Atlantis Studies in Mathematics for Engineering and Science

Volume 12

Aims and scope of the series

The series 'Atlantis Studies in Mathematics for Engineering and Science' (AMES) publishes high quality monographs in applied mathematics, computational mathematics, and statistics that have the potential to make a significant impact on the advancement of engineering and science on the one hand, and economics and commerce on the other. We welcome submission of book proposals and manuscripts from mathematical scientists worldwide who share our vision of mathematics as the engine of progress in the disciplines mentioned above.

For more information on this series and our other book series, please visit our website at: www.atlantis-press.com/publications/books

Atlantis Press
8, square des Bouleaux
75019 Paris, France

More information about this series at http://www.springer.com/series/10071

Radomir S. Stanković · Paul L. Butzer
Ferenc Schipp · William R. Wade
Authors/Co-Authors

Dyadic Walsh Analysis from 1924 Onwards Walsh-Gibbs-Butzer Dyadic Differentiation in Science Volume 1 Foundations

A Monograph Based on Articles of the Founding Authors, Reproduced in Full

in Collaboration with the Co-Authors
Weiyi Su
Yasushi Endow
Sándor Fridli
Boris I. Golubov
Franz Pichler
Kees (C.W.) Onneweer

ATLANTIS
PRESS

Authors
Radomir S. Stanković
Faculty of Electronic Engineering
Nis
Serbia

Paul L. Butzer
Lehrstuhl A für Mathematik
RWTH Aachen
Aachen
Germany

Ferenc Schipp
Numerical Analysis
Eötvös Loránd University
Budapest
Hungary

William R. Wade
Mathematics and Computer Science
BIOLA University
La Miranda, California
USA

Co-Authors
Weiyi Su
Mathematics
Nanjing University
Nanjing
China

Yasushi Endow
Chuo University
Tokyo
Japan

Sándor Fridli
Numerical Analysis
Eötvös Loránd University
Budapest
Hungary

Boris I. Golubov
Chair of Higher Mathematics
Moscow Institute of Physics
and Technology (State University)
Dolgoproudny, Moscow Region
Russia

Franz Pichler
System Theory
Johannes Kepler University, Linz
Linz
Austria

Kees (C.W.) Onneweer
Department of Mathematics and Statistics
University of New Mexico
Albuquerque
USA

ISSN 1875-7642 ISSN 2467-9631 (electronic)
Atlantis Studies in Mathematics for Engineering and Science
ISBN 978-94-6239-159-8 ISBN 978-94-6239-160-4 (eBook)
DOI 10.2991/978-94-6239-160-4

Library of Congress Control Number: 2015954975

Printed on acid-free paper

To the memory of Joseph Leonard Walsh and James Edmund Gibbs

Preface

The central theme of this book is the Gibbs derivative, a broad class of differential operators the most prominent member of which is the Butzer-Wagner derivative. The goal of the book is to introduce researchers in applied mathematics, information sciences, computing, and related areas to this particular concept in dyadic analysis and various generalizations of it.

The rationale, and hopefully an excuse, for bringing it to public notice, is a feeling that due to the present situation in information sciences and related technology, there can be a room for some concrete applications of these differential operators. This situation can be briefly summarized as follows.

The information and computing technologies are currently faced with problems in further exploiting approaches which were used in the last couple of decades for developing computing hardware based on semiconductor devices since their physical limits have been nearly reached. There can be observed some similarity with the situation in early seventies when Walsh dyadic analysis emerged as an answer to demands for improving performances in performing certain, primarily signal processing related, algorithms by changing the underlying mathematical foundations, since the limited computing power of hardware available at the time could not provide acceptable solutions based on classical theories.

The information sciences, with this term understood in a broad sense, are continually challenged to provide answers to demands for processing ever increasing amounts of data. Extracting and processing the content data contain is possible only by extensive use of computing hardware that is being constantly improved and performance enhanced. Therefore, for both computing technologies and information sciences, there is a need for essentially new discoveries possibly based on alternative approaches to those currently used. There is again a similarity with introducing the Walsh dyadic analysis in former times as an alternative to classical Fourier analysis in certain applications.

Differential operators, as introduced by Newton and Leibniz, can be generally viewed as a tool for estimating the rate and the direction of change of signals. The Gibbs derivatives were introduced over 45 years ago to serve the same purpose in information sciences as the Newton-Leibniz derivative does in mathematical physics and related areas [1]. Therefore, these differential operators can be viewed as an alternative to the existing differential operators tailored for application in Walsh dyadic analysis and extedning the classes of functions differentiable in certain sense.

At that time, Walsh (Fourier) analysis had attracted much attention for two reasons. First, it enjoyed peculiar properties (being a bounded, piecewise constant orthonormal system), and second, its coefficients and transforms could be easily and efficiently computed (since it took on only the values ± 1). These properties gave the Walsh system quite an important advantage over the classical Fourier analysis regarding the available computer hardware.

The Gibbs derivatives were introduced as differential operators having Walsh functions as their eigenfunctions. It might be said that these operators were intended to provide a mathematical tool offering advantages of differential operators primarily taking into account demands for processing of binary encoded signals, but also the computing power offered by the technology at the time. Therefore, interest in Walsh analysis, in general especially discrete Walsh analysis, and in the Gibbs derivatives in particular, can be explained by two reasons. First engineering practice required fast implementation of various algorithms for signal and information processing based on spectral (Walsh-Fourier) analysis, and second the limited computing power of the technology at the time ultimatively demanded something that took less memory than older, more established classical Fourier analysis.

For these reasons, regarding the present situation, we believe that the way of thinking that led to this concept can be equally useful in defining new concepts and related methods in computing with applications in signal and information processing now and in future.

To serve its goal, the book contains reprints of several papers setting the foundations of Gibbs differential operators in general, and the Butzer-Wagner derivative in particular, as well as some generalizations of these concepts.

Reprinting original former publications instead of rewriting the main theory, is motivated by the following considerations.

By taking the line of development of Walsh analysis as a computationally less demanding mathematical foundations for signal processing algorithms into account, we did not want to rephrase or rewrite the original statements in early papers on this subject for two reasons. First, we believe that the way these statement were presented originally by the authors, has a particular value for the reader. Second, the manner of writing and the way of formulating and discussing certain important concepts often reflects both the fashion of the times as well as the authors attitude to the subject. This, in a way, also expresses other circumstances at the time when the notions were introduced and concepts and theories formulated.

Therefore, we did not dare disturb and destroy the initial composition of the contents and their formulation, as done by the authors, by rewording the statements in present terminology. Thus, we restricted the contribution on our part to the selection of what we consider most interesting to present to the readers notice.

[1] Gibbs, J.E., "A contribution to a revolution?" *NPL News*, 1971 May, 1-4, (1971).
It is meant the informatics revolution that started at about that time (remark by editors)

Taking this into account, the reprinted papers are selected by the following criteria.

We wanted to reprint papers in which important results in this area were initially introduced and discussed. Most of these papers were presented at workshops which make them not so easily accessible. That was another motivation to reprint them rather than more elaborated versions that were possibly published latter in some journals. We also wanted to present papers which suggested alternative approaches to the Gibbs differentiation, proposed particular extensions and generalizations of the concept, or discussed certain applications.

The reprinted papers shuld serve as a basis for reading newly written chapters reviewing former and presenting recent and new results in the area. These chapters provide reviews of particular aspects of the theory of Gibbs differentiation and are written mostly by the founders of this theory. The book includes discussions of recent developments in the area as well as a collection of open and unsolved problems. Attempting to provide a rather complete picture of the development in the field, we asked the contributors to this book who started their research in the field in seventies and eighties for their reminiscences of personal involvement in the subject.

With a hope that the book will serve its purpose, we thank all the contributors as well as other friends without whose help its appearance would hardly be possible.

Acknowledgments

The four editors of this monograph, in particular Radomir S. Stanković and Paul L. Butzer, express their thanks for its preparation to the authors of the monograph together with their former students. The initiator of dyadic differentiation, James Edmund Gibbs, National Physical Laboratory, Teddington, UK, died much too early, in January 2007, and so could, most unfortunately, not participate in our present project.

Mrs. Merion Gibbs reported that according to her husbands diary, his first definition of his "logical derivative" was written in January 1967. Of his 27 publications on dyadic derivatives six were written together with Dr. Brian Ireland, Bath University, UK, who joined Dr. Gibbs in the work on this subject in 1971.

The joint work of authors towards the present monograph, which took place over a period of almost four years, was very constructive and unbelievably harmonious, and even during the time from 1969 until 1990 when the chief results of dyadic Walsh analysis were developed by them. This is by no means self-evident since the authors come from ten countries, namely Austria, Canada, China, England, Germany, Hungary, Japan, Russia, Serbia, USA, with quite different mathematical/scientific traditions.

The majority of these, together with some of their former students, only met for the first time at the Workshop "Theory and Applications of Gibbs Derivatives", conducted by Radomir S. Stanković at Kupari- Dubrovnik, Yugoslavia, on September 26-28, 1989.

All authors express their special thanks to Charles Chui, Stanford University, for accepting our monograph as Editor of the series "Atlantis Studies in Mathematics for Engineering and Science" of Atlantis Press, Paris, without any hesitations. It is quite common for selected or complete publications of one author (which could cover several fields) to appear in book-form. But this is by no means so for selected papers of all founding authors of one field, even together with reader-friendly reviews of their own papers, to appear in book-form. Further, the authors thank Dr. Keith Jones for handling all matters concerned with Atlantis Press, Paris, in an efficient and friendly manner, as well as Springer for production, marketing, and distribution of this book. In particular, thanks are due to Ms. Devi Ignasy, the Project Coordinator, and Ms. Gajalakshmi Sundaram, the Production Editor, from Springer, for excellent processing of the manuscript, especially for very good care of reprinted papers, the source versions of some of them which were not so clear and easily readable.

The Editors

Organization of the Book

The book is organized in two volumes each consisting of 12 chapters. The present Volume 1 comprises 6 parts, each consisting of a review of work in the area of dyadic analysis and generalizations by particular research groups acting in this field from its origins up to date.

Each part of this volume contains a chapter or couple of chapters written by experts in the field, most of them being founders of dyadic analysis as a mathematical discipline originating in demands from engineering applications. These chapters are followed by reprinted carefully selected former publications, which provide basic concepts, related definitions as they were initially introduced, and main results as originally presented by their inventors.

The book starts with a chapter on the development of Walsh analysis followed by two reprinted papers, the initial paper by Joseph Leonard Walsh introducing the Walsh system in 1923, and a paper by Henning F. Harmuth published in 1976 discussing related applications of Walsh functions in communications as viewed at that time. The presentation continues with reminiscences of the engagement of F. Pichler in the subject highlighting the situation in the research in this field in West Europe and USA. This part ends with a reprinted technical report by Franz Pichler discussing early ideas of the applications of Walsh analysis in linear system theory.

The second part contains a short chapter presenting some remarks about the origins of the Gibbs differentiation due to James Edmund Gibbs and basic remarks on further development of the concept, in particular, the Butzer-Wagner derivative. The main idea with this chapter is to introduce the reader to this subject area and make him well prepared for the continuing contents of the book. The chapter is followed by three reprinted research reports by Dr. Gibbs and his associates, in which the basic concepts were introduced.

The third part is devoted to the central concept in the subject area of the book, the dyadic derivative on infinite dyadic groups, as formulated by P.L. Butzer and his former student H.J. Wagner, and then accepted and further studied by F. Schipp and his associates. The part contains two reviews, four reprinted papers, and reminiscences of P.L. Butzer and F. Schipp of the early work in the area.

The part four starts with brief introductory remarks by C.K.W. Onneweer about different formulations of the definition of the dyadic derivative tailored to incorporate different particular features of the operator with respect to its eigenfunctions, and continues with reprinted five former related papers. Then, a new article and a reprinted paper by F. Weisz discuss the two-dimensional dyadic derivative.

Term by term differentiation of Walsh series and related issues is the central theme of the part five as discussed by W.R. Wade, followed by his reminiscences of the early work and establishing the relationships and cooperation with researchers from former Soviet Union and Hungary, in particular with Valentin Skvotsov from Moscow, and Ferenc Schipp from Budapest and his associates. This part also contains reprinted five publications prepared independently by P.L. Butzer, and F. Schipp, and their associates.

The sixth part discusses the relationships between the dyadic derivative and Walsh functions and transform as viewed by B.I. Golbov, followed by his reminiscences of the involvement in the subject.

This completes the Volume 1 which consists of 12 chapters and 22 reprinted papers.

Volume 2 is devoted to the generalizations and certain practical applications different from applications in mathematical analysis and consists of 12 chapters and 8 reprinted papers organized in five parts.

Regarding the circumstances, including political situation in the World in early seventies, under which the theory of dyadic differentiation was initially created, it is important to take into account the specificity of the situation at the time in China and former Soviet Union and Russia, therefore, the first two parts in Volume 2 present an overview and highlight particular aspects of early work in the area done by scholars in these countries.

In the first part, after a brief analysis of former work in China written by Weiyi Su, we reprinted three papers that were initially published in Chinese and are for the first time translated into English. This part also contains a new review of recent results on pseudo-differential operators in local fields by Weiyi Su and her associates.

The second part first presents a discussion of the work in dyadic analysis and generalizations in former Soviet Union as viewed by R.S. Stanković, and also the point of view to the work in this area in Russia by B.I. Golubov and his associate S. Volosivets.

Two new chapters in the part three are devoted to the generalization of the theory of dyadic derivative to Vilenkin groups by B. Golubov and S. Volosivets, and G. Gát and R. Toledo, respectively.

The fourth part is devoted to the generalizations of the concept of dyadic Gibbs differentiation to finite not necessarily Abelian groups. It contains a review chapter on this development of the theory of Gibbs derivatives due to Dr. Gibbs and his associates for Abelian groups, and extensions to finite non-Abelian groups by R.S. Stanković and his reminiscences of the involvement in the subject. Another chapter is devoted to methods of computing the Gibbs derivative on contemporary technological platforms. This part also contains three reprinted former publications related to Gibbs differentiation on groups, including a treatise of the relationships of the Newton-Leibniz derivative and the Gibbs derivative presented by Dr. Gibbs in 1989.

The main subject of the next part is the application of the Gibbs derivative in the theory of dyadic stationary processes. This parts contains a chapter presenting new results in the field by Y. Endow as well as his reminiscences of the former work. The reprinted paper by Endow published in 1984 presents the initial results in the subject area and serves as a good introduction to scholars and practitioners interested in this and related topics.

The volume ends with a chapter presenting open problems in theory and applications of the dyadic Gibbs derivatives and their generalizations.

Reprinted Papers

As mentioned above, the book contains reprinted early and classical papers on Gibbs differentiators with the main background idea to bring to the attention of contemporary researchers basic notions and properties of these operators as they were originally formulated by their inventors. The selection is based upon recommendations by several researchers in the field that are founders of this area and at the same time contributors to the present volume.

The first bibliography on Gibbs derivatives was prepared on the occasion of the *First International Workshop on Gibbs Derivatives*, held on September 26-28, 1989, at Kupari-Dubrovnik, former Yugoslavia, and consisted of 156 papers (Gibbs, J.E., Stanković, R.S., "Why IWGD-89? A look at the bibliography of Gibbs derivatives", in Butzer, P.L., Stanković, R.S. (edit.), *Theory and Applications of Gibbs Derivatives - Proceedings of the First International Workshop on Gibbs Derivatives* held September 26-28, 1989 at Kupari-Dubrovnik, Yugoslavia, Mathematical Institute, Belgrade, 1990, xi-xxiv.)

Collecting was continued by J. E. Gibbs and R.S. Stanković until 2007, and the present selection is based on the most recent bibliography on Gibbs derivatives consisting of more than 300 papers (Stanković, R.S., Astola, J, *Gibbs Derivatives - the First Forty Years* TICSP Series #39 2008, Tampere International Center for Signal Processing, Tampere University of Technology, Tampere, Finland, 2008, ISBN 978-952-15-1973-4.)

The main selection criteria determining papers to be reprinted can be briefly summarized as follows.

1. We selected papers in which particular concepts were initially introduced or certain generalizations proposed.
2. Attempts are made to cover different aspect of the theory of Gibbs differentiation, provide a review of generalizations of the initial concept, highlight basic properties, and point out some applications.
3. Focus is on papers that have been published before 1990, since more recent papers are easier to find [2].
4. For the same reason of hard availability and accessability, we mainly selected papers presented at some workshops or conferences but also included journal papers fundamental for the area.

The first reprinted publications is the Technical Report of the National Physical Laboratory, Middlesex, Teddington, United Kingdom, by James Edmund Gibbs from 1967 where the notion of Gibbs derivatives under the name logic differentiation was formulated for the first time. The Gibbs derivative is defined as the differ-

[2] An exception is the paper 16 by Prof. F. Weisz that is published in 1996. This paper is reprinted due to its importance since presents links between two-dimensional dyadic derivatives and Hardy spaces providing a basis for a particular theory of multidimensional dyadic derivatives. It should be noted that two-dimensional dyadic derivatives were discussed in 1982 in Butzer, P.L., Engels, W., "Dyadic calculus and sampling theorems for functions with multidimensional domain", *Inform. and Control*, Vol. 52, 1982, 333-351.

ential operator for functions on finite dyadic groups having the Walsh functions as eigenfunctions. The other reprinted papers are ordered by subjects as follows. The central part are papers introducing and discussing the Butzer-Wanger derivative, that is the dyadic differentiator on infinite dyadic groups. After that, we present different generalizations of this way of differentiation to functions defined on different domains, including the positive part of the real line R_+, cyclic groups, finite Abelian and non-Abelian groups.

Table 1 shows a classification of differential operators discussed in reprinted and translated papers by the domain of functions to be differentiated, which also determines the eigenfunctions of the operators.

List of Reprinted Papers

1. Walsh, J.L., "A closed set of orthogonal functions", *Amer. J. Math.*, 55, 1923, 5-24.

2. Harmuth, H.F., "Applications of Walsh functions in communications", *IEEE Spectrum*, Vol. 6, No. 11, 1969, 82-91.

3. Pichler, F., *Walsh Functions and Linear System Theory*, Technical Research Report, T-70-05, Dept. of Electrical Engineering, University of Maryland, College Park, Maryland 20742, April 1970, ii+46.

4. Gibbs, J.E., "Walsh spectrometry, a form of spectral analysis well suited to binary digital computation", *Nat. Phys. Lab.*, Teddington, Middx, UK, 1967, 24pp.

5. Gibbs, J.E., Millard, Margaret J., "Walsh functions as solutions of a logical differential equation", *NPL DES Rept.*, No. 1, 1969, 9 pp.

6. Gibbs, J.E., Ireland, B., "Some generalizations of the logical derivative", *DES Report No. 8, Nat. Phys. Lab.*, August 1971, 22+ii pages.

7. Butzer, P.L., Wagner, H.J., "Walsh-Fourier series and the concept of a derivative", *Applicable Anal.*, 3, 1973, 29-46.

8. Butzer, P.L., Wagner, H.J., "Approximation by Walsh polynomials and the concept of a derivative", *Proc. Sympos. Applic. Walsh Functions*, Washington, D.C., 1972, 388-392.

9. Butzer, P.L., Wagner, H.J., "On a Gibbs-type derivative in Walsh-Fourier analysis with applications", *Proc. Nat. Electron. Conf.*, 27, 1972, 393-398.

Table 1: Classification of reprinted papers by the domain of functions to be differentiated and eigenfunctions of related differential operators.

Domain and Eigenfunctions	Paper
Finite dyadic group Discrete Walsh functions	[4] Gibbs, J.E., 1967 [5] Gibbs, J.E., Millard, Margaret J., 1969
Infinite dyadic group Walsh functions	[7] Butzer, P.L., Wagner, H.J., 1973a [8] Butzer, P.L., Wagner, H.J., 1972a [9] Butzer, P.L., Wagner, H.J., 1972b [13] Butzer, P.L., Wagner, H.J., 1973b [11] Schipp, F., 1976 [21] Butzer, P.L., Wagner, H.J., 1975 [10] Schipp, F., 1974 [12] Onneweer, C.W., 1979 [17] Skvortsov, V.A., Wade, W.R., 1979 [18] Engels, W., 1985 [19] Schipp, F., 1986
Two-dimensional Walsh functions	[16] Weisz, F., 1996
Infinite dyadic group Generalized Walsh functions	[23] He, Zelin, 1983
Infinite *p*-adic group Generalized *p*-adic Walsh functions	[1] translated, Ren, Su, Zheng, 1978 [2] translated, Zheng, 1979 [3] translated, Zheng, Su, 1981
R_+ Generalized Walsh functions	[15] Butzer, P.L., Wagner, H.J., 1973 [2] Wagner, H.J., 1975 [20] Pàl, J., Schipp, F., 1978
$R = (-\infty, \infty)$ Haar functions	[21] Butzer, P.L., Splettstosser, W., Wagner, H.J., 1975
Finite Abelian group Cyclic group Group characters	[6] Gibbs, J.E., Ireland, B., 1971 [25] Gibbs, J.E., Simpson, J., 1974 [26] Gibbs, J.E., 1989
Finite non-Abelian groups Unitary irreducible representations	[27] Stanković, R.S., 1989
	Applications
Infinite dyadic group Walsh functions	[3] Pichler, F., 1970 [28] Endow, Y., 1984 [8], [9] Butzer, P.L., Wagner, H.J., 1972 [14], [15] Butzer, P.L., Wagner, H.J., 1973 [21] Butzer, P.L., Splettstosser, W., Wagner, H.J., 1975 [23] He, Zelin, 1983 [1] Ren, F.X., Su, W.Y., Zheng, W.X., 1978 (translated paper) [4] Gibbs, J.E., 1967, [5] Gibbs, J.E., Millard, Margaret J., 1969 [27] Stanković, R.S., 1989

10. Schipp, F., "Über einen Ableitungsbegriff von P.L. Butzer und H.J. Wagner", *Mathematica Balkanica*, 4.103, 1974, 541-546.

11. Schipp, F., "On the dyadic derivative", *Acta Math. Acad. Sci. Hungar.*, Vol. 28, 1976, 145-152.

12. Onneweer, C.W., "On the definition of dyadic differentiation", *Applicable Analysis*, Vol. 9, 1979, 267-278.

13. Butzer, P.L., Wagner, H.J., "On dyadic analysis based on the pointwise dyadic derivative", *Anal. Math.*, 1, 1975, 171-196.

14. Butzer, P.L., Wagner, H.J., "A new calculus for Walsh functions with applications", *Proc. Sympos. Theory & Applic. Walsh & Other Non-sinus. Functions*, Hatfield, England, 1973 June 28-29, 1973, ii + 16.

15. Butzer, P.L., Wagner, H.J., "A calculus for Walsh functions defined on R_+", *Proc. Sympos. Applic. Walsh Functions, Washington*, D.C., 1973, 75-81.

16. Weisz, F., "(H_p, L_p)-type inequalities for the two-dimensional dyadic derivative", *Studia Methematica*, Vol. 120, No. 3, 1996, 271-288.

17. Skvortsov, V.A., Wade, W.R., "Generalization of some results concerning Walsh series and the dyadic derivative", *Analysis Mathematica*, Vol. 5, 1979, 249-255.

18. Engels, W., "On the characterization of the dyadic derivative", *Acta Math. Hungar.*, Vol. 46, No. 1-2, 1985, 47-56.

19. Schipp, F., "Über gewisse Maximaloperatoren", *Annales Univ. Sci. Budapest, Sect. Math.*, Vol. 28, 1986, 145-152.

20. Pàl, J., Schipp, F., "On the dyadic differentiability of dyadic integral functions on R_+", *Annales Univ. Sci. Budapest., Sec. Comp.*, Vol. 8, 1978, 91-108.

21. Butzer, P.L., Splettstosser, W., Wagner, H.J., "On the role of Walsh and Haar functions in dyadic analysis", *Proc. Spec. Sess. Sequency Techniques (in cooperation with the 1st Sympos. & Tech. Exhib. EMC*, Montreux, 1975 May 20-22), Ed. H. Hubner, Darmstadt, 1975, 1-6.

22. Wagner, H.J., "On dyadic calculus for functions defined on R_+", Arbeitsber., Lehrstuhl A für Math., RWTH Aachen, ii+27 pp. 1974, also in *Proc. Sympos. Theory & Applic. Walsh & Other Non-sinus. Functions*, Hatfield, UK, 1975 July 1-3, 1975.

23. He, Zelin, ”The derivatives and integrals of fractional order in Walsh-Fourier analysis, with applications to approximation theory”, *J. Approx. Theory*, 39, 1983, 361-373.

24. Selected pages from the book
Agaev, G.N., Vilenkin, N.Ja., Dzhafarli, G.M., and Rubinstein, A.I., *Multiplicative Systems of Functions and Harmonic Analysis on Zero-dimensional Groups* (Russian), Izd. ”Elm”, Baku, 1981. 28

25. Gibbs, J.E., J. Simpson, ”Differentiation on finite Abelian groups”, *Nat. Phys. Lab.*, Teddington, Middx, UK, 1974, 34pp.

26. Gibbs, J.E., ”Local and global views of differentiation”, in Butzer, P.L., Stanković, R.S. (edit.), *Theory and Applications of Gibbs Derivatives: Proceedings of the First International Workshop on Gibbs Derivatives*, September 26-28, 1989, Kupari-Dubrovnik, Yugoslavia, Matematički Institut, Beograd, 1990, 1-18.

27. Stanković, R.S., ”Gibbs derivatives on finite non-Abelian groups”, in Butzer, P.L., Stanković, R.S. (edit.), *Theory and Applications of Gibbs Derivatives: Proceedings of the First International Workshop on Gibbs Derivatives*, September 26-28, 1989, Kupari-Dubrovnik, Yugoslavia, Matematički Institut, Beograd, 1990, 269-297.

28. Endow, Y., ”Analysis of dyadic stationary processes using the generalized Walsh functions”, *Tohoku Math. J., Ser.* 2, 36, No. 4, 1984, 485-503.

List of Translated Papers

1. Ren, F.X., Su, W.Y., Zheng, W.X., ”The generalized logical derivatives and its applications”, *J. of Nanjing University*, Vol. 3, 1978, 1-8.

2. Zheng, W.X., ”Generalized Walsh transform and an extremum problem”, *Acta Mathematica Sinica*, Vol. 22, No. 3, 1979, 362-374.

3. Zheng, W.X., Su, W.Y., ”The logical derivatives and integrals”, *J. Math. Res. Exposition*, Vol. 1, 1981, 79-90.

Contents

1 Early History of Walsh Analysis 1
William R. Wade
1.1 History of the American School of Walsh Analysis 3

2 My involvement in Walsh and Dyadic Analysis 37
Franz Pichler
References 39

3 The Origins of the Dyadic Derivative due to James Edmund Gibbs .. 89
Radomir S. Stanković, Ferenc Schipp
3.1 Definition of the Gibbs Derivative 91
3.2 Further Development 93
References 95

4 Early Contributions from the Aachen School to Dyadic Walsh Analysis with Applications to Dyadic PDEs and Approximation Theory 161
Paul L. Butzer, Heinrich Josef Wagner
4.1 Butzer-Wagner Dyadic Derivative 163
4.2 Further Developments 164
4.3 A Dyadic PDE, the Wave Equation 165
4.4 Applications to Approximation by Walsh Polynomials and Walsh-Fourier Series 168
4.5 Later Contributions from Aachen 170
4.6 Decimal and Dyadic Systems 171
4.7 Further Suggestions for Studying Our Monograph 172
References 173
4.8 Publications by Members of the Aachen School of Dyadic Analysis 176

5 Dyadic Derivative, Summation, Approximation 209
S. Fridli, F. Schipp
5.1 Introduction 209

5.2 Dyadic Derivative . . . 211
5.2.1 Dyadic derivative of functions of one variable . . . 211
5.2.2 Dyadic derivative in the multivariable case . . . 214
5.2.3 Dyadic derivative on the real line . . . 214
5.2.4 Dyadic derivative on groups . . . 215
5.3 Summation, Strong summation . . . 216
5.3.1 $(C,1)$ summability of Walsh–Fourier series . . . 216
5.3.2 Strong summability . . . 218
5.3.3 Summability of multivariable Walsh–Fourier series . . . 220
5.3.4 Dyadic Cesáro and Copson operators . . . 221
5.3.5 Multipliers . . . 222
References . . . 223

6 How I Started My Research in Walsh and Dyadic Analysis . . . 235
Ferenc Schipp
References . . . 237

7 My Involvement with the Dyadic Derivative . . . 247
Kees (C.W.) Onneweer
References . . . 249

8 Hardy Spaces in the Theory of Dyadic Derivative . . . 315
Ferenc Weisz
8.1 Introduction . . . 315
8.2 One-dimensional Hardy Spaces . . . 316
8.3 The One-dimensional Dyadic Derivative . . . 318
8.4 d-dimensional Hardy Spaces . . . 321
8.4.1 The Hardy Spaces $H_p^{\square}[0,1)^d$ . . . 322
8.4.2 The Hardy Spaces $H_p[0,1)^d$ . . . 323
8.5 More-dimensional dyadic derivative . . . 325
8.5.1 Restricted dyadic derivative . . . 326
8.5.2 Unrestricted dyadic derivative . . . 326
References . . . 327

9 Term by Term Dyadic Differentiation of Walsh Series . . . 347
William R. Wade
9.1 Introduction . . . 347
9.2 The original problem . . . 348
9.3 Gap series . . . 351
9.4 Rapidly converging series . . . 352
9.5 Term-by-term strong dyadic differentiation . . . 355
References . . . 358

10 Why I got Interested in Dyadic Differentiation . . . 359
William R. Wade

11 Dyadic Derivative and Walsh-Fourier Transform 443
Boris I. Golubov
11.1 Introduction 443
11.2 Notations and Definitions 444
11.3 Lemmas 444
11.4 Dyadic Differentiation and Integration of Walsh-Fourier Transform 445
References 446

12 How I started my research in Walsh and dyadic analysis 449
Boris I. Golubov
References 452

Index 453

List of Contributors

The names of the contributors to this book are given in alphabetical order.

Paul L. Butzer
Lahrstuhl A für Mathematik, RWTH Aachen, Germany
e-mail: butzer@rwth-aachen.de

Yasushi Endow
Department of Industrial and Systems Engineering, Chuo University, Tokyo, 112-8551 Japan, e-mail: endow@indsys.chuo-u.ac.jp

Sándor Fridli
Department of Numerical Analysis, Eötvös L. University, Budapest, Pázmány P. sétány 1\C, H-1117 Hungary, e-mail: fridli@numanal.inf.elte.hu

Dušan Gajić
Dept. of Computer Science, Faculty of Electrical Engineering, Niš, Serbia, e-mail: dule.gajic@gmail.com

György Gát
College of Nyíregyháza, Inst. of Math. and Comp. Sci., Nyíregyháza, P.O.Box 166., H–4400, Hungary, e-mail: gatgy@nyf.hu

Ushangi Goginava
Dept. of Mathematics, Faculty of Exact and Natural Science, Tbilisi State University, Tbilisi, Georgia, e-mail: ushangi.goginava@tsu.ge

Boris I.Golubov
Chair of Higher Mathematics, Moscow Institute of Physics and Technology (State University), 141700 Dolgoproudny, Moscow Region, Rusia, e-mail: golubov@mail.mipt.ru

Qiu Hua
Mathematics Department, Nanjing University, China, e-mail: qiuhua@nju.edu.cn

Kees (C.W.) Onneweer
Department of Mathematics and Statistics, University of New Mexico, Albuquerque, USA, e-mail: onneweer@math.unm.edu

Franz Pichler
Johannes Kepler University, Linz, Austria, e-mail: Franz.Pichler@jku.at

Ferenc Schipp
Department of Numerical Analysis, Eötvös L. University, Budapest, Pázmány P. sétány 1\C, H-1117 Hungary, e-mail: schipp@numanal.inf.elte.hu

Radomir S. Stanković
Dept. of Computer Science, Faculty of Electronic Engineering, University of Niš, Niš, Serbia, e-mail: Radomir.Stankovic@gmail.com

Weiyi Su
Mathematics Department, Nanjing University, China, e-mail: suqiu@nju.edu.cn

Rodolfo Toledo
College of Nyíregyháza, Inst. of Math. and Comp. Sci., Nyíregyháza, P.O.Box 166., H–4400, Hungary, e-mail: toledo@nyf.hu

Sergey S. Volosivets
Faculty of Mechanics and Mathematics, Saratov State University, 410012 Saratov, Russia, e-mail: volosivetsSS@mail.ru

William R. Wade
Mathematics and Computer Science, Biola University, La Mirada CA 90639, USA e-mail: william.wade@biola.edu

Hans Josef Wagner
Krauthausener Str. 6A, 52223 Stolberg-Dorff, Germany

Ferenc Weisz
Department of Numerical Analysis, Eötvös L. University H-1117 Budapest, Pázmány P. sétány 1/C., Hungary, e-mail: weisz@inf.elte.hu

Chapter 1
Early History of Walsh Analysis

William R. Wade

The history of Walsh series really begins with Haar's dissertation, "Zur Theorie der Orthogonalen Functionsysteme," which introduced the Haar system in 1909. Folklore has it that Hilbert, Haar's advisor at Göttingen University, asked Haar to find an orthogonal system on the interval $[0,1)$ whose Fourier series of continuous functions converged uniformly. Haar's clever solution was to restrict his attention to piecewise constant functions supported on dyadic intervals. He constructed his system so that the Dirichlet kernels took on values 2^n on intervals of length 2^{-n}. This allowed him to use the fundamental theorem of calculus to prove that Haar-Fourier series of continuous functions converge uniformly. By the same reasoning, Lebesgue's theorem shows that Haar-Fourier series of Lebesgue integrable functions converge almost everywhere. Here, then, is an orthonormal system with the sparkling convergence properties that had been envisioned for the trigonometric system by its creators.

The Haar functions had one huge drawback. They were unbounded. Since the theory of bounded orthonormal systems is much more accessible and much more like the classical trigonometric model, it was only a matter of time until someone would try to modify Haar's system to produce a bounded orthonormal system with good convergence properties. That someone turned out to be Joseph Leonard Walsh.

Walsh began working on a dissertation at Harvard University under Maxime Bôcher in 1917. When the United States entered World War I later that year, Walsh enlisted in the U.S. Navy for whom he served on troop transport vessels in the North Atlantic. By the time he returned to Harvard in 1919, Bôcher had died. Walsh finished his dissertation under G.D. Birkhoff, in 1920, on roots of a Jacobian of two binary forms. Walsh was awarded a Sheldon Traveling Fellowship for the academic year 1920-1921, and spent his time in Paris working with Montel. Where in

William R. Wade
Mathematics and Computer Science, Biola University, La Mirada CA 90639, USA, e-mail: william.wade@biola.edu

R.S. Stanković et al. (eds.), *Dyadic Walsh Analysis from 1924 Onwards Walsh-Gibbs-Butzer Dyadic Differentiation in Science Volume 1 Foundations*, Atlantis Studies in Mathematics for Engineering and Science 12, DOI 10.2991/978-94-6239-160-4_1

those travels he picked up the idea of modifying the Haar system is anyone's guess. Bôcher, Birkhoff, and Montel were broadly trained, well connected mathematicians who could have suggested it, or Walsh could have come up with it on his own. One thing is certain. It definitely was not part of his dissertation as has been widely reported (me included).

Walsh used linear combinations of Haar functions (actually, Hadamard transforms) to produce a new orthonormal system, now called the Walsh system. The linear combinations made the support of each Walsh function equal to the interval $[0,1)$ and the whole system was bounded. The Walsh functions took on the values ± 1 (and 0 since he insisted in averaging them at jumps) and, since the 2^nth partial sums of Walsh-Fourier series of a fixed function equaled the 2^nth partial sums of the corresponding Haar-Fourier series, had well behaved Fourier series as long as attention was restricted to dyadic partial sums. It would turn out that convergence properties of the full sequence of partial sums of Walsh-Fourier series was no better than those of the trigonometric system, but for many applications, dyadic partial sums are enough. Since Walsh's method of generating the Walsh functions can be done recursively, and generates functions whose zero crossings match those of the trigonometric system, this ordering is preferred for applications [1]

The Walsh functions were reordered by R.E.A.C. Paley in 1932. He showed that the Walsh system was the completion of the Rademacher system, and could be defined using products of Rademacher functions. This was the first example of a product system and would lead to a general theory of such systems. It also provided a link between the Walsh system and probability theory that would be used to define analogues of Hardy spaces without resorting to analytic function theory, unavailable for the non-continuous Walsh system. Using products of Rademacher functions to define the Walsh system provided a whole new set of tools for Walsh analysis, especially for finding closed "product" forms for various kernels. Unfortunately, Paley did not live long enough to exploit his creation. He was buried in an avalanche in 1933 while on vacation in the Canadian Rockies.

Paley was an enthusiastic, gregarious person who inspired and collaborated with many who came in contact with him. In his short professional career, 1927-1933, he wrote joint papers with Littlewood (his dissertation advisor at Cambridge), Wiener, and Zygmund which produced ground breaking results that are still being taught and used today: the Littlewood-Paley Lemma, the Paley-Wiener theorems, and the Paley-Zygmund inequality.

Paley met Antoni Zygmund when the latter had a Rockefeller Fellowship during the academic year 1930-1931. Zygmund spent a semester at Oxford working with Hardy, and a semester at Cambridge working with Littlewood. While staying at Cambridge, Paley and Zygmund produced what is now called the Paley-Zygmund inequality. They must have also discussed the Walsh system which Plaey was beginning to think about, because after this, Zygmund exhibited a life-long fascination for the Walsh functions.

---

[1] This concerns applications in the same areas as trigonometric systems. For applications in some other areas, as for instance Digital logic, other orderings of Walsh functions are preferred. The same is with respect to computational point of view.

Zygmund liked the Walsh functions because they were elegant and simple, yet had a rich theory that paralleled the trigonometric system. He predicted decades before it happened, that as Walsh theory matured, it would shed light on the older trigonometric system. He told me [2] early in my career not to abandon the Walsh system. He said it was roughly like the situation for double trigonometric series in the 1930's. The subject was not yet held in high esteem because most of the results were analogues of older one-dimensional trigonometric results. However, when we began to ask the right questions, the Walsh system would come into its own.

His enthusiasm must have been catching. Although Zygmund never worked on the Walsh system, he did direct two dissertations on the subject. Moreover, over the years, several of his other students directed dissertations on the Walsh system. This is the source of most of the American results in the area. Here, then, is a brief summary of what transpired.

1.1 History of the American School of Walsh Analysis

N.J. Fine, who worked under Zygmund at the University of Pennsylvania, wrote a dissertation, published in 1949, which introduced the dyadic group, a compact Abelian group whose characters can be identified with the Walsh functions. (This same observation had been made in 1947 by Vilenkin in a more general context, but the west's knowledge of what was going on in the Soviet Union was very sketchy in those days.) Since the Walsh functions are continuous on this group, we could now use many more trigonometric methods to study Walsh Fourier analysis.

G.W. Morgenthaler, who worked under Zygmund at the University of Chicago, wrote a dissertation, published in 1957, which studied summability of Walsh-Fourier series. His work reinforced the analogy between the trigonometric system and the Walsh system by proving that Walsh-Fourier series are summable in the L^p norm and that, on the average, Walsh-Fourier coefficients of absolutely continuous functions are $o(1/k)$ as $k \to \infty$.

Victor Shapiro, who worked under Zygmund at the University of Chicago, wrote a dissertation on double trigonometric series published in 1954, but got interested in Walsh series later. Besides writing two papers on Walsh series himself, three of his students wrote dissertations on them as well.

Richard Crittenden, who worked under Shapiro at the University of Oregon, wrote a dissertation on uniqueness of summable Walsh series which was published in 1963. He also wrote a paper, joint with Shapiro, about uniqueness of Walsh series which converge off a countable set.

Larry Harper, who worked under Shapiro at the University of Oregon, wrote a dissertation on sets of divergence and α-capacity on the dyadic group which was published in 1965.

[2] To Wiliam R. Wade

W.R. Wade, who worked under Shapiro at the University of California at Riverside, wrote a dissertation on uniqueness of Walsh series which converge almost everywhere but satisfy a growth condition off a countable set. This was published in 1969.

Daniel Waterman, who worked under Zygmund at the University of Chicago, wrote a dissertation on trigonometric series that was published in 1954. He got interested in Walsh analysis after he graduated and produced two students who furthered the field.

James McLaughlin, who studied under Waterman at Wayne State University in Detroit, wrote a dissertation on Haar and Walsh series which was published in 1968. In 1974, he published an interesting paper on growth of of Walsh-Fourier coefficients.

C.W. Onneweer, who also studied under Waterman at Wayne State University, wrote a dissertation on convergence of Fourier series on Vilenkin groups, a generalization of the dyadic group, which was published in 1968. He and Waterman wrote two highly regarded papers on uniform convergence of Vilenkin-Fourier series, and one on functions of bounded fluctuation. Onneweer went on to write many papers on the Walsh system, and its generalizations, on a wide variety of topics including uniform convergence, absolute convergence, sets of divergence, and the dyadic derivative. Regarding the latter, he managed to define an analogue of the dyadic derivative in a general context that included the p-adic and p-series setting.

Returning to students of Zygmund, Elias Stein, who worked under Zygmund at the University of Chicago, wrote a dissertation on linear operators in L^p spaces published in 1955. Although he never published anything on Walsh series, one of his students did a great deal to further the study of Walsh functions.

Mitch Taibleson, who worked under Stein at the University of Chicago, wrote a dissertation on smoothness and differentiability conditions which was published in 1962, got interested in Walsh functions later. He wrote a monograph "Fourier Analysis on Local Fields," published by Princeton University Press in 1975, which showed that the dyadic group is a special case of a local field. This gave the Walsh system another general context for interpretation, and another set of tools to use. Taibleson had a hand in training two students who had a huge influence on Walsh Fourier analysis. In reverse chronological order, they were Chao and Hunt.

John Chao, who worked under Taibleson at Washington University in Saint Louis, wrote a dissertation on conjugate systems in local fields which was published in 1972. Chao continued to produce highly original work for the next decade on the intersection of probability theory and the Walsh system via martingales. His promising career was cut short by cancer.

Richard Hunt, who worked under Taibleson and Guido Weiss (another Zygmund student at Chicago from the mid fifties), wrote a dissertation at Washington University at Saint Louis on Lorentz spaces which was published in 1965. Hunt wrote a paper on almost everywhere convergence of Walsh Fourier series in 1970 and directed several dissertations on Walsh functions.

John Gosselin, who worked under Hunt at Purdue University, wrote a dissertation on almost everywhere convergence of Fourier series in Vilenkin systems of bounded type, a generalization of the Walsh system, which was published in 1972. He went on to publish several papers on Walsh functions which explored convergence and the Littlewood-Paley theory.

K.H. Moon, who also worked under Hunt at Purdue University, wrote a dissertation on maximal functions published in 1972. A consequence of his work is that there exist functions in $L(\log^+ \log^+ L)^{1-\varepsilon}$ for $\varepsilon > 0$ whose Walsh-Fourier series diverge everywhere.

Wo-Sang Young, who worked under Hunt at Purdue University, wrote a dissertation on maximal inequalities and almost everywhere convergence of Fourier series published in 1973. Her papers on Walsh Fourier analysis, which were a model of elegance and compactness. She got results on Hardy-Littlewood maximal inequalities, mean convergence, and rearrangements of Walsh-Fourier series including the first result on the Kaczmarz ordering.

David Pankratz, who studied under Hunt at Purdue University, wrote a dissertation on almost everywhere convergence of Walsh-Fourier series published in 1974.

N.R. Ladhawala, who studied under Hunt at Purdue University, wrote a dissertation on a distribution inequality which was published in 1976. He and Pankratz published a paper that shows Walsh-Fourier series of H^1 functions can diverge almost everywhere.

Finally, John Coury, who studied under Edwin Hewitt at the University of Washington, wrote a dissertation on sets of uniqueness and multiplicity on product compact topological groups which was published in 1969. He then wrote a string of papers on Walsh series on a wide variety of topics including sets of multiplicity, lacunary series, and Walsh series with monotone coefficients.

A CLOSED SET OF NORMAL ORTHOGONAL FUNCTIONS.*

By J. L. Walsh.

Introduction.

A set of normal orthogonal functions $\{\chi\}$ for the interval $0 \leqq x \leqq 1$ has been constructed by Haar,† each function taking merely one constant value in each of a finite number of sub-intervals into which the entire interval (0, 1) is divided. Haar's set is, however, merely one of an infinity of sets which can be constructed of functions of this same character. It is the object of the present paper to study a certain new closed set of functions $\{\varphi\}$ normal and orthogonal on the interval (0, 1); each function φ has this same property of being constant over each of a finite number of sub-intervals into which the interval (0, 1) is divided. In fact each function φ takes only the values $+ 1$ and $- 1$, except at a finite number of points of discontinuity, where it takes the value zero.

The chief interest of the set φ lies in its similarity to the usual (e.g., sine, cosine, Sturm-Liouville, Legendre) sets of orthogonal functions, while the chief interest of the set χ lies in its *dissimilarity* to these ordinary sets. The set φ shares with the familiar sets the following properties, none of which is possessed by the set χ: the nth function has $n - 1$ zeroes (or better, sign-changes) interior to the interval considered, each function is either odd or even with respect to the mid-point of the interval, no function vanishes identically on any sub-interval of the original interval, and the entire set is uniformly bounded.

Each function χ can be expressed as a linear combination of a finite number of functions φ, so the paper illustrates the changes in properties which may arise from a simple orthogonal transformation of a set of functions.

In § 1 we define the set χ and give some of its principal properties. In § 2 we define the set φ and compare it with the set χ. In § 3 and § 4 we develop some of the properties of the set φ, and prove in particular that every continuous function of bounded variation can be expanded in terms of the φ's and that every continuous function can be so developed in the sense not of convergence of the series but of summability by the first Cesàro mean. In § 5 it is proved that there exists a continuous function which

---

* Presented to the American Mathematical Society, Feb. 25, 1922.

† *Mathematische Annalen*, Vol. 69 (1910), pp. 331–371; especially pp. 361–371.

cannot be expanded in a convergent series of the functions φ. In § 6 there is studied the nature of the approach of the approximating functions to the sum function at a point of discontinuity, and in § 7 there is considered the uniqueness of the development of a function.

§ 1. Haar's Set χ.

Consider the following set of functions:

$$f_0(x) \equiv 1, \qquad 0 \leqq x \leqq 1,$$

$$f_1^{(1)}(x) \equiv \begin{cases} 1, 0 \leqq x < \frac{1}{2}, \\ 0, \frac{1}{2} < x \leqq 1, \end{cases} \qquad f_1^{(2)}(x) \equiv \begin{cases} 1, \frac{1}{2} < x \leqq 1, \\ 0, 0 \leqq x < \frac{1}{2}, \end{cases}$$

. . . .

$$f_k^{(i)}(x) \equiv \begin{cases} 1, \frac{i-1}{2^k} < x < \frac{i}{2^k}, & i = 1, 2, 3, \cdots, 2^k, \\ 0, 0 \leqq x < \frac{i-1}{2^k}, \text{ or } \frac{i}{2^k} < x \leqq 1, & k = 1, 2, 3, \cdots, \infty; \end{cases}$$

these functions may be defined at a point of discontinuity to have the average of the limits approached on the two sides of the discontinuity.

If we have at our disposal all the functions $f_k^{(i)}$, it is clear that we can approximate to any continuous function in the interval $0 \leqq x \leqq 1$ as closely as desired and hence that we can expand any continuous function in a uniformly convergent series of functions $f_k^{(i)}$. For a continuous function $F(x)$ is uniformly continuous in the interval (0, 1), and thus uniformly in that entire interval can be approximated as closely as desired by a linear combination of the functions $f_k^{(i)}$ where k is chosen sufficiently large but fixed. The approximation can be made better and better and thus will lead to a uniformly convergent series of functions $f_k^{(i)}$.

Haar's set χ may be found by normalizing and orthogonalizing the set $f_k^{(i)}$, those functions to be ordered with increasing k, and for each k with increasing i. The set χ consists of the following functions:*

$$\chi_0(x) \equiv 1, \qquad 0 \leqq x \leqq 1, \qquad \chi_1(x) \equiv \begin{cases} 1, 0 \leqq x < \frac{1}{2}, \\ -1, \frac{1}{2} < x \leqq 1, \end{cases}$$

$$\begin{array}{lll} \chi_2^{(1)}(x) = \sqrt{2}, & \chi_2^{(2)} = 0, & 0 \leqq x < \frac{1}{4}, \\ \quad = -\sqrt{2}, & \quad = 0, & \frac{1}{4} < x < \frac{1}{2}, \\ \quad = 0, & \quad = \sqrt{2}, & \frac{1}{2} < x < \frac{3}{4}, \\ \quad = 0, & \quad = -\sqrt{2}, & \frac{3}{4} < x \leqq 1, \end{array}$$

. . . .

* L. c., p. 361.

$$\begin{aligned} \chi_n^{(k)} &= \sqrt{2^{n-1}}, & \frac{k-1}{2^{n-1}} < x < \frac{2k-1}{2^n}, & \quad k = 1, 2, 3, \cdots, 2^{n-1}, \\ &= -\sqrt{2^{n-1}}, & \frac{2k-1}{2^n} < x < \frac{k}{2^{n-1}}, & \quad n = 1, 2, 3, \cdots, \infty, \\ &= 0, & 0 < x < \frac{k-1}{2^{n-1}} \text{ or } \frac{k}{2^{n-1}} < x < 1. & \end{aligned}$$

The same convention as to the value of $\chi_n^{(k)}$ at a point of discontinuity is made as for the $f_n^{(k)}$, and $\chi_n^{(k)}(0)$ and $\chi_n^{(k)}(1)$ are defined as the limits of $\chi_n^{(k)}$ as x approaches 0 and 1.

For any particular value of N, all the functions $f_n^{(k)}$, $n < N$, can be expressed linearly in terms of the functions $\chi_n^{(k)}$, $n < N$, and conversely.

Let $F(x)$ be any function integrable and with an integrable square in the interval (0, 1); its formal development in terms of the functions χ is

$$\begin{aligned} F(x) \sim \chi_0(x) \int_0^1 F(y)\chi_0(y)dy + \chi_1(x) \int^1 F(y)\chi_1(y)dy + \cdots \\ + \chi_n^{(k)}(x) \int_0^1 F(y)\chi_n^{(k)}(y)dy + \cdots. \end{aligned} \quad (1)$$

This series (1) is formed with coefficients determined formally as for the Fourier expansions, and it is well known that $S_m(x)$, the sum of the first m terms of this series, is that linear combination $F_m(x)$ of the first m of the functions χ which renders a minimum the integral

$$\int_0^1 (F(x) - F_m(x))^2 dx.$$

That is, $S_m(x)$ is in the sense of least squares the best approximation to $F(x)$ which can be formed from a linear combination of the first m functions χ; it is likewise true that $S_m(x)$ is the best approximation to $F(x)$ which can be formed from a linear combination of those functions $f_n^{(k)}$ that are dependent on the first m functions χ.

Let $F(x)$ be continuous in the closed interval (0, 1). If ϵ is any positive number, there exists a corresponding number n such that

$$|F(x') - F(x'')| < \epsilon \qquad \text{whenever} \qquad |x' - x''| < \frac{1}{2^n}.$$

We interpret $S_{2^n}(x)$ as a linear combination of the functions $f_n^{(k)}$. The multiplier of the function $f_n^{(k)}$ which appears in $S_{2^n}(x)$ is chosen so as to furnish the best approximation in the interval $\left(\frac{k-1}{2^n}, \frac{k}{2^n}\right)$ to the function $F(x)$, so it is evident that $S_{2^n}(x)$ approximates to $F(x)$ uniformly in the entire interval (0, 1) with an approximation better than ϵ. The function

$S_{2^{n-1}}(x)$ cannot differ from $F(x)$ by more than ϵ at any point of the interval $(0, 1)$, and so for all the functions $S_{2^{n-1}}(x)$. Thus we have

THEOREM I. *If $F(x)$ is continuous in the interval $(0, 1)$, series (1) converges uniformly to the value $F(x)$ if the terms are grouped so that each group contains all the 2^{n-1} terms of a set $\chi_n^{(k)}$, $k = 1, 2, 3, \cdots, 2^{n-1}$.*

Haar proves that the series actually converges uniformly to $F(x)$ without the grouping of terms,* and establishes many other results for expansions in terms of the set χ; to some of these results we shall return later.

§ 2. The Set φ.

The set φ, which it is the main purpose of this paper to study, consists of the following functions:

$$\varphi_0(x) \equiv 1, \qquad 0 \leqq x \leqq 1, \qquad \varphi_1(x) \equiv \begin{cases} 1, & 0 \leqq x < \frac{1}{2}, \\ -1, & \frac{1}{2} < x \leqq 1, \end{cases}$$

$$\varphi_2^{(1)}(x) \equiv \begin{cases} 1, & 0 \leqq x < \frac{1}{4}, \ \frac{3}{4} < x \leqq 1, \\ -1, & \frac{1}{4} < x < \frac{3}{4}, \end{cases}$$

$$\varphi_2^{(2)}(x) \equiv \begin{cases} 1, & 0 \leqq x < \frac{1}{4}, \ \frac{1}{2} < x < \frac{3}{4}, \\ -1, & \frac{1}{4} < x < \frac{1}{2}, \ \frac{3}{4} < x \leqq 1, \end{cases}$$

. . . .

$$\varphi_{n+1}^{(2k-1)}(x) \equiv \begin{cases} \varphi_n^{(k)}(2x), & 0 \leqq x < \frac{1}{2}, \\ (-1)^{k+1}\varphi_n^{(k)}(2x-1), & \frac{1}{2} < x \leqq 1, \end{cases}$$
$$\varphi_{n+1}^{(2k)}(x) \equiv \begin{cases} \varphi_n^{(k)}(2x), & 0 \leqq x < \frac{1}{2}, \\ (-1)^{k}\varphi_n^{(k)}(2x-1), & \frac{1}{2} < x \leqq 1, \end{cases} \tag{2}$$
$$k = 1, 2, 3, \cdots, 2^{n-1}, \qquad n = 1, 2, 3, \cdots, \infty.$$

In general, the function $\varphi_n^{(1)}$, $n > 0$, is to be used, with the horizontal scale reduced one half and the vertical scale unchanged, to form the functions $\varphi_{n+1}^{(1)}$ and $\varphi_{n+1}^{(2)}$ in each of the halves $(0, \frac{1}{2})$, $(\frac{1}{2}, 1)$ of the original interval; the function $\varphi_{n+1}^{(1)}(x)$ is to be even and the function $\varphi_{n+1}^{(2)}$ odd with respect to the point $x = \frac{1}{2}$. Similarly the function $\varphi_n^{(k)}$ is to be used to form the functions $\varphi_{n+1}^{(2k-1)}$ and $\varphi_{n+1}^{(2k)}$, the former of which is even and the latter odd with respect to the point $x = \frac{1}{2}$. All the functions $\varphi_n^{(k)}$ are to be taken positive in the interval $\left(0, \frac{1}{2^n}\right)$. The function $\varphi_n^{(k)}$ is to be defined at points of discontinuity as were the functions f and χ, and at $x = 0$ to have the value 1, and at $x = 1$ to have the value $(-1)^{k+1}$.† The function

* L. c., p. 368.

† If it is desired to develop periodic functions by means of the set φ [or the similar sets f and χ] simultaneously in all the intervals $\cdots$, $(-2, -1)$, $(-1, 0)$, $(0, 1)$, $(1, 2)$, $\cdots$, it will be wise to change these definitions at $x = 0$ and $x = 1$ so that always the value of $\varphi_n^{(k)}(x)$ is the arithmetic mean of the limits approached at these points to the right and to the left.

$\varphi_n^{(k)}$ is odd or even with respect to the point $x = \frac{1}{2}$ according as k is even or odd.

The functions $\varphi_0, \varphi_1, \varphi_2^{(1)}, \varphi_2^{(2)}$ have 0, 1, 2, 3 zeroes (i.e., sign-changes) respectively interior to the interval (0, 1). The function $\varphi_{n+1}^{(2k-1)}(x)$ has twice as many zeroes as the function $\varphi_n^{(k)}$; and $\varphi_{n+1}^{(2k)}(x)$ has one more zero, namely at $x = \frac{1}{2}$, than has $\varphi_{n+1}^{(2k-1)}(x)$. Thus the function $\varphi_n^{(k)}$ has $2^{n-1} + k - 1$ zeroes; this formula holds for $n = 2$ and follows for the general case by induction. Hence each function $\varphi_n^{(k)}$ has one more zero than the preceding; the zeroes of these functions increase in number precisely as do the zeroes of the classical sets of functions—sine, cosine, Sturm-Liouville, Legendre, etc. We shall at times find it convenient to use the notation $\varphi_0, \varphi_1, \varphi_2, \cdots$ for the functions $\varphi_n^{(k)}$; the subscript denotes the number of zeroes.

The orthogonality of the system φ is easily established. Any two functions $\varphi_n^{(k)}$ are orthogonal if $n < 3$, as may be found by actually testing the various pairs of functions. Let us assume this fact to hold for $n = 1, 2, 3, \cdots, N - 1$; we shall prove that it holds for $n = N$. By the method of construction of the functions φ, each of the integrals

$$\int_0^{1/2} \varphi_N^{(k)}(x)\varphi_m^{(l)}(x)dx, \qquad \int_{1/2}^1 \varphi_N^{(k)}(x)\varphi_m^{(l)}(x)dx, \qquad m \leqq N,$$

is the same except possibly for sign as an integral

$$\int_0^1 \varphi_{N-1}^{(j)}(y)\varphi_{m-1}^{(i)}(y)dy$$

after the change of variable $y = 2x$ or $y = 2x - 1$. Each of these two integrals [in fact, they are the same integral] whose variable is y has the value zero, so we have the orthogonality of $\varphi_N^{(k)}(x)$ and $\varphi_m^{(l)}(x)$:

$$\int_0^1 \varphi_N^{(k)}(x)\varphi_m^{(l)}(x)dx = 0.$$

This proof breaks down if the two functions $\varphi_{N-1}^{(j)}(y)$, $\varphi_{m-1}^{(i)}(y)$ are the same, but in that case either $\varphi_N^{(k)}(x)$ and $\varphi_m^{(l)}(x)$ are the same and we do not wish to prove their orthogonality, or one of the functions $\varphi_N^{(k)}(x)$, $\varphi_m^{(l)}(x)$ is odd and the other even, so the two are orthogonal.

Each of the functions $\varphi_n^{(k)}(x)$ is normal, for we have

$$|\varphi_n^{(k)}(x)| \equiv 1$$

except at a finite number of points.

Each of the functions $\chi_0, \chi_1, \chi_2^{(1)}, \chi_2^{(2)}, \cdots, \chi_{n+1}^{(2^n)}$ can be expressed linearly in terms of the functions $\varphi_0, \varphi_1, \varphi_2^{(1)}, \varphi_2^{(2)}, \cdots, \varphi_{n+1}^{(2^n)}$. Thus for $n = 1$ we have

$$\chi_0 = \varphi_0, \qquad \chi_1 = \varphi_1, \qquad \chi_2^{(1)} = \tfrac{1}{2}\sqrt{2}(\varphi_2^{(1)} + \varphi_2^{(2)}), \qquad \chi_2^{(2)} = \tfrac{1}{2}\sqrt{2}(-\varphi_2^{(1)} + \varphi_2^{(2)}).$$

It is true generally that except for a constant normalizing factor $\sqrt{2}$, the function $\chi_{n+1}^{(k)}$, $k \leqq 2^{n-1}$, is the same linear combination of the functions $\frac{1}{2}[\varphi_{n+1}^{(2k-1)} + \varphi_{n+1}^{(2k)}]$ as is $\chi_n^{(k)}$ of the functions $\varphi_n^{(k)}$, and the function $\chi_{n+1}^{(k)}$, $k > 2^{n-1}$, is the same linear combination of the functions $\frac{1}{2}(-1)^{k+1}[\varphi_{n+1}^{(2k-1)} - \varphi_{n+1}^{(k)}]$ as is $\chi_n^{(k-2^{n-1})}$ of the functions $\varphi_n^{(k)}$.

It is similarly true that all the functions $\varphi_0, \varphi_1, \cdots, \varphi_{n+1}^{(2^n)}$ can be expressed linearly in terms of the functions $\chi_0, \chi_1, \cdots, \chi_{n+1}^{(2^n)}$. Thus we have for $n = 2$,

$$\varphi_0 = \chi_0, \qquad \varphi_1 = \chi_1, \qquad \varphi_2^{(1)} = \tfrac{1}{2}\sqrt{2}(\chi_2^{(1)} - \chi_2^{(2)}), \qquad \varphi_2^{(2)} = \tfrac{1}{2}\sqrt{2}(\chi_2^{(1)} + \chi_2^{(2)}).$$

The general fact appears by induction from the very definition of the functions φ.

The set χ is known to be closed;* it follows from the expression of the χ in terms of the φ that the set φ is also closed.

The definition of the functions $\varphi_n^{(k)}$ enables us to give a formula for $\varphi_n^{(k)}(x)$. Let us set, in binary notation,

$$x = \frac{a_1}{2^1} + \frac{a_2}{2^2} + \frac{a_3}{2^3} + \cdots, \qquad a_i = 0 \text{ or } 1.$$

If x is a binary irrational or if in the binary expansion of x there exists $a_i \neq 0$, $i > n$, the following formulas hold for $\varphi_n^{(k)}$:

$$\begin{aligned}
\varphi_0 &= 1, & \varphi_1 &= (-1)^{a_1},\\
\varphi_2^{(1)} &= (-1)^{a_1+a_2}, & \varphi_2^{(2)} &= (-1)^{a_2},\\
\varphi_3^{(1)} &= (-1)^{a_2+a_3}, & \varphi_3^{(2)} &= (-1)^{a_1+a_2+a_3},\\
\varphi_3^{(3)} &= (-1)^{a_1+a_3}, & \varphi_3^{(4)} &= (-1)^{a_3},\\
\varphi_4^{(1)} &= (-1)^{a_3+a_4}, & \varphi_4^{(2)} &= (-1)^{a_1+a_3+a_4},\\
\varphi_4^{(3)} &= (-1)^{a_1+a_2+a_3+a_4}, & \varphi_4^{(4)} &= (-1)^{a_2+a_3+a_4},\\
\varphi_4^{(5)} &= (-1)^{a_2+a_4}, & \varphi_4^{(6)} &= (-1)^{a_1+a_2+a_4},\\
\varphi_4^{(7)} &= (-1)^{a_1+a_4}, & \varphi_4^{(8)} &= (-1)^{a_4},\\
&\cdots\cdots\cdots\cdots\cdots\cdots,\\
&\cdots\cdots\cdots\cdots\cdots\cdots.
\end{aligned} \tag{3}$$

The general law appears from these relations; always we have

$$\begin{aligned}
\varphi_n^{(1)} &= (-1)^{a_{n-1}+a_n},\\
\varphi_n^{(k)} &= \varphi_{k-1}\varphi_n^{(1)}.
\end{aligned} \tag{4}$$

A general expression for $\varphi_n^{(k)}(x)$ when x is a binary rational can readily be computed from formulas (3), for we have expressions for the values of $\varphi_n^{(k)}$ for neighboring larger and smaller values of the argument than x.

* That is, there exists no non-null Lebesgue-integrable function on the interval (0, 1) which is orthogonal to all functions of the set; l. c., p. 362.

§ 3. Expansions in Terms of the Set $\{\varphi\}$.

The following theorem results from Theorem I by virtue of the remark that all the functions $\varphi_n^{(k)}$ can be expressed in terms of the functions $\chi_n^{(i)}$ and conversely, and from the least squares interpretation of a partial sum of a series of orthogonal functions:

THEOREM II. *If $F(x)$ is continuous in the interval* (0, 1), *the series*

$$F(x) \sim \varphi_0(x) \int_0^1 F(y)\varphi_0(y)dy + \varphi_1(x) \int_0^1 F(y)\varphi_1(y)dy + \cdots \varphi_i^{(j)}(x) \int_0^1 F(y)\varphi_i^{(j)}(y)dy + \cdots, \tag{5}$$

converges uniformly to the value $F(x)$ if the terms are grouped so that each group contains all the 2^{n-1} terms of a set $\varphi_n^{(k)}$, $k = 1, 2, 3, \cdots, 2^{n-1}$.

Series (5) after the grouping of terms is precisely the same as series (1) after the grouping of terms.

Theorem II can be extended to include even discontinuous functions $F(x)$; we suppose $F(x)$ to be integrable in the sense of Lebesgue. Let us introduce the notation

$$F(a+0) = \lim_{\epsilon=0} F(a+\epsilon), \qquad F(a-0) = \lim_{\epsilon=0} F(a-\epsilon), \qquad \epsilon > 0,$$

and suppose that these limits exist for a particular point $x = a$. We introduce the functions

$$F_1(x) = \begin{cases} F(x), & x < a, \\ F(a-0), & x \geqq a, \end{cases} \qquad F_2(x) = \begin{cases} F(a+0), & x \leqq a, \\ F(x), & x > a, \end{cases} \tag{6}$$

The least squares interpretation of the partial sums $S_{2^n}(x)$ of the series (1) or (5) as expressed in terms of the $f_i^{(j)}$ gives the result that if $h_1 < F(x) < h_2$ in any interval, then also $h_1 < S_{2^n}(x) < h_2$ in any completely interior interval if n is sufficiently large. It follows that $F_1(x)$ is closely approximated at $x = a$ by its partial sum S_{2^n} if n is sufficiently large, and that this approximation is uniform in any interval about the point $x = a$ in which $F_1(x)$ is continuous. A similar result holds for $F_2(x)$.

The function $F_1(x) + F_2(x)$ differs from the original function $F(x)$ merely by the function

$$G(x) = \begin{cases} F(a+0), & x < a, \\ F(a-0), & x > a. \end{cases}$$

The representation of such functions by sequences of the kind we are considering will be studied in more detail later (§ 6), but it is fairly obvious that such a function is represented uniformly except in the neighborhood of the point a. If $F(x)$ is continuous at and in the neighborhood of a, or if a is dyadically rational, the approximation to $G(x)$ is uniform at the point a as well. Thus we have

THEOREM III. *If $F(x)$ is any integrable function and if $\lim_{x=a} F(x)$ exists for a point a, then when the terms of the series (5) are grouped as described in Theorem II, the series so obtained converges for $x = a$ to the value $\lim_{x=a} F(x)$. If $F(x)$ is continuous at and in the neighborhood of a, then this convergence is uniform in a neighborhood of a.*

If $F(x)$ is any integrable function and if the limits $F(a - 0)$ and $F(a + 0)$ exist for a dyadically rational point $x = a$, then the series with the terms grouped converges for $x = a$ to the value $\frac{1}{2}[F(a + 0) + F(a - 0)]$; this convergence is uniform in the neighborhood of the point $x = a$ if $F(x)$ is continuous on two intervals extending from a, one in each direction.

It is now time to study the convergence of series (5) when the terms are not grouped as in Theorems II and III. We shall establish

THEOREM IV. *Let the function $F(x)$ be of limited variation in the interval $0 \leqq x \leqq 1$. Then the series (5) converges to the value $F(x)$ at every point at which $F(a + 0) = F(a - 0)$ and at every point at which $x = a$ is dyadically rational. This convergence is uniform in the neighborhood of $x = a$ in each of these cases if $F(x)$ is continuous in two intervals extending from a, one in each direction.*

Since $F(x)$ is of limited variation, $F(a + 0)$ and $F(a - 0)$ exist at every point a. Theorem IV tacitly assumes $F(x)$ to be defined at every point of discontinuity a so that $F(a) = \frac{1}{2}[F(a + 0) + F(a - 0)]$.

Any such function $F(x)$ can be considered as the difference of two monotonically increasing functions, so the theorem will be proved if it is proved merely for a monotonically increasing function. We shall assume that $F(x)$ is such a function, and positive. We are to evaluate the limit of

$$\int_0^1 F(y) K_n^{(k)}(x, y) dy,$$

$$K_n^{(k)}(x, y) = \varphi_0(x)\varphi_0(y) + \varphi_1(x)\varphi_1(y) + \cdots + \varphi_n^{(k)}(x)\varphi_n^{(k)}(y).$$

We have already evaluated this limit for the sequence $k = 2^{n-1}$, so it remains merely to prove that

$$\lim_{n=\infty} \int_0^1 F(y) Q_n^{(k)}(x, y) dy = 0, \tag{7}$$

$$Q_n^{(k)}(x, y) = \varphi_n^{(1)}(x)\varphi_n^{(1)}(y) + \varphi_n^{(2)}(x)\varphi_n^{(2)}(y) + \cdots + \varphi_n^{(k)}(x)\varphi_n^{(k)}(y),$$

whatever may be the value of k.

We shall consider the function $F(x)$ merely at a point $x = a$ of continuity; that is, we study essentially the new functions F_1 and F_2 defined by equations (6). In the sequel we suppose a to be dyadically irrational; the necessary modifications for a rational can be made by the reader.

The following formulas are easily found by the definition of the $Q_n^{(k)}$; both x and y are supposed dyadically irrational:

$$Q_2^{(1)}(x, y) = \pm 1,$$

$$Q_2^{(2)}(x, y) = \begin{cases} 0 \text{ if } x < \tfrac{1}{2}, y > \tfrac{1}{2} \text{ or if } x > \tfrac{1}{2}, y < \tfrac{1}{2}, \\ \pm 2 \text{ if } x < \tfrac{1}{2}, y < \tfrac{1}{2} \text{ or if } x > \tfrac{1}{2}, y > \tfrac{1}{2}, \end{cases}$$

. . . .

$$Q_n^{(1)}(x, y) = \pm 1,$$

$$Q_n^{(2)}(x, y) = \begin{cases} 0 \text{ if } x < \tfrac{1}{2}, y > \tfrac{1}{2} \text{ or if } x > \tfrac{1}{2}, y < \tfrac{1}{2}, \\ 2Q_{n-1}^{(1)}(2x, 2y) \text{ if } x < \tfrac{1}{2}, y < \tfrac{1}{2}, \\ 2Q_{n-1}^{(1)}(2x-1, 2y-1) \text{ if } x > \tfrac{1}{2}, y > \tfrac{1}{2}, \end{cases}$$

. . . .

$$Q_n^{(2k)}(x, y) = \begin{cases} 0 \text{ if } x < \tfrac{1}{2}, y > \tfrac{1}{2} \text{ or if } x > \tfrac{1}{2}, y < \tfrac{1}{2}, \\ 2Q_{n-1}^{(k)}(2x, 2y) \text{ if } x < \tfrac{1}{2}, y < \tfrac{1}{2}, \\ 2Q_{n-1}^{(k)}(2x-1, 2y-1) \text{ if } x > \tfrac{1}{2}, y > \tfrac{1}{2}, \end{cases}$$

$$Q_n^{(2k+1)}(x, y) = \begin{cases} \pm 1 \text{ if } x < \tfrac{1}{2}, y > \tfrac{1}{2} \text{ or if } x > \tfrac{1}{2}, y < \tfrac{1}{2}, \\ \dfrac{Q_n^{(2k)} + Q_n^{(2k+2}}{2} \text{ if } x < \tfrac{1}{2}, y < \tfrac{1}{2} \text{ or if } x > \tfrac{1}{2}, y > \tfrac{1}{2}. \end{cases}$$

The integral in (7) for $x = a$ is to be divided into three parts. Consider an interval bounded by two points of the form $x = \frac{\rho}{2^\nu}$, $x = \frac{\rho+1}{2^\nu}$, where ρ and ν are integers and such that

$$\frac{\rho}{2^\nu} < a < \frac{\rho+1}{2^\nu}.$$

Then we have

$$\int_0^1 F_1(y)Q_n^{(k)}(a, y)dy = \int_0^{\rho/2^\nu} F_1(y)Q_n^{(k)}(a, y)dy$$
$$+ \int_{\rho/2^\nu}^{(\rho+1)/2^\nu} F_1(y)Q_n^{(k)}(a, y)dy + \int_{(\rho+1)/2^\nu}^1 F_1(y)Q_n^{(k)}(a, y)dy. \tag{8}$$

These integrals on the right need separate consideration.

Let us set

$$\frac{\rho}{2^\nu} = \frac{\mu_1}{2^1} + \frac{\mu_2}{2^2} + \frac{\mu_3}{2^3} + \cdots + \frac{\mu_\nu}{2^\nu}, \qquad \mu_i = 0 \text{ or } 1.$$

The first integral in the right-hand member of (8) can be written

$$\int_0^{\mu_1/2^1} + \int_{\mu_1/2^1}^{(\mu_1/2^1)+(\mu_2/2^2)} + \cdots \int_{(\rho/2^\nu)-(\mu_\nu/2^\nu)}^{\rho/2^\nu} F_1(y)Q_n^{(k)}(a, y)dy. \tag{9}$$

Each of these integrals is readily treated. Thus, on the interval $0 \leqq y \leqq \frac{\mu_1}{2^1}$, $Q_n^{(k)}(a, y)$ takes only the values ± 1 or 0, is 0 if k is even and has the value $\pm \varphi_n^{(k)}(y)$ if k is odd. It is of course true that

$$\lim_{n=\infty} \int_0^1 \Phi(y)\varphi_n^{(k)}(y)dy = 0 \tag{10}$$

no matter what may be the function $\Phi(y)$ integrable in the sense of Lebesgue and with an integrable square.* Hence we have

$$\lim_{n=\infty} \int_0^{\mu_1/2^1} F_1(y)Q_n^{(k)}(a, y)dy = 0.$$

On the interval $\frac{\mu_1}{2^1} \leqq y \leqq \frac{\mu_1}{2^1} + \frac{\mu_2}{2^2}$, the function $Q_n^{(k)}(a, y)$ takes only the values 0, ± 1, ± 2, and except for one of these numbers as constant factor, has the value $\varphi_n^{(k)}(y)$. It is thus true that

$$\lim_{n=\infty} \int_{\mu_1/2^1}^{(\mu_1/2^1)+(\mu_2/2^2)} F_1(y)Q_n^{(k)}(a, y)dy = 0.$$

From the corresponding result for each of the integrals in (9) and a similar treatment of the last integral in the right-hand member of (8), we have

$$\begin{aligned} \lim_{n=\infty} \int_0^{p/2^\nu} F_1(y)Q_n^{(k)}(a, y)dy &= 0, \\ \lim_{n=\infty} \int_{(p+1)/2^\nu}^1 F_1(y)Q_n^{(k)}(a, y)dy &= 0. \end{aligned} \tag{11}$$

We shall obtain an upper limit for the second integral in (8) by the second law of the mean. We notice that

$$\left| \int_\xi^{(p+1)/2^\nu} Q_n^{(k)}(a, y)dy \right| \leqq \tfrac{1}{2},$$

whatever may be the value of ξ. In fact, this relation is immediate if n

---

* This well-known fact follows from the convergence of the series

$$\Sigma\, (a_n^{(k)})^2,$$

proved from the inequality

$$\int_0^1 (\Phi(x) - a_0\varphi_0 - a_1\varphi_1 - a_2^{(1)}\varphi_2^{(1)} - \cdots - a_n^{(k)}\varphi_n^{(k)})^2 dx \geqq 0,$$

where $a_n^{(k)} = \int_0^1 \Phi(y)\varphi_n^{(k)}(y)dy$.

is small and it follows for the larger values of n by virtue of the method of construction of the $Q_n^{(k)}$. Moreover, if $n \geqq \nu$ and if $\xi = \frac{\rho}{2^\nu}$, this integral has the value zero. We therefore have from the second law of the mean, $n \geqq \nu$,

$$\int_{\rho/2^\nu}^{(\rho+1)/2^\nu} F_1(y) Q_n^{(k)}(a, y)dy = F_1\left(\frac{\rho}{2^\nu}\right) \int_{\rho/2^\nu}^{\xi} Q_n^{(k)}(a, y)dy$$

$$+ F_1\left(\frac{\rho+1}{2^\nu}\right) \int_{\xi}^{(\rho+1)/2^\nu} Q_n^{(k)}(a, y)dy$$

$$= \left[F_1(a) - F_1\left(\frac{\rho}{2^\nu}\right)\right] \int_{\xi}^{(\rho+1)/2^\nu} Q_n^{(k)}(a, y)dy.$$

By a proper choice of the point $\frac{\rho}{2^\nu}$ we can make the factor of this last integral as small as desired; the entire expression will be as small as desired for sufficiently large n. The relations (11) are independent of the choice of $\frac{\rho}{2^\nu}$, so (7) is completely proved for the function F_1. A similar proof applies to F_2, so (7) can be considered as completely proved for the original function $F(x)$.

The uniform convergence of (5) as stated in Theorem IV follows from the uniform continuity of $F(x)$ and will be readily established by the reader.

§ 4. Further Expansion Properties of the Set φ.

The least square interpretation already given for the partial sums and the expression of the φ's in terms of the f's show that if the terms of (5) are grouped as in Theorems II and III, the question of convergence or divergence of the series at a point depends merely on that point and the nature of the function $F(x)$ in the neighborhood of that point. This same fact for series (5) when the terms are not grouped follows from (8) and (10) if $F(x)$ is integrable and with an integrable square. We shall further extend this result and prove:

THEOREM V. *If $F(x)$ is any integrable function, then the convergence or divergence of the series (5) at a point depends merely on that point and on the behavior of the function in the neighborhood of that point. If in particular $F(x)$ is of limited variation in the neighborhood of a point $x = a$, and if a is dyadically rational or if $F(a - 0) = F(a + 0)$, then series (5) converges for $x = a$ to the value $\frac{1}{2}[F(a - 0) + F(a + 0)]$. If $F(x)$ is not only of limited variation but is also continuous in two neighborhoods one on each side of a, and if a is dyadically rational or if $F(a - 0) = F(a + 0)$, the convergence of (5) is uniform in the neighborhood of the point a.*

Theorem V follows immediately from the reasoning already given and from (10) proved without restriction on Φ; we state the theorem for any bounded normal orthogonal set of functions ψ_n:

THEOREM VI. *If* $\{\psi_n(x)\}$ *is a uniformly bounded set of normal orthogonal functions on the interval* (0, 1), *and if* $\Phi(x)$ *is any integrable function, then*

$$\lim_{n=\infty} \int_0^1 \Phi(x)\psi_n(x)dx = 0. \qquad (12)$$

Denote by E the point set which contains all points of the interval for which $|\Phi(x)| > N$; we choose N so large that

$$\int_E |\Phi(x)|dx < \epsilon,$$

where ϵ is arbitrary. Denote by E_1 the point set complementary to E; then we have

$$\int_0^1 \Phi(x)\psi_n(x)dx = \int_E \Phi(x)\psi_n(x)dx + \int_{E_1} \Phi(x)\psi_n(x)dx.$$

It follows from the proof of (10) already indicated that the second integral on the right approaches zero as n becomes infinite. The first integral is in absolute value less than $M\epsilon$ whatever may be the value of n, where M is the uniform bound of the ψ_n. It therefore follows that these two integrals can be made as small as desired, first by choosing ϵ sufficiently small and then by choosing n sufficiently large.*

It is interesting to note that Theorem VI breaks down if we omit the hypothesis that the set ψ_n is uniformly bounded. In fact Theorem VI does not hold for Haar's set χ. Thus consider the function

$$\Phi(x) = (x - \tfrac{1}{2})^{-\nu}, \qquad \nu < 1.$$

We have

$$\int_0^1 \Phi(x)\chi_n^{(2^{n-2}+1)}(x)dx = \sqrt{2^{n-1}} \int_{1/2}^{1/2+1/2^n} (x - \tfrac{1}{2})^{-\nu}dx$$

$$- \sqrt{2^{n-1}} \int_{1/2+1/2^n}^{1/2+1/2^{n-1}} (x - \tfrac{1}{2})^{-\nu}dx = \frac{(2^{n-1})^{\nu-(1/2)}}{1-\nu}[2^\nu - 1].$$

Whenever $\nu \geqq \frac{1}{2}$, it is clear that (12) cannot hold, and if $\nu > \frac{1}{2}$, there is a sub-sequence of the sequence in (12) which actually becomes infinite.

* Theorem VI is proved by essentially this method for the set $\psi_n(x) = \sqrt{2}\sin n\pi x$ by Lebesgue, *Annales scientifiques de l'école normale supérieure*, ser. 3, Vol. XX, 1903. See also Hobson, *Functions of a Real Variable* (1907), p. 675, and Lebesque, *Annales de la Faculté des Science de Toulouse*, ser 3, Vol. I (1909), pp. 25–117, especially p. 52.

We turn now from the study of the convergence of such a series expansion as (5) to the study of the summability of such expansions, and are to prove

THEOREM VII. *If* $F(x)$ *is continuous in the closed interval* (0, 1), *the series* (5) *is summable uniformly in the entire interval to the sum* $F(x)$.

If $F(x)$ *is integrable in the interval* (0, 1), *and if* $F(a-0)$ *and* $F(a+0)$ *exist, and if either* $F(a-0) = F(a+0)$ *or* a *is dyadically rational, then the series* (5) *is summable for* $x = a$ *to the value* $\frac{1}{2}[F(a-0) + F(a+0)]$. *If* $F(x)$ *is continuous in the neighborhood of the point* $x = a$, *or if* a *is dyadically rational and* $F(x)$ *continuous in the neighborhood of* a *except for a finite jump at* a, *the summability is uniform throughout a neighborhood of that point.*

In this theorem and below, the term *summability* indicates summability by the first Cesàro mean.

We shall find it convenient to have for reference the following

LEMMA. *Suppose that the series*

$$(b_1 + b_2 + \cdots + b_{n_1}) + (b_{n_1+1} + b_{n_1+2} + \cdots + b_{n_2}) + \cdots + (b_{n_k+1} + b_{n_k+2} + \cdots + b_{n_{k+1}}) + \cdots \tag{13}$$

converges to the sum B and that the sequence

$$\begin{gathered} b_1, \quad \frac{2b_1 + b_2}{2}, \quad \frac{3b_1 + 2b_2 + b_3}{3}, \quad \cdots \quad \frac{(n_1-1)b_1 + (n_1-2)b_2 + \cdots + b_{n_1-1}}{n_1 - 1}, \\ \frac{(n_1-1)b_1 + \cdots + b_{n_1-1}}{n}, \quad \frac{(n_1-1)b_1 + (n_1-2)b_2 + \cdots + b_{n_1-1} + b_{n_1+1}}{n_1+1}, \\ \frac{(n_1-1)b_1 + \cdots + b_{n_1-1} + 2b_{n_1+1} + b_{n_1+2}}{n_1+2}, \quad \cdots \\ \frac{(n_1-1)b_1 + \cdots + b_{n_1-1} + (n_2-n_1-1)b_{n_1+1} + (n_2-n_1-2)b_{n_1+2} + \cdots + b_{n_2-1}}{n_2-1}, \\ \cdots, \end{gathered} \tag{14}$$

converges to zero. Then the series

$$b_1 + b_2 + b_3 + \cdots \tag{15}$$

is summable to the sum B.

This lemma involves merely a transformation of the formulas involving the limit notions. Insert zeroes in series (13) so that the parentheses are respectively the n_1-th, n_2-th, n_3-th terms of the new series; this new series

2

converges to the sum B and hence is summable to the sum B. The term-by-term difference of the new series and (15) is the series

$$b_1 + b_2 + \cdots + b_{n_1-1} - (b_1 + b_2 + \cdots + b_{n_1-1}) + b_{n_1+1} + b_{n_1+2} + \cdots + b_{n_2-1} - (b_{n_1+1} + b_{n_1+2} + \cdots + b_{n_2-1}) + \cdots, \tag{16}$$

which is to be shown to be summable to the sum zero. The sequence corresponding to the summation of (16) is precisely (14).

A sufficient condition for the convergence to zero of (14) is that we have, independently of m,

$$\lim_{k=\infty} \frac{mb_{n_k+1} + (m-1)b_{n_k+2} + \cdots + b_{n_k+m}}{m} = 0, \qquad m \leqq n_{k+1} - n_k, \tag{17}$$

for from a geometric point of view each term of the sequence (14) is the center of gravity of a number of terms such as occur in (17), each term weighted according to the number of b_i that appear in it. An (ϵ, δ)-proof can be supplied with no difficulty.

For the case of Theorem VII let us assume $F(x)$ integrable and that $F(a-0)$ and $F(a+0)$ exist. The series (15) is to be identified with the series (5), and (13) with (5) after the terms are grouped as in Theorem III. The sum that appears in (17) is, then, for $x = a$,

$$\frac{1}{m}\int_0^1 [m\varphi_n^{(1)}(a)\varphi_n^{(1)}(y) + (m-1)\varphi_n^{(2)}(a)\varphi_n^{(2)}(y) + \cdots + \varphi_n^{(m)}(a)\varphi_n^{(m)}(y)]F(y)dy, \quad m \leqq 2^{n-1}. \tag{18}$$

We shall prove that (18) formed for the function $F_1(y)$ defined in (6) and for a dyadically irrational has the limit zero as n becomes infinite.

Let us notice that

$$\frac{1}{m}\int_0^1 |m\varphi_n^{(1)}(a)\varphi_n^{(1)}(y) + (m-1)\varphi_n^{(2)}(a)\varphi_n^{(2)}(y) + \cdots + \varphi_n^{(m)}(a)\varphi_n^{(m)}(y)|dy = 1. \tag{19}$$

This follows directly from (3) and (4). The value of the integral in (19) is unchanged if we replace a by any dyadic irrational b. Choose $0 < b < 2^{-n}$, so that all the functions $\varphi_0, \varphi_1, \varphi_2, \cdots, \varphi_{m-1}$ are positive for $x = b$. Then the integrand in (19) can be reduced merely to $m\varphi_0(y)$, so (19) is proved.

Let us consider the integral (18) formed for the function $F_1(y)$ to be divided as in (8), where as before

$$\frac{\rho}{2^\nu} < a < \frac{\rho+1}{2^\nu},$$

and let us denote by (20), (21), (22), (23) respectively the entire integral and its three parts. Then (22) can be made as small as desired simply by proper choice of the point $\frac{\rho}{2^\nu}$, for in the interval $\left(\frac{\rho}{2^\nu}, \frac{\rho+1}{2^\nu}\right)$ we can make $|F_1(y) - F_1(a)|$ uniformly small, we have established (19), and we have also

$$\int_{\rho/2^\nu}^{(\rho+1)/2^\nu} [m\varphi_n^{(1)}(a)\varphi_n^{(1)}(y) + (m-1)\varphi_n^{(2)}(a)\varphi_n^{(2)}(y) + \cdots + \varphi_n^{(m)}(a)\varphi_n^{(m)}(y)]F_1(a)dy = 0$$

if merely $n > \nu$.

The integral (21) is the average of m integrals of the type that appear in (8):

$$\int_0^{p/2^\nu} F_1(y)Q_n^{(k)}(a, y)dy, \qquad k = 1, 2, \cdots, m.$$

Thus the entire integral (21) approaches zero as n becomes infinite. Treatment in a similar way of the integral (23) proves that (20) approaches zero. It is likewise true that (18) formed for the function $F_2(y)$ also approaches zero as n becomes infinite. This completes the proof of the second sentence in Theorem VII for a dyadic irrational; we omit the proof for a dyadic rational. The uniformity of the continuity of $F(x)$ gives us readily the remaining parts of Theorem VII.

§ 5. Not Every Continuous Function Can Be Expanded in Terms of the φ.

The summability of the expansions of continuous functions in terms of the functions φ is another point of resemblance of those functions to the Fourier sine and cosine functions. Still another point of resemblance which we shall now establish is that there exists a continuous function whose expansion in terms of the φ's does not converge at every point of the interval.

Our proof rests on a beautiful theorem due to Haar,* by virtue of which the existence of such a continuous function will be shown if we prove merely that

$$\int_0^1 |K_n^{(k)}(a, y)| dy \tag{24}$$

is not bounded uniformly for all n and k. The point a is a point of divergence of the expansion of the continuous function and for our particular case may be chosen any point of the interval (0, 1). We shall study (24) in detail merely for a dyadically irrational; the integral (24) is independent of the point a chosen if a is dyadically irrational.

* L. c., p. 335. This condition holds for any set of normal orthogonal functions and is necessary as well as sufficient, if a slight restriction is added.

The integral (24) is bounded uniformly for all the values n if $k = 2^{n-1}$, so it will be sufficient to consider the integral

$$c_n^{(k)} = \int_0^1 |Q_n^{(k)}(a, y)|\,dy.$$

The following table shows the value of $c_n^{(k)}$ for small values of n and for each value of k:

$n = 2$				1								1				
$n = 3$			1				1				$1\frac{1}{2}$				1	
$n = 4$		1		1		$1\frac{1}{2}$		1		$1\frac{3}{4}$		$1\frac{1}{2}$		$1\frac{3}{4}$		1
$n = 5$	1,	1,	$1\frac{1}{2}$,	1,	$1\frac{3}{4}$,	$1\frac{1}{2}$,	$1\frac{3}{4}$,	1,	$1\frac{7}{8}$,	$1\frac{3}{4}$,	$2\frac{1}{8}$,	$1\frac{1}{2}$,	$2\frac{1}{8}$,	$1\frac{3}{4}$,	$1\frac{7}{8}$,	1,
	.	.	.	.	.	.	.	.	.	.	.	.	.	.	.	.

We have the general formulas

$$c_n^{(1)} = c_n^{(2n+1)} = 1,$$
$$c_n^{(k)} = c_{n+1}^{(2k)},$$
$$c_{n+1}^{(2k+1)} = \tfrac{1}{2}\left[c_n^{(k)} + c_n^{(k+1)}\right] + \tfrac{1}{2},$$

so the $c_n^{(k)}$ are not uniformly bounded.

THEOREM VIII. *If a point a is arbitrarily chosen, there will exist a continuous function whose φ-development does not converge at a.*

§ 6. The Approximation to a Function at a Discontinuity.

We have considered in § 3 and § 4 with a fair degree of completeness the nature of the approach to $F(x)$ of the formal development of an arbitrary function $F(x)$ in the neighborhood of a point of continuity of $F(x)$. We shall now consider the approach to $F(x)$ of this formal development in the neighborhood of a point of discontinuity of $F(x)$. We study this problem merely for a function which is constant except for a single discontinuity, a finite jump, but this leads directly to similar results for any function $F(x)$ at an isolated discontinuity which is a finite jump, if $F(x)$ is of such a nature that the expansion of $F(x)$ would converge uniformly in the neighborhood of the point of discontinuity were that discontinuity removed by the addition of a function constant except for a finite jump.

Let us consider the function

$$f(x) = \begin{cases} 1, & 0 \leqq x < a, \\ 0, & a < x \leqq 1. \end{cases}$$

If a is dyadically rational, $f(x)$ can be expressed as a finite sum of functions φ,* and thus is represented uniformly, if we make the definition $f(a)$

* A discontinuity at $x = 0$ or $x = 1$ is slightly different [compare the first footnote of § 2]. Under the present definition of the φ's it acts like an artificial discontinuity in the interior of the interval and has no effect on the sequence representing the function.

$= \frac{1}{2}[f(a-0)+f(a+0)]$; this follows from the evident possibility of expanding $f(x)$ in terms of the functions $f_0, f_1, f_2^{(1)}, \cdots$.

If the point a is dyadically irrational, $f(x)$ *cannot be expanded in terms of the* φ. The formal development of $f(x)$ converges in fact for every value of x other than a and diverges for $x = a$.* The convergence for $x \neq a$ follows, indeed, from Theorem IV. We proceed to demonstrate the divergence.

Use the dyadic notation

$$a = \frac{a_1}{2^1} + \frac{a_2}{2^2} + \frac{a_3}{2^3} + \cdots, \qquad a_n = 0 \text{ or } 1.$$

The partial sum

$$S_n^{(k)}(x) = \varphi_0(x)\int_0^1 f(y)\varphi_0(y)dy + \varphi_1(x)\int_0^1 f(y)\varphi_1(y)dy + \cdots + \varphi_n^{(k)}(x)\int_0^1 f(y)\varphi_n^{(k)}(y)dy$$

is in the sense of least squares the best approximation to $f(x)$ that can be formed from the functions $\varphi_0, \varphi_1, \cdots, \varphi_n^{(k)}$. It is therefore true that when $k = 2^{n-1}$, on every subinterval $\left(\frac{r}{2^n}, \frac{r+1}{2^n}\right)$ on which $f(x)$ is constant, $S_n^{(k)}(x)$ is also constant and equal to $f(x)$. On that subinterval $\left(\frac{m}{2^n}, \frac{m+1}{2^n}\right)$ which contains the point a, $S_n^{(k)}$ has the value

$$2^n a - m = \frac{a_{n+1}}{2^1} + \frac{a_{n+2}}{2^2} + \frac{a_{n+3}}{2^3} + \cdots, \tag{25}$$

which lies between zero and unity. Thus $S_n^{(k)}(x)$ $[n > 1]$ is a function with two points of discontinuity and which takes on three distinct values at its totality of points of continuity.

The infinite series corresponding to the sequence (25) is

$$\left(\frac{a_2}{2^1} + \frac{a_3}{2^2} + \frac{a_4}{2^3} + \cdots\right) + \left(\frac{a_3}{2^2} + \frac{a_4}{2^3} + \cdots - \frac{a_2}{2}\right) + \left(\frac{a_4}{2^2} + \frac{a_5}{2^3} + \frac{a_6}{2^4} + \cdots - \frac{a_3}{2}\right) + \left(\frac{a_5}{2^2} + \frac{a_6}{2^3} + \frac{a_7}{2^4} + \cdots - \frac{a_4}{2}\right) + \cdots. \tag{26}$$

Not all the numbers a_n after a certain point can be zero and not all of them

* This was pointed out for the set χ by Faber, *Jahresbericht der deutschen Mathematiker-Vereinigung*, Vol. 19 (1910), pp. 104–112.

can be unity, so the general term of the series (26) cannot approach zero and the sequence (25) cannot converge.

It is likewise true that the sequence (25) is not always summable and if summable may not be summable to the value $\frac{1}{2}$. Thus if we choose

$$a = \frac{1}{2} + \frac{1}{2^2} + \frac{0}{2^3} + \frac{1}{2^4} + \frac{1}{2^5} + \frac{0}{2^6} + \frac{1}{2^7} + \cdots,$$

the sequence (25) is summable to the sum $\frac{2}{3}$. Likewise the sequence $S_n^{(k)}(x)$ for $x = a$ and where we consider all values of n and k, is summable to the value $\frac{2}{3}$.

The general behavior of $S_n^{(k)}(x)$ for $f(x)$ where we do not make the restriction $k = 2^{n-1}$ is quite easily found from the behavior for $k = 2^{n-1}$ and the relation

$$\varphi_n^{(i)}(a) \int_0^1 f(y)\varphi_n^{(i)}(y)dy = \varphi_n^{(k)}(a) \int_0^1 f(y)\varphi_n^{(k)}(y)dy,$$

which holds for all values of i, k, and n.

In fact there occurs a phenomenon quite analogous to Gibbs's phenomenon for Fornier's series. For the set φ, the approximating functions are uniformly bounded. The peaks of the approximating function $S_n^{(k)}$ disappear entirely for $k = 2^{n-1}$ but reappear (usually altered in height) for for larger values of n.

It is clear that the facts concerning the approximating curves for $f(x)$ hold without essential modification for a function of limited variation at a simple finite discontinuity, and that the facts for the summation of the approximating sequence hold without essential modification for a function continuous except at a simple finite discontinuity.

§ 7. The Uniqueness of Expansions.

We now study the possibility of a series of the form

$$a_0\varphi_0(x) + a_1\varphi_1(x) + \cdots + a_n\varphi_n(x) + \cdots \tag{27}$$

which converges on $0 \leqq x \leqq 1$ to the sum zero, with the possible exception of a certain number of points x. Faber has pointed out* that there exists a series of the functions $\chi_n^{(k)}(x)$ which converges to zero except at one single point, and the convergence is uniform except in the neighborhood of that point.

We state for reference the easily proved

Lemma. *If the series* (27) *converges for even one dyadically irrational value of x, then* $\lim_{n=\infty} a_n = 0$.

* L. c., p. 111.

This lemma results immediately from the fact that $\varphi_n^{(k)}(x) = \pm 1$ if x is dyadically irrational.*

We shall now use this lemma to establish

THEOREM IX. *If the series* (27) *converges to the sum zero uniformly except in the neighborhood of a single value of x, then $a_n = 0$ for every n.*

We phrase the argument to apply when this exceptional value x_1 is dyadically irrational. If $x_1 > \frac{1}{2}$, we have for $0 \leqq x \leqq \frac{1}{2}$,

$$a_0\varphi_0(x) + a_1\varphi_1(x) + \cdots + a_n\varphi_n(x) + \cdots = 0,$$
$$(a_0 + a_1)\varphi_0(y) + (a_2 + a_3)\varphi_1(y) + (a_4 + a_5)\varphi_2(y) + \cdots = 0,$$

for every value of $y = 2x$. Then we have from the uniformity of the convergence,

$$a_0 + a_1 = 0, \qquad a_2 + a_3 = 0, \qquad a_4 + a_5 = 0, \qquad \cdots. \tag{28}$$

If $x_1 < \frac{3}{4}$, we have for $\frac{3}{4} \leqq x \leqq 1$,

$$a_0\varphi_0(x) + a_1\varphi_1(x) + \cdots + a_n\varphi_n(x) + \cdots = 0,$$

or for $0 \leqq y \leqq 1$, $y = 4x - 3$,

$$(a_0 - a_1 + a_2 - a_3)\varphi_0(y) + (a_4 - a_5 + a_6 - a_7)\varphi_1(y) + (a_{4n} - a_{4n+1} + a_{4n+2} - a_{4n+3})\varphi_n(y) + \cdots = 0.$$

From the uniformity of the convergence we have

$$a_0 - a_1 + a_2 - a_3 = 0,$$
$$a_4 - a_5 + a_6 - a_7 = 0,$$
$$\cdot \quad \cdot \quad \cdot \quad \cdot \quad \cdot \quad \cdot,$$

or from (28),

$$a_0 = -a_1 = -a_2 = a_3,$$
$$a_4 = -a_5 = -a_6 = a_7,$$
$$\cdot \quad \cdot \quad \cdot \quad \cdot \quad \cdot \quad \cdot.$$

If $x_1 > \frac{5}{8}$, we have for $\frac{5}{8} \leqq x \leqq \frac{3}{4}$,

$$a_0\varphi_0(x) + a_1\varphi_1(x) + \cdots = 0,$$

or for $0 \leqq y \leqq 1$, $y = 8x - 5$,

$$(a_0 - a_1 - a_2 + a_3 - a_4 + a_5 + a_6 - a_7)\varphi_0(y) + (a_8 - a_9 - a_{10} + a_{11} - a_{12} + a_{13} + a_4 - a_{15})\varphi_1(y) + \cdots = 0.$$

Then each of these coefficients must vanish, and hence

$$a_0 = -a_1 = -a_2 = a_3 = a_4 = -a_5 = -a_6 = a_7.$$

* This lemma is closely connected with a general theorem due to Osgood, *Transactions of the American Mathematical Society*, Vol. 10 (1909), pp. 337–346.

See also Plancherel, *Mathematische Annalen*, Vol. 68 (1909–1910), pp. 270–278.

Continuation in this way together with the Lemma shows that every a_n must vanish. This reasoning is typical and does not essentially depend on our numerical assumptions about x_1. Then Theorem IX is proved.

The reasoning is precisely similar if instead of the hypothesis of Theorem IX we admit the possibility of a finite number of points in the neighborhood of each of which the convergence is not assumed uniform:

THEOREM X. *If the series*

$$a_0\varphi_0(x) + a_1\varphi_1(x) + \cdots + a_n\varphi_n(x) + \cdots$$

converges to the sum zero uniformly, $0 \leqq x \leqq 1$, *except in the neighborhood of a finite number of points, then* $0 = a_1 = a_2 = \cdots = a_n = \cdots$.

HARVARD UNIVERSITY,
May, 1922.

Applications of Walsh functions in communications

It is—with our advanced technology—no longer necessary to design communications equipment around a concept of trigonometric-based functions. A possible sophistication now is neatly netted with Walsh functions

Henning F. Harmuth University of Maryland

Communication theory was founded on the system of sine-cosine functions. A more general theory has become known more recently; it replaces the sine-cosine functions by other systems of orthogonal functions, and the concept of frequency by that of sequency. Of these systems, the Walsh functions are of great practical interest since they lead to equipment that is easily implemented by semiconductor technology. Filters, multiplexing equipment, and a voice analyzer/synthesizer have been built successfully for Walsh functions. Some interesting applications of electromagnetic Walsh waves have been found theoretically.

Traditionally, the theory of communication has been based on the complete, orthogonal* system of sine and cosine functions. The concept of frequency is a consequence of these functions, since frequency is defined as the parameter f in $\sin 2\pi ft$ and $\cos 2\pi ft$. The question arises whether there are other systems of functions on which theories of similar scope can be based, and that lead to equipment of practical interest. Since sine and cosine form a system of orthogonal functions, it is reasonable to investigate other systems of orthogonal functions.

Figure 1 shows three orthogonal systems: sine-cosine functions, Walsh functions, and block pulses, for which the normalized time $\theta = t/T$ is the variable. The block pulses are representative of several pulse shapes used for time multiplexing. The notations sal (i, θ) and cal (i, θ) are used here for the Walsh functions. (The letters *s* and *c* allude to the sine and cosine functions which are closely related to Walsh functions; the letters *al* are derived from the name Walsh.)[1-8]

Block pulses form an incomplete system; sine–cosine and Walsh functions form a complete system. Explicitly, the difference is that additional sine–cosine or Walsh functions may be drawn in Fig. 1 for $i = 5, 6, \ldots$ in the interval $-\frac{1}{2} \leq \theta < \frac{1}{2}$, but no other block pulses are orthogonal to the eight shown. Practically, the difference between complete and incomplete systems of functions is shown by the existence of elaborate theories based on sine–cosine functions for antennas, waveguides, and filters; no such theories exist for block pulses although used in communications much longer.

The Walsh functions in Fig. 1 assume the values +1 and −1 only—a useful feature if circuits are to be constructed with binary digital components. The functions cal (i, θ) and $\sqrt{2}\cos 2i\gamma\theta$ are symmetric (even). The functions sal (i, θ) and $\sqrt{2}\sin 2i\pi\theta$ are asymmetric.

In Fig. 1, the parameter i in $\sqrt{2}\sin 2i\pi\theta$ and $\sqrt{2}\cos 2i\pi\theta$ gives the number of oscillations in the interval $-\frac{1}{2} \leq \theta < \frac{1}{2}$ (that is, the normalized frequency $i = fT$). One may interpret i as "one half the number of zero crossings per unit time" rather than as "oscillations per unit time." (The zero crossing at the left side, $\theta = -\frac{1}{2}$, but not the one at the right side, $\theta = +\frac{1}{2}$, of the time interval is counted for sine functions.)

The parameter i also equals one half the number of zero crossings in the interval $-\frac{1}{2} \leq \theta < \frac{1}{2}$ for Walsh functions. In contrast to sine–cosine functions the sign changes are not equidistant.† If, unlike Fig. 1, i is not an integer, then it equals "one half the average number of zero crossings per unit time." The term "normalized sequency" has been introduced for i, and $\varphi = i/T$ is called the nonnormalized sequency:

Sequency in zps = ½ (average number of zero crossings per second)

The general form of a sine function $V \sin(2\pi ft + \alpha)$ contains the parameters amplitude V, frequency f, and phase angle α. The general form of a Walsh function V sal $(\varphi T, t/T + t_0/T)$ contains the parameters amplitude V, sequency φ, the delay t_0, and time base T. The normalized delay, t_0/T, corresponds to the phase angle. The time base T is an additional parameter and it causes a major part of the differences in the applications of sine-cosine and Walsh functions.

* The two functions $f(j, x)$ and $f(k, x)$ in Fig. 1 are orthogonal in the interval $-\frac{1}{2} \leq x \leq \frac{1}{2}$ if the integral

$$\int_{-1/2}^{1/2} f(j, x) f(k, x)\, dx$$

is zero for $j \neq k$. They are orthogonal and normal or orthonormal if the integral equals 1 for $j = k$. A system of functions $\{f(j, x)\}$, orthogonal in a certain interval, is called complete if any function $F(x)$ quadratically integrable in that interval can be represented by a superposition of the functions $f(j, x)$ with a vanishing mean-square error.

† The first five Walsh functions look like heavily amplitude-clipped sine or cosine functions and have equidistant sign changes. This does not in general hold for functions with i larger than 2.

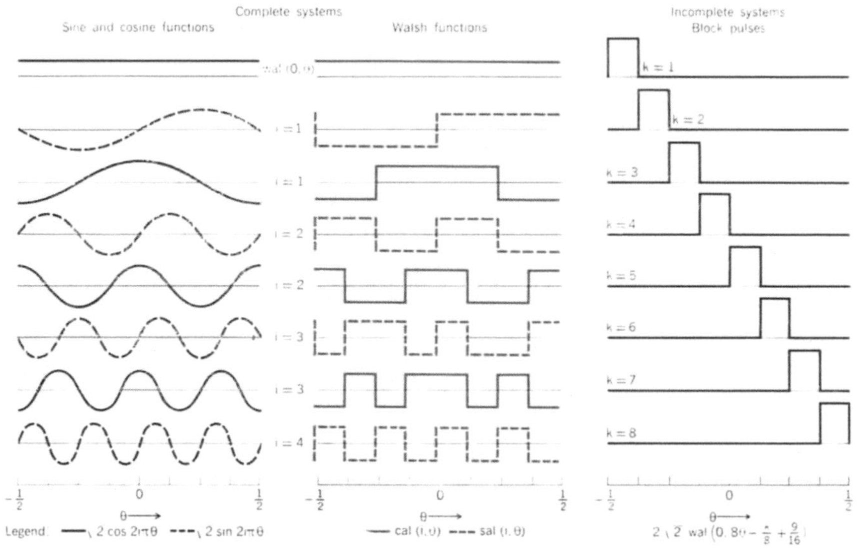

FIGURE 1. Orthonormal systems of functions.

Another alternative

So far, Walsh functions are the only known functions with desirable features comparable to sine-cosine functions for use in communications.* Development of semiconductor technology has imparted practical interest in them at this time. As an example of how this development has changed the approach to filter synthesis, consider the role of the capacitor and coil, which until recently were the most desirable components of filters. Such frequency-selective filters are linear, time-invariant, and thus a theory based on sine-cosine functions has indisputable advantages. But filters based on Walsh functions are linear and periodic time-variable. Generally speaking, the transition from sine-cosine functions to other complete systems of orthogonal functions means a transition from linear, time-invariant components and equipment to linear, time-variable components and equipment, which, of course, constitute a much larger class.

Figure 2 lists features of sine-cosine functions, Walsh functions, and block pulses. The mathematical theory of Walsh-Fourier analysis corresponds to the Fourier analysis used for sine-cosine functions. There is no theory of similar scope for block pulses, because they are incomplete.

* Walsh functions are closely related to Hadamard matrices. These matrices are orthogonal, consisting of square arrays of plus and minus ones, and are of the order 2^n. Other complete systems can be derived from Hadamard matrices of different rank.

Sine and cosine transforms of a function $F(\theta)$ are†

$$a_s'(\mu) = \int_{-\infty}^{\infty} F(\theta)\sqrt{2}\sin 2\pi\mu\theta \, d\theta$$
$$a_c'(\mu) = \int_{-\infty}^{\infty} F(\theta)\sqrt{2}\cos 2\pi\mu\theta \, d\theta \tag{1}$$

The corresponding sal and cal transforms of Walsh-Fourier analysis are defined by

$$a_s(\mu) = \int_{-\infty}^{\infty} F(\theta)\,\mathrm{sal}(\mu, \theta)\, d\theta$$
$$a_c(\mu) = \int_{-\infty}^{\infty} F(\theta)\,\mathrm{cal}(\mu, \theta)\, d\theta \tag{2}$$

$$F(\theta) = \int_{-\infty}^{\infty} [a_s(\mu)\,\mathrm{sal}(\mu, \theta) + a_c(\mu)\,\mathrm{cal}(\mu, \theta)]\, d\mu \tag{3}$$

where $\mu = \varphi T$ and $\theta = t/T$.

Walsh-function filters

Figure 3 shows the block diagram, the time diagram, and a practical circuit of a sequency low-pass filter based

† The functions of Fig. 1 are defined in a finite interval but may be continued periodically to infinity. The parameter i is an integer and assumes denumerably many values. These functions are used for Fourier and Walsh-Fourier series expansions. The functions used for the Fourier and Walsh-Fourier transforms in Eqs. (1) and (2) are defined in the infinite interval. The parameter μ may be any real number and assumes nondenumerably many values.

Harmuth—Applications of Walsh functions in communications

	Sine Functions	Walsh Functions	Block Pulses
Parameters	Amplitude Frequency Phase angle ...	Amplitude Sequency Delay Time base	Amplitude ... Pulse position Pulse width
Mathematical theory	Fourier analysis	Walsh–Fourier analysis	...
Power spectrum	Frequency spectrum	Sequency spectrum	...
Filters	Time-invariable, linear	Periodically time variable, linear	Time variable, linear
Characterization	Frequency response of attenuation and phase shift	Sequency response of attenuation and delay	Attenuation as function of time
Multiplex	Frequency division	Sequency division	Time division
Modulation	Amplitude, phase, frequency modulation	Amplitude, time position, time base, code modulation	Amplitude, pulse position, pulse width modulation
Radiable	$\sin 2\pi ft$, $\cos 2\pi ft$	$\mathrm{sal}\left(i, \frac{t}{T}\right)$, $\mathrm{cal}\left(i, \frac{t}{T}\right)$	...

FIGURE 2. List of features and applications of sine–cosine functions, Walsh functions, and block pulses.

FIGURE 3. Sequency low-pass filter. Top to bottom: practical circuit, block circuit, time diagrams.

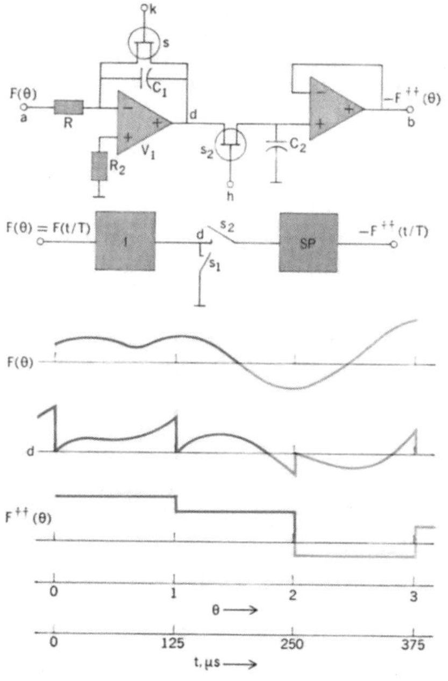

on Walsh functions. The input signal, $F(\theta)$, is transformed into a step function, $F\dagger\dagger(\theta)$, with steps of a certain width, by integrating $F(\theta)$ during an interval equal to the step width (see line d). The amplitudes of the steps are chosen so that $F\dagger\dagger(\theta)$ yields a least-mean-square approximation of $F(\theta)$. In addition, $F\dagger\dagger(\theta)$ is delayed with respect to $F(\theta)$ by one step width. The voltage obtained at the end of the interval is sampled by the switch s_2 and stored in the holding circuit SP. Immediately after sampling, the integrator is reset by s_1. If the width of the steps is 125 μs, $F\dagger\dagger(\theta)$ will have 8000 independent amplitudes per second. $F\dagger\dagger(\theta)$ may be considered to consist of a superposition of Walsh functions having 0 to 8000 zero crossings per second or a sequency between 0 and 4 kzps. The output signal of a frequency low-pass filter with 4-kHz cutoff frequency also has 8000 independent amplitudes per second. Hence, the sampling theorems of Fourier and Walsh–Fourier analysis permit the comparison of frequency and sequency filters.

Consider the multiplication theorems of sine and cosine shown in Fig. 4. The product of these two functions always yields a sum of two functions with argument $(k - i)\theta$ and $(k + i)\theta$. Let $\cos i\theta$ and $\sin i\theta$ represent carriers and let $\cos k\theta$ and $\sin k\theta$ represent Fourier components of a signal. The terms on the right sides of the multiplication theorems of Fig. 4 represent "lower" and "upper components" produced by amplitude modulation. Lower and upper sidebands are obtained if a carrier is amplitude-modulated by many rather than by one Fourier component. Hence, double-sideband modulation is a result of the multiplication theorems of sine–cosine functions.

Figure 4 also shows multiplication theorems for Walsh functions. The symbol $\oplus$ indicates an addition modulo 2: The numbers are written in binary form and added

$$2\cos k\theta \ \cos i\theta = \cos(k-i)\theta + \cos(k+i)\theta$$
$$2\sin k\theta \ \cos i\theta = \sin(k-i)\theta + \cos(k+i)\theta$$
$$2\cos k\theta \ \sin i\theta = -\sin(k-i)\theta + \sin(k+i)\theta$$
$$2\sin k\theta \ \sin i\theta = \cos(k-i)\theta - \cos(k+i)\theta$$

$$\mathrm{cal}(k,\theta)\,\mathrm{cal}(i,\theta) = \mathrm{cal}\{k \oplus i, \theta\}$$
$$\mathrm{sal}(k,\theta)\,\mathrm{cal}(i,\theta) = \mathrm{sal}\{[i \oplus (k-1)]+1, \theta\}$$
$$\mathrm{cal}(k,\theta)\,\mathrm{sal}(i,\theta) = \mathrm{sal}\{[k \oplus (i-1)]+1, \theta\}$$
$$\mathrm{sal}(k,\theta)\,\mathrm{sal}(i,\theta) = \mathrm{cal}\{(k-1) \oplus (i-1), \theta\}$$
$$\mathrm{cal}(0,\theta) = \mathrm{wal}(0,\theta)$$

FIGURE 4. Multiplication theorems of sine-cosine and Walsh functions.

FIGURE 5. Sequency bandpass filter (A) and multiplier for Walsh functions (B).

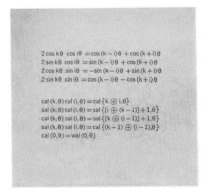

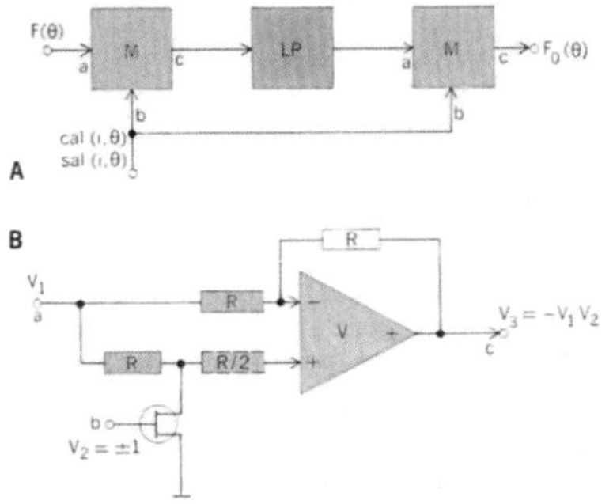

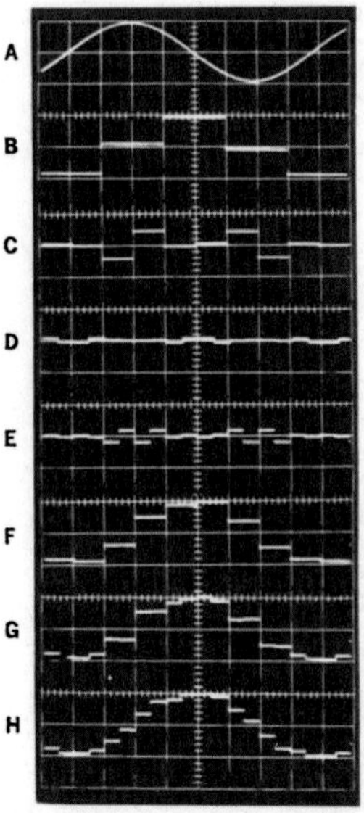

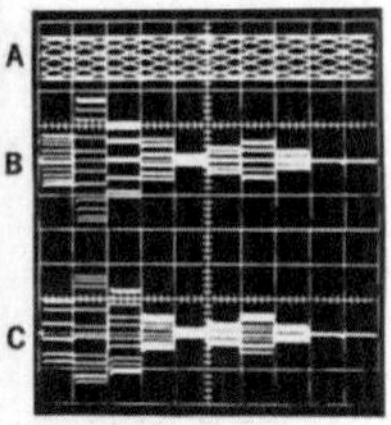

FIGURE 6 (left). Filtering of a sinusoidal voltage by various sequency filters; (A) is a sinusoidal function, frequency 250 Hz. Time base T = 1 ms; horizontal scale 0.5 ms/div. The following functions pass through the filters: (B) cal (φT, t/T), 0 ≤ φ < 1 kzps; (C) sal (φT, t/T), 0 < φ ≤ 1 kzps; (D) cal (φT, t/T), 1 kzps ≤ φ < 2 kzps; (E) sal (φT, t/T), 1 kzps < φ ≤ 2 kzps; (F) sum of B and C; (G) sum of B, C, and D; (H) sum of B, C, D, and F. (Courtesy Boeswetter and Klein)

FIGURE 7 (right). Amplitude spectra of sinsoidal voltages. Line (A) represents sinusoidal voltages, frequency 1 kHz, various phases; horizontal scale 0.1 ms/div. Lines (B) and (C) are amplitude spectra $a_c(\varphi T)$ and $a_s(\varphi T)$; time base T = 1.6 ms; horizontal scale 625 zps/div. (Courtesy Boeswetter and Klein)

according to the rules 1 ⊕ 0 = 0 ⊕ 1 = 1, 0 ⊕ 0 = 1 ⊕ 1 = 0 (no carry). The point is, the product of two Walsh functions yields only one Walsh function and not two. Let cal (*i*, *θ*) and sal (*i*, *θ*) represent carriers and let cal (*k*, *θ*) and sal (*k*, *θ*) represent Walsh-Fourier components of a signal. The amplitude modulation of a Walsh carrier yields only one component or only one (sequency) sideband.

A typical application of the multiplication theorems of Walsh functions is in the design of sequency-bandpass filters. Figure 5 shows the operating principle of such a bandpass. The input signal *F(θ)* is shifted in sequency by multiplication with a Walsh carrier, cal (*i*, *θ*) or sal (*i*, *θ*), and then passed through a sequency low-pass filter *LP* shown in Fig. 3. The filtered signal is subsequently shifted to its original position in the sequency domain by multiplication with the same Walsh carrier used to shift the input signal. Figure 5 also shows a typical multiplier for Walsh functions. Note that a signal *F(θ)* is multiplied by only +1 or −1. Multiplication by +1 leaves the signal unchanged; −1 reverses amplitude.

Sequency filters based on Walsh functions have been built by Boeswetter (Technische Hochschule, Darmstadt, W. Germany) for a voice analyzer and synthesizer, and also by Lueke and Maile (AEG-Telefunken, Research Department, Ulm, W. Germany) for a telephony multiplex system. In the latter application, a minimum crosstalk attenuation of about −60 dB was achieved in the stop bands. Such high-quality filters differ, of course, from the circuits shown in Figs. 3 and 5. Vandivere (Telcom Inc., McLean, Va.) has developed sequency filters for a signal analyzer.

Decomposition of voice signals by Walsh functions first seems to have been investigated theoretically by Sandy[9] in 1962. Another early investigator of sequency power spectra was Ohnsorg (Honeywell Inc., St. Paul, Minn.) whose work has not yet been published. Klein (Technische Hochschule, Darmstadt, W. Germany) has

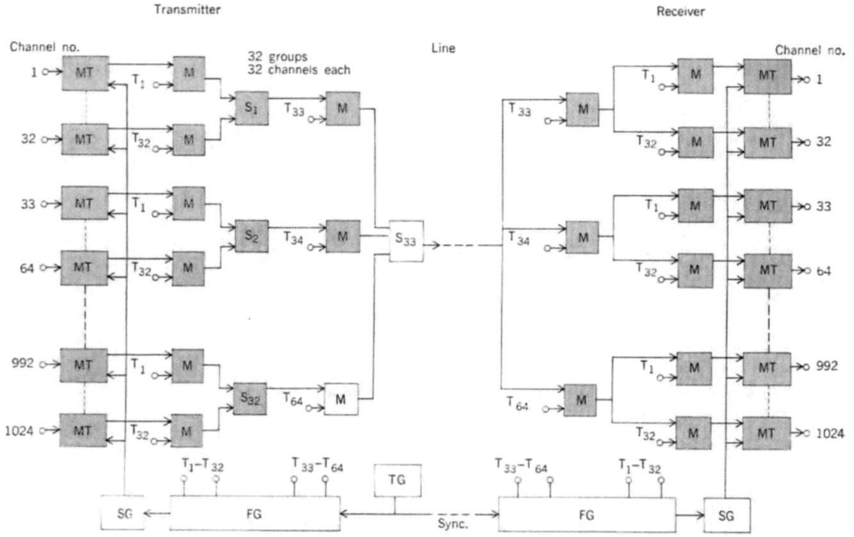

FIGURE 8. Block diagram of a sequency-multiplex system for 1024 telephony channels. Designations are: MT, sequency low-pass filter; M, multipliers; S, adder; FG, function generator; TG, clock pulse generator; SG, timing generator for the filter.

shown experimentally for some simple cases that voice signals have sequency formants just as they have frequency formants. Work on voice signals was also recently started by Elsner (Technische Hochschule, Braunschweig, W. Germany) and Strum (Mitre Corp., McLean, Va.). C. Brown (Systems Research Labs., Dayton, Ohio) is investigating signal processing techniques using sequency spectra for the purpose of detection and recognition of signals in noise, signal sorting, and signal parameter identification.

Figure 6 shows oscillograms of a sinusoidal voltage (A) and the voltages it produces at the output of various sequency low-pass filters (B, F, G, H) and sequency bandpass filters (C, D, E). Figure 7 shows sinusoidal voltages of fixed frequency and their sequency amplitude spectra $a_c(\mu) = a_c(\varphi T)$ and $a_s(\mu) = a_s(\varphi T)$.

Signal multiplexing

The multiplication theorems of the Walsh functions make signal multiplexing an attractive application. Figure 8 shows the principle of a sequency-multiplex system for 1028 telephony channels. Analog or digital signals are fed through sequency low-pass filters *MT* to multipliers *M*. For voice signals the bandwidth of these low-pass filters is $\Delta\varphi = 4$ kzps. Thirty-two carriers, T_1 to T_{32}, consisting of Walsh functions cal (i, θ) and sal (i, θ) are fed to the multipliers.* The time base of these functions equals $\Delta\varphi/2 = 125\ \mu$s. Each of the output voltages of the 32 multipliers is summed by the adders S_1 to S_{32}.

* Rules for the selection of carriers that avoid crosstalk and waste of sequency bandwidth can be derived from the multiplication theorems of the Walsh functions.[10]

These summed voltages may further be multiplied by another set of multipliers with other Walsh carriers, denoted T_{33} to T_{64} in Fig. 8. A sequency-multiplex system permits repeated sequency shifting just as a frequency multiplex system permits repeated frequency shifting.

The signals are separated at the receiver by multiplication with the same synchronized Walsh carriers used in the transmitter. The block diagram of Fig. 8 differs from that of a frequency-multiplex system only by the missing single-sideband filters. The circuitry inside the blocks is, of course, very different.

A sequency-multiplex system according to Fig. 8 has been developed by Lueke and Maile. The system is designed for 256 voice channels, of which three are fully completed.† More channels are presently being added by Huebner[12] of the West German Post Office Department in preparation for tests on post office lines. One of the tests will be the transmission of PCM voice signals via a scatter link, since, compared with time-division PCM, a gain of some 3 dB is predicted for this application. A Walsh-function tracking filter is used in the equipment of Lueke and Maile to establish synchronization between transmitter and receiver. The synchronization is good enough to yield a crosstalk attenuation of −57 dB, or better, in back-to-back operation. Crosstalk due to all

† Development of sequency-multiplex equipment was also started at ETH Zurich (Swiss Federal Institute of Technology, Institute for Advanced Electrical Engineering) and at the Research Institute of the West German Post Office Department in Darmstadt. Considerable theoretical work on PCM transmission by Walsh functions was done by Taki and Hatori[11] of Tokyo University, Japan. Experimental work has also been reported by Cox of the M.I.T. Instrument Laboratory, Cambridge, Mass.

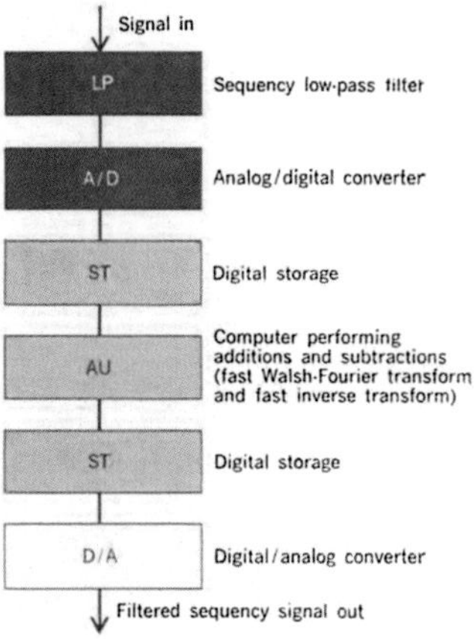

FIGURE 9. Digital sequency filter.

causes is −53 dB or better in back-to-back operation. Using a compandor (compressor plus expander), the apparent crosstalk attenuation could be made some −75 dB. These figures are, of course, in excess of that required for good PCM transmission.

Sequency-multiplex systems do not need single-sideband filters. The low-pass and bandpass filters required do not cause attenuation or delay distortions. These features are highly important for data transmission. All filters can be implemented by integrated circuit techniques. No individual tuning of the filters is required. The bandwidth of the filters is determined by the timing of the pulses that drive the switches s_1 and s_2 in Fig. 3; correct timing of the pulses replaces the tuning of filters: no temperature compensation is needed. The Walsh carriers can be produced by means of binary counters and gates. The only part in a sequency-multiplex system that requires tuning and temperature compensation is the clock pulse-generator.

Time-multiplex systems do not need single-sideband filters either, and they are also well suited for implementation by semiconductor technology. The advantage of a sequency-multiplex system here rests with reduced sensitivity to disturbances—brought about by two causes:

1. Only a fraction of all channels is active in a multiplex system at any one time. For instance, the activity factor of a telephone system does not exceed one quarter even during peak traffic.* Hence, the power amplifiers are not used three quarters of the time and the average useful signal power is reduced correspondingly in a time-multiplex system. Sequency-multiplex systems yield a higher average signal power for equal peak power. Particularly advantageous is that the average power can be maintained approximately constant by means of automatic gain-control amplifiers when the activity factor drops during low-traffic hours. It is also noteworthy that the sequency bandwidth of a sequency-multiplex signal is not changed by a nonlinear compressor or expander characteristic; hence, sequency-multiplex signals can readily be passed through instantaneous compandors.

2. Digital signal errors during transmission through telephone lines are mainly caused by pulse-type disturbances. Time division is more susceptible to these disturbances than frequency or sequency division since a particular block pulse may be changed appreciably by a disturbing pulse, although the preceding and the following pulses are not changed at all. In the case of frequency or sequency multiplexing, many sine–cosine or Walsh pulses are transmitted simultaneously. The energy of a disturbing pulse is thus spread over many signal pulses. Only considerable energy can disturb signal pulses that are quantized. Measurements with binary sine–cosine pulses have yielded—and theoretically Walsh pulses should yield—error rates 100 times smaller than for block pulses having the same average power.

There are several additional—although not so important—differences between time- and sequency-multiplex systems. For example, equipment for time division is often less expensive; sequency multiplexing is more adaptable to the problems of communication networks and makes it easier to mix voice and data signals with different average power.

Digital filtering and multiplexing

One of the most promising aspects of Walsh functions is the ease with which filters and multiplex equipment can be implemented as digital circuits. The reason is that numerical Walsh–Fourier transformation and numerical sequency shifting of signals require summations and subtractions only. In the case of sine–cosine functions, the corresponding operations require multiplications with irrational numbers. The simplification of numerical computations by the use of Walsh functions has been recognized by many scientists.[12–17]

Figure 9 shows the block diagram of a digital filter based on Walsh functions. The input signal passes first through a sequency low-pass filter *LP* that transforms it into a step function. This step function is sampled and the samples are transformed into numbers by an analog/digital converter. A series of these numbers is stored in a digital storage *ST*. A Walsh–Fourier transform of this series is obtained by performing certain additions and subtractions in the arithmetic unit *AU*. Some or all of the obtained coefficients, that represent sequency components, may be suppressed or altered—in effect, a filtering process. An inverse Walsh–Fourier transform yields the filtered signal as a series of numbers. These numbers are stored in a second digital storage *ST* and transformed into an analog signal by digital/analog converter *D/A*. Since there is a fast Walsh–Fourier transform just as there is a fast Fourier transform, the arithmetic operations in a digital sequency filter are not only simpler than in a digital frequency filter but can be performed faster.[10, 18–21]

Voice signals are functions of the one variable, time. A black-and-white photograph is a function of two space variables, and a black-and-white television picture is a function of two space variables and time. Digital filters may be applied to signals that are functions of two or three variables by using a two- or three-dimensional

* The activity factor gives the number of channels actually carrying signals. It is smaller than one during peak traffic because one party must listen while the other one talks, because there are idle periods in conversations, because a channel becoming available is not immediately used, etc.

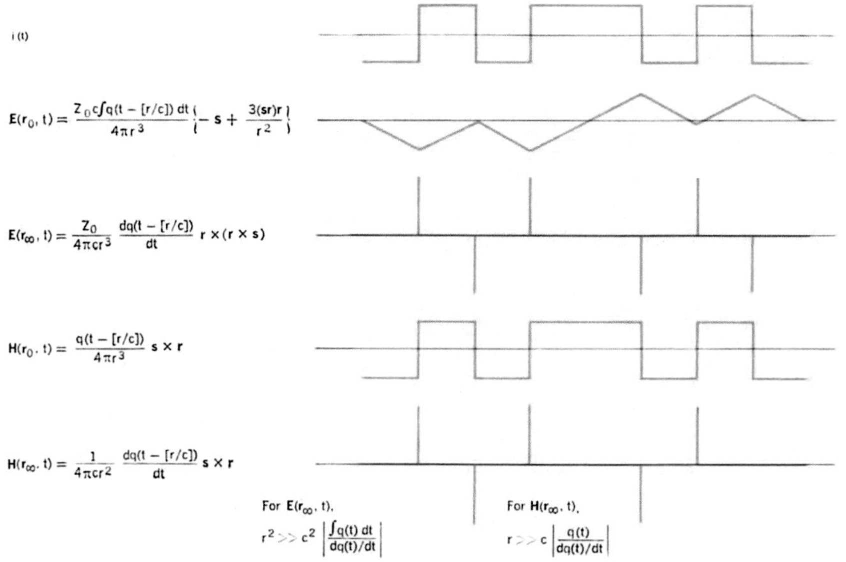

FIGURE 10. Electric and magnetic field strengths in the near and wave zones due to a current q(t) fed into a Hertzian dipole. The functions on the right show the time variations caused by a Walsh-shaped current q(t).

Walsh-Fourier transform.* This has been done by Pratt, Kane, and Andrews[18] for functions of two variables. Roth and Lueg (Technische Hochschule, Aachen, W. Germany) as well as Held (Technische Hochschule, Darmstadt, W. Germany) and Klein have also recently started to develop digital sequency filters. A very general method for digital signal filtering—not restricted to Walsh functions—was devised by Andrews[22] and Caspari (International Telephone and Telegraph Corporation, Electro-Physics Laboratories, Hyattsville, Md.).

Digital multiplexing is a straightforward extension of digital filtering. Rather than multiplying signals and Walsh carriers represented by voltages, one multiplies signals and Walsh carriers represented by series of numbers. Multiplexed signals are again represented by a series of numbers that may be transmitted by any digital communication equipment. The promising feature of such a digital-sequency-multiplex system is that it has essentially the same immunity to disturbances as the previously discussed analog-sequency-multiplex equipment, but is compatible with existing transmission equipment.

Electromagnetic Walsh waves

At the present time, sinusoidal electromagnetic waves exclusively are used for radio communication. Such waves are characterized by a sinusoidal variation with time of the electric and magnetic field strengths $\mathbf{E}(\mathbf{r}, t)$ and $\mathbf{H}(\mathbf{r}, t)$. These field strengths are produced by feeding a sinusoidal current into the antenna.

Figure 10 shows the time variation of electric and magnetic field strengths for a Walsh-shaped current $q(t)$ fed into a Hertzian dipole. A typical current is shown in the first line. $\mathbf{E}(\mathbf{r}_0, t)$ and $\mathbf{H}(\mathbf{r}_0, t)$ are the field strengths in the near zone, $\mathbf{E}(\mathbf{r}_\infty, t)$ and $\mathbf{H}(\mathbf{r}_\infty, t)$ the field strengths in the wave zone. Z_0 (377 ohms) is the wave impedance of free space, c the velocity of light, r the distance between dipole and observation point, **r** the vector from dipole to the observation point, and **s** the dipole vector. The spatial variation of **E** and **H** depends on the vectors **s** and **r**; it is the same for sine, Walsh, and other waves. The time variation depends solely on the current $q(t)$ fed into the Hertzian dipole. This time variation is plotted in Fig. 10 on the left for the near zone and the wave zone of **E** and **H**. In the wave zone, one obtains Dirac delta functions for **E** and **H** that are located at the jumps of $q(t - r/c)$. The time variation of **H** in the near zone equals $q(t - r/c)$; the time variation of **E** equals $\int q(t - r/c)\, dt$.

The wave zone of **E** and **H** is the region in which the distance r between dipole and observation point satisfies the conditions shown in the bottom line of Fig. 10. In the near zone, the "much larger" signs in the inequalities are replaced by "much smaller" signs. These conditions have the more familiar form $r >> \lambda$ or $r << \lambda$ for sinusoidal currents $q(t) = I \sin 2\pi f t = I \sin (2\pi c t/\lambda)$.

A receiver always receives the sum of the field strengths

* Analog filters based on sine-cosine functions can be implemented for functions of two space variables by optical means. Analog filters based on Walsh functions can be implemented with relative ease for functions of two space variables by resistors and operational amplifiers; their implementation for functions of two space variables and time, or three space variables with and without the time variable, is possible but expensive.

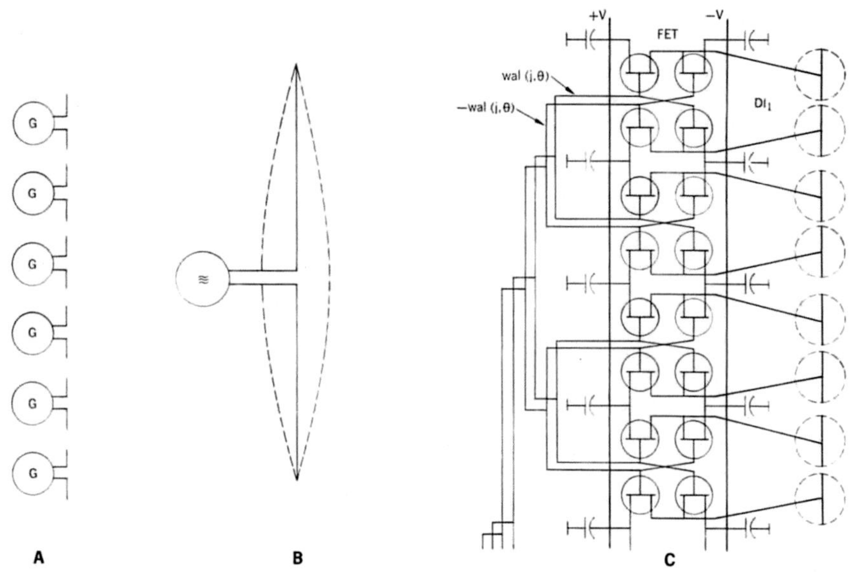

FIGURE 11. A—General array of Hertzian dipoles. B—λ/4 dipole for sine waves. C—Array of Hertzian dipoles and power amplifiers for Walsh waves [wal (2j, θ) = cal (j, θ)].

FIGURE 12. Beam width of an antenna for Walsh waves.

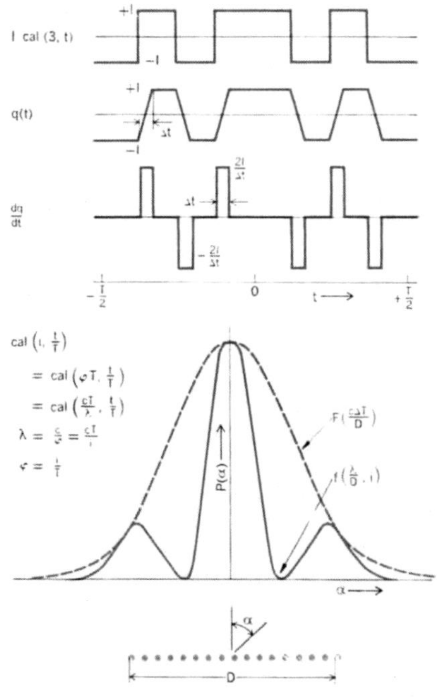

Harmuth—Applications of Walsh functions in communications

from the near and wave zones. $\mathbf{E}(\mathbf{r}_\infty, t)$ and $\mathbf{H}(\mathbf{r}_\infty, t)$ decrease proportionally with $1/r$, whereas $\mathbf{E}(\mathbf{r}_0, t)$ and $\mathbf{H}(\mathbf{r}_0, t)$ decrease proportionally with $1/r^3$ and $1/r^2$. Hence, a comparison of the field strengths of the near and the wave zones allows the distance of the transmitter to be determined. This determination of course, requires the time variation of the near- and the wave-zone components to be different in order to distinguish between them. A sinusoidal current $q(t)$ produces sinusoidal near-zone and wave-zone components since the differential as well as the integral of a sine function is simply a time-shifted sine function.

A Hertzian dipole radiates vanishingly little power. Many such dipoles may be used simultaneously to radiate more power. Figure 11(A) shows several Hertzian dipoles, each fed by one generator. In the case of sinusoidal waves, one may feed many Hertzian dipoles by a single high-power generator using standing waves in a half-wave or similar dipole [Fig. 11(B)]. This is also possible but not economical for Walsh waves. Since a generator for Walsh-shaped currents consists of switches that feed positive or negative currents into the dipole, it is better to feed many Hertzian dipoles through many switches as shown in Fig. 11(C). These switches must feed constant currents and not constant voltages to the dipoles. Current feed is automatic if the switches are semiconductors.

The Hertzian dipoles *DI* in Fig. 11(C) are circular areas of conducting material. They do not have to be arranged along a line as in Fig. 11(C), but may be arranged in two dimensions. Such a two-dimensional arrangement, however, cannot be used for antennas that use the standing-wave principle such as the one in Fig. 11(B).

The deviations of the antenna current $q(t)$ from the ideal shape shown in Fig. 10 must be taken into account

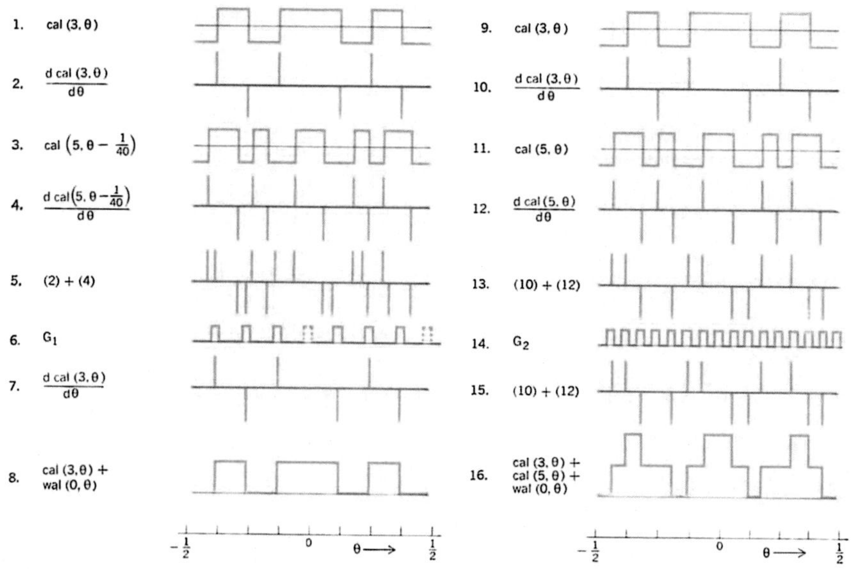

FIGURE 13. Principle of the separation of two Walsh waves in mobile radio communication.

if the directional characteristic of an antenna, consisting of many Hertzian dipoles, is to be determined. For example the idealized current, I cal (3, t) shown on the first line of Fig. 12 is not achievable since it is not possible to switch back and forth between $+I$ and $-I$ in zero time. The current $q(t)$ with finite switching times Δt is more realistic. The time differential dq/dt of the current $q(t)$ is shown in the third line. Narrow, rectangular pulses with finite amplitude—not delta pulses—result.

A row of Hertzian dipoles is shown at the bottom of Fig. 12. The length of this row is D; each dipole is short compared to $c\Delta t$. Let a current $q(t)$, as shown in the second line of Fig. 12, be fed into each dipole. A radiation diagram $P(\alpha)$ gives the average radiated power in the wave zone as function of the angle α. A typical radiation diagram is shown in Fig. 12 by the solid line above the row of dipoles. The dashed line is the envelope of the main lobe and the sidelobes. This dashed line is a function of $c\Delta t/D$, where c denotes the velocity of light and Δt the switching time of the current $q(t)$. A reduction of Δt makes the dashed line approach zero for angles $\alpha \neq 0$, and the antenna radiates into a very narrow angle for small Δt.

The location of the zeros of the solid curve in Fig. 12 depends on the ratio λ/D and the normalized sequency i of the Walsh waves. (λ is the average wavelength.)

The following definitions apply to cal (i, t/T):

$$\text{cal}\,(i, t/T) = \text{cal}\,(\varphi T, t/T) = \text{cal}\,(cT^2/\lambda, t/T)$$

where $\lambda = c/\varphi = cT/i$ = average wavelength, $\varphi = i/T$ = sequency, and $i = 1, 2, \cdots$ = normalized sequency.

The directivity of an antenna is determined by the ratio λ/D for sine waves. In the case of Walsh waves it is determined by the ratios λ/D and $c\Delta t/D$, and by the normalized sequency i. The important point is that a reduction of Δt yields a significantly better directivity without a need to decrease λ or increase D. These results not only hold true for a row of dipoles in Fig. 12, but also for other antenna shapes, particularly parabolic reflectors. (D represents the reflector diameter.)

Maintaining orthogonality after time shifting

It is known that orthogonal functions—for example, voltages or field strengths varying with time—can be separated. It is not necessary that this separation be by frequency or time division; the more general orthogonal division is sufficient.[10] Hence, a point-to-point transmission is perfectly possible with Walsh waves. Mobile communication is more difficult, since waves radiated from various transmitters show various time shifts at the receiver due to the propagation times of the waves. These unknown time shifts generally destroy the orthogonality of the received functions. Up to now, the only known exception has been sine waves. A time shift destroys only the orthogonality between sine and cosine functions of the same—not different frequency.*

Walsh waves in the wave zone are a second exception† and may also be separated regardless of any time shift.

* The relation $\sin(\omega t + \gamma) = \sin\gamma\cos\omega t + \cos\gamma\sin\omega t$ shows that a time-shifted sine function consists of not-shifted sine and cosine functions with the same frequency. Any function orthogonal to $\sin\omega t$ and $\cos\omega t$ is thus orthogonal to $\sin(\omega t + \gamma)$.

† These waves have the shape of differentiated Walsh functions according to Fig. 10 rather than the shape of the Walsh functions.

Figure 13, line one, shows the function cal (3, θ) = cal (3, t/T); line two shows its differential, which represents the time variation of the electric and magnetic field strength in the wave zone—Dirac pulses at the jumps of cal (3, θ). (The deviation from this theoretical shape is neglected here, just as it is usual to neglect other than idealized, infinitely long sine functions.)

Line three shows another Walsh function, cal (5, θ − 1/40), time-shifted relative to cal (3, θ); line four shows its differential; line five shows the sum of lines two and four. This sum is received if one transmitter radiates the wave cal (3, θ) and another the wave cal (5, θ). The voltage produced in the receiving antenna also varies with time according to line five. As shown by line six, a gate permits those pulses to pass that arrive at the proper time; the pulses of line seven are thus derived from those of line five. These are the same pulses as those of line two. Hence, the desired signal is separated from the non-desired one.* Integration of the pulses of line seven yields the function cal (3, θ) and a superimposed dc component in line eight. The dc component is of no consequence.

Lines nine to 16 in Fig. 13 show what happens if cal (3, θ) and cal (5, θ) are not time-shifted relative to each other. This case cannot occur if the Dirac pulses in lines two and four as well as the gating intervals in line six are infinitely short, but it is important for the practical pulses of finite duration. (More precisely, the probability of this case to occur is zero.) Lines nine to 13 correspond to one to five except for the time shift. The gate, however, opens 16 times according to line 14. Integration of the pulses passed (line 15) yields the sum, cal (3, θ) + cal (5, θ) + dc component wal (0, θ) [equals cal (0, θ)] in line 16. Correlation of this sum with a sample function cal (3, θ) suppresses the components cal (5, θ) and wal (0, θ). The general rule for the opening times of the gate is as follows:

Let an arbitrary number of transmitters radiate Walsh waves cal (i_1, t/T), cal (i_2, t/T), $\cdots$, and let the normalized sequencies i_1, i_2, $\cdots$ assume values from 1 to 2^k only (k = integer). The gate must open periodically 2×2^k times during the time T and allow pulses to pass. Example: i_1 equals 3 and i_2 equals 5 in Fig. 13, line nine. These numbers are between 1 and 8 = 2^3 = 2^k. Hence, the gate of line 14 must open periodically $2 \times 8 = 16$ times during the time T (the interval $-T/2 \leq t < +T/2$ corresponds to the interval $-\frac{1}{2} \leq \theta = t/T < +\frac{1}{2}$). The time interval during which the gate stays open should be about as wide as the received pulses.

* The problem of separation is essentially the same as for synchronous reception of sinusoidal waves. Walsh waves of different sequency can be distinguished and thus separated like sine waves of different frequency. On the other hand, two waves caused by antenna currents I sal (i, θ) and I cal (i, θ) cannot be distinguished without a synchronization signal just as two waves caused by antenna currents $I \sin \omega t$ and $I \cos \omega t$ cannot be distinguished without a synchronization signal.

REFERENCES

1. Walsh, J. L., "A closed set of orthogonal functions," *Am. J. Math.*, vol. 55, pp. 5–24, 1923.
2. Fine, N. J., "The generalized Walsh functions," *Trans. Am. Math. Soc.*, vol. 69, pp. 66–77, 1950.
3. Selfridge, R. G., "Generalized Walsh transform," *Pacific J. Math.*, vol. 5, pp. 451–480, 1955.
4. Watari, C., "Mean convergence of Walsh–Fourier series," *Tôhoku Math. J.*, vol. 16 (2), pp. 183–188, 1964.
5. Yano, S., "On Walsh–Fourier series," *Tôhoku Math. J.*, vol. 3 (2), pp. 223–242, 1951.
6. Price, J. J., "A density theorem for Walsh functions," *Proc. Am. Math. Soc.*, vol. 18, pp. 209–211, 1967.
7. Harrington, W. J., and Cell, J. W., "A set of square wave functions orthogonal and complete in $L_2(0, 2)$," *Duke Math. J.*, pp. 393–407, 1961.
8. Pichler, F. "Das System der sal- und cal-Funktionen als Erweiterung des Systems der Walsh–Funktionen und die Theorie der sal- und cal-Fourier Transformation," Ph.D. thesis, Innsbruck University, Austria, 1967.
9. Sandy, G. F., "Square wave (Rademacher–Walsh functions) analysis," Mitre report WP-1585, Oct. 1968.
10. Harmuth, H. F., *Transmission of Information by Orthogonal Functions.* New York/Berlin 1969: Springer–Verlag. (This book contains an extensive list of references on Walsh functions.)
11. Taki, Y., and Hatori, M., "PCM communication system using Hadamard transformation," *Electron. Commun. Japan*, vol. 49, no. 11, pp. 247–267, 1966.
12. Polyak, B. T., and Shreider, Yu. A., "The application of Walsh functions in approximate calculations," *Voprosy Teorii Matematicheskix Mashin*, vol. 2, Yu. Ya. Bazilevskii, ed. Moscow: Fizmatgiz, 1962 (in Russian).
13. Howe, P. W., "The use of Laguerre and Walsh functions in materials problems of variable loading at high temperature," rept. AD-434122, 1964.
14. Corrington, M. S., "Advanced analytical and signal processing techniques," rept. AD-277942, Apr. 1962.
15. Ito, T., "A way of approximating functions by computer—an application of the Walsh functions," IEEE Computer Group, paper depository R-69-49.
16. Gibbs, J. E., "Walsh spectrometry, a form of spectral analysis well suited to binary digital computation," (to be published).
17. Meltzer, B., Searle, N. H., and Brown, R., "Numerical specification of biological form," *Nature*, vol. 216, pp. 32–36, Oct. 1967.
18. Pratt, W. K., Kane, J., and Andrews, H. C., "Hadamard transform image coding," *Proc. IEEE*, vol. 57, pp. 58–67, Jan. 1969.
19. Shanks, J. L., "Computation of the fast Walsh–Fourier transform," *IEEE Trans. Computers*, vol. C-18, pp. 457–459, May 1969.
20. Whelchel, J. E., and Guinn, D. F., "The fast Fourier–Hadamard transform and its use in signal representation and classification," *1968 EASCON Record*, pp. 561–571.
21. Green, R. R., "A serial orthogonal decoder," Space Programs Summary, Jet Propulsion Laboratory, Pasadena, Calif., No. 37–39, vol. IV, pp. 247–251, 1966.
22. Andrews, H. C., and Caspari, K. L., "A generalized spectrum analyzer," *IEEE Trans. Computers*, vol. C-18, in print.

BIBLIOGRAPHY

Crowther, W. R., and Rader, C. M., "Efficient coding of Vocoder channel signals using linear transformation," *Proc. IEEE*, vol. 54, pp. 1594–1595, Nov. 1966.

Huebner, H., "Walsh-Funktionen und ihre Anwendung," *Gesellschaft No Fuer Ortung und Navigation.* Dusseldorf: Amwehrhahn 94, W. Germany, 1969.

Nambiar, K. K., "A note on the Walsh functions," *IEEE Trans. Electronic Computers*, vol. EC-13, pp. 631–632, Oct. 1964.

Swick, D. A., "Walsh function generation," *IEEE Trans. Information Theory*, vol. IT-15, p. 167, Jan. 1969.

Szok, W. G., "Waveform characterization in terms of Walsh functions," master's thesis, Syracuse University, June 1968.

Weiser, F. E., "Walsh function analysis of instantaneous nonlinear stochastic problems," Ph.D. thesis, Polytechnic Institute of Brooklyn, June 1964.

Henning F. Harmuth (M) left West Germany—where he was a consulting engineer—a year ago to assume a staff position with the Electrical Engineering Department, University of Maryland, College Park, Md., as visiting associate professor. A native of Austria, he earned the degree of Diplom-Ingenieur in 1951 and the doctor's degree of technical science in 1953 from the Technische Hochschule in Vienna. An active author, Dr. Harmuth has written some 40 papers and one book, "Transmission of Information by Orthogonal Functions."

Chapter 2
My involvement in Walsh and Dyadic Analysis

Franz Pichler

The following description of involvement of F. Pichler in Walsh and dyadic analysis is an excerpt from *Reprints from the Early Days of Information Sciences Reminiscences of the Early Work in Walsh Functions Interviews with Franz Pichler, William R. Wade, Ferenc Schipp*, Radomir S. Stanković, Jaakko T. Astola, (eds.), TICSP Series # 58, ISBN 978-952-15-2598-8, ISSN 1456-2744.

In the above interview, Prof. Pichler said the following.

It is so that the inventor of the Walsh functions for Innsbruck was Roman Liedl [1], he is still there a Professor, maybe he retired already, a mathematician, and he invented as many other researchers also on his own the Walsh functions. Later he found out that the concept exist already, but Liedl already saw also the group relations, group theoretical relations and topological group relations [1], [2], [3]. Then research started and I think that in Innsbruck about twenty PhD theses on Walsh functions were made. Many theses were defended. For example, Peter Weiss, he is still at Linz, was one of the first, and they were mainly devoted to generalized Walsh functions. At that time we looked at the work of Lèvy [4], and Rice and Selfdrige [10] and others. Selfridge, these were names that passed, and also Vilenkin, the Russian important Walsh function researcher.

Then Harmuth discovered that there was a Walsh researcher in Innsbruck and he contacted the Innsbruck people. He finally wanted to develop some theoretical framework for his meander functions, which were essentially what have been called *cal* and *sal* functions [5].

Since I was already starting a PhD work, and they knew, the mathematicians there, Roman Liedl knew that I had a communication background, I was the right one to get interested in that, and so it started, and it was interesting. But of course,

Franz Pichler
Johannes Kepler University Linz, A-4040 Linz, Austria, e-mail: Franz.Pichler@jku.at

[1] Roman Rudolf Liedl, Professor of Mathematics, Institute of Mathematics University of Innsbruck, Austria

R.S. Stanković et al. (eds.), *Dyadic Walsh Analysis from 1924 Onwards Walsh-Gibbs-Butzer Dyadic Differentiation in Science Volume 1 Foundations*, Atlantis Studies in Mathematics for Engineering and Science 12, DOI 10.2991/978-94-6239-160-4_2

at that time we had no overview, for example, I had to discover also the concept of the dyadic filter and this is dyadic convolution, I did not know that before. So this is all separate, you are a student, you do not know, and so it started.

How I came to Maryland is interesting, because Harmuth was not an easy man, and he is until today not easy, but he is a devoted scientist and so on, so I had some struggle with him in Innsbruck already. But he was so concentrated to push forward Walsh functions and research, and then, I think we split. We had no much contacts, but I continued to make my work, especially the PhD thesis, and also papers. Yes, and in 1968 I think I published my first paper in the *AEU* (*Archiv der Elektrischen Übertragung* abbreviated as *Archiv eiektr. Übertragung*), this is the journal where Hansi Piesch was an editor, or co-editor [6]. You see, it was not so easy at that time to publish about Walsh functions. AEU had already published the papers of Harmuth, and I was still a student, and not experienced. So I had my doubts if they would accept it, but I had a promoter in the East Germany, in the DDR. This was Franz Heinrich Lange, a Professor of communication engineering in Rostock. He was very well known, and he was fan of Harmuth and of Walsh functions, and so on, and he knew about my work, and when I wrote and sent my paper to the AEU, I do not remember the main editor there, the secretary so to say, I mentioned that if they would not publish it, I could publish it in the DDR, because Lange would have liked to take it. I was already clever, I think, to mention this and they finally reviewed the paper and one of the reviewers was Hansi Piesch. And so I really brought the final version of the paper to Vienna. I drove with my Fiat 600 from Innsbruck, with a friend of mine, we drove to Vienna, and I went to her apartment, yes, at the Gürtel near Sud Bahnhof, and was friendly welcomed and I gave the paper for publication.

I continued publishing in this area, let me mention just first papers of mine [7], [8], [9].

I went to Linz in 1968, and Harmuth again contacted me, and needed me for this first conference as a mathematician, because he was always criticized that he could not define exactly the Walsh function in the continuous case of sequency as he called it. Yes, so he needed me, Harmuth was able to define Walsh functions just as a limes, yes, if n goes to infinity then this is the function, yes, so he was picked by some people when they said *Tell me, how does the Walsh function with sequency π, yes, 3.14 and so on, does look like?*, and he could not answer. He could not really answer, he was, and really is a gifted intuitive working scientist. A kind of engineer with mathematic intellect, Harmuth, there is no doubt yes. Like also Gibbs, they would make formulas without knowing how they can derive these formulas. I was just the opposite. I was, say, educated as a step-wise, going further, operating mathematician, so he needed me in Washington (for the conference on Walsh functions in 1970), and I got the invitation as a visiting research assistant professor.

My stay at the Laboratory of Professor Harmuth at Maryland University resulted in two reports that are reprinted in this book [2]

---

[2] *Comment by Editors* More on the early work of Prof. Franz Pichler in this area and his cooperation with Dr. J.E. Gibbs, can be found in F. Pichler, "Remembering J. Edmund Gibbs", in *Walsh and Dyadic Analysis*, R.S. Stanković, (ed.), *Proc. Workshop on Walsh and Dyadic Analysis*, October 2007, Faculty of Electronic Engineering, Niš, Serbia, XXI-XXVI, 2008.

References

1. Liedl, R., *Vollständige orthonormierte Furiktionen-folgen des Hilbertraumes L_2, deren Elemente bezglich der Multiplikation eine Gruppe bilden*, Dissertation, Philosophische Fakultät Universität Innsbruck, 1964.
2. Liedl, R., *Eine algebraische Herleitung und eine Verallgemeinerung des Satzes von Fine ber Gruppen von orthonormalen Funktionen und eine Beschreibung der vielfait-erhaltenden Transformationen des Intervalles* $[0,1]$ *audsich selbst*, Habilitationsschrift, Universität, Innsbruck, Austria.
3. Liedl, R., "Über eine spezielle Klasse von stark multiplikativ orthogonalen Funktionensystemen", *Monatshefte für Mathematik*, 68, 1964, 130-137.
4. Levy, P., "Sur une generalization de fonctions orthogonales de M. Rademacher", *Comment. Math. Helv.*, 16, 1944, 146-152.
5. Ohnsorg, F. R., "Application of Walsh functions to complex signals", *Proc. of the Walsh Functions Symposium*, Naval Research Laboratory, Washington, DC, USA, 1970, 123-127.
6. Pichler, F., "Synthese lin earer periodisch zeitvariabler Filter mit vorgschriebenem Sequenzverhalten", *Archiv elektr. Übertragung*, 22, 1968, 150-161.
7. Pichler, F., "Walsh-Fourier Synthese optimaler Filter", *Archiv elektr. Übertragung*, 24, 1970, 350-360.
8. $*$Pichler , F., *Walsh Functions and Linear Systems Theory*, Techn. Report, Department of Electrical Engineering, University of Maryland, Washington D.C., Appendix B-l to 8-1, "Sampling Theorem with Respect to Walsh-Fourier Analysis", April 1970.
9. Pichler, F., "Walsh Functions-Introduction to the Theory", in *Proceedings Nato Advanced Study Institute on Signal Processing*, Loughbourough, U.K., 1972, 18 pages.
10. Selfridge, R.G., "Generalized Walsh transforms", *Pacific J. Math.*, 5, 1955, 451-480.

The $*$ sign in the citations indicates that the paper is reprinted in this book.

REPORT T-70-05

WALSH FUNCTIONS AND LINEAR SYSTEM THEORY

TECHNICAL
RESEARCH
REPORT

BY

FRANZ PICHLER

DEPARTMENT OF
ELECTRICAL ENGINEERING

UNIVERSITY OF MARYLAND

COLLEGE PARK, MARYLAND 20742

Walsh Functions and Linear System Theory*

Franz Pichler

Department of Electrical Engineering
University of Maryland

and

Institut für Mathematik
Hochschule Linz
Linz, Austria

* Lecture to be presented at the workshop on "Applications of Walsh Functions", Naval Research Laboratory and University of Maryland April 2, 1970.

CONTENTS

	page
Introduction	1
Walsh Functions	1
Linear Dyadic-Invariant Systems	9
On a State Space Approach for Linear Dyadic-Invariant Systems	13
References	21
Appendix A: Figure 1-8	
Appendix B: Sampling Theorem with Respect to Walsh-Fourier Analysis	

I. Introduction

In this paper we present some ideas in the applications of Walsh functions to the analysis and synthesis of linear systems.

To do this we first consider the mathematical background of Walsh functions. We define the Walsh functions $\psi(y, \cdot)$, $\psi^*(y, \cdot)$ and cal $(s, \cdot)$, sal $(s, \cdot)$ and consider some of the mathematical results. Next, we define a special class of linear systems, which we call dyadic-invariant systems. During the course of examining the dyadic-invariant systems we outline the synthesis procedure of sequency bandpasses and of optimal-filters. Finally, we shall attempt the first steps of a state-space approach for dyadic-invariant systems.

2. Walsh Functions

The mathematical theory of Walsh functions is well developed. Since the fundamental paper of J. L. Walsh [1] was published in 1923, many additional papers concerning this theory have been published. Among these are Paley [2], Levy [3], Fine [4], [5], Vilenkin [6], [7], Chrestenson [8], Watari [9], Weiss [10] and Liedl [11].

There are two common aspects to these papers: The one is, that they search for a theory similar to the theory of the trigonometric functions. Questions concerning summability and convergence of Walsh-Fourier series and Walsh-Fourier integrals are often the most interesting ones. The other aspect is an attempt to embed the theory of Walsh-functions in a more general one: the theory of abstract harmonic analysis. There the Walsh-functions can be derived from the characters of a certain locally compact topological group: the Dyadic Group F of Fine [5]. Often it happens, that theoretical questions concerning Walsh-functions can be solved by means of the theory of abstract harmonic analysis.

2.1 Definition of the Walsh Functions

There are two different definitions in common use. The first gives us the Walsh-functions wal $(i, \cdot)$. For these functions the parameter i represents a count of the sign changes of the function per unit interval. This principle of ordering the Walsh-functions was also used originally by Walsh [1]. In the application of the functions it is often of practical interest to use particular symbols for even and odd Walsh-functions. For the even Walsh-functions we use the symbol cal $(i, \cdot)$, and for the odd we use sal $(i, \cdot)$. We have:

$$\text{cal}\,(i, \cdot) = \text{wal}\,(2i, \cdot) \quad \text{for } i = 0, 1, 2, \ldots$$

and (1)

$$\text{sal}\,(i, \cdot) = \text{wal}\,(2i-1, \cdot) \quad \text{for } i = 1, 2, 3, \ldots$$

The number i represents the generalized frequency, which has been called "sequency" (Harmuth [12], [13]).

The second method of defining the Walsh-functions was introduced by Paley [2]. This method presents the Walsh-functions $\psi(n, \cdot)$ as products of Rademacher-functions $\phi(k, \cdot)$. If the number n has the dyadic representation.

$$n = \sum_{k=-N}^{0} n_k 2^{-k} \tag{2}$$

then the Walsh-function $\psi(n, \cdot)$ is defined by

$$\psi(n, t) = \prod_{k=-N}^{0} \left[\phi(-k, t)\right]^{n_k} \tag{3}$$

The Rademacher-functions $\phi(-k, \cdot)$, $k \in Z$ (Z denotes the set of integers), can be defined as the functions given by

$$\phi(-k, t) = \exp \pi i t_{1-k} \tag{4}$$

where t is a nonnegative real number given by

$$t = \sum_{k=-\infty}^{\infty} t_k 2^{-k}$$

For negative real numbers-t the Rademacher functions are defined by

$$\phi(-k, -t) = -\phi(-k, t) \tag{5}$$

It should be mentioned that the number V(n), given by

$$V(n) = \sum_{k=-N}^{0} n_k \tag{6}$$

has been called the "Vielfalt" of the Walsh-function $\psi(n, \cdot)$. V(n) is the number of Rademacher-functions $\phi(-k, \cdot)$ which generates $\psi(n, \cdot)$. There are many interesting mathematical results associated with the "Vielfalt" and with the problem of approximating a polynomial with a Walsh-Fourier series [10], [11].

The relationship between the Walsh-functions wal $(i, \cdot)$ and $\psi(n, \cdot)$ is given by the formula

$$\text{wal}\ (i, \cdot) = \begin{cases} \psi\,(i/2) \oplus i, \cdot) & i = 0, 2, 4, \ldots \\ \psi\,[(i-1)/2] \oplus i, \cdot) & i = 1, 3, 5, \ldots \end{cases} \qquad (7)$$

where $\oplus$ denotes the addition modulo 2 of the integers written as binary numbers.

2.2 Walsh-Fourier Analysis

It is well known that the Walsh functions form a complete orthonormal system of functions for the real Hilbert-space $L_2\ [a, a+1]$ of functions defined on a interval of Length 1. Therefore, we have a theory to analyze and synthesize functions $f \in L_2\ [a, a+1]$. The discrete Walsh-Fourier transform $\hat{f}$ is defined by

$$\hat{f}\,(n) = \langle f, \psi\,(n, \cdot)\rangle \quad n = 0, 1, 2, \ldots \qquad (8)$$

where $\langle \cdot, \cdot \rangle$ denotes the inner-product of the space $L_2[a, a+1]$. Using Walsh functions cal $(i, \cdot)$ and sal $(i, \cdot)$, we will denote the discrete Walsh-Fourier transform of $f \in L_2\ [a, a+1]$ by the symbols F_c and F_s:

$$\begin{aligned} F_c(i) &= \langle f,\ \text{cal}\ (i, \cdot)\rangle \quad i = 0, 1, 2, \ldots \\ F_s(i) &= \langle f,\ \text{sal}\ (i, \cdot)\rangle \quad i = 1, 2, 3, \ldots \end{aligned} \qquad (9)$$

Next, we must have a theory of Walsh-Fourier integrals to represent nonperiodic functions as superpositions of Walsh functions. To formulate such a theory we must define the Walsh functions for continuous parameters. With respect to the Walsh functions $\psi\,(n, \cdot)$ this work was done in a paper by Fine [5]. According to Fine the Walsh functions $\psi(y,\cdot)$, $y \in R_+$ *), are defined in a very natural way as the functions given by

$$\psi(y,t) = \exp \pi i \sum_{k=-N}^{M+1} y_k t_{1-k} \tag{10}$$

where $y, t \in R_+$ have the dyadic representation

$$y = \sum_{k=-N}^{\infty} y_k 2^{-k} \quad \text{and} \quad t = \sum_{k=-M}^{\infty} t_k 2^{-k} \tag{11}$$

If $y \in D_+$ or $t \in D_+$ where D_+ denotes the set of nonnegative dyadic rational numbers, we use for (11) the finite dyadic representation. The Walsh functions $\psi^*(y,\cdot)$, $y \in D_+$, Fine defines as the functions given by

$$\psi^*(y,t) = \exp \pi i \sum_{k=-N}^{M+1} y_k t_{1-k} \tag{12}$$

where y is now represented by (11) by an infinite sum. Note that the Walsh functions $\psi(y,\cdot)$ are only defined for $t \in R_+$.

The Walsh functions cal $(s,\cdot)$ and sal $(s,\cdot)$ can be defined for $s \in R_+$ in a similar manner. If $s \in R_+$ and $t \in R_+$ have the dyadic representation

$$s = \sum_{k=-N}^{\infty} s_k 2^{-k} \quad \text{and} \quad t = \sum_{k=-M}^{\infty} t_k 2^{-k} \tag{13}$$

then the functions cal $(s,\cdot)$ and sal $(s,\cdot)$ are given for $t \in R_+$ by

$$\text{cal}\,(s,t) = \exp \pi i \sum_{k=-N}^{M+1} (s_k + s_{k+1}) t_{1-k} \tag{14}$$

*) R denotes the set of real numbers, R_+ denotes the nonnegative real numbers.

6

and

$$\text{sal}(s,t) = \exp \pi i \sum_{k=-N}^{M+1} (s_k + s_{k+1}) t_{1-k} \qquad (15)$$

The difference between (14) and (15) is only that if $s \in D_+$ we have to use in (14) the finite dyadic representation of s but in (15) the infinite representation.

To define the Walsh functions cal (s, ·) and sal (s, ·) on the entire real axis we determine the cal functions to be even and the sal functions to be odd functions of the variable t.

The connection of the cal and sal functions to the functions ψ and ψ^* is given by the formula

$$\begin{aligned} \text{cal}(s,t) &= \psi(s,t)\psi(2s,t) \qquad && s,t \in R_+ \\ \text{sal}(s,t) &= \psi^*(s,t)\psi^*(2s,t) \qquad && s \in D_+,\ t \in R_+ \end{aligned} \qquad (16)$$

Now one can derive a theory of Walsh-Fourier integrals.

If $f \in L_2[0,\infty)$, the Walsh-Fourier transform $\hat{f}$ of f is defined as the function given by the integral

$$\hat{f}(y) = \int_0^\infty f(t)\psi(y,t)dt \qquad (17)$$

where the integral converges with respect to norm of the space $L_2[o,\infty)$. If $f \in L_2(-\infty,\infty)$, we define the cal transform F_c and the sal transform F_s of f as the functions given by

$$F_c(s) = \int_{-\infty}^{\infty} f(t)\,\text{cal}(s,t)dt \qquad (18)$$

and

$$F_s(s) = \int_{-\infty}^{\infty} f(t)\ \mathrm{sal}\ (s,t)dt \tag{19}$$

There are theorems concerning the transforms of a function $f \varepsilon L_1[0,\infty)$ and $f \varepsilon L_1(-\infty,\infty)$, respectively. For theorems such as Plancherel theorem and convolution theorem the reader is advised to consult the papers of Fine [5], Vilenkin [7], Selfridge [14] and Pichler [15].

A general transform theory defined on an arbitrary locally compact Abelian group has been presented in a paper by Falb and Friedman [16].

2.3 Dyadic Correlation Analysis

Next, we are concerned with a generalization of correlation methods. A generalization may be obtained by using "addition modulo 2" $\oplus$ instead of the usual addition of real numbers. Let us first define what we mean by "addition modulo 2" of real numbers. Let $u \varepsilon R_+$ and $v \varepsilon R_+$ with, if possible, u and v having a finite dyadic representation

$$u = \sum_{i=-N}^{\infty} u_i 2^{-i}, \quad v = \sum_{i=-M}^{\infty} v_i 2^{-i}, \tag{21}$$

then the real number $u \oplus v$ is given by

$$u \oplus v = \sum_{i=-L}^{\infty} (u_i \oplus v_i) 2^{-i} \tag{22}$$

where $L = \max(N, M)$ and $u_i \oplus v_i$ denotes addition modulo of numbers 0 or 1 in the usual sense ($0 \oplus 1 = 1 \oplus 0 = 1, 0 \oplus 0 = 1 \oplus 1 = 0$). For negative real numbers, addition modulo can be defined by

8

$$u \oplus (-v) = (-u) \oplus v = -(u \oplus v) \tag{23}$$

and

$$(-u) \oplus (-v) = u \oplus v \tag{24}$$

Let $W_2(-\infty, \infty)$ denote the space of functions $f: R \rightarrow R$ for which the integral

$$\lim_{t \to \infty} \frac{1}{2T} \int_{-T}^{+T} f^2(t) dt \tag{25}$$

exists and has a finite value. Let $f, g \in W_2(-\infty, \infty)$. The dyadic cross-correlation function $R(f, g, \cdot)$ of f and g then is defined as the function given by

$$R(f, g, \tau) = \lim_{T \to \infty} \frac{1}{2T} \int_{-T}^{+T} f(t) g(t \oplus \tau) d\tau \tag{26}$$

For g = f we write for $R(f, f, \cdot)$ easier $R(f, \cdot)$. $R(f, \cdot)$ is called the dyadic autocorrelation function of f.

For dyadic correlation there is a theory similar to the classical theory of correlation. The initial development of this theory was presented in a thesis by Weiser [17] and a paper by this author [18]*). The possibility of such a theory was mentioned in a paper of Wiener and Paley [19] as early as 1932, but there have been no reports of further development. There is, however, hope that progress will be made in this direction. Like the generalized harmonic analysis of N. Wiener, one can embed this theory in a more general theory of certain stochastic processes on topological groups and now there are many efforts to complete this theory.

*) See also, J. E. Gibbs and H. A. Gebbie: "Application of Walsh Functions to Transform Spectroscopy", Nature, pp. 1012-1013, Dec. 1969.

3. Linear Dyadic-Invariant Systems

We shall now define a certain class of linear systems, which have an invariant behavior against dyadic translations of the input-functions.

Let I denote the space of input-functions of a scalar continuous linear system S. Each $x \in I$ should be a real-valued stepwise integrable time functions. Let O denote the corresponding space of output-functions of the system S. We define then S as the system, which has an input-output relation $R(S) \subset I \times O$ given by the integral

$$y(t) = \int_{-\infty}^{\infty} h(t \oplus \tau)\, x(\tau) d\tau \tag{27}$$

where y denotes an output-function and $h: R \to R$ is assumed to be absolutely integrable on R. So $(x, y) \in R(S)$ if x and y fulfill the equation (27). The function $h^{*}: R \times R \to R$ given by $(t, \tau) \to h(t \oplus \tau)$ is the impulse response of S and h is the impulse response for $\tau = 0$.

The analogy of our system S given by (27) to a linear time-invariant system given by its steady-state representation is apparent. Equation (27) also has a convolution form: it defines the dyadic convolution of the functions h and x. It is easy to show that S is invariant against dyadic translations of the input functions. With that we mean that from $(x, y) \in R(S)$ follows that for all $\lambda \in R$, also $(x_\lambda, y_\lambda) \in R(S)$. Here x_λ and y_λ denotes the λ-dyadic translations of x and y defined by

$$x_\lambda(t) = x(t \oplus \lambda)$$

and (28)

$$y_\lambda(t) = y(t \oplus \lambda)$$

If x is a suited function (e.g. if $x \in L_1(-\infty, \infty)$) applying the convolution theorem of Walsh-Fourier transform theory one can obtain from (27) the equations

$$\begin{aligned} Y_c(s) &= H_c(s)X_c(s) \\ Y_s(s) &= H_s(s)X_s(s) \end{aligned} \tag{29}$$

where $Y_c, Y_s, H_c, H_s, X_c, X_s$, denotes the sal and cal transforms of y, h and x given by integrals of the form (18) and (19) respectively.

In (29) we have a description of the system S in terms of "sequency". The functions H_c and H_s are called the transfer functions of the system S.

We have assumed that $h \in L_1(-\infty, \infty)$. From that follows, that S is "bounded-input bounded-output stable". For S to be nonanticipative (causal), it is necessary that

$$h(t) = 0 \quad \text{for all } t < 0 \tag{30}$$

But observe that the condition (30) is only necessary and, in general, not sufficient.

3.1 Synthesis of Sequency Bandpasses

We will now outline a synthesis procedure for a certain class of dyadic-invariant system which are bandpasses. A detailed treatment of these has been given in a paper by this author [20].

Let the transfer functions H_c and H_s of a dyadic-invariant linear system S defined by (27) be given by

$$H_c(s) = \begin{cases} 1 & \text{for all } s \in [n2^k, (n+1)2^k) \\ 0 & \text{elsewhere} \end{cases}$$

and

$$H_c(s) = \begin{cases} 1 & \text{for all } s \in (n2^k, (n+1)2^k] \\ 0 & \text{elsewhere} \end{cases}$$

where k is an integer and n a nonnegative integer. The system S, defined by (31), we shall call a sequency-bandpass with a normalized bandwith $\Delta s = 2^k$ and a cutoff-sequency $s = n2^k$. To obtain the impulse response h of this sequency-bandpass, we apply the inverse Walsh-Fourier transform to H_c and H_s and we have then generally

$$h(t) = \int_0^\infty [H_c(s)\,\mathrm{cal}(s,t) + H_s(s)\,\mathrm{sal}(s,t)]\,ds \tag{32}$$

With H_c and H_s defined in (31) h becomes

$$h(t) = \begin{cases} 2^{k+1}\mathrm{cal}(n2^k, t) & \text{for } t \in [0, 2^{-k-1}) \\ 0 & \text{elsewhere} \end{cases} \tag{33}$$

We see that h is a Walsh impulse given on the interval $[0, 2^{-k-1})$. It can be shown that the filters we get in this way are b.i.b.o. stable and nonanticipative. To make these filters "technically realizable" we have to allow a constant delay β with $\beta \geq 2^{-k-1}$. It happens that these filters are exactly the same as the sequency-bandpasses of Harmuth [12], [21], derived by a different approach.

3.2 Synthesis of Optimal Filters

The development of a theory of optimal filtering is now near at hand, based on a Walsh-Fourier decomposition of signals and systems. One can formulate such a theory in a manner similar to the classical theory of optimal filtering of Wiener and Kolmogoroff. In a method similar to Wiener, one can do this without the theory of probability. Assuming that the signal u, as also the noise v, is an element of the space $W_2(-\infty, \infty)$ we obtain relations for the transfer functions H_c and H_s of the optimal sequency filter.

Let the input x of a linear dyadic-invariant system be the sum of a signal u and noise v; $x = u + v$. Both u and v are assumed to be elements of the space $W_2(-\infty, \infty)$. We want to find the impulse response h of the system so that the mean-square deviation $\overline{\varepsilon^2}$ defined by

$$\overline{\varepsilon^2} = \lim_{T \to \infty} \frac{1}{2T} \int_{-T}^{T} (y(t)-u(t))^2 dt \qquad (34)$$

is minimal. Here y denotes the outputsignal related with the inputsignal x. A linear dyadic-invariant system with such an impulse response, h, could be called an optimal sequency filter.

A solution to the problem of finding h is given in a paper by the author [18]. It can be shown that the transfer function H_c and H_s of the optimal filter are related to the one-sided Walsh-Fourier transforms $S_{co}(u, x, \cdot)$, $S_{so}(u, x, \cdot)$, $S_{co}(x, \cdot)$ and $S_{so}(x, \cdot)$ of the dyadic correlation functions $R(u, x, \cdot)$ and $R(x, \cdot)$ respectively, in the following expressions:

$$\begin{aligned} S_{co}(u, x, s) &= H_c(s) S_{co}(x, s) \\ S_{so}(u, x, s) &= H_s(s) S_{so}(x, s) \end{aligned} \qquad (35)$$

Further it can be shown, that the derived optimal filter is nonanticipative at the time $\tau = 0$. For applications in communciations it would be of interest to find optimal sequency filters which could easily be built up with electronic elements.

With a slight modification of this theory for other classes of signals and noise, one can get the Harmuth sequency bandpasses as optimal filters [18].

4. On a State Space Approach for Linear Dyadic-Invariant Systems

Now we shall deal with some of the concepts of a state-space approach for dyadic-invariant systems. The development of such a theory seems to be of interest. It could give us a deeper insight into the internal working of these systems. Further it could be a bridge to software systems and to applications of dyadic invariant systems in control problems.

4.1 Dyadic Differentiation

First, we are concerned with the concept of a generalized differentiation. The fundamental work for this theory was done in a paper by Gibbs and Millard [22]. There the main interest was that of finite discrete sturctures. Some slight modifications must be made to obtain a theory of generalized differentiation defined for real valued functions of a continuous nonnegative real variable. Let f be such a function. To f we attach, if possible, a function $f^{[1]}$, given by

$$f^{[1]}(t) = \sum_{k=-\infty}^{\infty} [f(t) - f(t \oplus 2^{-k})]\, 2^{k-2} \qquad (36)$$

If the function $f^{[1]}$ exists, we shall call it the first dyadic derivative of f. *) The function f we call in this case dyadic differentiable. The term "dyadic differentiation" is used since this linear operation has certain qualities, which one associates with differentiation. As the first result of dyadic differentiation we have for the Walsh functions $\psi(y, \cdot)$

$$\psi^{[1]}(y,t) = y\psi(y,t) \qquad \forall y, t \in R_+ \tag{37}$$

Equation (37) shows us, that the Walsh functions $\psi(y, \cdot)$ are eigenfunctions of the dyadic differential operator. This is easy to prove:

From (36) we have

$$\psi^{[1]}(y,t) = \sum_{k=-\infty}^{\infty} [\psi(y,t) - \psi(y, t\oplus 2^{-k}]\, 2^{k-2} \tag{38}$$

With the formula

$$\psi(y, t\oplus 2^{-k}) = \psi(y,t)\psi(y, 2^{-k}) \tag{39}$$

we get

$$\psi^{[1]}(y,t) = \sum_{k=-\infty}^{\infty} [1-\psi(y,2^{-k})]\, 2^{k-2}\psi(y,t) \tag{40}$$

Due to the fact that $\psi(y, 2^{-k}) = \exp \pi i y_{1-k}$

and so $[1-\psi(y,2^{-k})]\,2^{-1} = y_{1-k}$ we get from (40)

$$\psi^{[1]}(y,t) = \sum_{j=-\infty}^{\infty} y_{1-k} 2^{k-1}\psi(y,t) \tag{41}$$

*) Gibbs and Millard use the terms "logical differentiation" and "logical derivative".

and with the substitution $l-k \rightarrow j$ we obtain

$$\psi^{[1]}(y,t) = \sum_{j=-\infty}^{\infty} y_j 2^{-j} \psi(y,t) = y\psi(y,t) \tag{42}$$

So the Walsh functions $\psi(y,\cdot)$ can be seen to be the solutions of the following dyadic differential equation of the first order for different values of $y \in R_+$.

$$f^{[1]} - yf = 0 \tag{43}$$

To get this property of the Walsh functions originally was the motive of Gibbs to define a "logical differentiation".

Another interesting result concerns the Walsh-Fourier transforms of dyadic derivatives. Let f be an n-time dyadic-differentiable function. Let $\hat{f}$ denote the Walsh-Fourier transform of f given by (17). We have then the following theorem:

$$\widehat{f^{[n]}}(y) = y^n \hat{f}(y) \quad \forall y \in R_+ \tag{44}$$

The proof of this theorem is straight forward and may here be omitted. This theorem is analogous to the well-known theorem concerning the Laplace transforms of derivatives. The difference is the absence of initial values.

4.2 Dyadic Differential Equations and Dyadic Invariant Systems

The concept of dyadic differentiation directly leads to the concept of dyadic differential equations. We define an ordinary linear dyadic differential equation with constant coefficients as a relation given by

$$\sum_{k=o}^{n} a_k f^{[k]} = \sum_{k=o}^{m} b_k u^{[k]} \tag{45}$$

where us is assumed to be k-time dyadic differentiable; a_k and b_k should be real numbers; f denotes the general solution of (45). The relation $R(S_d)$ given by (45) can be interpreted as an input-output relation of a system S_d. It is clear that the system S_d is linear.

We get the general solution, f, (the general output function) of equation (45) as the sum of the solution f_h of the homogeneous equation (the zero-input response of the system) and the particular solution, f_p, of the inhomogeneous equation (the zero-state response)

$$f = f_h + f_p \tag{46}$$

f_h is given by a linear combination of Walsh functions, $\psi(\alpha, \cdot)$. If we assume that the characteristic equation of (45) given by

$$\sum_{k=o}^{n} a_k \alpha^k = 0 \tag{47}$$

has n real and distinct solutions, $\alpha_1, \ldots, \alpha_n$, then f_h is given by

$$f_h = \sum_{k=1}^{n} c_k \psi(\alpha_k, \cdot) \tag{48}$$

Hereby the coefficients $c_1, \ldots, c_n$ result from the initial values connected with equation (45). If the initial values of f are given at time 0 by

$$f(0),\ f^{[1]}(0), \ldots, f^{[n-1]}(0), \tag{49}$$

then we can get the coefficients $c_1, \ldots, c_n$, as the solutions of the following regular system of linear equations

$$\sum_{k=1}^{n} c_k \alpha_k^i = f^{[1]}(0) \quad i = 0, 1, \ldots, n-1 \tag{50}$$

To get the particular solution of the dyadic differential equation given by (45) we apply the Walsh-Fourier transform on both sides of (45) and get with regard to (44)

$$\sum_{k=o}^{n} a_k y^k \hat{f}_p(y) = \sum_{k=o}^{m} b_k y^k \hat{u}(y) \tag{51}$$

Here we have assumed that both f_p and u have a Walsh-Fourier transform. With the polynomials p and q given by

$$p(y) = \sum_{k=o}^{m} b_k y^k \quad \text{and}$$

$$q(y) = \sum_{k=o}^{n} a_k y^k \tag{52}$$

we get from (51)

$$\hat{f}_p(y) = \frac{p(y)}{q(y)} \hat{u}(y) \tag{53}$$

The function $\hat{h}$ given by

$$\hat{h} = \frac{p}{q} \tag{54}$$

is called the transfer function of the system S_d. We shall see that the system S_d associated with the dyadic differential equation (45) is a linear dyadic invariant system.

From (53), the inverse Walsh-Fourier transform produces the particular solution f_p

$$f_p(t) = \int_o^\infty \hat{h}(y)\hat{u}(y)\psi(y, t)dy \tag{55}$$

Assuming that we can apply a convolution theorem, we get from (55)

$$f_p(t) = \int_o^\infty h(t \oplus \tau)u(\tau)d\tau \tag{56}$$

where h denotes the inverse Walsh-Fourier transform of $\hat{h}$. We see that (56) has a form similar to (27). The difference is only that the lower limit of the integral is 0 rather than $-\infty$. This is the result of assuming that the functions are only defined on R_+. It seems possible to extend this theory to functions defined on the whole real axis R.

So we have the following result: The ordinary linear dyadic differential equation given by

$$\sum_{k=o}^{n} a_k f^{[k]} = \sum_{k=o}^{m} b_k u^{[k]} \tag{57}$$

where u is a k-time dyadic-differentiable function and a_k, $k = o, \ldots, n$, and b_k, $k = o, \ldots, m$, are real numbers, has a general solution f of the following form

$$f(t) = \sum_{k=1}^{n} c_k \psi(\alpha_k, t) + \int_o^\infty h(t \oplus \tau)u(\tau)d\tau \tag{58}$$

if the following assumptions are fulfilled:

a) The characteristic equation

$$\sum_{k=o}^{n} a_k \alpha^k = 0 \tag{59}$$

of (57) has n distinct real roots,

$$\alpha_1, \ldots, \alpha_n.$$

b) The transfer dunction $\hat{h}$ of (57) given by

$$\hat{h}(y) = \frac{\sum_{k=o}^{m} b_k y^k}{\sum_{k=o}^{n} a_k y^k} \quad \forall y \in R_+ \tag{60}$$

has an inverse Walsh-Fourier transform h.

c) The real numbers c_k, $k = 1, \ldots, n$ of (58) are given by following regular system of linear equations

$$\sum_{k=1}^{n} c_k \alpha_k^i = f^{[i]}(0) \quad i = 0, 1, \ldots, n-1 \tag{61}$$

where $f^{[i]}(0)$, $i = 0, 1, \ldots, n-1$, are the initial values of f at time 0.

For the scalar linear system S_d connected with the dyadic differential equation (57) the solution (58) represents an input-output-state relation [23]. Hereby determine the constant numbers $c_1, \ldots, c_n$ the state x(o) at time 0; $x(o)' = (c_1, \ldots, c_n)$. The vector $\psi(\alpha, \cdot) = (\psi(\alpha_1, \cdot), \ldots, \psi(\alpha_n, \cdot))$ is a basis vector whose components $\psi(\alpha_1, \cdot), \ldots, \psi(\alpha_n, \cdot)$ represents the zero-input response of S_d starting in the initial states $(1, 0, \ldots, 0), \ldots, (0, \ldots, 1)$ respectively.

h is the impulse response of S_d, that is, the zero-state response of S_d to the unit impulse $\delta(t)$. h is the inverse Walsh-Fourier transform of $\hat{h}$, the transfer function of S_d.

To get the linear dyadic invariant system S_d nonanticipative we have to assume that

$$h(t \oplus \tau) = 0 \text{ for all } t < \tau \qquad (62)$$

With (62) and expressing the sum in (58) as the scalarproduct $<\psi(\alpha, t), x(o)>$ of the vectors $\psi(\alpha, t)$ and $x(o)$ we get for a linear dyadic invariant nonanticipative system S_d the input-output-state relation in the form

$$f(t) = <\psi(\alpha, t), x(o)> + \int_{o}^{t} h(t \oplus \tau) u(\tau) d\tau \qquad (63)$$

The formula (63) has an analogous counterpart in the theory of linear time-invariant systems.

These are first steps to a general theory. Much work must yet be invested in this theory.

Acknowledgements

I should like to thank Mrs. Barbara Aycock and Mrs. Olivia Goetz for typing the manuscript and for their help in correcting the grammar.

REFERENCES

[1] Walsh, J. L., "A closed set of normal orthogonal functions", Amer. Math., vol. 45, pp. 5-24, 1923.

[2] Paley, R. E. A. C., "A remarkable series of orthogonal functions", Proc. London Math. Soc., vol. 34, pp. 241-279, 1932.

[3] Levy, P., "Sur une generalisation de fonctions orthogonales de M. Rademacher", Comment. Math. Helv., vol. 16, pp. 146-152, 1944.

[4] Fine, N. J., "On the Walsh functions", Trans. Amer. Math. Soc., vol. 65, pp. 372-414, 1949.

[5] Fine, N. J., "The generalized Walsh functions", Trans. Amer. Math. Soc., vol. 69, pp. 66-77, 1950.

[6] Vilenkin, N. Y., "On a class of complete orthogonal systems", Izv. Akad. Nauk Ser. Math., vol. 11, pp. 363-400, 1947 (in Russian).

[7] Vilenkin, N. Y., "On the theory of Fourier integrals on topological groups", Mat. Sbornik (N.S.), vol. 30 (72), pp. 233-244, 1952 (in Russian).

[8] Chrestenson, H. E., "A class of generalized Walsh functions", Pacific J. Math., vol. 5, pp. 17-31, 1955.

[9] Watari, Ch., "On generalized Walsh-Fourier series", Tohoku Math. J. (2), vol. 10, pp. 211-241, 1958.

[10] Weiss, P., "Zusammenhang von Walsh-Fourierreihen mit Polynomen", Monatshefte für Mathematik, vol. 71, pp. 165-179, 1967.

[11] Liedl, R., "Eine algebraische Herleitung und eine Verallgemeinerung des Satzes von FINE über Gruppen von orthonormalen Funktionen und eine Beschreibung der vielfalterhaltenden Transformationen des Intervalls [0,1] auf sich selbst, "Monatshefte für Mathematik, vol. 72, pp. 45-80, 1968.

[12] Harmuth, H. F., "Transmission of information by orthogonal functions", Springer-Verlag, New York, 1969.

[13] Harmuth, H. F., "A generalized concept of frequency and some applications", IEEE Trans. Information Theory, vol. IT-14, No. 3, 375-382, May, 1968.

[14] Selfridge, R. G., "Generalized Walsh transform", Pacific J. Math., vol. 5, pp. 451-480, 1955.

[15] Pichler, F., "Das system der sal-und cal-Funktionen als Erweiterung des Systems der Walsh-Funktionen und die Theorie der sal-und cal-Fouriertransformation", Dissertation, Philosophische Fakultät der Universität Innsbruck, 1967.

[16] Falb, B. L. and M. I. Friedmann, "A generalized transform theory for causal operators", Siam J. Control, vol. 7, pp. 452-471, 1969.

[17] Weiser, F. E., "Walsh function analysis of instantaneous nonlinear stochastic processes", Dissertation, Polytechnic Institute of Brooklyn, June, 1964.

[18] Pichler, F., "Walsh-Fourier Synthese optimaler Filter" to appear in A. E. Ü., 1970.

[19] Wiener, N. and R. E. A. C. Paley, "Analytic properties of infinite abelian groups", Verhandlungen des Internationalen Mathematiker-Kongresses Zürich, 1932, II. Band, p. 95.

[20] Pichler, F., "Synthese linearer periodisch zeitvariabler Filter mit vorgeschriebenem Sequenzverhalten", A. E. Ü., vol. 22, pp. 150-161, 1968.

[21] Harmuth, H. F., "Grundzüge einer Filtertheorie für die Mäanderfunktionen $A_n(\theta)$," A. E. Ü., vol. 18, pp. 544-554, 1964.

[22] Gibbs, J. E., and Margaret J. Millard, "Some methods of solution of linear ordinary logical differential equations", National Physical Laboratory, DES Report No. 2, December, 1969.

[23] Zadeh, L. A., and Ch. A. Desoer, "Linear System Theory, The State Space Approach", McGraw-Hill Book Company, New York, 1963.

APPENDIX A

A-1

(a)

$t \longrightarrow$, $\tau \downarrow$

$t \oplus \tau$	0	1	2	3	4	5	6	7
0	0	1	2	3	4	5	6	7
1	1	0	3	2	5	4	7	6
2	2	3	0	1	6	7	4	5
3	3	2	1	0	7	6	5	4
4	4	5	6	7	0	1	2	3
5	5	4	7	6	1	0	3	2
6	6	7	4	5	2	3	0	1
7	7	6	5	4	3	2	1	0

(b)

$t \longrightarrow$, $\tau \downarrow$

τ^*	0	1	2	3	4	5	6	7
0	0	0	0	0	0	0	0	0
1	1	-1	1	-1	1	-1	1	-1
2	2	2	-2	-2	2	2	-2	-2
3	3	1	-1	-3	3	1	-1	-3
4	4	4	4	4	-4	-4	-4	-4
5	5	3	5	3	-3	-5	-3	-5
6	6	6	2	2	-2	-2	-6	-6
7	7	5	3	1	-1	-3	-5	-7

Figure 1: (a) Addition modulo 2 table for the integers between 0 and 7.

(b) Time shift τ^*, given by $t + \tau^* = t \oplus \tau$ for $t, \tau = 0, 1, \ldots, 7$.

A-2

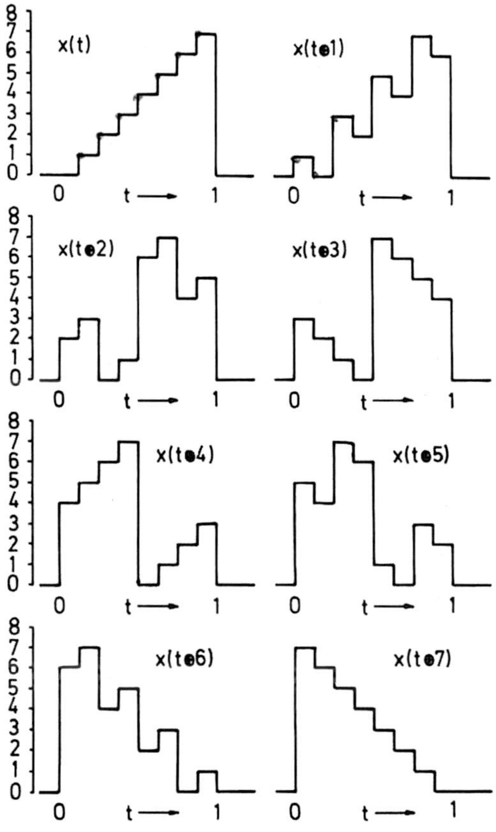

Figure 2: Graph of a stepfunction x given by $x(t) = [8t]$ for $t \in [0,1)$ and $x(t) = 0$ for $t \notin [0,1)$ and its dyadic translations x_λ, $\lambda = 1, 2, \ldots, 7$.

A - 3

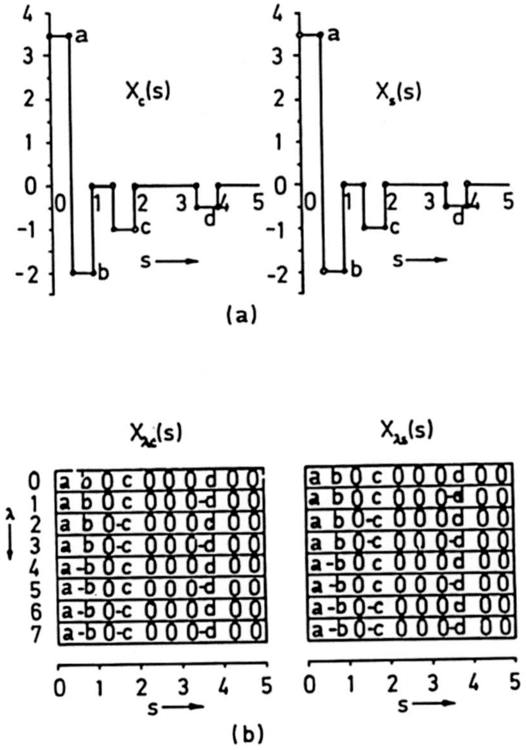

Figure 3: (a) Walsh-Fourier transforms X_c and X_s of the function x given in figure 2

(b) table for the values of the Walsh-Fourier transforms $X_{\lambda c}$ $X_{\lambda s}$ of the functions x_λ of figure 2.

Notice that $X_{\lambda c}^2(s) = X_c^2(s)$ and $X_{\lambda s}^2(s) = X_s^2(s)$ for all λ.

A-4

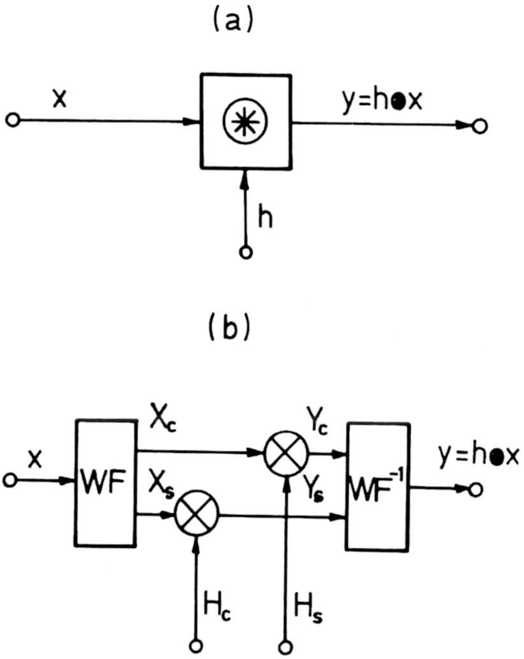

Figure 4: (a) Block-diagram for a relation defined by dyadic convolution. x inputsignal, h inpulse response at time t = 0, y outputsignal.

(b) Dyadic convolution using Walsh-Fourier transformation and dyadic convolution theorem.

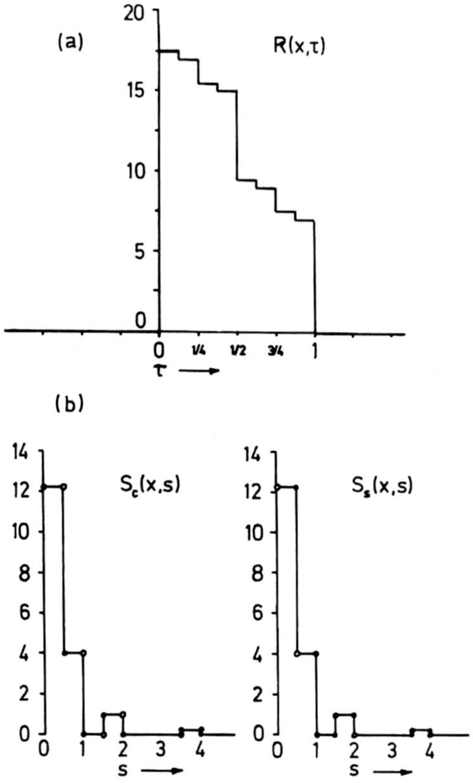

Figure 5: (a) Dyadic autocorrelation function $R(x, \cdot)$ of the functions x_λ of figure 2.

(b) Walsh-Fourier transforms $S_c(x, \cdot)$ and $S_s(x, \cdot)$ of $R(x, \cdot)$

A comparison with $X_{\lambda c}$ and $X_{\lambda s}$ of figure 3 shows, that we have $S_c(x, s) = X^2_{\lambda c}(s)$ and $S_s(x, s) = X^2_{\lambda s}(s)$ for all $\lambda = 0, 1, \dots, 7$ and all s. We call therefore the functions $S_c(x, \cdot)$ and $S_s(x, \cdot)$ the sequency power spectra of the functions x_λ.

A-6

(a)

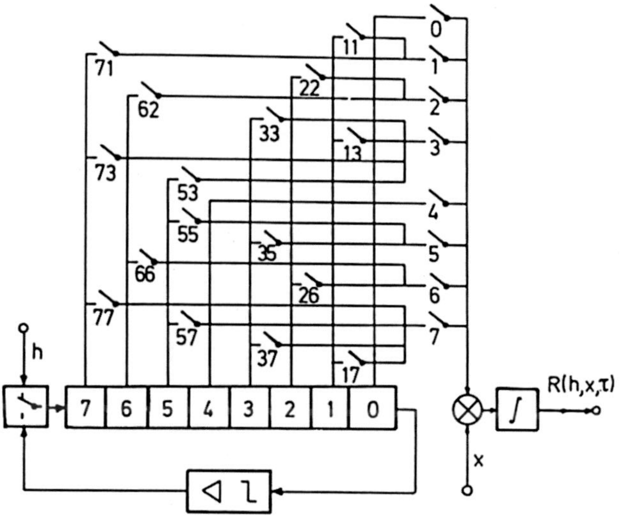

Figure 6: Principles of dyadic correlators.

(a) Dyadic correlator using mod 8 shift register for a stepfunction h and a signal x. The switches perform the values $h(t \oplus \tau)$, $t \in [0,1)$ and $\tau = 0, 1, \ldots, 7$. The integrator computes the values

$$R(h, x, \tau) = \int_0^1 h(t \oplus \tau) x(t) dt$$

where $\tau = 0, 1, \ldots, 7$.

This principle of a dyadic correlator seems to be interesting for the design of dyadic invariant systems (sequency filters) with prescribed impulse response h.

A-7

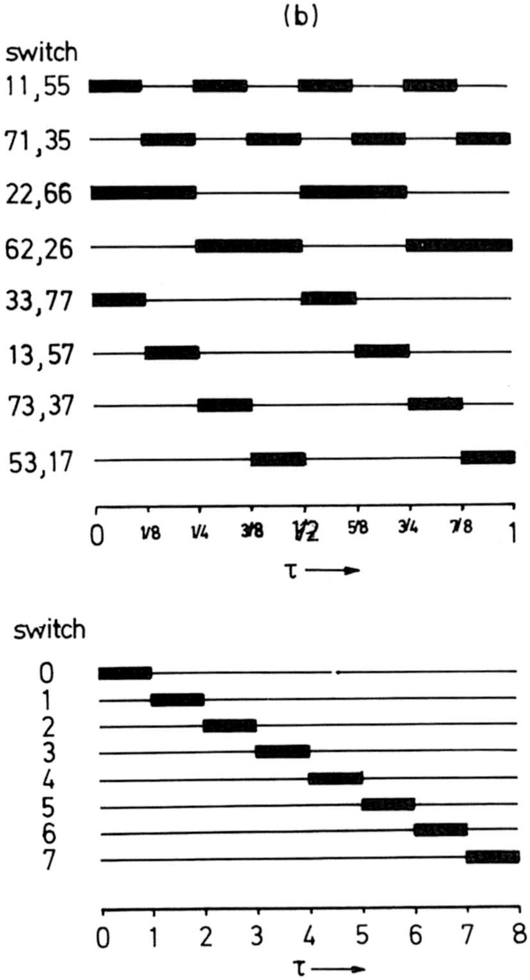

(b) Time table for the switches of the dyadic correlator of figure 6(a).

A-8

(c)

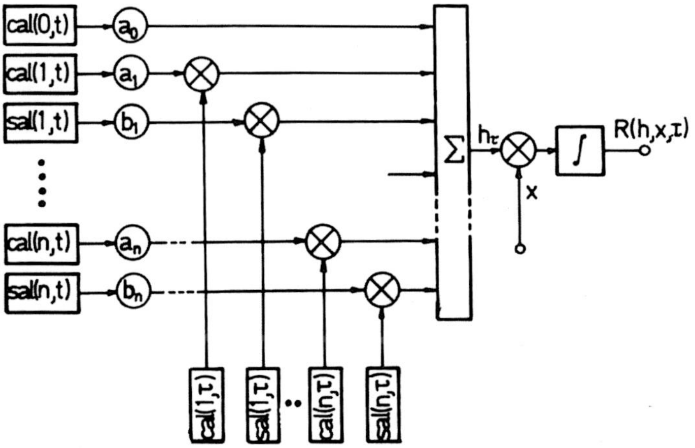

(c) Dyadic correlator using the Walsh-Fourier representation of h. The signal h is assumed to be given by

$$h(t) = \sum_{i=0}^{n} a_i \mathrm{cal}(i,t) + \sum_{i=1}^{n} b_i \mathrm{sal}(i,t)$$

With the multiplication theorem for Walsh functions we get

$$h_\tau(t) = h(t \oplus \tau) = \sum_{i=0}^{n} a_i \mathrm{cal}(i,t)\mathrm{cal}(i,\tau) + \sum_{i=1}^{n} b_i \mathrm{sal}(i,t)\mathrm{sal}(i,\tau)$$

This signal appears as output of the summator. Multiplication with x and integration of the product $h_\tau x$ gives us the dyadic correlation functions $R(h, x, \cdot)$.

A-9

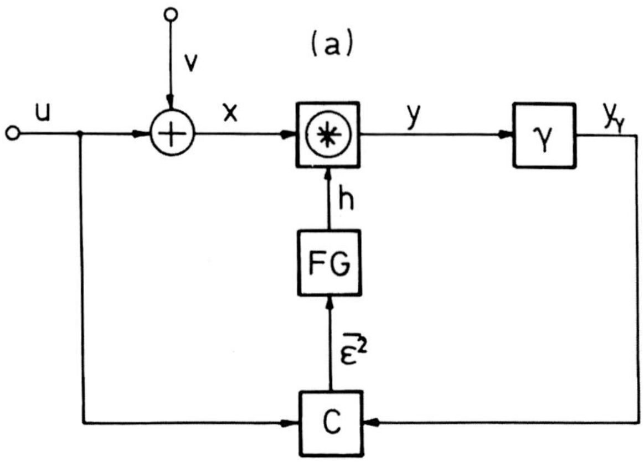

Figure 7: Principles to determine the impulse response h of the optimal linear dyadic invariant system.

(a) minimization of $\overline{\varepsilon^2}$. γ denotes a constant delayor with delay γ. For $\gamma = 0$ we have the case of an optimal filter.

FG function generator which is controlled by $\overline{\varepsilon^2}$.

C correlator performing $\overline{\varepsilon^2}$.

A-10

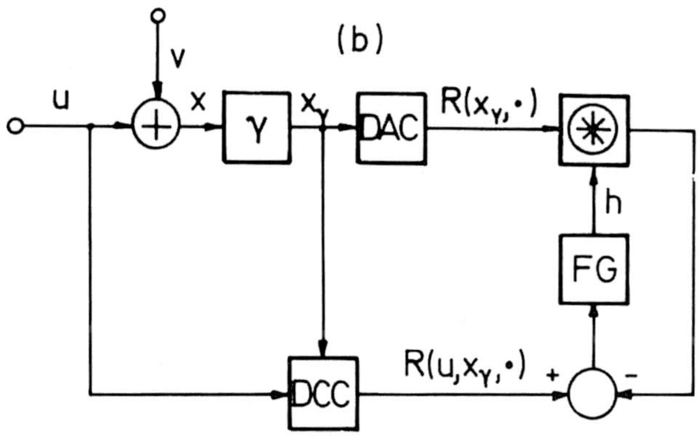

(b) impulse response h of the optimal system as solution of the generalized Wiener-Hopf integral equation

$$R(u, x_\gamma, \tau) - \int_0^\infty R(x_\gamma, \tau \oplus \lambda) h(\lambda) d\lambda = 0$$

for all $\tau > 0$.

DAC dyadic autocorrelator

DCC dyadic cross correlator

FG function generator

A-11

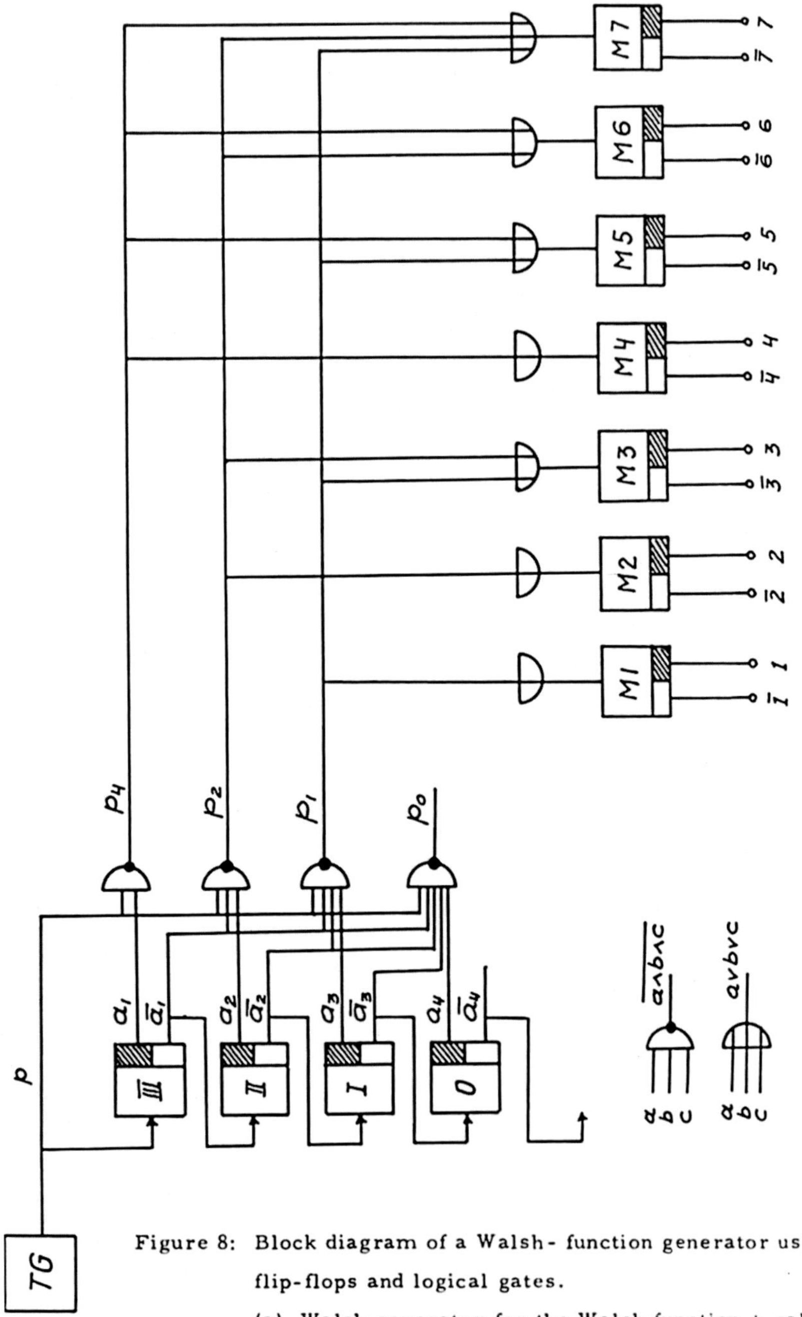

Figure 8: Block diagram of a Walsh- function generator using flip-flops and logical gates.

(a) Walsh generator for the Walsh-function $\pm$ cal(i, t), i = 1, . . . , 7.

A-12

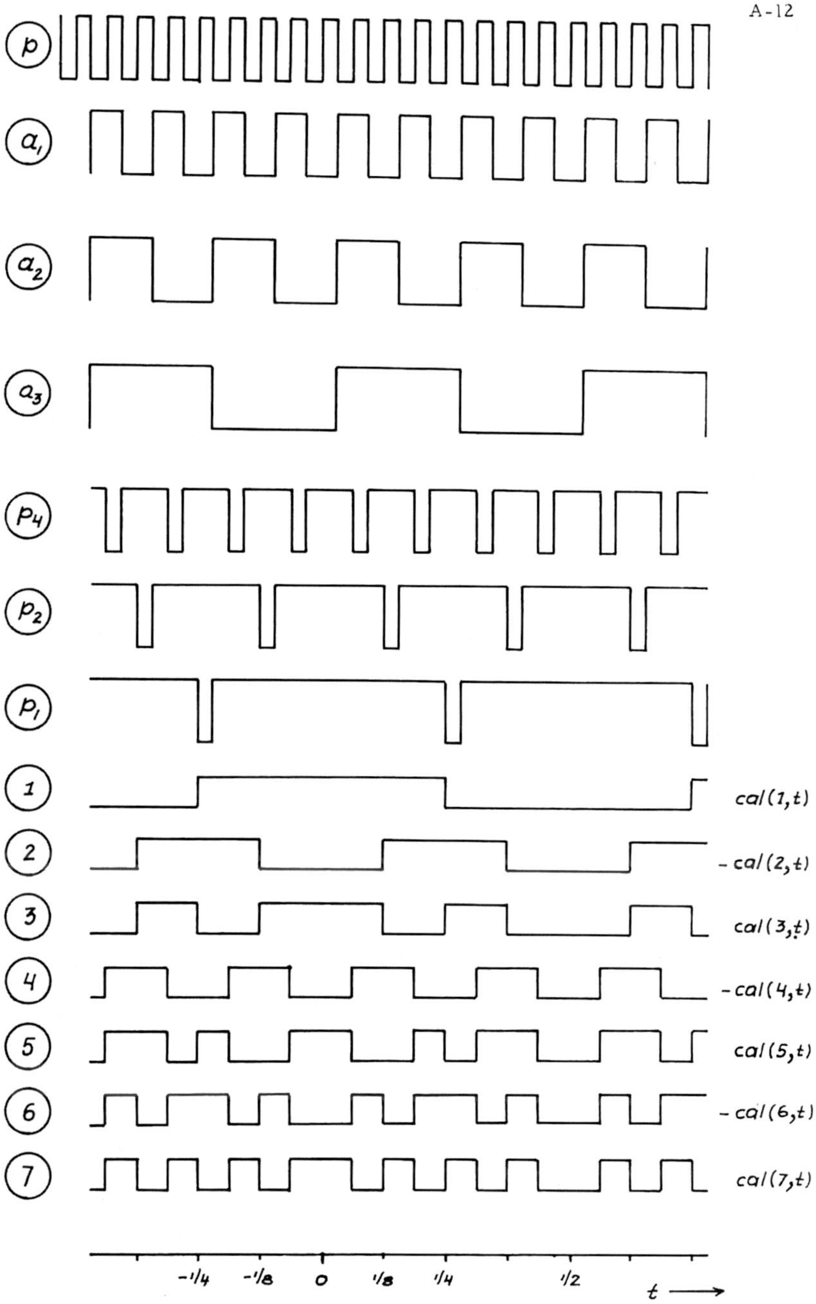

(b) pulse diagram to the Walsh generator of fig. 8 (a).

A-13

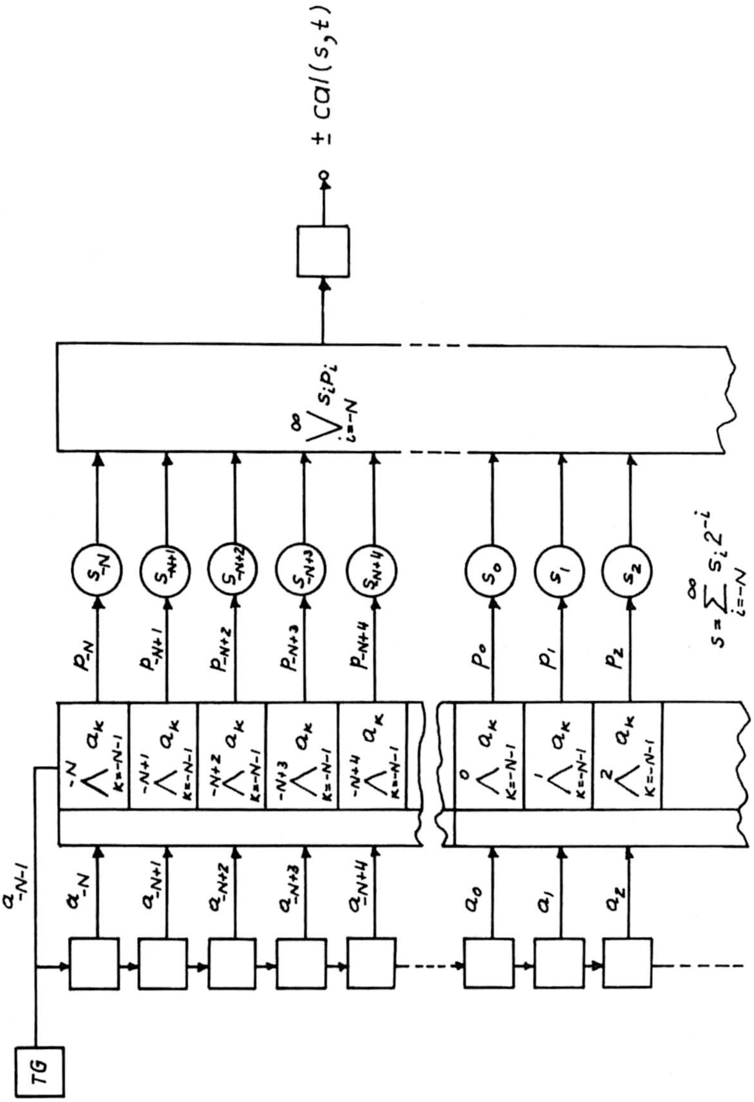

(c) principles of the generation of a arbitrary Walsh function cal(s, t), s > 0.

APPENDIX B

Sampling Theorem with Respect to Walsh-Fourier Analysis

1. Introduction

In the following we are concerned with a generalization of the sampling theorem for band-limited signals. The origin of this classical theorem of communications can hardly be traced. In the mathematical literature it, or some of its analogues, is connected to several authors (e.g. Cauchy [1], Whittaker [2]). The same situation seems to be in the field of communications (e.g. Kotelnikov [3], Shannon [4], Raabe [5]).

Here we shall deal with a formulation of a sampling theorem for the case of representing signals as superpositions of Walsh functions. We shall use results of a paper by Kluvànec [6], in which the sampling theorem has been formulated in terms of the theory of abstract harmonic analysis. We specialize the case of an arbitrary locally compact abelian group to the case of the dyadic group of Fine [7] and get the sampling theorem in terms of Walsh-Fourier analysis. It turns out, that this theorem is a very trivial one if one pays attention to the fact that a function, band limited in the sense of Walsh-harmonic analysis, is equal almost everywhere to a stepfunction. A slight modification of this theorem, depending on the use of the Walsh functions sal(s, t) and cal(s, t), has been presented in a former paper of this author [8].

B-2

2. Sampling Theorem in Abstract Harmonic Analysis

We follow Kluvánek [6]. Let G be a locally compact abelian group (written additively) and $\hat{G}$ the dual group. The value of a character $y \varepsilon G$ at a point $x \varepsilon G$ will be written as usual as (x, y).

Suppose H be a discrete subgroup of G with discrete annihilator Λ given by $\Lambda = \{ y \varepsilon \hat{G}: (x, y) = 1 \text{ for all } x \varepsilon H \}$. Let $[y]$ denote the coset of Λ which contains the point $y \varepsilon \hat{G}$; i.e. $[y] = y + \Lambda$. Let further the set Ω be defined as a measurable subset of $\hat{G}$, $\Omega \subset \hat{G}$, which contains exactly one point from every coset $[y]$ of Λ, i.e. $\Omega \cap [y]$ is a singleton for all $y \varepsilon \hat{G}$.

The Haar measure on G is denoted by m, that of $\hat{G}$ by $\hat{m}$. The Haar measure $[\hat{m}]$ of the factorgroup $\hat{G}/\Lambda$ should be normalized, so that $[\hat{m}](\hat{G}/\Lambda) = 1$. The Haar measure on H and Λ respectively have at each point $x \varepsilon H$ and $y \varepsilon \Lambda$ respectively the value 1. Then $\hat{m}$ can be normalized, so that

$$\int_{\hat{G}} \hat{f}(y)\, d\hat{m}(y) = \int_{\hat{G}/\Lambda} \sum_{z \varepsilon \Lambda} \hat{f}(y+z)\, d[\hat{m}]([y]) \tag{1}$$

holds for every integrable function $\hat{f}$ on $\hat{G}$. Finally let the Haar measure m on G be adjusted so that the inversion formulas for Fourier transforms holds, i.e. by the relations

$$\hat{f}(y) = \int_G (-x, y) f(x)\, dm(x) \tag{2}$$

and

$$f(x) = \int_{\hat{G}} (x, y) \hat{f}(y)\, d\hat{m}(y) \tag{3}$$

B-3

Let further the function φ be defined by

$$\varphi(x) = \int_{\Omega} (x, y)d\hat{m}(y) \tag{4}$$

Due to Kluvanek [6] we have the following lemma and theorem:

Lemma: The function φ is defined for all $x \in G$. It is continuous, positive-definite and belongs to $L^2(G)$. Its norm $||\varphi||$ in $L^2(G)$ is equal 1 and $\varphi(0) = 1$. For all $z \in H$ with $z \neq 0$ we have $\varphi(z) = 0$ and

$$\int_G \varphi(x)\overline{\varphi(x-z)}\, dm(x) = 0 \tag{5}$$

Sampling theorem: Suppose $f \in L^2(G)$ and $\hat{f}(y) = 0$ for almost all $y \notin \Omega$. Then f is equal almost everywhere to a continuous function. If f itself is continuous then

$$f(x) = \sum_{z \in H} f(z)\varphi(x-z) \tag{6}$$

uniformly on G and in the sense of the convergence in $L^2(G)$. Furthermore

$$||f||^2 = \sum_{z \in H} |f(z)|^2 \tag{7}$$

If $G = T = (-\infty, \infty)$, the set of real numbers, and H is given by $H = \{mT: m = 0, \pm 1, + 2, \ldots\}$ where $T = 1/2f_o$, then we have $\Omega = (-2\pi f_o, 2\pi f_o)$ and φ is given by

$$\varphi(x) = \frac{\sin 2\pi f_o(x - mT)}{2\pi f_o(x - mT)} \tag{8}$$

and formula (6) becomes the classical form

$$f(x) = \sum_{m=-\infty}^{\infty} f(mT) \frac{\sin 2\pi f_o(x - mT)}{2\pi f_o(x - mT)} \tag{9}$$

3. Sampling Theorem in Dyadic Harmonic Analysis

Our intention is now, to formulate the sampling theorem of above with respect to the case that $G = \mathfrak{F}$, the dyadic group of Fine [7]. According to Fine the dyadic group $\mathfrak{F}$ is given by the set of infinite sequences $\bar{x}$ of the form

$$\bar{x} = (x_i) = (\ldots 00x_{-M}x_{-M+1}\ldots x_0x_1x_2\ldots)$$

the components x_i being 0 or 1 and each $\bar{x}\epsilon\mathfrak{F}$ being 0-periodic to the left side. It is helpful to identify the set $\mathfrak{F}$ with the set of binary representations of nonnegative real numbers. The composition $\oplus$ in $\mathfrak{F}$ is defined via addition modulo 2 of the components. $(\mathfrak{F}, \oplus)$ is then obviously an abelian group, each $x\epsilon\mathfrak{F}$ having order 2; $\bar{x}\oplus\bar{x} = 0$. A topology can be found, such that $(\mathfrak{F}, \oplus)$ becomes a locally compact topological group [7], [9]. The character group $\hat{\mathfrak{F}}$ of $\mathfrak{F}$ is algebraically isomorph to $\mathfrak{F}$. So each character $\bar{y}\epsilon\hat{\mathfrak{F}}$ can be represented by a sequence of the form

$$\bar{y} = (y_i) = (\ldots 00y_{-N}y_{-N+1}\ldots y_0y_1\ldots)$$

The value $(\bar{x}, \bar{y})$ of a character $\bar{y}\epsilon\hat{\mathfrak{F}}$ at a point $\bar{x}$ is given by

$$(\bar{x}, \bar{y}) = \exp \pi i \sum_{i+k=1} y_i x_k \tag{10}$$

From (10) we can see that the characters $\bar{y}\epsilon\hat{\mathfrak{F}}$ are real valued functions which maps $\mathfrak{F}$ onto the set $\{+1, -1\}$. The Haar measure on $\mathfrak{F}$ and $\hat{\mathfrak{F}}$ respectively should be normalized such that the subgroup $\mathfrak{D}$ and $\hat{\mathfrak{D}}$

B-5

consisting of all sequences $\bar{x} = (x_i)$ and $\bar{y} = (y_i)$ respectively, where $x_i = y_i = 0$ for all $i \leq 0$ have the measure 1. We can now establish the sampling theorem for the dyadic group $\mathfrak{F}$. The subgroup H should be given by $H = \{\bar{x} \varepsilon G: x_i = 0 \text{ for all } i > k\}$, where k is an integer. Then Λ is given by the subgroup $\Lambda = \{\bar{y} \varepsilon \hat{G}: y_i = 0 \text{ for all } i > -k\}$. The set Ω consists of all sequences $\bar{y} \varepsilon \hat{\mathfrak{F}}$ with $y_i = 0$ for all $i \leq -k$. The functions φ we define (4) by

$$\varphi(\bar{x}) = 2^{-k} \int_{\Omega} (\bar{x}, \bar{y})\, dm(\bar{y}) \tag{11}$$

The difference of (11) to (4) comes from the fact, that we normalized the Haar measure m so that $m(\Omega) = 2^k$ rather than 1. Integration gives us

$$\varphi(\bar{x}) = \chi_{1/\Omega}(\bar{x}) \tag{12}$$

where $\chi_{1/\Omega}$ denotes the characteristic function of the set $1/\Omega$ given by $1/\Omega = \{\bar{x}\varepsilon \;|\, x_i = 0 \text{ for all } i \leq k\}$. Therefore formula (6) of the sampling theorem gets the form

$$f(\bar{x}) = \sum_{\bar{z}\varepsilon H} f(\bar{z})\, \chi_{1/\Omega}(\bar{x} \oplus \bar{z}) \tag{13}$$

4. Sampling Theorem in Walsh-Fourier Analysis

Walsh-Fourier analysis is a Fourier analysis for functions defined on the nonnegative real line $[0, \infty)$. It can be derived from dyadic Fourier analysis in the following way: The map $\lambda: \mathfrak{F} \to [0, \infty)$ is according to Fine [7] defined as the map, which takes a point $\bar{x} = (x_i) \varepsilon \mathfrak{F}$

B-6

into a point $x \in [0, \infty)$ given by

$$x = \sum_i x_i 2^{-i} \tag{14}$$

For $x \in [0, \infty)$ the inverse mapping μ is defined by (11) choosing the finite dyadic representation if x is a dyadic rational. The map μ neglects only a set $\mathfrak{E}$, consisting of sequences which are 1-periodic to the right side; μ: $[0, \infty) \to \mathfrak{F} \setminus \mathfrak{E}$. The set $\mathfrak{E}$ has Haar measure zero. So we have

$$\lambda(\mu(x) = x \qquad \text{for all } x \in [0, \infty) \tag{15}$$

and

$$\mu(\lambda(\overline{x})) = \overline{x} \qquad \text{for all } \overline{x} \in \mathfrak{F} \setminus \mathfrak{E} \tag{16}$$

he mappings $\hat{\lambda}$: $\hat{\mathfrak{F}} \to [0, \infty)$ and $\hat{\mu}$: $[0, \infty) \to \hat{\mathfrak{F}} \setminus \hat{\mathfrak{E}}$ where $\hat{\mathfrak{E}} \cong \mathfrak{E}$, are efined in a analogous way as the maps λ and μ.

The Walsh functions $\psi(y, \cdot)$: $[0, \infty) \to \{+1, -1\}$ are defined as the functions given for all $x, y \in [0, \infty)$ by

$$\psi(y, x) = (\mu(x), \hat{\mu}(y)) \tag{17}$$

and the transforms formula given by (2) and (3) becomes the form

$$\hat{f}(y) = \int_0^\infty \psi(y, x) f(x)\, dx \tag{18}$$

and

$$f(x) = \int_0^\infty \psi(y, x) \hat{f}(y)\, dy \tag{19}$$

where $f \in L^2[0, \infty)$. This comes from the fact that the map λ^*: $P(\mathfrak{F}) \to P([0, \infty))$,

B-7

which maps the power set $P(\mathfrak{F})$ of $\mathfrak{F}$ into the power set $P([0, \infty))$ of $[0, \infty)$ and which is induced by λ, is a Haar-Lebesgue measure preserving map. The same is true for $\hat{\lambda}^*$. From that it turns out, that the algebraic and measure theoretic results of the sampling theorem formulated above for $G = \mathfrak{F}$ are invariant against the map λ^* and $\hat{\lambda}^*$. However, the topological results concerning continuity etc. are altered. We have so

$$\lambda^*(\mathfrak{F}) = [0, \infty), \quad \hat{\lambda}^*(\hat{\mathfrak{F}}) = [0, \infty),$$

$$\lambda^*(H) = \{m2^{-k}: \; m = 0, 1, 2, \ldots\}$$

$$\hat{\lambda}^*(\Lambda) = \{n2^{k} : n = 0, 1, 2, \ldots\}$$

$$\hat{\lambda}^*(\Omega) = [0, 2^{k}), \; \hat{\lambda}^*(1/\Omega) = [0, 2^{-k})$$

$$dm(\bar{x}) = dx$$

$$dm(\bar{y}) = dy$$

The function $\lambda^* \varphi$: $[0, \infty) \to R$ given for all $x \in [0, \infty)$ by

$$\lambda^*\varphi(x) = 2^{-k} \int_0^{2^k} \psi(y, x)dy \tag{20}$$

is the characteristic function $\chi_{[0, 2^{-k})}$ of the interval $\lambda^*(1/\Omega) = [0, 2^{-k})$. From above the sampling theorem becomes for Walsh-Fourier analysis the form:

<u>Theorem</u>: Suppose f is a real valued function of the space $L^2[0, \infty)$ and $\hat{f}(y) = 0$ for almost all $y \notin [0, 2^{k})$. Then f is equal almost everywhere to a stepfunction, continuous from the left which jumps only at points x of the form $x = m2^{-k}$, $m = 0, 1, 2, \ldots$. If f itself is of that kind then

B-8

$$f(x) = \sum_{m=0}^{\infty} f(m2^{-k})\chi_{[0,\,2^{-k})}(x - m2^{-k}) \tag{21}$$

uniformly on $[0, \infty)$ and in the sense of the convergence in $L^2[0, \infty)$.

Furthermore

$$2^k \|f\|^2 = \sum_{m=0}^{\infty} f^2(m2^{-k}) \tag{22}$$

Knowing that f is equal almost everywhere to a stepfunction of that kind described above, this theorem is a trivial one. It is obvious that such a stepfunction can be generated by characteristic functions as shown in equation (21).

B-9

REFERENCES

[1] Cauchy A., "Memoire sur diverses formules d'analyse", C. R. Acad. Sci. Paris 12 (1841), 283-298.

[2] Whittaker E. T., "On the functions which are represented by the expansions of the interpolation-theory", Proc. Roy. Soc. Edinburgh. Section A. 35 (1915), 181-194.

[3] Kotelnikov, V. A., "The Theory of Optimum Noise Immunity", Doctoral dissertation, 1947, Translation into English by R. A. Silverman, McGraw-Hill, New York, 1959.

[4] Shannon C. E., "Communication in the Presence of Noise", Proc. IRE 37, (1949), 10-21.

[5] Raabe, H., Untersuchungen an der wechselzeitigen Mehrfachueberträgung (Multiplexübertragung), Elektrische Nachrichten-Technik 16 (1939), 213-228.

Kluvánek, I., "Sampling Theorem in Abstract Harmonic Analysis", Matematicko-fyzikalny Casopis. Sloven. Akad. Vied., 15 (1965), 43-48.

[7] Fine, N. J., "The generalized Walsh functions", Trans. Amer. Math. Soc. 69, (1950), 66-77.

[8] Pichler, F., "Synthese linearer periodisch zeitvariabler Filter mit vorgeschriebenem Sequenzverhalten", Archiv der Elektr. Übertr. 22 (1968), 150-161.

B-10

[9] Liedl, R., "Eine algebraische Herleitung und eine Verallgemeinerung des Satzes von Fine über Gruppen von orthonormalen Funktionen und eine Beschreibung der vielfalterhaltenden Transformationen des Intervalls [0,1] auf sich selbst", Monatshefte fuer Mathematik, 72, (1968), 45-60.

Chapter 3
The Origins of the Dyadic Derivative due to James Edmund Gibbs

Radomir S. Stanković, Ferenc Schipp

At the *First International Workshop on Gibbs Derivatives*, held on September 26-28, 1989, at Kupari-Dubrovnik, former Yugoslavia, under auspices of the Mathematical Institute, Belgrade, Serbia, it was agreed that the term *Gibbs derivative* will be used to denote the class of differential operators that can be viewed as descendants of the initial concept of the logical derivative introduced by James Edmund Gibbs in 1967. The related concepts introduced later have their own names to provide a distinction among these different concepts due to their peculiar features. Probably the most important and most widely investigated among them is the Butzer-Wagner derivative. Various other generalizations, for instance p-adic derivative and fractional p-adic derivative, have also attracted a lot of attention and have certain interesting applications.

In the sixties of the last century, the cold war was at its peak and this situation was an important factor providing motivations for developing various signal processing algorithms. In many cases, algorithms with good performance had been developed, however, their practical application was still unfeasible due to complex computations and limited computing powers. In this context, there was high interest in fast algorithms for various kinds of spectral (Fourier) analysis of different signals including radar, sonar, seismic, speech, and other signals produced by different sources.

The relatively limited computing power of the computing technology available at that time had imposed some restrictions in selecting mathematical foundations that

Radomir S. Stanković
Dept. of Computer Science, Faculty of Electronic Engineering, University of Niš, Niš, Serbia, e-mail: Radomir.Stankovic@gmail.com

Ferenc Schipp
Department of Numerical Analysis, Eötvös L. University, Budapest, Pázmány P. sétány 1\C, H-1117 Hungary,e-mail: schipp@numanal.inf.elte.hu

R.S. Stanković et al. (eds.), *Dyadic Walsh Analysis from 1924 Onwards Walsh-Gibbs-Butzer Dyadic Differentiation in Science Volume 1 Foundations*, Atlantis Studies in Mathematics for Engineering and Science 12, DOI 10.2991/978-94-6239-160-4_3

could be used to achieve both the desired features and properties of these algorithms, and their efficient implementation. That was one of the reasons which motivated the interest in various applications of Walsh and other non-sinusoidal functions. Another reason was the beautiful and very useful mathematical properties these systems of functions exhibited. The Walsh functions take values ± 1 and, therefore, computing with them reduces to addition and subtraction with no complex multiplication as in the case of the Discrete Fourier transform (DFT). Walsh (Fourier) analysis, the kernel of which are the Walsh functions, possesses most of the useful properties of the classical Fourier analysis. Actually, the Fourier and Walsh-Fourier analysis can be viewed as mathematical concepts with the same origins, just defined on different domain groups, the group of real numbers and the dyadic group, with both groups viewed as particular locally compact Abelian groups. In both cases, the group characters are used as functional kernels of the related spectral transforms. The differences between the structure of these groups results in the differences in features of the transforms, and due to that, each of them has certain advantages as well as disadvantages in concrete particular applications.

This high interest in practical applications of Walsh functions that spread in early seventies of the former century was unexpected and even surprising also for the Professor Captain Joseph Leonard Walsh who introduced this system of orthogonal functions in 1923 [1] The main ideas of the research were presented in his talk at the American Mathematical Society, the abstract of which is published as

Walsh, J.L., "A closed set of normal orthogonal functions", Abstract of presentation on February 25, 1922, *Bull. Amer. Math. Soc.*, 28, 1922, 241

resulting later in the paper

Walsh, J.L., "A closed set of orthogonal functions", *Amer. J. Math.*, 55, 1923, 5-24.

Due primarily to his work leading to this publication J.L. Walsh is honoured as the father of dyadic (Walsh-Fourier) analysis as it is clearly stated in *Applications of Walsh Functions and Sequency Theory*, (Eds.) H. Schreiber, G.F. Sandy, IEEE, New York, p.iii (1974).

In the realms of dyadic analysis, the Gibbs derivatives and their various extensions and generalizations can be viewed as counterparts of the Newton-Leibniz derivative in classical mathematical analysis and mathematical physics.

When thinking about the initial ideas and origins of the Gibbs derivative it is important to recall the following. While one of us was gathering information on

[1] As pointed out by W.R. Wade, it is a common myth that Walsh functions were introduced in his PhD Thesis. The thesis was *On Roots of the Jacobian of Two Binary Forms* based on two papers published under the same title "On the location of the roots of the Jacobian of two binary forms, and of the derivative of a rational function", *Trans. of the Amer. Math. Soc.*, Vol. 19, No. 3, 1918, 291-298, and *Trans. of the Amer. Math. Soc.*, Vol. 22, 1921, 101-116. See, also *Trans. of the Amer. Math. Soc.*, Vol. 23, No. 1, 1922, 31-69, and articles by J.L. Walsh in *Annals of Mathematics*, Vol. 22, No. 2, 1920, 128-144, *Comptes Rendus du Congres International des Mathematiciens*, Strasbourg, 1920, 349-352.

updating and preparing the bibliography on Gibbs derivatives resulting eventually in [19], J.E. Gibbs pointed out that Ms. H.A. Gebbie, the Director of the National Physical Laboratory, Teddington, Middlesex, UK, had participated at a Christmas dinner in 1966 at the home of Prof. Walsh where a relevant conversation encouraged her in the idea of applying the Walsh functions in spectroscopy [4]. See, also [8].

In the abstract of [8], it is explicitly stated that the choice of Walsh functions is motivated by their compatibility with binary digital computers

THIS article suggests that Walsh functions might be used in transform Spectroscopy to replace the sinusoidal functions appearing in the Fourier transform. We think this might be the case because Walsh functions are a complete orthonormal set, and therefore give rise to an integral transform of Fourier type; and they take only the values $+1$ *and* -1 *and are therefore likely to be well suited to the binary digital computer.*

Ms. Gebbie asked J.E. Gibbs to investigate the possibility of defining a differential operator related to Walsh functions and one of the results was in the limited circulation, on 1967 January 13, of the first document defining a finite dyadic derivative (or the logical derivative, as it was initially called). A page from the personal diary of Dr. Gibbs containing the handwritten first explicit formulation of the definition of the logical derivative is reproduced in [18] by the curtesy of Mrs. Merion Gibbs. This book is a collection of papers presented at the *Workshop on Walsh and Dyadic Analysis*, devoted to the memory of James Edmund Gibbs held on October 2007 in Niš, Serbia.

The Gibbs derivative as introduced in 1967 was purposely defined as an operator designed so that the discrete Walsh functions can be generated as the solutions of an eigenvalue problem.

Recall that the classical (Newton-Leibniz) differentiator is a closed linear operator on the linear space of differentiable functions on the real group. It may be defined as that operator that maps each group character $\exp(2\pi ikx)$ to the scalar multiple $2\pi ik\exp(2\pi ikx)$ of that character. In the same way, the Gibbs differentiator is that operator (on the space of complex-valued functions on a finite dyadic group) that maps each character (discrete Walsh function of index w) to w times itself.

To present a formal definition of the logical Gibbs derivative, we need to first introduce some basic concepts and notations.

3.1 Definition of the Gibbs Derivative

The discrete Walsh functions are group characters of the finite dyadic group G_n whose elements are the n-tuples $x = (x_0, x_1, \ldots, x_{n-1})$ with $x_i \in \{0,1\}$. The group operation $\oplus$ is coordinatewise addition modulo 2. Each element of G_n can be associated with a unique element of the set of non-negative integers less than 2^n, $B_n = \{0,1,\ldots,2^{n-1}\}$, by means of a function $V_n : G_n \rightarrow B_n$ defined by

$$V_n(x) = \sum_{i=0}^{n-1} 2^{n-i-1} x_i.$$

The space of all bounded complex-valued functions f on G_n (or on B_n) will be denoted by L_n, or, since it is not usually necessary to express the dependence on n, simply by L.

Definition 3.1 *To each function $f \in L$ we assign a function $f^{[1]} \in L$ defined by*

$$f^{[1]}(x) = -\frac{1}{2}\sum_{r=0}^{n-1}(f(x \oplus 2^r) - f(x))2^r, \quad (x \in B_n).$$

We call $f^{[1]}$ the (first-order) Gibbs derivative of f. The corresponding operator $D_n : L \to L$, defined by $D_n f = f^{[1]}$, $(f \in L)$, is called the Gibbs differentiator.

Nomenclature of this type enables us to distinguish each of the various species of Gibbs differentiator, by indicating the domain and range of the functions on which it operates.

In [9], the increment of the variable is written as e_r defined as the basis vector such that $(e_r)_{(s)} = \delta(r,s)$, where δ is the Kronecker delta.

Thus defined the logical differential operator satisfies

$$f_w^{[1]}(x) = w f_w,$$

for $f_w = wal(w,\cdot)$, where $wal(w,\cdot)$ is the discrete Walsh function of index $w \in B_n$.

In other words, the discrete Walsh functions are the 2^n solutions of the eigenvalue problem

$$D_n f = wf,$$

of which the eigenvalues are $w = 0, 1, \ldots, 2^n - 1$, and the corresponding eigenfunctions $wal(0,\cdot), wal(1,\cdot), \ldots, wal(2^n - 1,\cdot)$, respectively [7].

In this context, we distinguish Gibbs derivatives of the first and second kinds, corresponding respectively to the Walsh-Paley [10] and Walsh-Kaczmarz [11] orderings of the Walsh functions [5].

The Gibbs derivative can be extended to arbitrary complex order p by way of the definition of the discrete δ-function

$$\delta(x) = 2^{-n}\sum_{w=0}^{2^n-1} wal(w,x), \quad (x \in B_n).$$

The function thus defined has the properties

$$\delta(x) = \begin{cases} 1, & (x = 0), \\ 0, & (x \neq 0), \end{cases}$$

and

$$f = \delta * f, \quad (f \in L).$$

The derivative of order $p \in K$ of the δ-function is defined by

$$\delta^{[p]}(x) = 2^{-n} \sum_{w=0}^{2^n-1} w^p wal(w,x).$$

The derivative of order p of an arbitrary function $f \in L$ may be conveniently obtained as the dyadic convolution product of $\delta^{[p]}$ and f. Thus,

$$D^p f = f^{[p]} = \delta * f^{[p]} = \delta^{[p]} * f.$$

The Gibbs derivative of fractional order is obtained in the case that p is real.

For more details on the first formulation of the Gibbs derivative, see [12].

3.2 Further Development

The interest in Walsh and related functions resulted in the organization of conferences *Applications of Walsh Functions* held in 1970 to 1974 in Washington DC, as well as *Symposium on Theory & Applications of Walsh & Other Non-sinusoidal Functions*, Hatfield, England, June 28-29, 1973, and July 1-2, 1975.

The conferences organized in Washington DC by the American Navy in the military complex of Naval Labs were not an environment particularly convenient for a free exchange of information and informal scientific discussions. For this reason, during the conference in 1970, Prof. Walsh invited on April 4 several participants to a dinner at this home. It is believed that discussions at this stately diner were very influential for further development of theory and applications of Walsh functions. This belief was reconfirmed by Prof. P.L. Butzer in March 2009 in some informal talks at the occasion of the conference *Approximation Theory and Signal Analysis* held in Lindau (Lake Constance), Germany, devoted to the celebration of his 80th birthday. In particular, as said by Prof. Butzer, this dinner was important for further increasing his interest in dyadic analysis and dyadic derivatives, since other guests at this dinner include J.E. Gibbs and F. Pichler.

In his report from April 1970, in Section 4.1 entitled *Dyadic differentiation*, Pichler refers for the definition of the dyadic derivative to the report of Gibbs and Millard [9]. After a comment that this is the differentiator for functions on discrete structures, Pichler wrote that *Slight modifications must be made to obtain a theory of generalized differentiation defined for real valued functions of a continuous nonnegative real variable. Let f be such a function. To f we attach, if possible, a function $f^{[1]}$, given by*

$$f^{[1]}(t) = \sum_{k=-\infty}^{\infty} (f(t) - f(t \oplus 2^{-k}))2^{k-2}.$$

If the function $f^{[1]}$ exists, we shall call it the first dyadic derivative of f.

In a footnote, Pichler remarked that Gibbs and Millard used the terms *logical differentiation* and *logical derivative*. The same definition and terminology for the dyadic derivative Pichler used in [13]. Thus, it can be concluded that the term dyadic differentiation was coined by Pichler in 1970.

Two papers introducing the concept of the Butzer-Wagner derivative, [1] and [3], were completed in preprint form in the autumn of 1971. Thus, P.L. Butzer, a participant together with J. Edmund Gibbs and F. Pichler at the famous dinner at the home of Walsh in 1970, and his student H. Wagner were foremost in recognizing the value of the Gibbs derivative in analysis.

In the first publication by Butzer and Wagner discussing dyadic differentiation [1], the following definition is used for functions of period 1 which are p-th power integrable $1 \leq p < \infty$, in the norm

$$\|f\|_p = \left(\int_0^1 |f(x)|^p dx \right)^{1/p}.$$

The space of such functions is denoted as $L^p(0,1)$.

If for $f \in L^p(0,1)$, there exists $f \in L^p(0,1)$ such that

$$\lim_{m \to \infty} \|\frac{1}{2} \sum_{j=0}^{m} 2^j (f(\cdot) - f(\cdot \oplus 2^{-j-1})) - g(\cdot)\| = 0,$$

then g is called the strong derivative of f denoted by $D^{[1]} f$.

Note that this definition differs from that given by J. E. Gibbs in [6] essentially in the sense that the factor 2^j is replaced by 2^{-j}, however, Gibbs' definition is taken in the pointwise sense.

In [2], the increment of variable in the derivative is expressed as h_j which is defined as $h_j = \{h_{j,s}\}$, $s \in N$, where

$$h_{j,s} = \begin{cases} 0, \text{ for } 1 \leq s \leq j, \\ 1, \text{ for } j < s. \end{cases}$$

while in [3], this increment is written as e_{j+1}, where $e_j = \{x_s^j\}$ with $x_s^j = 0$ for $j \neq s$, and $x_s^j = 1$ for $j = s$.

It should be mentioned that from 1970 onwards, Prof. Franz Pichler, led the way in applications to system theory [13], [14]. This was the reason for reprinting the research report by Prof. Pichler prepared when he was a visiting scholar at Maryland University working with Prof. Henning F. Harmuth who championed the application of Walsh functions in different areas of information transmission and communications. These considerations were extended in another report written by Prof. Pichler [15].

It should be noticed that the report at Maryland University by F. Pichler [13] contains the formulation of the sampling theorem in dyadic analysis. For a wider-ranging historical account of the mathematical and engineering background to Walsh functions and the Gibbs derivative, reference may be made to the text of an admirable talk by Pichler [16] and also discussing the same subject from a different engineering area point of view [17].

In the forty five years since the concept of Gibbs derivative was first introduced in the restricted context of a finite dyadic group, the scope of the definition has been greatly extended, resulting in the Butzer-Wagner derivative as the differential operator on infinite dyadic groups, and further generalizations of differentiators on various other not-necessarily Abelian groups, including also differentiation of different classes of signals and two-dimensional signals. In this area, many results have been obtained, and certain areas of application have been established as it can be seen in from the bibliography on the subject [19] in connection with a quantitative discussion of the rise of interest in the subject. The study of Gibbs differentiation may therefore be said to have reached a certain level of maturity, and this book aims to give an overall view of the present state of knowledge, in the hope that it will encourage and facilitate further developments.

References

1. ∗Butzer, P.L., Wagner, H.J., "Approximation by Walsh polynomials and the concept of a derivative", *Proc. Sympos. Applic. Walsh Functions*, Washington, D.C., 1972, 388-392.
2. ∗Butzer, P.L., Wagner, H.J., "On a Gibbs-type derivative in Walsh-Fourier analysis with applications", *Proc. Nat. Electron. Conf.*, 27, 1972, 393-398.
3. ∗Butzer, P.L., Wagner, H.J., "Walsh-Fourier series and the concept of a derivative", *Applicable Anal.*, 3, 1973, 29-46.
4. Gebbie, H.A., "Walsh functions and the experimental spectroscopist", *Proc. Sympos. Applic. Walsh Functions*, Washington, D.C., 1970, 99-100.
5. Gibbs, J.E., "Sine waves and Walsh waves in physics", *Proc. Sympos. Applic. Walsh Functions*, Washington, D.C., 1970 Mar. 31-Apr. 3, 1970, 260-274.
6. ∗Gibbs, J.E., Ireland, B., "Some generalizations of the logical derivative", *DES Report No. 8, Nat. Phys. Lab.*, August 1971, 22+ii pages.
7. Gibbs, J.E., "Walsh functions and the Gibbs derivative", *NPL DES Memo.*, No. 10, ii + 13, 1973.
8. Gibbs, J. E., Gebbie, H. A., "Application of Walsh functions to transform spectroscopy", *Nature*, Vol. 224, No. 5223, 1969, 1012-1013, publication date 12/1969.
9. ∗Gibbs, J.E., Millard, Margaret J., "Walsh functions as solutions of a logical differential equation", *NPL DES Rept.*, No. 1, 1969, 9 pp.
10. Paley, R.E.A.C., "A remarkable series of orthogonal functions", *Proc. London Math. Soc.*, 34, 1932, 241-279.
11. Kaczmarz, S., "Uber ein Orthogonalsysteme", *Comptes rendus du premiercongres des mathematiciens des pays slaves*, (Varsovie, 1929), (1930), 189-192.
12. Pichler, F., *Reminiscences*, in R.S. Stanković, (ed.), *Walsh and Dyadic Analysis*, Faculty of Electronic Engineering, Niš, Serbia, 2008, 31-36, ISBN 978-86-85195-47-1.
13. Pichler, F.R., "Walsh functions and linear system theory", *Proc. Sympos. Applic. Walsh Functions*, Washington, D.C., 1970, 175-182.

14. Pichler, F.R., *Walsh Functions and Linear System Theory*, Univ. Maryland Tech. Res. Rept T-70-05, 1970.
15. Pichler, F.R., *Some Aspects of a Theory of Correlation with Respect to Walsh Harmonic Analysis*, Univ. Maryland Tech. Res. Rept R-70-11, 1970.
16. Pichler, F.R., "Some historical remarks on the theory of Walsh functions and their applications in information engineering", in Butzer, P.L., Stanković, R.S. (eds.), *Theory and Applications of Gibbs Derivatives: Proceedings of the First International Workshop on Gibbs Derivatives* September 26-28, 1989, Kupari-Dubrovnik, Yugoslavia, Mathematical Institute, Belgrade, 1990, xxv-xxx.
17. Pichler, F., "Walsh functions - early ideas on their application in signal processing and communication engineering", *Proc. The 2004 International TICSP Workshop on Spectral Methods and Multirate Signal Processing Proceedings*, SMMSP2004, Vienna, Austria, September 11-12, 2004, ISBN 952-15-1229-6.
18. Stanković, R.S., (ed.), *Walsh and Dyadic Analysis, Proceedings of the Workshop devoted to the memory of James Edmund Gibbs*, held on October 18-19, 2007, Niš, Serbia, Faculty of Electronics, Niš Serbia, 3-14, (2008), (ISBN 978-86-85195-47-1).
19. Stanković, R.S., Astola, J, *Gibbs Derivatives - the First Forty Years* TICSP Series #39 2008, Tampere International Center for Signal processing,Tampere University of Technology, Tampere, Finland, 2008, ISBN 978-952-15-1973-4.

The $*$ sign in the citations indicates that the paper is reprinted in this book.

Biography of James Edmund Gibbs written by Merion Gibbs

J. E. Gibbs: born 19 January 1928 in Norwood, SE, London; died 31 January 2007

Education

1939-1946 Surbiton Boy's High School	
1949-1952 Imperial College, London	BSc in Physics BSc in Mathematics
1949-1952 Imperial College Technical Optics Section	PhD Stelar photometry (supervisor Prof. W.D. Wright)

Work history

1952-1953 National Research Council of Canada; Carleton College, Ottawa	
1953-1955 Royal Observatory Edinburgh Cormack Research Fellow of the Royal Society of Edinburgh	Stellar scintillation
1955-1961 NPL Light Division	Primary standard of luminance; Colour-temperature standards
1961-1968 NPL Basic Physics Division Theoretician	Infra-red Fourier-transform spectroscopy; Walsh functions (from 1966) Gibbs differentiation (from 1967)
1968-1976 NPL Division of Electrical Science Principal Scientific Officer Mathematical Support Group	Dyadic analysis; Involvement in international Walsh-function community (from 1970) including preparation and examination of PhD students
1976-1988 NPL Division of Electrical Science Time and Frequency Section	A hypothetico-eductive theory of time: Development of algorithms for generating a unifrom time-scale from data from atomic clocks.

Letter of Merion Gibbs to Prof. Paul L. Butzer

Marion Gibbs
9 Merryfield Road
Petersfield
GU31 4BS

9 October 2013

Prof Emeritus P L Butzer
RWTH Aachen
Lehrstuhl A für Mathematik
D-52056 Aachen
Germany

Dear Professor Butzer

Walsh Analysis and Gibbs Differentiation

I passed a copy of your letter of 27 September to Bryan Ireland. Bryan regrets that he cannot assist in the project. He has by now discarded all reports, journals, papers and diaries of the seventies era, and says his memory is not sufficiently accurate to recall much of what took place then. He recalls attending two Walsh functions symposia at Hatfield Polytechnic, but only remembers that Edmund (Dr J E Gibbs) was a major speaker at both. These symposia would have been those held in June 1971.

I do not have any reference to Hatfield symposia in 1974.

Looking through Edmund's papers and reports, I note that Bryan Ireland, in the University of Bath School of Mathematics, was Industrial Training Tutor for a succession of sandwich course students from there to the NPL where they were supervised by Edmund. Margaret Millard was the first of these students, January to August 1969.

Bryan Ireland had one Vacation Consultantship at the NPL, in 1971. Franz Pichler had a Vacation Consultantship in spring 1971, and John Marshall, from the University of Bath, had a Vacation Consultantship in summer 1972.

Thus the early work in Gibbs differentiation was carried out by a floating team. All Edmund's reports include an abstract or summary.

The NPL became more averse to pure research and in 1976 Edmund got transferred to the less innovative field of time and frequency standards. He was unable to keep up with developments in Walsh analysis and Gibbs differentiation but he worked on some ideas in retirement— a short paper appearing in the proceedings of the 2007 Niš workshop.

I enclose a CV for Edmund.

Best wishes,
Yours sincerely

Marion Gibbs

Marion Gibbs

- 1 -

Walsh spectrometry, a form of spectrol analysis well suited to binary digital computation

J. E. Gibbs, 1967

Summary

A kind of spectral analysis is proposed in which the role of the sinusoidal functions is taken over by the Walsh functions. The coefficients in the binary expansions of integers expressible in not more than n bits are regarded as co-ordinates in an n-dimensional space. The "logical derivative" of a function of such an integer is identified with the "logical gradient" of a function of the space variables. The Walsh functions of order n are defined as the eigensolutions of a certain first-order logical differential equation. Some of their elementary algebraic properties are derived. The one-dimensional Walsh transform of order n is shown to be the same as the n-dimensional finite discrete Fourier transform modulo 2.

The basis of Walsh spectrometry is the logical analogue of the (arithmetic) Wiener-Khintchine theorem, which states that the Walsh covariance spectrum is the Walsh transform of the logical autocovariance function. The logical is related to the arithmetic autocovariance by an arithmetico-logical transform. The relationship between the wave-number and the eigenvalue of a Walsh function is established, and a simple recursive algorism is obtained for computing Walsh functions ordered by wave-number. Two algorisms are stated for effecting the AIU transform, defined as the resultant of the arithmetico-logical and Walsh transforms. The more sophisticated of these is based on a remarkable identity governing the kernel of the AIU transform.

Transform spectroscopy is a method of studying electromagnetic radiation in which a two-beam interferometer is used to measure a function of path-difference (or time-delay), which is simultaneously or subsequently transformed by the action of a digital computer into a function of wave-number (or frequency). This process of inverting the independent variable does not, ideally, change the information obtained from the experiment, but presents it in a more intelligible form. The inversion process is made much more efficient by basing it on a set of orthogonal functions intimately related to the binary arithmetic used by the computer, rather than on the quite unrelated sinusoidal functions. A measure of the improved speed is obtained by noting that a computer of word-length n bits can, in principle, compute an n th order Walsh function in a single "short" hardware operation. Whereas a sinusoidal function requires a sub-routine of appreciable complexity.

* * * * *

The covariance spectrum (power spectrum)[1] of a stationary random process expresses the power associated with each of the sinusoidal components into which the process may be resolved. A familiar example is the spectrum of electromagnetic radiation. Such a spectrum may be determined in the infra-red region by the well-established method of Fourier spectrometry[2], in which the interferogram of a Michelson two-beam interferometer is transformed by digital computation into the power spectrum. This method is based on the fact that the interferogram is the autocovariance function of the radiation; and on the Wiener-Khintchine theorem that the power spectrum is the Fourier transform of the autocovariance function.

The availability of electronic digital computers has made the numerical computation of Fourier transforms a practical proposition, and considerable ingenuity has been devoted to devising fast methods of performing such calculations. Particularly notable is the Cooley-Tukey algorism[3], recently modified by Gentleman and Sande[4]. A fundamental limitation on the speed of digital Fourier transformation lies in the unavoidability of calculating, directly or indirectly, values of the Fourier kernel $\exp(-2\pi ikx)$, and of multiplying them by the data. The formation of such functions as the exponential is a relatively very slow operation on a digital computer, and even multiplication of two numbers is much slower than addition or logical operations. It is clear that a profound improvement in speed may be expected if the transformation process can be made to depend almost exclusively on addition and logical operations.

A complete description of the statistical properties of a given (stationary Gaussian) radiation is given either by the interferogram function or by its Fourier transform, the power spectrum. Both carry the same information, but in general the spectrum is more readily intelligible than the interferogram to the spectroscopist. Wave-number (or frequency) appears more natural as an independent variable than path-difference (or time delay): hence the labour expended on inverting the variable in Fourier spectrometry.

These considerations have led us to examine the possibility of inverting the variable by means of a transformation of Fourier type in which the kernel shall be much more readily formed in a binary digital computer than the familiar complex exponential function. In particular, we assume that the kernel $u_k(x)$ sought may be calculated by logical operations on the binary expansions of its arguments.

Since we are concerned with a realisable numerical process, we shall go straight to the case of the finite discrete transform, and assume that the independent variable x is restricted to the non-negative integers expressible in n bits, that is,

$$x = 0, 1, \ldots 2^n-1 .$$

We shall use the notation

$$x = \sum_{r=0}^{n-1} x_r 2^r ,$$

with $x_r = 0$ or 1, $r = 0, 1, \ldots n-1$.

The logical operations that we shall consider are logical addition and logical multiplication. Logical addition is defined as bit-by-bit addition (without carry) under the rules $0 \oplus 0 = 1 \oplus 1 = 0$, $0 \oplus 1 = 1 \oplus 0 = 1$. Logical multiplication is defined as bit-by-bit multiplication under the rules $0 \wedge 0 = 0 \wedge 1 = 1 \wedge 0 = 0$, $1 \wedge 1 = 1$.

Since these operations treat each bit independently, we are led to regard the powers $2^0, 2^1, \ldots 2^{n-1}$ as orthogonal unit vectors, and the bits $x_0, x_1, \ldots x^{n-1}$ as co-ordinates in a space of n dimensions. This space is cyclic in the sense that two successive identical displacements x restore a point a to its original position; that is,

$$a \oplus x \oplus x = a .$$

In particular, $x \oplus x = 0$.

The following lemma on logical addition is important: the sum of an arbitrary function $f(x)$ is invariant under logical addition of a constant c to its argument, that is,

$$\sum_{x=0}^{2^n-1} f(x \oplus c) = \sum_{x=0}^{2^n-1} f(x) ,$$

$$c = 0, 1, \ldots 2^n-1.$$

- 5 -

For the terms of the sum on the left are the same as those of that on the right, but in a different order.

The gradient $u'(x)$ of a function $u(x)$ of a position vector

$$x = \sum_{r=0}^{n-1} x_r i_r ,$$

where the i_r are unit vectors, is defined as

$$u'(x) = \sum_{r=0}^{n-1} \frac{\partial u}{\partial x_r} i_r$$

By analogy, we define the logical derivative $u^1(x)$ of a function $u(x)$ c the n-bit integer x as

$$u^1(x) = \sum_{r=0}^{n-1} \frac{\delta u_r}{\delta x_r} 2^r ,$$

where δu_r is the increment in u due to a logical increment $\delta x_r = 1$ (the only possible non-zero logical increment) in x_r, that is,

$$\delta u_r = u\,(x \oplus 2^r) - u(x) .$$

Thus

$$u^1(x) = \sum_{r=0}^{n-1} \{u\,(x \oplus 2^r) - u(x)\}\, 2^r .$$

The orthogonal functions $v_k(x) = \exp(-2\pi ikx)$, $k = 0, \pm 1, \pm 2, \ldots,$ may t regarded as eigensolutions of the ordinary differential equation

$$v'(x) + 2\pi ik\, v(x) = 0$$

- 6 -

under the boundary conditions

$$v(0) = v(1) = 1 .$$

The functions we seek may likewise be regarded as eigensolutions of the logical differential equation

$$u^1(x) + 2k\, u(x) = 0$$

under the boundary conditions

$$u(0) = u(2^n) = 1 .$$

Thus far we have assumed that the integer argument x is expressible in n bits. If the argument is allowed to be <u>any</u> integer, we may clearly extend the function periodically by reducing the argument to its least non-negative residue modulo 2^n. With this understanding, the boundary conditions are equivalent to

$$u(0) = 1 .$$

To solve the boundary problem, we note that, if $u_k(x)$ is the eigenfunction corresponding to the eigenvalue k, then

$$\sum_{r=0}^{n-1} \{u_k(x \oplus 2^r) - u_k(x)\}2^r + 2k\, u_k(x) = 0, \qquad x = 0,1,\ldots,2^n-1.$$

By equating to zero the coefficient of each power of 2 on the left, we see that k is of the form

$$k = \sum_{r=0}^{n-1} k_r\, 2^r;$$

and that

$$u_k(x \oplus 2^r) = (1 - 2k_r)\, u_k(x), \qquad r = 0,1,\ldots,n-1.$$

- 7 -

By putting first $x = 0$ and then $x = 2^r$ in this relation, and using the boundary condition, we get

$$u_k(2^r) = 1 - 2k_r = (1 - 2k_r)^{-1} .$$

Hence $k_r = 0$ or 1; and the eigenvalues are $k = 0,1,\dots 2^n-1$. Also

$$u_k(2^r) = \exp \pi i k_r .$$

If $\xi_r = 0$ or 1, it follows that

$$u_k(\xi_r 2^r) = \exp \pi i k_r \xi_r .$$

We have

$$u_k(x \oplus 2^r) = u_k(x)\, u_k(2^r);$$

or, again using the boundary condition,

$$u_k(x \oplus \xi_r 2^r) = u_k(x)\, u_k(\xi_r 2^r) = u_k(x) \exp \pi i k_r \xi_r .$$

Putting $x = \xi_0 2^0, r = 1$, we get

$$u_k(\xi_0 2^0 + \xi_1 2^1) = \exp \pi i (k_0 \xi_0 + k_1 \xi_1);$$

putting $x = \xi_0 2^0 + \xi_1 2^1$ and $r = 2$, we get

$$u_k(\xi_0 2^0 + \xi_1 2^1 + \xi_2 2^2) = \exp \pi i (k_0 \xi_0 + k_1 \xi_1 + k_2 \xi_2);$$

and so on. After a finite number of such steps we obtain the eigenfunction in the form (replacing ξ by x)

$$u_k(x) = u_k\left(\sum_{r=0}^{n-1} x_r 2^r\right) = \exp \pi i \sum_{r=0}^{n-1} k_r x_r .$$

Evidently $$u_k(x) = u_x(k) .$$

The notation $u(k,x)$ expressing this symmetry is often more convenient.

It may be shown that these functions are closely related to the functions ϕ constructed by Walsh [5] in 1922, further studied and generalized by

Paley [6] and Fine [7], and popularized by Polyak and Shreider [8] in 1962. The difference between the functions $u_k(x)$ and the Walsh functions is that the latter are defined on the interval $[0,1)$, and, in particular, for the binary rationals $0/2^n$, $1/2^n$, ... $(2^n-1)/2^n$, while our functions are defined only for the binary integers $0,1,\ldots 2^n-1$.

This restriction enables us to avoid analytic considerations in connexion with the "Walsh transform"

$$S(k) = 2^{-n/2} \sum_{x=0}^{2^n-1} f(x)\, u_k(x) ,$$

whose properties are thus merely a matter of algebra. There is a close analogy with the finite discrete Fourier transform

$$S(k) = N^{-1/2} \sum_{x=0}^{N-1} f(x) \exp \frac{2\pi ikx}{N}$$

discussed by Gentleman and Sande [4].

The orthogonality of the functions $u_k(x)$ may be proved from the defining logical differential equation. If $u(x)$, $v(x)$ are eigensolutions corresponding to distinct eigenvalues k_i, k_j, we have

$$u^1(x) = -2k_i\, u(x) ,$$
$$v^1(x) = -2k_j\, v(x) ;$$

and hence

$$vu^1 - uv^1 = -2(k_i-k_j)\, uv .$$

- 9 -

Summing this equation over x, we find

$$-2(k_i-k_j)\sum_{x=0}^{2^n-1} u(x)\,v(x) = \sum_{x=0}^{2^n-1} (vu^1 - uv^1)$$

$$= \sum_{x=0}^{2^n-1}\left[v(x)\sum_{r=0}^{n-1}\{u(x\oplus 2^r) - u(x)\}2^r - u(x)\sum_{r=0}^{n-1}\{v(x\oplus 2^r) - v(x)\}2^r\right]$$

$$= \sum_{r=0}^{n-1} 2^r \sum_{x=0}^{2^n-1}\{u(x\oplus 2^r)\,v(x) - u(x)\,v(x\oplus 2^r)\} = 0 ,$$

by the lemma on logical addition.

The functions $u_k(x)$ are also normal, for

$$\sum_{x=0}^{2^n-1} u_k^2(x) = \sum_{x=0}^{2^n-1} 1 = 2^n ;$$

thus

$$\sum_{x=0}^{2^n-1} u_k(x)\,u_h(x) = 2^n\,\delta(k,h) .$$

The functions $u_k(x)$ are furthermore complete, in the sense that any bounded function $f(x)$ defined for $x = 0,1,\ldots 2^n-1$ may be expressed linearly in terms of the $u_k(x)$. Thus

$$f(x) = 2^{-n/2}\sum_{k=0}^{2^n-1} S(k)\,u_k(x), \qquad x = 0,1,\ldots 2^n-1 ,$$

where the coefficients $S(k)$ are determined by

$$S(k) = 2^{-n/2}\sum_{x=0}^{2^n-1} f(x)\,u_x(k), \qquad k = 0,1,\ldots 2^n-1 ,$$

- 10 -

The identity of form of these two relations shows that the Walsh transform, as we have defined it, is self-inverse. The inversion theorem is easily proved; thus

$$2^{-n/2} \sum_{k=0}^{2^n-1} S(k)\, u_k(x)$$

$$= 2^{-n/2} \sum_{k=0}^{2^n-1} 2^{-n/2} \sum_{\xi=0}^{2^n-1} f(\xi)\, u_\xi(k)\, u_k(x)$$

$$= 2^{-n} \sum_{\xi=0}^{2^n-1} f(\xi) \sum_{k=0}^{2^n-1} u_\xi(k)\, u_x(k)$$

$$= \sum_{\xi=0}^{2^n-1} f(\xi)\, \delta(\xi,x) = f(x) .$$

The leading algebraic property of the functions $u_k(x)$ is the multiplication theorem

$$u_k(x)\, u_k(y) = u_k(x \oplus y) .$$

A proof is obtained by writing the product on the left as

$$\exp \pi i \sum_{r=0}^{n-1} k_r(x_r + y_r) = \exp \pi i \sum_{r=0}^{n-1} k_r(x_r \oplus y_r) ;$$

for replacing x_r+y_r by $x_r \oplus y_r$ can at worst add an integral multiple of $2\pi i$ to the argument of the exponential function.

A Walsh function of order n can be expressed as the product of n Walsh functions of the first order. For

$$u_n(k,x) = \prod_{r=0}^{n-1} \exp \pi i k_r\, x_r$$

$$= \prod_{r=0}^{n-1} u_n(k_r 2^r, x)$$

$$= \prod_{r=0}^{n-1} u_1(k_r, x_r) .$$

– 11 –

This property enables us to recognize the one-dimensional Walsh transform of order n as the n-dimensional finite discrete Fourier transform modulo 2. For

$$2^{-n/2}\sum_{x=0}^{2^n-1} = \prod_{r=0}^{n-1}\left(2^{-1/2}\sum_{x_r=0}^{1}\right),$$

$$u_n(k,x) = \prod_{r=0}^{n-1} \exp\frac{2\pi i k_r x_r}{2},$$

$$f_n(x) = F_n(x_0,x_1,\dots x_{n-1});$$

so

$$S_n(k) = 2^{-n/2}\sum_{x=0}^{2^n-1} f_n(x)\, u_n(k,x)$$

$$= \prod_{r=0}^{n-1}\left(2^{-1/2}\sum_{x_r=0}^{1}\exp\frac{2\pi i k_r x_r}{2}\right) F_n(x_0,x_1,\dots x_{n-1}).$$

The formula

$$\sum_{r=0}^{n-1} 2^r\, u_n(2^r,x) = 2^n-1-2x$$

is easily proved by writing $u_n(2^r,x) = 1-2x_r$. It is the simplest of a family of Walsh expansions of zig-zag functions.

The logical convolution $f(x) \circledast g(x)$ of order n of two functions $f(x)$, $g(x)$ given for $x = 0,1,\dots 2^n-1$ is defined as

$$f(x) \circledast g(x) = 2^{-n/2}\sum_{y=0}^{2^n-1} f(x\oplus y)\, g(y).$$

Let the Walsh transforms of $f(x)$, $g(x)$ be $S(k)$, $T(k)$. The logical convolution theorem asserts that the Walsh transform of $f(x)\, g(x)$ is

- 12 -

$S(k) \circledast T(k)$, and that that of $f(x) \circledast g(x)$ is $S(k)\,T(k)$. This follows at once from the definitions with the aid of the multiplication theorem and the lemma on logical addition.

It is convenient to introduce the function

$$\delta_n(x,a) \;=\; 2^{n/2}\,\delta(x,a), \qquad x,a \;=\; 0,1,\ldots\, 2^n-1\,,$$

where $\delta(x,a)$ is the Kronecker delta. We distinguish $\delta_n(x,a)$ as the Walsh delta function. It is the Walsh transform of the Walsh function $u_k(a)$, as is easily seen from the orthogonality property.

Logical convolution of a function $f(x)$ with $\delta_n(x,a)$ yields the "logically shifted" function $f(x \oplus a)$. We notice, by applying the logical convolution theorem, that the Walsh transform of $f(x \oplus a)$ is $S(k)\,u_k(a)$.

Using the multiplication theorem and the orthogonality property, we can show that the logical convolution of two Walsh functions vanishes, unless those functions have the same eigenvalue. Thus

$$u_k(x) \circledast u_h(x) \;=\; 2^{-n/2} \sum_{y=0}^{2^n-1} u_k(x \oplus y)\, u_h(y)$$

$$=\; 2^{-n/2}\, u_k(x) \sum_{y=0}^{2^n-1} u_k(y)\, u_h(y)$$

$$=\; u_k(x)\, \delta_n(k,h)\,.$$

The logical derivative $f^1(x)$ and the Walsh transform $S(k)$ of a function $f(x)$ are related by the formula

$$f^1(x) \;=\; 2^{-n/2} \sum_{k=0}^{2^n-1} \{-2k\, S(k)\}\, u_k(x)\,.$$

- 13 -

For term-by-term logical differentiation of the Walsh expansion of $f(x)$ gives

$$f^{1}(x) = 2^{-n/2} \sum_{k=0}^{2^n-1} S(k)\, u_k^{1}(x) \;;$$

and the derivative $u_k^{1}(x)$ is given by the defining logical differential equation as

$$u_k^{1}(x) = -2ku_k(x) \,.$$

By repeated logical differentiation, we can define logical derivatives $f^{11}(x)$, $f^{111}(x)$, ... $f^{1^p}(x)$ of higher order. These are evidently related to the Walsh transform by

$$f^{1^p}(x) = 2^{-n/2} \sum_{k=0}^{2^n-1} (-2k)^p\, S(k)\, u_k(x), \qquad p = 0,1,2,\ldots$$

If we put $S(k) = 1$ in this, so that

$$f(x) = 2^{-n/2} \sum_{k=0}^{2^n-1} u_k(x) = \delta_n(x,0) \,,$$

we get

$$2^{-n/2} \sum_{k=0}^{2^n-1} k^p\, u_k(x) = (-2)^{-p}\, \delta_n^{1^p}(x,0) \,.$$

Thus the Walsh transforms of the positive integral powers are related to the logical derivatives, of appropriate orders, of the Walsh delta function supported at the origin.

The logical derivative of a logical convolution $f(x) \circledast g(x)$ is equal to the product of either function and the logical derivative of the other. For

$$\{f(x) \circledast g(x)\}^1 = 2^{-n/2} \sum_{k=0}^{2^n-1} S(k)\, T(k)\, u_k^1(x)$$

$$= 2^{-n/2} \sum_{k=0}^{2^n-1} S(k)\, T(k)\, (-2k)\, u_k(x);$$

and, by grouping the factors appropriately and applying the convolution theorem, we obtain either $f^1(x) \circledast g(x)$ or $f(x) \circledast g^1(x)$.

The arithmetic mean value of the logical derivative of an arbitrary function is zero. For

$$\sum_{x=0}^{2^n-1} f^1(x) = \sum_{r=0}^{n-1} \sum_{x=0}^{2^n-1} \{f(x \oplus 2^r) - f(x)\}2^r = \sum_{r=0}^{n-1} 0.2^r = 0 ,$$

by the lemma on logical addition.

The formula relating higher-order logical derivatives to the Walsh transform permits an obvious generalization to define logical derivatives of fractional and negative order. In particular, by setting $p = -1$, we can define a logical indefinite integral $g(x) = f^{1^{-1}}(x)$ of a function $f(x)$. A necessary condition for the existence of $g(x)$ is given by the mean value theorem above as

$$\sum_{x=0}^{2^n-1} f(x) = 0 .$$

We shall show, by actually constructing $g(x)$, that this condition is also sufficient. It can be anticipated that $g(x)$ will not be unique, but will be arbitrary to the extent of a constant, from the fact that the logical derivative of a constant is zero.

Given f(x), of zero mean, our algorism yields

$$g(x) = 2^{-n/2} \sum_{k=0}^{2^n-1} (-2k)^{-1} S(k)\, u_k(x) ,$$

where

$$S(k) = 2^{-n/2} \sum_{x=0}^{2^n-1} f(x)\, u_x(k) ,$$

with S(0) = 0. Every term in the summation for g(x) is determinate except the first, k = 0. This represents the expected arbitrary constant. To allow a further integration, we must clearly determine this constant, as is always possible, so that

$$\sum_{x=0}^{2^n-1} g(x) = 0 .$$

The process may evidently be repeated any number of times, provided that the arbitrary constant arising at each stage is fixed so as to give an integral of zero mean.

The basis of Walsh spectrometry is the logical analogue of the Wiener-Khintchine theorem. This logical theorem asserts that the Walsh covariance (power) spectrum of a stationary stochastic function f(x) is the Walsh transform of the logical autocovariance function of f(x).

The Walsh power spectrum $P_n(k)$ of order n of f(x) is defined as the mean square Walsh transform of 2^n consecutive values f(t), f(t+1), ... $f(t+2^n-1)$ of f(x). Since f(x) is stationary, the mean may be taken over t. Thus

$$P_n(k) = \lim_{T\to\infty} \frac{1}{T} \sum_{t=0}^{T-1} S_n^2(k,t), \qquad k = 0,1,\ldots 2^n-1 ,$$

where

$$S_n(k,t) = 2^{-n/2} \sum_{x=0}^{2^n-1} f(x+t)\, u_k(x) .$$

The logical autocovariance $L_n(x)$ of order n of $f(x)$ is defined as the mean logical convolution of order n of $f(x)$ with itself (it is assumed that $f(x)$ is of zero mean). Thus

$$L_n(x) = \lim_{T\to\infty} \frac{1}{T} \sum_{t=0}^{T-1} f(x,t) \circledast f(x,t)$$

$$= \lim_{T\to\infty} \frac{1}{T} \sum_{t=0}^{T-1} 2^{-n/2} \sum_{y=0}^{2^n-1} f(x \oplus y,t)\, f(y,t) ,$$

$$x = 0,1,\ldots 2^n-1.$$

Here we have replaced $f(x+t)$ by $f(x,t)$ to avoid confusion in applying the definition of logical convolution.

The theorem is proved by substituting

$$f(x,t) = 2^{-n/2} \sum_{k=0}^{2^n-1} S_n(k,t)\, u_k(x)$$

in the last expression, and using the formula

$$u_k(x) \circledast u_h(x) = 2^{n/2}\, u_k(x)\, \delta(k,h)$$

for the logical convolution of two Walsh functions. Thus

$$L_n(x) = \lim_{T\to\infty} \frac{1}{T} \sum_{t=0}^{T-1} 2^{-n} \sum_{k=0}^{2^n-1} \sum_{h=0}^{2^n-1} S_n(k,t)\, S_n(h,t)\, 2^{-n/2} \sum_{y=0}^{2^n-1} u_k(x \oplus y)\, u_h(y)$$

$$= 2^{-n/2} \sum_{k=0}^{2^n-1} \lim_{T\to\infty} \frac{1}{T} \sum_{t=0}^{T-1} S_n^2(k,t)\, u_k(x)$$

$$= 2^{-n/2} \sum_{k=0}^{2^n-1} P_n(k)\, u_k(x), \quad \text{which proves the theorem.}$$

– 17 –

The logical autocovariance $L_n(x)$ is easily expressed in terms of the arithmetic autocovariance $A_n(x)$, defined in the present context by

$$A_n(x) = \lim_{T\to\infty} \frac{1}{T} \sum_{t=0}^{T-1} f(x,t)\, f(0,t), \qquad x = 0,1,\ldots 2^n-1.$$

For

$$L_n(x) = 2^{-n/2} \sum_{y=0}^{2^n-1} \lim_{T\to\infty} \frac{1}{T} \sum_{t=0}^{T-1} f(x \oplus y, t)\, f(y,t)$$

$$= 2^{-n/2} \sum_{y=0}^{2^n-1} \lim_{T\to\infty} \frac{1}{T} \sum_{t=0}^{T-1} f\{(x \oplus y) - y + t\}\, f\{t\}$$

$$= 2^{-n/2} \sum_{y=0}^{2^n-1} A_n\{(x \oplus y) - y\} .$$

The Walsh power spectrum $P_n(k)$ expresses the power associated with each of the Walsh functions $u_k(x)$ for $k = 0,1, \ldots 2^n-1$. The power spectrum may be expressed as a function of wave-number if the concept of wave-number is suitably generalised. We define the wave-number ν of the Walsh function $u_k(x)$ as $2^{-(n+1)}\ell$, where ℓ is the number of changes of sign of $u_k(x)$ as x takes the values $0,1,\ldots 2^n-1$. As k takes the values $0,1,\ldots 2^n-1$, the number ℓ of sign-changes of $u_k(x)$ takes the same values $0,1,\ldots 2^n-1$, but not in that order. In fact, the binary expansion of ℓ is related to that of k by

$$k_r = \ell_{n-r} \oplus \ell_{n-r-1}, \qquad r = 0,1,\ldots n-1.$$

This expression enables the power spectrum ordinates $P_n(k)$ to be re-arranged in natural order as $S_n(\ell)$, the power associated with the Walsh function

$$\phi_n(\ell,x) = u_n(k,x) .$$

- 18 -

On account of the relation between ℓ and k, we have

$$\phi_n(\ell,x) = \exp \pi i \sum_{r=0}^{n-1} (\ell_{n-r} + \ell_{n-r-1})x_r$$

$$= \exp \pi i \sum_{r=0}^{n-1} (x_r + x_{r+1})\, \ell_{n-r-1}$$

$$= \exp \pi i \sum_{r=0}^{n-1} (x_{n-r} + x_{n-r-1})\ell_r = \phi_n(x,\ell) .$$

Also

$$\phi_n(\ell,x) = \exp \pi i \left[\ell_{n-1}\, x_0 + \sum_{r=1}^{n-1} (\ell_{n-r} + \ell_{n-r-1})x_r\right]$$

$$= \exp \pi i \left[\ell_{n-1}\, x_0 + \sum_{r=1}^{n-1} \{(1 - \ell_{n-r}) + (1 - \ell_{n-r-1})\}x_r\right]$$

$$= \exp \pi i \left[\{1 - (2^n-1-\ell)_{n-1}\}x_0 + \sum_{r=1}^{n-1} \{(2^n-1-\ell)_{n-r} + (2^n-1-\ell)_{n-r-1}\}x_r\right]$$

$$= (-)^x\, \phi_n(2^n-1-\ell,x) = (-)^\ell\, \phi_n(\ell,2^n-1-x) .$$

By analogy with sinusoidal Fourier spectrometry, the wave-number $2^{-(n+1)}2^n = \frac{1}{2}$ is called the cut-off wave-number. Likewise, the nominal resolution limit is the difference $2^{-(n+1)}$ between consecutive wave-numbers ν. We have so far used dimensionless units, the sampling period for the interferogram being taken as unity. If the actual sampling period is denoted by q, then the cut-off wave-number for Walsh spectrometry is $p = 1/(2q)$, and the nominal resolution limit $r = 1/(2^{n+1}q) = p/2^n$, just as for (unapodized) sinusoidal Fourier spectrometry.

- 19 -

The formula

$$k_r = \ell_{n-r} \oplus \ell_{n-r-1}$$

enables us to express each of the $\phi_n(\ell,x)$ as a product of two of the $u_n(k,x)$ with subscripts given in terms of the binary expansion of ℓ. Thus

$$\phi_n(\ell,x) = u_n(k,x) = \prod_{r=0}^{n-1} u_n(k_r 2^r,x)$$

$$= \prod_{r=0}^{n-1} u_n\{(\ell_{n-r} \oplus \ell_{n-r-1})2^r,x\}$$

$$= \left\{\prod_{r=0}^{n-1} u_n(\ell_{n-r}2^r,x)\right\}\left\{\prod_{r=0}^{n-1} u_n(\ell_{n-r-1}2^r,x)\right\}$$

$$= u_n\left(\sum_{r=0}^{n-1} \ell_{n-r}2^r,x\right) u_n\left(\sum_{r=0}^{n-1} \ell_{n-r-1}2^r,x\right).$$

In particular,

$$\phi(2^\rho,x) = u(2^{n-\rho},x)\, u(2^{n-\rho-1},x)$$

$$= \exp \pi i\, (x_{n-\rho} + x_{n-\rho-1}).$$

The multiplication theorem for the functions $\phi_\ell(x)$ may be derived as follows:

$$\phi_n(\ell,x)\, \phi_n(\ell',x) = u_n(\Sigma\, \ell_{n-r}2^r,x)\, u_n(\Sigma\, \ell_{n-r-1}2^r,x)$$

$$u_n(\Sigma\, \ell'_{n-r}2^r,x)\, u_n(\Sigma\, \ell'_{n-r-1}2^r,x)$$

$$= u_n\{\Sigma(\ell \oplus \ell')_{n-r}2^r,x\}$$

$$u_n\{\Sigma(\ell \oplus \ell')_{n-r-1}2^r,x\}$$

$$= \phi_n(\ell \oplus \ell',x).$$

In particular,

$$\phi_n\left(2^\rho + \sum_{r=0}^{\rho-1} \ell_r 2^r,x\right)$$

$$= \phi_n(2^\rho,x)\, \phi_n\left(\sum_{r=0}^{\rho-1} \ell_r 2^r,x\right),$$

$$\rho = 1,2,\dots n-1.$$

- 20 -

or, by using the previous result,

$$\phi_n\left(2^\rho + \sum_{r=0}^{\rho-1} \ell_r 2^r, x\right) = \phi_n\left(\sum_{r=0}^{\rho-1} \ell_r 2^r, x\right) \exp \pi i \,(x_{n-\rho} + x_{n-\rho-1}) .$$

This expression provides us with a simple recursive scheme for calculating the 2^n Walsh functions

$$\phi_n(\ell,x), \qquad \ell = 0,1,\ldots 2^n-1,$$

without explicit reference to the definition. We have

$$\begin{aligned}
\phi_n(0,x) &= 1, \\
\phi_n(1,x) &= \exp \pi i x_{n-1}, \\
\phi_n(2,x) &= \exp \pi i \,(x_{n-1} + x_{n-2}), \\
\phi_n(4,x) &= \exp \pi i \,(x_{n-2} + x_{n-3}), \\
&\ldots\ldots\ldots\ldots\ldots\ldots\ldots\ldots, \\
\phi_n(2^{n-1},x) &= \exp \pi i \,(x_1 + x_0) .
\end{aligned}$$

If the remaining Walsh functions $\phi_n(3,x)$, $\phi_n(5,x)$, $\phi_n(6,x)$, ... $\phi_n(2^n-1,x)$ are calculated from

$$\phi_n\left(2^\rho + \sum_{r=0}^{\rho-1} \ell_r 2^r, x\right) = \phi_n(2^\rho,x) \; \phi_n\left(\sum_{r=0}^{\rho-1} \ell_r 2^r, x\right)$$

in ascending order of ℓ, then values of the two factors on the right will always be available at the stage they are required.

Up to this point, the computation of the Walsh power spectrum has been envisaged in two stages: first, the transformation of the arithmetical autocovariance $A(\xi)$ into the logical autocovariance $L(\xi)$; and, second, the Walsh transform proper. Such a method requires the storing initially of all the interferogram data (or at least the storing of the logical autocovariance), and therefore limits the resolution in the Walsh spectrum that can be obtained using a computer with a small main store. This limitation can be avoided if

- 21 -

the contributions of each interferogram ordinate to all the spectrum ordinates are evaluated completely just after reading in that ordinate, which in this case need not be stored. For this form of computation, we require to know the kernel $\psi(\ell,x)$ of the overall transform (A.I.U. transform)

$$S(\ell) = \sum_{\xi=0}^{2^n-1} \Lambda(\xi)\, \psi(\ell,\xi) .$$

To obtain an expression for $\psi(\ell,\xi)$, we note first that

$$S(\ell) = 2^{-n/2} \sum_{x=0}^{2^n-1} L(x)\, \phi_\ell(x) .$$

Secondly, it may be shown that

$$L(x) = 2^{n/2} \sum_{0\leq\xi<2^n} \Lambda(\xi) \exp\left\{\left(1 - \delta(x,0) - \sum_{r=0}^{n-1} x_r\right) \ln 2\right\},$$

where the summation extends over all non-negative integers ξ less than 2^n such that

$$\xi = \sum_{r=0}^{n-1} (\pm x_r 2^r)$$

for some combination of + and - signs.

The A.I.U. transform is therefore

$$S(\ell) = \sum_{x=0}^{2^n-1} \sum_{0\leq\xi<2^n} \Lambda(\xi) \exp\left\{\left(1 - \delta(x,0) - \sum_{r=0}^{n-1} x_r\right) \ln 2\right\} \phi_\ell(x)$$

$$= \sum_{\xi=0}^{2^n-1} \Lambda(\xi)\, \psi(\ell,\xi) ,$$

where

$$\psi(\ell,\xi) = \sum_{0\leq x<2^n} \phi_\ell(x) \exp\left\{\left(1 - \delta(x,0) - \sum_{r=0}^{n-1} x_r\right) \ln 2\right\},$$

with the summation on the right extending over non-negative integers x less than 2^n such that

$$\xi = \sum_{r=0}^{n-1} (\pm x_r 2^r)$$

The solutions of this equation in x may be written

$$x = \xi + 2c ,$$

where c takes those values out of

$$0,1,\dots \left[\tfrac{1}{2}(2^n - 1 - \xi)\right]$$

that are consistent with

$$(\xi + 2c) \wedge c = c .$$

The last expression for $\psi(\ell,\xi)$ is the basis of a computer program that has enabled us to tabulate $\psi(\ell,\xi)$ for $\ell,\xi = 0,1,\dots 2^n-1$, with $n = 2, 3, 4, 5,$ and 6. By inspection, the following symbolic identity has been discovered:

$$\psi(\ell,\xi) = \psi(2\ell + \tfrac{1}{2}, \tfrac{1}{2}\xi) .$$

Combined with certain rules of interpretation, this identity enables us to express ψ, for arbitrary ℓ and ξ, in terms of $\psi(\ell,0)$ and $\psi(\ell,1)$, for which we have the expressions

$$\psi(\ell,0) = 1,$$
$$2^{n-1}\psi(\ell,1) = 2^n - 1 - 2\ell .$$

- 23 -

The interpretative rules are the following:

(a) If ξ is an integer, and $\ell \geq 2^{n-1}$, then $\psi(\ell,\xi)$ is to be replaced by $(-)^{\xi}\,\psi(2^n-1-\ell,\xi)$.

(b) Fractional arguments are to be understood by linear interpolation; thus

$$\begin{aligned}\psi(\ell+\theta,\ \xi+\eta) &= (1-\theta)\ (1-\eta)\ \psi(\ell,\xi) + \theta(1-\eta)\ \psi(\ell+1,\ \xi) \\ &\quad + (1-\theta)\eta\ \psi(\ell,\ \xi+1) + \theta\eta\psi(\ell+1,\ \xi+1)\ ,\end{aligned}$$

where θ,η are non-negative proper fractions.

(c) $\psi(\ell,0)$ is to be replaced by 2, <u>not</u> by its value, 1.

By way of illustration, we append an example of reduction of $\psi(\ell,\xi)$ to the form $p_0\ \psi(\ell,0) + p_1\ \psi(\ell,1)$. If $n = 4$,

$$\begin{aligned}\psi(13,3) &= -\ \psi(2,3) \\ &= -\ \psi(\tfrac{9}{2},\ \tfrac{3}{2}) \\ &= -\ \psi(\tfrac{19}{2},\tfrac{3}{4}) \\ &= -\ \tfrac{1}{4}\ \{\psi(\tfrac{19}{2},0) + 3\ \psi(\tfrac{19}{2},1)\} \\ &= -\ \tfrac{1}{4}\ \{\psi(\tfrac{11}{2},0) - 3\ \psi(\tfrac{11}{2},1)\} \\ &= -\ \tfrac{1}{4}(2 - \tfrac{3}{2}) \ = \ -\ \tfrac{1}{8}\end{aligned}$$

- 24 -

References

[1] Blackman, R.B., and Tukey, J.W., 1959, "The Measurement of Power Spectra" (New York : Dover).

[2] Gebbie, H.A., in Singer, J.R., edit., "Advances in Quantum Electronics", 155-163 (Columbia Univ. Press, 1961).

[3] Cooley, J.W., and Tukey, J.W., 1965, Math. Computation, 19, 297-301.

[4] Gentleman, W.M., and Sande, G., 1966, Proc. Fall Joint Computer Conf., 563-578.

[5] Walsh, J.L., 1923, Amer. J. Math., 55, 5-24.

[6] Paley, R.E.A.C., 1932, Proc. Lond. Math. Soc., 34, 241-279.

[7a] Fine, N.J., 1949, Trans. Amer. Math. Soc., 65, 372-414.

[7b] Fine, N.J., 1950, Trans. Amer. Math. Soc., 69, 66-77.

[8] Polyak, B.T., and Shreider, Yu.A., 1966, Problems in Theory of Mathematical Machines. Collection II, pp.174-190. Fizmatgiz, Moscow.

See note inside cover

DES REPORT No. 1

December 1969

Gi 8

NATIONAL PHYSICAL LABORATORY

DIVISION OF ELECTRICAL SCIENCE

WALSH FUNCTIONS AS SOLUTIONS OF A LOGICAL DIFFERENTIAL EQUATION

by

J. E. Gibbs and Margaret J. Millard

NATIONAL PHYSICAL LABORATORY 1900

A Station of the
Ministry of Technology

Approved on behalf of Director, NPL by
Mr. A.E. Bailey, Superintendent, Division of Electrical Science

Walsh functions as solutions of a logical differential equation

by

J.E. Gibbs and Margaret J. Millard*

Abstract

Most well-known sets of orthogonal functions are solutions of linear ordinary differential equations. The Walsh functions, important in the sequency theory of communications [4], are apparently an exception. An analogue, for these functions, of the defining differential equation is formulated *via* an appropriate generalization of the concept of derivative.

It has been remarked by Harmuth [1] that most of the well-known complete orthonormal systems of functions are defined by linear ordinary differential equations of the second order, but that the Walsh functions [2 - 4] are defined by a difference equation rather than a differential equation. We believe that Harmuth's so-called difference equation is more correctly regarded as a recursion formula, and as analogous to the recursion formulae known for the classical orthonormal systems. On this view, an analogue for Walsh functions of the defining ordinary differential equation remains to be discovered. The interest that would attach to finding such an analogue is hinted at by Meltzer [5].

The Walsh functions take the values +1 and -1 everywhere except at a finite number of discontinuities on any unit interval. It seems likely from this that, before we may discover a 'differential equation' that they satisfy, we shall have to arrive at a fresh conception of 'derivative'.

The points of discontinuity of the Walsh functions are at the binary rational numbers, of the form $p/2^q$, where p and q are integers. This

---

*Now at School of Mathematics, Bath University, Bath, Somerset.

suggests that a domain of definition for these functions more natural than the real line would be the set of non-negative integers less than 2^n, namely $\{0, 1, \ldots, 2^n-1\}$, where n is a positive integer. We shall therefore study some properties of functions defined on this domain, and see whether Walsh functions do not emerge naturally in this context.

The domain $\{0, \ldots, 2^n-1\}$ is ripe for investigation, since it corresponds, for example, to the set of all locations in a random-access computer store addressable by a binary word of n bits. The range of the function may be thought of as the set of data stored in the locations mentioned, and the function itself as the mapping from addresses to contents. We do not wish to imply, however, that the range of the function considered is necessarily a set of computer-representable data.

The computer analogy is a happy one, since, as Polyak and Schreider [6] pointed out in 1962, one of the distinguishing features of Walsh functions is that they can be computed by a single short hardware operation from their arguments expressed in binary notation. This at once implies that the Walsh functions are well 'impedance-matched' to the binary digital computer, in the sense that, for example, of all finite discrete [7] numerical transforms of Fourier type, the computer displays by far the 'maximum power' in respect of the Walsh-Fourier transform.

The methods of modern numerical analysis have of course been considerably influenced by the peculiar aptitudes of the digital computer. The hardware generation of Walsh functions seems, however, to be the first non-trivial influence of the computer's peculiarity of representing numbers in binary notation. Computer people have recognized that an n-bit word may be interpreted arithmetically or as an n-tuple of boolean variables; hence the distinction between arithmetical and logical orders within the computer's order code. These two subsets of orders are generally kept conceptually well apart;

- 3 -

but there may be more legitimate opportunities for mixing them than we suspect.

As an example of the judicious mixing of arithmetical and logical operations, we shall define a 'logical derivative' of a function defined on the set

$$B_n = \{0, \ldots, 2^n - 1\}.$$

It will then be possible to write down a 'logical differential equation' of very simple form having the Walsh functions as solutions.

We begin by establishing a one-to-one correspondence between the elements of B_n and the elements of the n-dimensional vector space B_1^n over the field $B_1 = \{0, 1\}$, in which the field operations are logical addition $\oplus$ defined by

$$u \oplus v = 1 - \delta(u,v)$$

and logical multiplication $\wedge$ defined by

$$u \wedge v = uv$$

for all u, v in B_1. (We use $\delta(u,v)$ for the Kronecker delta).
The correspondence, in other words, is between non-negative integers less than 2^n on the one hand; and, on the other hand, vectors of the form

$$x = (x_{(n-1)}, x_{(n-2)}, \ldots, x_{(0)}),$$

where $x_{(r)}$ is either 0 or 1 for $0 \leqslant r \leqslant n-1$. To each vector x in the space B_1^n corresponds a scalar ξ in B_n given by

$$\xi = \sum_{r=0}^{n-1} x_{(r)} 2^r;$$

- 4 -

thus the components $x_{(r)}$ of the vector x are the coefficients in the direct dyadic expansion of the integer ξ. The above expression for ξ is in fact a norm [8] for the space $B_1{}^n$.

Unlike the ordinary Euclidean norm

$$\xi = \sqrt{\sum_{r=0}^{n-1} x_{(r)}{}^2}$$

of a vector $(x_{(n-1)}, \ldots, x_{(0)})$, our norm ξ is one-to-one; while many vectors have the same Euclidean norm, our norm determines a unique vector. This is merely to say that the direct dyadic expansion is unique. To emphasize that the norm ξ is one-to-one, we write down an algorithm to obtain (uniquely) x from ξ:

$$\sigma_0 = \xi; \ \sigma_{r+1} = [\tfrac{1}{2}\sigma_r], \ x_{(r)} = \sigma_r - 2\sigma_{r+1}, \ r = 0, \ldots, n-1,$$

where $[\sigma]$ denotes the greatest integer not exceeding σ.

As a matter of notational convenience, we shall reserve Greek letters to represent scalars obtained from the corresponding vectors (represented by italic letters) by the algorithm

$$\xi = \sum_{r=0}^{n-1} x_{(r)} 2^r.$$

The same algorithm $f(x) = (f_{(n-1)}(x), \ldots, f_{(0)}(x)) \mapsto \phi(x)$ may be used to define a many-to-one mapping

$$c: \ (\mathbb{K}^n)^{B_1{}^n} \to \mathbb{K}^{B_1{}^n}$$

- 5 -

from the space of all complex n-vector-valued functions on $B_1{}^n$ to the space of all complex scalar-valued functions on $B_1{}^n$:

$$\phi(x) = \sum_{r=0}^{n-1} f_{(r)}(x)2^r, \quad \forall\, x \in B_1{}^n.$$

We shall be concerned with a subspace of $(\mathbb{K}^n)^{B_1{}^n}$, namely the set S of functions $f(x)$ such that, for some $\phi(x) \in \mathbb{K}^{B_1{}^n}$,

$$f_{(r)}(x) = (2^n-1)^{-1}\bar{\phi}$$

$$+2^{-n} \sum_{\kappa=1}^{2^n-1} \kappa^{-1} k_{(r)} \sum_{\xi'=0}^{2^n-1} (\phi(x') - \phi) \exp \pi i \sum_{s=0}^{n-1} k_{(s)} (x_{(s)} \oplus x'_{(s)}),$$

where

$$\bar{\phi} = 2^{-n} \sum_{\xi=0}^{2^n-1} \phi(x),$$

and, according to convention,

$$\kappa = \sum_{r=0}^{n-1} k_{(r)}2^r, \quad \xi' = \sum_{r=0}^{n-1} x'_{(r)} 2^r.$$

The restriction of the mapping c to the domain S is a bijection from S to $\mathbb{K}^{B_1{}^n}$, and the inverse of this bijection is $\phi(x) \mapsto f(x)$, where $f_{(r)}(x)$ is given by the expression above.

It is easy to verify that this inverse mapping is a generalization of the inversion of the norm ξ of a vector x in $B_1{}^n$: setting $\phi(x) = \xi$ yields $f_{(r)}(x) = x_{(r)}$. Thus we may regard $f(x)$ as the 'direct dyadic expansion' of $\phi(x)$.

- 6 -

The logical derivative will now be introduced by way of the 'logical gradient'. The gradient of a scalar function ϕ on Cartesian n-space is a vector of which each component is equal to the rate of change of ϕ with respect to the corresponding space co-ordinate. In $B_1{}^n$, each 'space co-ordinate' has only two possible values, 0 and 1. The 'rate of change' of ϕ at a given point with respect to a particular co-ordinate, if we may evaluate it at all, can only be the algebraic difference between: (a) the value of ϕ when that co-ordinate alone is given its only other value, and (b) the value of ϕ at the point of evaluation. More precisely, if ϕ is a scalar function on $B_1{}^n$, we shall define the logical gradient of ϕ as the vector function on $B_1{}^n$

$$f^{[1]}(x) = (f^{[1]}{}_{(n-1)}(x), \ldots, f^{[1]}{}_{(0)}(x)),$$

where

$$f^{[1]}{}_{(r)}(x) = -\tfrac{1}{2}(\phi(x \oplus e_r) - \phi(x)).$$

Here e_r is the basis vector such that

$$(e_r)_{(s)} = \delta(r,s);$$

and logical addition of vectors is defined by

$$(x \oplus y)_{(r)} = x_{(r)} \oplus y_{(r)}.$$

The numerical coefficient $-\frac{1}{2}$ is introduced to obviate later occurrence of an otiose -2.

- 7 -

The bijection c connects the logical gradient $f^{[1]}(x)$ with the logical derivative $\phi^{[1]}(x)$:

$$\phi^{[1]}(x) = \sum_{r=0}^{n-1} f^{[1]}{}_{(r)}(x)2^r = -\tfrac{1}{2}\sum_{r=0}^{n-1} (\phi(x \oplus e_r) - \phi(x))2^r.$$

The (discrete [9-11]) Walsh functions may now be defined as the solutions of a certain first-order linear ordinary logical differential equation, namely

$$\eta^{[1]}(x) = \kappa\eta(x), \quad \forall\ x \in B_1{}^n.$$

This is a characteristic-value problem that may be solved by matrix methods. The characteristic values are

$$\kappa = \sum_{r=0}^{n-1} k_{(r)}2^r, \quad k_{(r)} \in B_1.$$

and the characteristic solutions

$$\eta_n(\kappa, \xi) \triangleq \exp\ \pi i \sum_{r=0}^{n-1} k_{(r)}\ x_{(r)}.$$

The functions thus defined are related to Paley's [12] form ψ of the Walsh functions by

$$\eta_n\ (\kappa, \xi) = \psi_\kappa\ (2^{-n}\ \xi) = \psi_\xi\ (2^{-n}\kappa).$$

By repeating the operation of logical differentiation, we can obtain

- 8 -

logical derivatives $\phi^{[2]}(x)$, $\phi^{[3]}(x)$, ... of higher order. More generally, the order of a logical derivative $\phi^{[\rho]}(x)$ may be any complex number ρ, and

$$\phi^{[\rho]}(x) = 2^{-n} \sum_{\kappa=1}^{2^n-1} \kappa^{\rho} \sum_{\xi'=0}^{2^n-1} \phi^{[0]}(x')\eta_n(\kappa, \xi \oplus \xi'),$$

where

$$\phi^{[0]}(x) = \phi(x) - \bar{\phi}.$$

With this notation, the algorithm for $f_{(r)}(x)$ may be written simply

$$f_{(r)}(x) = (2^n-1)^{-1}\,\bar{\phi} - \tfrac{1}{2}(\phi^{[-1]}(x \oplus e_r) - \phi^{[-1]}(x)), \quad \forall\, x \in B_1^{\,n}.$$

References

1 HARMUTH, H.F.: 'A generalized concept of frequency and some applications', IEEE Trans., 1968, IT-14, pp. 375-382

2 WALSH, J.L.: 'A closed set of normal orthogonal functions', Amer. J. Math., 1923, 45, pp. 5-24

3 FINE, N.J.: 'Walsh functions', article to be published in 'Encyclopaedic Dictionary of Physics' (Pergamon Press)

4 HARMUTH, H.F.: 'Transmission of information by orthogonal functions' (Springer, 1969)

5 MELTZER, B.: 'Speculations on the use of orthonormal functions in the study of morphogenesis', Nature, 1968, 217, p. 176

6 POLYAK, B.T., and SCHREIDER, Yu.A.: 'The application of Walsh functions in approximate calculations', Voprosy Teor. Matem. Mashin, 1962, Coll. II, pp. 174-190

7 GENTLEMAN, W.M., and SANDE, G.: 'Fast Fourier transforms - for fun and profit', AFIPS Conf. Proc., 1966, 29, pp. 563-578

- 9 -

8 PICHLER, F.: 'Das System der sal- und cal-Funktionen als Erweiterung des Systems der Walsh-Funktionen und die Theorie der sal- und cal-Fouriertransformation', Dissertation, Universität Innsbruck (1967)

9 GOLOMB, S.W.: 'On the classification of Boolean functions', IRE Trans., 1959, CT-6, pp. 176-186

10 ITO, T.: 'A note on a general expansion of functions of binary variables', Information and Control, 1968, 12, pp. 206-211

11 WONG, E., and EISENBERG, E.: 'Iterative synthesis of threshold functions', J. Math. Anal. Appl., 1965, 11, pp. 226-235

12 PALEY, R.E.A.C.: 'A remarkable series of orthogonal functions', Proc. Lond. Math. Soc., 1932, 34, 241-279

ET

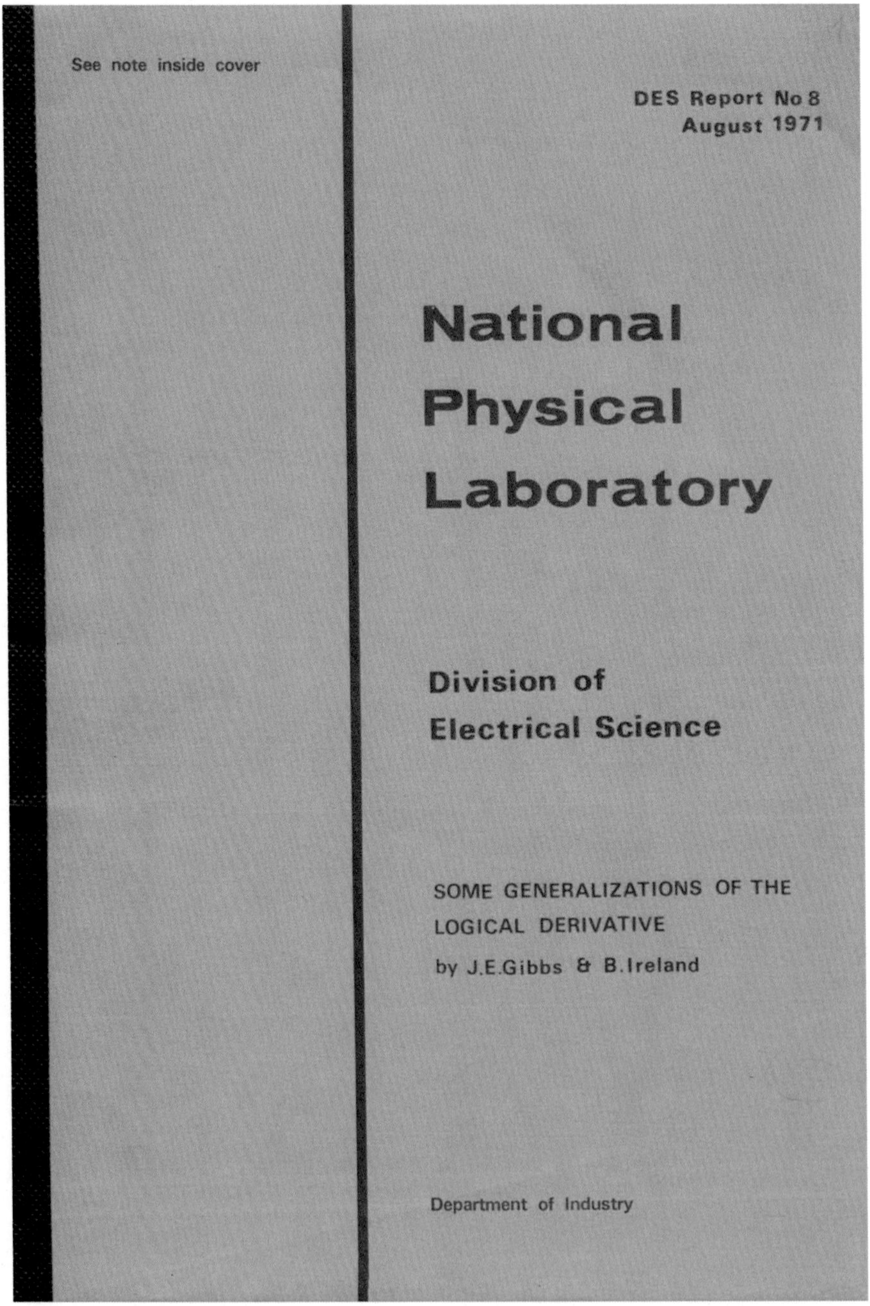
See note inside cover

DES Report No 8
August 1971

National
Physical
Laboratory

Division of
Electrical Science

SOME GENERALIZATIONS OF THE
LOGICAL DERIVATIVE
by J.E.Gibbs & B.Ireland

Department of Industry

Approved on behalf of Director, NPL by
Mr. A.E. Bailey, Superintendent, Division of Electrical Science

DES Report No 8

August 1971

NATIONAL PHYSICAL LABORATORY

SOME GENERALIZATIONS OF THE LOGICAL DERIVATIVE

by

J.E. Gibbs and B. Ireland
DIVISION OF ELECTRICAL SCIENCE

Abstract

Sets of functions, mutually orthogonal on an interval, and generated as solutions of differential equations, play an important part in the physical sciences. Sets of functions, mutually orthogonal with respect to a Boolean vector space, and generated as solutions of logical differential equations, have a similar role in the information sciences. Extensions of the existing definition of the logical derivative suffice to interpret logical differential equations in ways that give rise to various generalizations of the Walsh functions having values on the unit circle; the extensions leading, in particular, to the Walsh-Paley functions on the unit interval and to the Walsh-Lévy-Chrestenson functions are given in this paper. The Takayasu Ito functions, in which the Walsh functions are embedded as a special case, have values, in general, off the unit circle. By analogy with mathematical physics, the logical differential equation that generates the two-valued Walsh functions is regarded as a dichotomous wave equation; by a suitable generalization of the concept of dichotomy, we obtain a more general dichotomous wave equation whose solutions are the Ito functions.

This report was written during the tenure of a Vacation Consultantship in the Division of Electrical Science by one of us (B. Ireland) while on leave from the School of Mathematics of the University of Bath.

The Argument of the report was presented to the Symposium on Digital Filtering organized by Mr. R.E. Bogner in the Electrical Engineering Department of the Imperial College of Science and Technology on 1971 August 31-September 2.

SOME GENERALIZATIONS OF THE LOGICAL DERIVATIVE

CONTENTS

Argument		1
Appendix 1 -	The dichotomous logical derivative for functions on the unit interval	6
Appendix 2 -	Discrete Walsh functions as solutions of a dichotomous wave equation	8
Appendix 3 -	Discrete Walsh-Lévy-Chrestenson functions as solutions of a polychotomous wave equation	11
Appendix 4 -	The polychotomous logical derivative for functions on the unit interval	15
Appendix 5 -	Takayasu Ito functions as solutions of a generalized dichotomous wave equation	17
Epitome		20
References		21

Argument

Filtering, digital or other, makes use of the basic fact that any one of a wide class of complex-valued functions defined on the real line can be expressed almost everywhere as a linear combination of Euler functions (exponential functions of imaginary argument) of all frequencies (Gibbs 1970 d). This is a consequence of the fact that the Euler functions form a complete orthonormal set on the real line. Such sets of functions are generated as sets of eigensolutions of differential equations under suitable boundary conditions.

Just as orthogonal functions on the real line are important in the physical sciences, so orthogonal functions on a finite dyadic group (or on an n-dimensional Boolean vector space (Gibbs 1970 c)) are important in the information sciences (Gibbs 1971). The Takayasu Ito functions (Ito 1968) appear in this role in problems of pattern recognition, and (their special case) the Walsh functions have gained wide application in problems of

2

communication (Harmuth 1970) and calculation (Polyak and Schreider 1962) using two-level signals and two-state devices (Bass 1970, Zeek and Showalter 1971, Barrett and Lines 1971).

Orthogonal functions on the dyadic group can be generated as eigen-solutions of a differential equation, under a suitable generalization of the meaning of "differential equation". This generalization has been achieved in the case of discrete Walsh functions by defining "logical differentiation" for functions on a finite dyadic group (Gibbs 1967, 1969). With this definition, it is easy to write a first-order linear "logical differential equation" that has the discrete Walsh functions as eigensolutions (Gibbs and Millard 1969 a, 1969 b).

A number of extensions of this idea enable us to generate, in a similar way, certain orthogonal functions closely related to the discrete Walsh functions.

First, by modifying the definition of "logical derivative" (to "logical derivative of the second kind") in a certain way, one can generate, from a logical differential equation of the same form, the discrete Walsh functions in a new ordering-parameterization (ordered by generalized frequency ("sequency")) (Gibbs 1970 b).

Secondly, by using known mappings between the dyadic group and the real line, one can extend the definition of the logical derivative to functions defined, for example, on the unit interval, and thus (Appendix 1) generate the Walsh functions as defined by Paley (1932).

Thirdly, by using a linear logical differential equation of the second order (Gibbs 1970 b), one can generate "complex Walsh functions" (Gibbs 1970 a) taking the values $\pm i$ as well as ± 1 (and no other values).

Fourthly, by generalizing the definition of the logical derivative in such a way that it applies to functions on the same domain as the discrete analogue of the Walsh-Lévy-Chrestenson functions (Walsh 1923, Lévy 1944, Chrestenson 1955, Selfridge 1955), one can generate these functions from a first-order linear ("polychotomous") logical differential equation identical in form with that generating the two-valued Walsh functions.

These extensions can obviously be combined in various ways: for example, in the fourth extension, the Walsh-Lévy-Chrestenson functions will be generated in discrete form or on the unit interval according to the domain of the functions for which the polychotomous derivative is defined (Appendices 3 and 4).

In short, by suitable interpretations of the first logical derivative entering a certain linear logical differential equation of fixed form, it is possible to generate all manner of Walsh functions: discrete Walsh-Kaczmarz, Walsh-Paley, discrete Walsh-Lévy-Chrestenson, Walsh-Lévy-Chrestenson on the unit interval, and so on. The complex-valued Walsh functions of Gibbs appear as an isolated case, being directly generated by a second-order linear logical differential equation. All these kinds of Walsh functions are united, however, in having values on the unit circle in the complex plane.

The Takayasu Ito functions have a straightforward appearance in this context. Their domain, like that of the discrete Walsh functions, is an n-dimensional Boolean vector space; indeed, they include the discrete Walsh functions as a special case. But the Ito functions have hitherto resisted any attempt to generate them from a logical differential equation; the difficulty appears to be associated with the fact that their values are not confined to the unit circle.

The difficulty may be circumvented by generalizing not the idea of "logical derivative" but the idea of "logical differential equation."

The first-order linear logical differential equation that generates the discrete Walsh functions may be regarded as a "dichotomous wave equation": just as the physicist may choose to interpret a certain linear (partial) differential equation as a wave equation, so the "dichotomist" (information scientist) may choose to interpret a certain linear logical differential equation as a (dichotomous) wave equation.

The physicist arrives at a differential equation such as the wave equation by expressing the laws of physics relevant to his problem in a microscopic or differential form. Typically, the values of a physical quantity (dependent variable) at an arbitrary point in space and time (independent variables) is related, using physical laws, to the values of that quantity at neighbouring points and to the corresponding increments in the independent variables. By causing those increments to approach zero, one obtains, in the limit, a microscopic expression of the physical laws in the form of a differential equation.

The physicist sets himself a boundary-value problem by taking a certain domain of the independent variables; by demanding that his microscopic law shall be obeyed at all interior points of the domain, and that the dependent variable shall take assigned values (consistent with the laws of physics) on the boundary; and by seeking to determine the dependent variable throughout the domain. In short, he seeks a solution (global expression) of the differential equation (microscopic expression) subject to the boundary conditions.

The dichotomist arrives at a logical differential equation such as the dichotomous wave equation in an analogous way, by expressing certain laws of dichotomy in a microscopic form. The value of a dichotomous quantity (we mean, in the first instance, one that takes only the values +1 and -1) at an arbitrary point in an n-dimensional Boolean vector space is related to the values of that quantity at neighbouring points and to the corresponding (logical) increments in the independent variable. The relationship obtained in this way is a microscopic expression of the dichotomous laws in the form of a logical differential equation.

The dichotomist sets himself a boundary-value problem by taking a certain domain of the independent variable (a certain subset of the Boolean vector space: we consider, in the first instance, the whole space); by demanding that his microscopic law shall be obeyed at all points of the domain, and that the dependent variable shall take assigned values (consistent with the laws of dichotomy) on the boundary; and by seeking to determine the dependent variable throughout the domain. In short, he seeks a solution (global expression) of the logical differential equation (microscopic expression) subject to the boundary conditions.

The dichotomous wave equation (Appendix 2) is the microscopic expression of two very simple laws of dichotomy. The first is a mere matter of definition: at every point of the domain, the dependent variable takes one or the other of two possible values (conveniently denoted by +1 and −1). The second law is a hypothesis whose consequences we shall examine: for each of the n "dimensions" of the Boolean space, one or the other of the following two statements holds: (0) the dependent variable takes *the same* value at any two points of the space whose co-ordinates differ only in the relevant dimension; (1) the dependent variable takes *different* values at any two points of the space whose co-ordinates differ only in the relevant dimension.

We set ourselves a boundary-value problem by requiring that these two laws, expressed in microscopic form (ultimately, as a logical differential equation), shall be obeyed throughout the Boolean space; and that the dependent variable shall take assigned values (consistent with the dichotomous laws assumed) at the origin and at n points linearly independent with respect to the field GF(2). The general solution is a linear combination (of the 2^n discrete Walsh functions) in which only one of the coefficients (*which* one is at arbitrary choice) is non-zero (and equal to ±1). The boundary conditions are sufficient to determine the solution uniquely.

The dichotomous wave equation just discussed may be re-arranged in a form recognizable as our standard first-order linear logical differential equation: from our present point of view, it may be regarded as a vehicle for arriving heuristically at a definition of the (dichotomous) logical derivative.

Though the extended definition of the logical derivative required to generate the discrete Walsh-Lévy-Chrestenson functions from the standard logical differential equation may be obtained by purely technical means, it is instructive to develop this definition (Appendix 3) by generalizing our discussion of the dichotomous wave equation to the case of a dependent variable that can take any one of p different values (conveniently denoted by the complex p-th roots of unity). We speak of the wave equation in this case, and the logical derivative to whose definition it leads, as polychotomous.

The polychotomous logical derivative may be defined for functions on the unit interval in much the same way (Appendix 4) as the dichotomous logical derivative. With this definition, the standard first-order linear logical differential equation generates the Walsh-Lévy-Chrestenson functions on the unit interval in an ordering-parameterization analogous to that of the Walsh-Paley functions.

The generalized dichotomous wave equation associated with the Takayasu Ito functions is the microscopic expression of two laws of dichotomy somewhat more complicated than those which lead to the discrete Walsh functions. We postulate that, for each of the n "dimensions" of the Boolean space, one or the other of the following two assertions shall hold: (a0) the sign of the (real) dependent variable is the same at any two points of the space whose co-ordinates differ only in the relevant dimension; (a1) the sign of the dependent variable is different at any two points of the space whose co-ordinates differ only in the relevant dimension. We postulate further that, for each of the n "dimensions" of the space, there exists a non-zero real number b such that, if (a0) [respectively (a1)] holds, then the absolute value of the dependent variable is b times [respectively b^{-1} times] greater at each point having the relevant co-ordinate equal to unity than at the point whose co-ordinates differ only in the relevant dimension. The consequences of these postulates are worked out in Appendix 5.

6

Appendix 1: The dichotomous logical derivative for functions on the unit interval

Let $\mathbb{N}$ denote the set of natural numbers $\{1,2,\ldots\}$; $\forall p \in \mathbb{N}$, let $\mathbb{Z}_p$ denote the set $\{0,\ldots, p-1\}$ of non-negative integers less than p. Let $\mathbb{K}$ denote the complex plane; $\mathbb{R}$ the real line; and $\mathbb{D}$ the dyadic rationals in the unit interval $[0, 1)$.

We define a mapping γ: $[0, 1) \to \mathbb{Z}_2^{\mathbb{N}}$ by

$$\forall \xi \in [0, 1), \qquad \gamma(\xi) = x \triangleq (x_{(1)}, x_{(2)}, \ldots),$$

$$\text{with } \forall r \in \mathbb{N}, \qquad x_{(r)} = [2^r \xi] - 2[2^{r-1}\xi],$$

where $[q]$ means the greatest integer not exceeding q. We note that the range of γ is a proper subset of $\mathbb{Z}_2^{\mathbb{N}}$, namely, the complement in $\mathbb{Z}_2^{\mathbb{N}}$ of the set of all elements corresponding to the recurring expansions of dyadic rationals (Fine 1949, 1950; Pichler 1967). γ is a bijection of $[0, 1)$ onto $\operatorname{rng} \gamma \subset \mathbb{Z}_2^{\mathbb{N}}$:

$$\forall x \in \operatorname{rng} \gamma, \qquad \gamma^{-1}(x) = \sum_{r=1}^{\infty} x_{(r)} 2^{-r}.$$

With the operations of logical addition $\oplus$ defined by

$$\forall (x, y) \in \mathbb{Z}_2^{\mathbb{N}} \times \mathbb{Z}_2^{\mathbb{N}}, \qquad \forall r \in \mathbb{N},$$

$$\mathbb{Z}_2 \ni (x \oplus y)_{(r)} \equiv x_{(r)} + y_{(r)} \bmod 2,$$

and multiplication by an element of $\mathbb{Z}_2$ defined by

$$\forall (\alpha, x) \in \mathbb{Z}_2 \times \mathbb{Z}_2^{\mathbb{N}}, \qquad \forall r \in \mathbb{N},$$

$$(\alpha x)_{(r)} = \alpha x_{(r)},$$

the set $\mathbb{Z}_2^{\mathbb{N}}$ is a linear vector space over $\mathbb{Z}_2$. A basis for this space $\mathbb{Z}_2^{\mathbb{N}}$ is formed by the vectors $\{e_r : r \in \mathbb{N}\}$ defined by

$$\forall r \in \mathbb{N}, \qquad e_r = \gamma(2^{-r});$$

$$\text{or} \quad \forall (r, s) \in \mathbb{N} \times \mathbb{N}, \qquad e_{r(s)} = \delta_{rs},$$

where δ_{rs} is a Kronecker symbol.

We define logical differentiation for complex-valued functions on the unit interval $[0, 1)$ as a mapping $\phi \mapsto \phi^{[2;\,1]}$ from $\mathbb{K}^{[0,\,1)}$ into $\mathbb{K}^{[0,\,1)}$ such that,

$$\forall \phi \in \mathbb{K}^{[0,\,1)}, \quad \forall \xi \in [0, 1),$$

$$\phi^{[2;\,1]}(\xi) = \tfrac{1}{2}\phi(\xi) - \tfrac{1}{2} \sum_{r=1}^{\infty} \phi(\gamma^{-1}(\gamma(\xi) \oplus e_r))2^{-r}.$$

The superscript brackets will be used uniformly to denote a logical derivative (of the first kind); the number $p \in \mathbb{N}-\{1\}$ before the semi-colon refers to the modulus of the ring $\mathbb{Z}_p$ of residue classes; the number $\zeta \in \mathbb{K}$ after the semi-colon indicates the order of the logical derivative. Here $p=2$, $\zeta=1$; and we have a dichotomous first derivative.

A function $\phi \in \mathbb{K}^{[0,\ 1)}$ is logically differentiable at the point $\xi \in [0, 1)$ (possesses a first logical derivative at the point ξ) if, and only if, the series

$$\sum_{r=1}^{\infty} \phi(\gamma^{-1}(\gamma(\xi) \oplus e_r))2^{-r}$$

converges. This series converges (uniformly and absolutely) if ϕ is bounded on $[0, 1)$; for $\Sigma 2^{-r}$ is convergent. Thus, a sufficient condition that ϕ be logically differentiable (at all points of $[0, 1)$) is that ϕ be bounded on $[0, 1)$; this condition is clearly also necessary.

Further, if M is a bound for $\phi(\xi)$ on $[0, 1)$, we have, $\forall \xi \in [0, 1)$,

$$\begin{aligned} |\phi^{[2;\ 1]}(\xi)| &= |\tfrac{1}{2}\phi(\xi) - \tfrac{1}{2}\Sigma\phi(\gamma^{-1}(\gamma(\xi) \oplus e_r))2^{-r}| \\ &\leq \tfrac{1}{2}|\phi(\xi)| + \tfrac{1}{2}\Sigma|\phi(\gamma^{-1}(\gamma(\xi) \oplus e_r))|2^{-r} \\ &\leq M. \end{aligned}$$

It follows that, if ϕ is bounded on $[0, 1)$, not only ϕ, but also $\phi^{[2;\ 1]}$ is logically differentiable. By an inductive argument, we see that a bounded function on $[0, 1)$ possesses logical derivatives of all positive integer orders.

We define the Walsh-Paley functions on the unit interval $[0, 1)$ as a mapping $u\colon \mathbb{D} \times [0, 1) \to \mathbb{K}$

$\forall(\kappa, \xi) \in \mathbb{D} \times [0, 1)$,

$$u(\kappa, \xi) = \prod_{r=1}^{\infty} \exp \pi i k_{(r)} x_{(r)},$$

where $k = \gamma(\kappa)$, $x = \gamma(\xi)$.

With the foregoing definitions, we have

$\forall(\kappa, \xi) \in \mathbb{D} \times [0, 1)$,

$$\begin{aligned} &u^{[2;\ 1]}(\kappa, \xi) \\ &= \tfrac{1}{2} \prod_{s=1} \exp \pi i k_{(s)} x_{(s)} - \tfrac{1}{2} \sum_{r=1}^{\infty} 2^{-r} \prod_{s=1}^{\infty} \exp \pi i k_{(s)} (x_{(s)} \oplus \delta_{rs}) \\ &= \tfrac{1}{2} u(\kappa, \xi) \left[1 - \sum_{r=1}^{\infty} 2^{-r} \exp \pi i k_{(r)} \right] \\ &= \tfrac{1}{2} u(\kappa, \xi) \left[1 - \sum_{r=1}^{\infty} 2^{-r} (1 - 2k_{(r)}) \right] \\ &= \kappa u(\kappa, \xi). \end{aligned}$$

8

Appendix 2: Discrete Walsh functions as solutions of a dichotomous wave equation

By analogy with γ of Appendix 1, we define a mapping $\Gamma: \mathbf{Z}_{2^n} \to \mathbf{Z}_2^{\mathbf{Z}_n}$ by

$$\forall \xi \in \mathbf{Z}_{2^n}, \qquad \Gamma(\xi) = x \overset{\Delta}{=} (x_{(n-1)}, \ldots, x_{(0)}),$$

$$\text{with} \quad \forall r \in \mathbf{Z}_n, \qquad x_{(r)} = [2^{-r}\xi] - 2[2^{-r-1}\xi].$$

We note that Γ is a bijection of $\mathbf{Z}_{2^n}$ onto $\mathbf{Z}_2^{\mathbf{Z}_n}$:

$$\forall x \in \mathbf{Z}_2^{\mathbf{Z}_n}, \qquad \Gamma^{-1}(x) = \sum_{r=0}^{n-1} x_{(r)} 2^r.$$

With the operations of logical addition $\oplus$ defined by

$$\forall (x, y) \in \mathbf{Z}_2^{\mathbf{Z}_n} \times \mathbf{Z}_2^{\mathbf{Z}_n}, \qquad \forall r \in \mathbf{Z}_n,$$

$$\mathbf{Z}_2 \ni (x \oplus y)_{(r)} \equiv x_{(r)} \oplus y_{(r)} \bmod 2,$$

and multiplication by an element of $\mathbf{Z}_2$ defined by

$$\forall (\alpha, x) \in \mathbf{Z}_2 \times \mathbf{Z}_2^{\mathbf{Z}_n}, \qquad \forall r \in \mathbf{Z}_n,$$

$$(\alpha x)_{(r)} = \alpha x_{(r)},$$

the set $\mathbf{Z}_2^{\mathbf{Z}_n}$ is a vector space. A basis for this space is formed by the vectors $\{e_r : r \in \mathbf{Z}_n\}$ defined by

$$\forall r \in \mathbf{Z}_n, \qquad e_r = \Gamma(2^r);$$

$$\text{or} \quad \forall (r, s) \in \mathbf{Z}_n \times \mathbf{Z}_n, \qquad e_{r(s)} = \delta_{rs}.$$

In view of the first simple "law of dichotomy" expressed in the text, let

$$y: \mathbf{Z}_2^{\mathbf{Z}_n} \to \{+1, -1\}$$

be a (dichotomous) function to be determined. The hypothesis (second "law of dichotomy") expressed in the text may be written

$$\forall r \in \mathbf{Z}_n, \qquad \exists k_{(r)} \in \mathbf{Z}_2, \quad \forall x \in \mathbf{Z}_2^{\mathbf{Z}_n},$$

$$y(x \oplus e_r) = y(x) \exp \pi i k_{(r)}.$$

By assuming a separable solution for y in the form

$$\forall x \in \mathbf{Z}_2^{\mathbf{Z}_n}, \qquad y(x) = \prod_{r=0}^{n-1} y_r(x_{(r)}),$$

we get

$$\forall r \in \mathbf{Z}_n, \qquad \forall x_{(r)} \in \mathbf{Z}_2,$$

$$\left[\prod_{s=0}^{n-1} y_s(x_{(s)})\right] y_r(x_{(r)} \oplus 1)/y_r(x_{(r)}) = \left[\prod_{s=0}^{n-1} y_s(x_{(s)})\right] \exp \pi i k_{(r)},$$

or

$$y_r(x_{(r)} \oplus 1) = y_r(x_{(r)}) \exp \pi i k_{(r)}.$$

It is easily verified that the general solution of the last equation is (within a constant factor)

9

$$y_r(x_{(r)}) = \exp \pi i(k_{(r)} x_{(r)} \oplus m_{(r)}),$$

with $m_{(r)} \in \mathbb{Z}_2$. Hence, the general solution for y is

$$y(x) = \prod_{r=0}^{n-1} \exp \pi i(k_{(r)} x_{(r)} \oplus m_{(r)})$$

$$= M \prod_{r=0}^{n-1} \exp \pi i k_{(r)} x_{(r)},$$

with $M \in \{+1, -1\}$. By taking $M = +1$, and allowing

$$k \stackrel{\Delta}{=} (k_{(n-1)}, \ldots, k_{(0)}) \in \mathbb{Z}_2^{\mathbb{Z}_n}$$

to range over its 2^n possible values, we obtain a complete set of 2^n orthogonal, a fortiori linearly independent solutions

$$y(x) = \prod_{r=0}^{n-1} \exp \pi i k_{(r)} x_{(r)}.$$

These are the discrete Walsh-Paley functions as defined by Gibbs (1969).

The arbitrary constants $M \in \{+1, -1\}$ and $k \in \mathbb{Z}_2^{\mathbb{Z}_n}$ are determined uniquely if $y(0)$ and, $\forall s \in \mathbb{Z}_n$, $y(x_s)$ are given, provided that the $x_s \in \mathbb{Z}_2^{\mathbb{Z}_n}$ are linearly independent with respect to GF(2). For the given conditions may be written

$$M = y(0),$$

and

$$\forall s \in \mathbb{Z}_n, \qquad M \prod_{r=0}^{n-1} \exp \pi i k_{(r)} x_{s(r)} = y(x_s).$$

The first equation determines M; the remaining n equations may then be written

$$\sum_{r=0}^{n-1} k_{(r)} x_{s(r)} = \tfrac{1}{2}(1 - y(0)\, y(x_s)) = u_{(s)},$$

say, where $\sum$ denotes summation modulo 2. In the notation of matrices over the field GF(2), we have

$$XK = U,$$

where
$$X = \begin{bmatrix} x_0 \\ \cdot \\ \cdot \\ \cdot \\ x_{n-1} \end{bmatrix}, \quad K = \begin{bmatrix} k_{(0)} \\ \cdot \\ \cdot \\ \cdot \\ k_{(n-1)} \end{bmatrix}, \quad U = \begin{bmatrix} u_{(0)} \\ \cdot \\ \cdot \\ \cdot \\ u_{(n-1)} \end{bmatrix}.$$

The rows of X are linearly independent; so X^{-1} exists; so $K = X^{-1}U$.

10

The n conditions

$$\forall r \in \mathbb{Z}_n, \qquad y(x \oplus e_r) = y(x) \exp \pi i k_{(r)}$$

can be combined into a single condition by writing them

$$-\tfrac{1}{2}(y(x \oplus e_r) - y(x)) = k_{(r)} y(x),$$

multiplying by 2^r, and summing over r:

$$-\tfrac{1}{2} \sum_{r=0}^{n-1} (y(x \oplus e_r) - y(x))2^r = \kappa y(x),$$

where

$$\kappa = \gamma^{-1}(k) = \sum_{r=0}^{n-1} k_{(r)} 2^r .$$

We are thus led to define logical differentiation for complex-valued functions on the Boolean vector space $\mathbb{Z}_2^{\mathbb{Z}_n}$ as a mapping $\phi \mapsto \phi^{[n,\ 2;\ 1]}$ from $K^{\mathbb{Z}_2^{\mathbb{Z}_n}}$ into $K^{\mathbb{Z}_2^{\mathbb{Z}_n}}$ such that,

$$\forall \phi \in K^{\mathbb{Z}_2^{\mathbb{Z}_n}}, \qquad \forall x \in \mathbb{Z}_2^{\mathbb{Z}_n},$$

$$\phi^{[n,\ 2;\ 1]}(x) = -\tfrac{1}{2} \sum_{r=0}^{n-1} (\phi(x \oplus e_r) - \phi(x))2^r$$

$$= \tfrac{1}{2}(2^n - 1)\, \phi(x) - \tfrac{1}{2} \sum_{r=0}^{n-1} \phi(x \oplus e_r)2^r ;$$

for, with this definition, the dichotomous wave equation takes the form

$$y^{[n,\ 2;\ 1]} = \kappa y.$$

The number $n \in \mathbb{N}$ in the superscript of the form $[n,\ p;\ \zeta]$ indicates the dimension of the vector space implied by the definition.

The choice of n points in the vicinity of the arbitrary point x could be made in other ways: in particular, if we replace e_r in

$$y(x \oplus e_r) = y(x) \exp \pi i k_{(r)}$$

by $\Gamma(2^{n-r}-1)$, instead of $\Gamma(2^r)$, then we are led to a logical derivative $\phi \mapsto \phi^{\{n,\ 2;\ 1\}}$ of the second kind, such that

$$y^{\{n,\ 2;\ 1\}} = \lambda y$$

has as eigensolutions the discrete Walsh-Kaczmarz functions as defined by Gibbs (1970 b). These are a re-ordering of the discrete Walsh-Paley functions in ascending order of generalized frequency.

Appendix 3: Discrete Walsh-Lévy-Chrestenson functions as solutions of a polychotomous wave equation

As a generalization of the vector space $\mathbb{Z}_2^{\mathbb{Z}_n}$, we consider the $\mathbb{Z}_p$-module $\mathbb{Z}_p^{\mathbb{Z}_n}$, where p is an integer not less than 2; with logical addition $\oplus$ defined by

$$\forall (x, y) \in \mathbb{Z}_p^{\mathbb{Z}_n} \times \mathbb{Z}_p^{\mathbb{Z}_n}, \qquad \forall r \in \mathbb{Z}_n,$$

$$\mathbb{Z}_p \ni (x \oplus y)_{(r)} \equiv x_{(r)} + y_{(r)} \bmod p;$$

and multiplication by an element of $\mathbb{Z}_p$ by

$$\forall (\alpha, x) \in \mathbb{Z}_p \times \mathbb{Z}_p^{\mathbb{Z}_n}, \qquad \forall r \in \mathbb{Z}_n,$$

$$\mathbb{Z}_p \ni (\alpha x)_{(r)} \equiv \alpha x_{(r)} \bmod p.$$

As a generalization of the bijection Γ of Appendix 2, we define $\Gamma_p: \mathbb{Z}_{p^n} \to \mathbb{Z}_p^{\mathbb{Z}_n}$ by

$$\forall \xi \in \mathbb{Z}_{p^n}, \qquad \Gamma_p(\xi) = x \stackrel{\Delta}{=} (x_{(n-1)}, \ldots, x_{(0)}),$$

$$\text{with} \quad \forall r \in \mathbb{Z}_n, \qquad x_{(r)} = [p^{-r}\xi] - p[p^{-r-1}\xi];$$

we have inversely

$$\forall x \in \mathbb{Z}_p^{\mathbb{Z}_n}, \qquad \Gamma_p^{-1}(x) = \sum_{r=0}^{n-1} x_{(r)} p^r.$$

A basis for the module $\mathbb{Z}_p^{\mathbb{Z}_n}$ is formed by the elements $\{e_r : r \in \mathbb{Z}_n\}$ defined by

$$\forall r \in \mathbb{Z}_n, \qquad e_r = \Gamma_p(p^r);$$

$$\text{or} \quad \forall (r, s) \in \mathbb{Z}_n \times \mathbb{Z}_n, \qquad e_{r(s)} = \delta_{rs}.$$

Let $y: \mathbb{Z}_p^{\mathbb{Z}_n} \to \{1, \omega, \omega^2, \ldots, \omega^{p-1}\}$ be a p-chotomous function to be determined, where $\omega \stackrel{\Delta}{=} \exp 2\pi i/p$ is the principal p-th root of unity. The condition (second "law of p-chotomy") to be invoked is

$$\forall r \in \mathbb{Z}_n, \qquad \exists k_{(r)} \in \mathbb{Z}_p, \quad \forall x \in \mathbb{Z}_p^{\mathbb{Z}_n},$$

$$y(x \oplus e_r) = y(x)\, \omega^{k_{(r)}}.$$

By assuming a separable solution for y in the form

$$\forall x \in \mathbb{Z}_p^{\mathbb{Z}_n}, \qquad y(x) = \prod_{r=0}^{n-1} y_r(x_{(r)}),$$

we get

$$\forall r \in \mathbb{Z}_n, \qquad \forall x_{(r)} \in \mathbb{Z}_p,$$

$$y_r(x_{(r)} \oplus 1) = y_r(x_{(r)})\, \omega^{k_{(r)}};$$

of which the general solution is (within a constant factor)

12

$$y_r(x_{(r)}) = \omega^{k_{(r)} x_{(r)} \oplus m_{(r)}},$$

with $m_{(r)} \in \mathbf{Z}_p$. Hence,

$$y(x) = \prod_{r=0}^{n-1} \omega^{k_{(r)} x_{(r)} \oplus m_{(r)}}$$

$$= M \prod_{r=0}^{n-1} \omega^{k_{(r)} x_{(r)}},$$

with $M \in \{1, \ldots, \omega^{p-1}\}$. By taking $M = 1$, and allowing

$$k \stackrel{\Delta}{=} (k_{(n-1)}, \ldots, k_{(0)}) \in \mathbf{Z}_p^{\mathbf{Z}_n}$$

to range over its p^n possible values, we obtain a complete set of p^n orthogonal, <u>a fortiori</u> linearly independent solutions

$$y(x) = \prod_{r=0}^{n-1} \omega^{k_{(r)} x_{(r)}}.$$

These are the discrete Walsh-Lévy-Chrestenson functions.

At this point, we shall not insist upon sufficient conditions for determining the arbitrary constants M and k. If p is a prime, then $\mathbf{Z}_p^{\mathbf{Z}_n}$ is a vector space, and the discussion is analogous to that given, in Appendix 2, for the case $p = 2$.

Our immediate object is to determine an expression of the generalized (polychotomous) logical derivative $\phi \mapsto \phi^{[n,\ p;\ 1]}$ for functions defined on $\mathbf{Z}_p^{\mathbf{Z}_n}$ such that the Walsh-Lévy-Chrestenson functions are the eigensolutions of

$$y^{[n,\ p;\ 1]} = \kappa y.$$

To do this, we note, that, $\forall r \in \mathbf{Z}_n$, the condition

$$y(x \oplus e_r) = y(x)\,\omega^{k_{(r)}}$$

is a generator of p similar conditions

$$\forall s \in \mathbf{Z}_p, \qquad y(x \oplus se_r) = y(x)\,\omega^{sk_{(r)}}.$$

These may be written

$$\forall s \in \mathbf{Z}_p, \qquad y(x \oplus se_r) - y(x) = (\omega^{sk_{(r)}} - 1)\,y(x).$$

Now, $\forall s \in \mathbf{Z}_p - \{0\}$, $\quad \omega^{-s} - 1 \neq 0$; so,

$\forall r \in \mathbf{Z}_n$,

$$\sum_{s=1}^{p-1} \frac{y(x \oplus se_r) - y(x)}{\omega^{-s} - 1} = \sum_{s=1}^{p-1} \frac{\omega^{sk_{(r)}} - 1}{\omega^{-s} - 1}\, y(x).$$

Further, $\forall k_{(r)} \in \mathbb{Z}_p - \{0\}$,

$$\sum_{s=1}^{p-1} \frac{\omega^{sk_{(r)}} - 1}{\omega^{-s} - 1} = - \sum_{s=1}^{p-1} \sum_{t=1}^{k_{(r)}} (\omega^{s})^{t}$$

$$= - \sum_{t=1}^{k_{(r)}} \sum_{s=1}^{p-1} (\omega^{t})^{s}$$

$$= - \sum_{t=1}^{k_{(r)}} (-1)$$

$$= k_{(r)};$$

and this identity obviously holds also if $k_{(r)} = 0$. Thus,

$\forall r \in \mathbb{Z}_n$,

$$\sum_{s=1}^{p-1} \frac{y(x \oplus se_r) - y(x)}{\omega^{-s} - 1} = k_{(r)} y(x).$$

Multiplying each of these n relations by p^r, and summing over r, we get

$$\sum_{r=0}^{n-1} \sum_{s=1}^{p-1} \frac{y(x \oplus e_r) - y(x)}{\omega^{-s} - 1} p^r = \kappa y(x),$$

where

$$\kappa = \Gamma_p^{-1}(k) = \sum_{r=0}^{n-1} k_{(r)} p^r .$$

We are thus led to define logical differentiation for complex-valued functions on the $\mathbb{Z}_p$-module $\mathbb{Z}_p^{\mathbb{Z}_n}$ as a mapping $\phi \mapsto \phi^{[n,\ p;\ 1]}$ from $K^{\mathbb{Z}_p^{\mathbb{Z}_n}}$ into $K^{\mathbb{Z}_p^{\mathbb{Z}_n}}$ such that,

$\forall \phi \in K^{\mathbb{Z}_p^{\mathbb{Z}_n}}$, $\qquad \forall x \in \mathbb{Z}_p^{\mathbb{Z}_n}$,

$$\phi^{[n,\ p;\ 1]}(x) = \sum_{r=0}^{n-1} \sum_{s=1}^{p-1} \frac{\phi(x \oplus se_r) - \phi(x)}{\omega^{-s} - 1} p^r ;$$

for, with this definition, the p-chotomous wave equation (satisfied by the discrete Walsh-Lévy-Chrestenson functions) takes the form

$$y^{[n,\ p;\ 1]} = \kappa y.$$

The expression for $\phi^{[n,\ p;\ 1]}(x)$ is simplified by summing the terms in $\phi(x)$. Replacing s by $p-s$ in

14

$$S(p) \stackrel{\Delta}{=} \sum_{s=1}^{p-1} \frac{1}{\omega^{-s} - 1},$$

we have

$$S(p) = \sum_{s=1}^{p-1} \frac{1}{\omega^{s} - 1};$$

therefore

$$2S(p) = \sum_{s=1}^{p-1} \left(\frac{1}{\omega^{-s} - 1} + \frac{1}{\omega^{s} - 1} \right) = \sum_{s=1}^{p-1} (-1) = -(p-1).$$

Hence,

$$\phi^{[n,\ p;\ 1]}(x) = \tfrac{1}{2}(p^{n}-1)\ \phi(x) + \sum_{r=0}^{n-1} \sum_{s=1}^{p-1} \frac{\phi(x \oplus se_{r})}{\omega^{-s} - 1} p^{r}.$$

Appendix 4: The polychotomous logical derivative for functions on the unit interval

Combining the extensions of Appendices 1 and 3, we consider, $\forall p \in \mathbb{N} - \{1\}$, the $\mathbb{Z}_p$-module $\mathbb{Z}_p^{\mathbb{N}}$; with logical addition defined by

$\forall (x, y) \in \mathbb{Z}_p^{\mathbb{N}} \times \mathbb{Z}_p^{\mathbb{N}}$, $\forall r \in \mathbb{N}$,

$$\mathbb{Z}_p \ni (x \oplus y)_{(r)} \equiv x_{(r)} + y_{(r)} \mod p;$$

and multiplication by an element of $\mathbb{Z}_p$ by

$\forall (\alpha, x) \in \mathbb{Z}_p \times \mathbb{Z}_p^{\mathbb{N}}$, $\forall r \in \mathbb{N}$,

$$\mathbb{Z}_p \ni (\alpha x)_{(r)} \equiv \alpha x_{(r)} \mod p.$$

We define a mapping $\gamma_p : [0, 1) \to \mathbb{Z}_p^{\mathbb{N}}$ by

$\forall \xi \in [0, 1)$, $\gamma_p(\xi) = x \overset{\Delta}{=} (x_{(1)}, x_{(2)}, \ldots)$,

with $\forall r \in \mathbb{N}$, $x_{(r)} = [p^r \xi] - p[p^{r-1}\xi]$.

We note that γ_p is a bijection of $[0, 1)$ onto $\operatorname{rng} \gamma_p \subset \mathbb{Z}_p^{\mathbb{N}}$:

$\forall x \in \operatorname{rng} \gamma_p$,

$$\gamma_p^{-1}(x) = \sum_{r=1}^{\infty} x_{(r)} p^{-r};$$

A basis for the module $\mathbb{Z}_p^{\mathbb{N}}$ is formed by the elements $\{e_r : r \in \mathbb{N}\}$ defined by

$\forall r \in \mathbb{N}$, $e_r = \gamma_p(p^{-r})$;

or $\forall (r, s) \in \mathbb{N} \times \mathbb{N}$, $e_{r(s)} = \delta_{rs}$.

We define p-chotomous logical differentiation for complex-valued functions on the unit interval $[0, 1)$ as a mapping $\phi \mapsto \phi^{[p;\,1]}$ from $\mathbb{K}^{[0,\,1)}$ into $\mathbb{K}^{[0,\,1)}$ such that,

$\forall \phi \in \mathbb{K}^{[0,\,1)}$, $\forall \xi \in [0, 1)$,

$$\phi^{[p;\,1]}(\xi) = \sum_{r=1}^{\infty} \sum_{s=1}^{p-1} \frac{\phi(\gamma_p^{-1}(\gamma_p(\xi) \oplus se_r)) - \phi(\xi)}{p^r(\exp(-2\pi is/p) - 1)}$$

$$= \tfrac{1}{2}\phi(\xi) + \sum_{r=1}^{\infty} \sum_{s=1}^{p-1} \frac{\phi(\gamma_p^{-1}(\gamma_p(\xi) \oplus se_r))}{p^r(\exp(-2\pi is/p) - 1)}.$$

A function $\phi \in \mathbb{K}^{[0,\,1)}$ is p-chotomously logically differentiable at the point $\xi \in [0, 1)$ (possesses a first p-chotomous logical derivative $\phi^{[p;\,1]}$ at ξ) if, and only if, the series

$$\sum_{r=1}^{\infty} \frac{1}{p^r} \sum_{s=1}^{p-1} \frac{\phi(\gamma_p^{-1}(\gamma_p(\xi) \oplus se_r))}{\exp(-2\pi is/p) - 1}$$

16

converges. This series converges (uniformly and absolutely) if ϕ is bounded on $[0, 1)$; for Σp^{-r} is convergent. Thus, a necessary and sufficient condition that ϕ be logically differentiable on $[0, 1)$ is that ϕ be bounded on $[0, 1)$. Further, if M is a bound for ϕ on $[0, 1)$, we have, $\forall \xi \in [0, 1)$,

$$|\phi^{[p;\,1]}(\xi)| \leq \tfrac{1}{2}|\phi(\xi)| + \sum_{r=1}^{\infty} \frac{1}{p^r} \left| \sum_{s=1}^{p-1} \frac{\phi(\gamma_p^{-1}(\gamma_p(\xi) \oplus se_r))}{\exp(-2\pi is/p) - 1} \right|$$

$$\leq \tfrac{1}{2}M + M \sum_{r=1}^{\infty} \frac{1}{p^r} \sum_{s=1}^{p-1} \left| \frac{1}{\exp(-2\pi is/p) - 1} \right|$$

$$= \frac{M}{2(p-1)} \sum_{s=1}^{p-1} \left(1 + \left| \operatorname{cosec} \frac{\pi s}{p} \right| \right).$$

It follows that a bounded function on $[0, 1)$ possesses polychotomous logical derivatives of all positive integer orders.

Let $\mathbb{P}$ denote the set of all p-adic rationals in $[0, 1)$. For any positive integer p, we define the Walsh-Lévy-Chrestenson functions on $[0, 1)$ as a mapping $u_p : \mathbb{P} \times [0, 1) \to \mathbb{K}$ such that,

$\forall (\kappa, \xi) \in \mathbb{P} \times [0, 1)$,

$$u_p(\kappa, \xi) = \prod_{r=0}^{\infty} \exp \frac{2\pi i}{p} k_{(r)} x_{(r)},$$

where $k = \gamma_p(\kappa)$, $x = \gamma_p(\xi)$.

With the foregoing definitions, we have

$\forall (\kappa, \xi) \in \mathbb{P} \times [0, 1)$,

$u_p^{[p;\,1]}(\kappa, \xi)$

$$= \sum_{r=1}^{\infty} \sum_{s=1}^{p-1} \frac{\prod_{t=0}^{\infty} \exp(2\pi i k_{(t)} (x_{(t)} \oplus s\delta_{tr})/p) - \prod_{t=0}^{\infty} \exp(2\pi i k_{(t)} x_{(t)}/p)}{p^r (\exp(-2\pi is/p) - 1)}$$

$$= \left(\prod_{t=0}^{\infty} \exp \frac{2\pi i}{p} k_{(t)} x_{(t)} \right) \sum_{r=1}^{\infty} \sum_{s=1}^{p-1} \frac{\exp(2\pi i k_{(r)} s/p) - 1}{p^r (\exp(-2\pi is/p) - 1)}$$

$$= u_p(\kappa, \xi) \sum_{r=1}^{\infty} k_{(r)} p^{-r}$$

$$= \kappa\, u_p(\kappa, \xi).$$

Appendix 5: Takayasu Ito functions as solutions of a generalized dichotomous wave equation

Let $y: \mathbb{Z}_2^{\mathbb{Z}_n} \to \mathbb{R}$ be a (generalized-dichotomous) function to be determined in accordance with the microscopic laws of dichotomy postulated in the text for the Takayasu Ito functions. We shall use the symbols $\oplus$ and e_r in the same senses as in Appendix 2. The hypothesis expressed verbally in the text may be written

$$\forall r \in \mathbb{Z}_n, \qquad \exists b_r \in (0, \infty), \quad \forall x \in \mathbb{Z}_2^{\mathbb{Z}_n},$$

$$\text{either} \quad \left\{ \begin{array}{lll} \operatorname{sgn}(y(x \oplus e_r)) & = & \operatorname{sgn}(y(x)), \\ |y(x \oplus e_r)| & = & b_r^{(-1)^{x_{(r)}}} |y(x)|; \end{array} \right\}$$

$$\text{or} \quad \left\{ \begin{array}{lll} \operatorname{sgn}(y(x \oplus e_r)) & = & -\operatorname{sgn}(y(x)), \\ |y(x \oplus e_r)| & = & b_r^{(-1)^{x_{(r)} \oplus 1}} |y(x)|. \end{array} \right\}$$

We note that the second conditions in these two alternatives would be interchanged upon replacing b_r by b_r^{-1}; but $b_r \in (0, \infty)$ is arbitrary; so there is no loss of symmetry in pairing the conditions as shown.

The hypothesis may be expressed more concisely thus:

$$\forall r \in \mathbb{Z}_n, \qquad \exists (b_r, k_{(r)}) \in (0, \infty) \times \mathbb{Z}_2, \quad \forall x \in \mathbb{Z}_2^{\mathbb{Z}_n},$$

$$y(x \oplus e_r) = y(x)(-1)^{k_{(r)}} b_r^{(-1)^{k_{(r)} \oplus x_{(r)}}}.$$

This is our generalized dichotomous wave equation.

By assuming a separable solution for y in the form

$$\forall x \in \mathbb{Z}_2^{\mathbb{Z}_n}, \qquad y(x) = \prod_{r=0}^{n-1} y_r(x_{(r)}),$$

we get

$$\forall r \in \mathbb{Z}_n, \qquad \forall x_{(r)} \in \mathbb{Z}_2,$$

$$y_r(x_{(r)} \oplus 1) = y_r(x_{(r)})(-1)^{k_{(r)}} b_r^{(-1)^{k_{(r)} \oplus x_{(r)}}}.$$

It is easily verified that the general solution of the last equation is

$$y_r(x_{(r)}) = A(-1)^{k_{(r)} x_{(r)}} b_r^{-\frac{1}{2}(-1)^{k_{(r)} \oplus x_{(r)}}},$$

where $A \in \mathbb{R}$ is arbitrary. Hence, the general solution for y is

18

$$y(x) = B \prod_{r=0}^{n-1} (-1)^{k_{(r)} x_{(r)}} b_r^{-\frac{1}{2}(-1)^{k_{(r)}} \oplus x_{(r)}},$$

with $B \in \mathbb{R}$.

By taking $B = +1$, fixing

$$b \stackrel{\Delta}{=} (b_{(n-1)}, \ldots, b_{(0)}) \in \mathbb{R}^{\mathbb{Z}_n},$$

and allowing

$$k \stackrel{\Delta}{=} (k_{(n-1)}, \ldots, k_{(0)}) \in \mathbb{Z}_2^{\mathbb{Z}_n}$$

to range over its possible 2^n values, we obtain a complete set of 2^n orthogonal, a fortiori linearly independent solutions

$$y(x) = \prod_{r=0}^{n-1} (-1)^{k_{(r)} x_{(r)}} b_r^{-\frac{1}{2}(-1)^{k_{(r)}} \oplus x_{(r)}}.$$

Apart from a normalizing factor, these are the Takayasu Ito functions.

We append here a proof of the orthogonality not depending, as Ito's suggested proof does, upon mathematical induction. Since the Ito functions are real-valued, we omit complex conjugation.

Let h and j be distinct elements of $\mathbb{Z}_2^{\mathbb{Z}_n}$; then there is an $s \in \mathbb{Z}_n$ such that $h_{(s)} \neq j_{(s)}$. Let y_h and y_j be the corresponding solutions of the generalized dichotomous wave equation; then we have, in particular,

$$y_h(x \oplus e_s) = y_h(x)(-1)^{h_{(s)}} b_s^{(-1)^{h_{(s)}} \oplus x_{(s)}},$$

$$y_j(x \oplus e_s) = y_j(x)(-1)^{j_{(s)}} b_s^{(-1)^{j_{(s)}} \oplus x_{(s)}}.$$

Taking the product of these two equations, and recalling that $h_{(s)} \neq j_{(s)}$, we have

$$y_h(x \oplus e_s) y_j(x \oplus e_s) = -y_h(x) y_j(x) b_s^{\left((-1)^{h_{(s)}} + (-1)^{j_{(s)}}\right)(-1)^{x_{(s)}}}$$

$$= -y_h(x) y_j(x).$$

For all $c \in \mathbb{Z}_2^{\mathbb{Z}_n}$, the mapping $\mathbb{Z}_2^{\mathbb{Z}_n} \ni x \mapsto x \oplus c$ is a bijection of $\mathbb{Z}_2^{\mathbb{Z}_n}$ onto itself; so, for all functions f on $\mathbb{Z}_2^{\mathbb{Z}_n}$, $\forall c \in \mathbb{Z}_2^{\mathbb{Z}_n}$,

$$\sum_{x \in \mathbb{Z}_2^{\mathbb{Z}_n}} f(x) = \sum_{x \in \mathbb{Z}_2^{\mathbb{Z}_n}} f(x \oplus c).$$

In particular,

$$\sum_{x\in \mathbb{Z}_2^n} y_h(x)y_j(x) = \sum_{x\in \mathbb{Z}_2^n} y_h(x \oplus e_s)y_j(x \oplus e_s).$$

Combined with the previous result, this gives

$$\sum_x y_h(x)y_j(x) = -\sum_x y_h(x)y_j(x);$$

or

$$\sum_x y_h(x)y_j(x) = 0.$$

To obtain the normalizing factor, we note that,

$\forall k \in \mathbb{Z}_2^n$,

$$\sum_x (y_k(x))^2 = \sum_x \prod_{r=0}^{n-1} b_r^{(-1)^{k_{(r)} \oplus x_{(r)} \oplus 1}}$$

$$= \sum_{x_{(n-1)}=0}^{1} \cdots \sum_{x_{(0)}=0}^{1} \prod_{r=0}^{n-1} b_r^{(-1)^{x_{(r)}}}$$

$$= \prod_{r=0}^{n-1} (b_r^{-1} + b_r).$$

We therefore define the Takayasu Ito functions as the mapping

$$t_n : \mathbb{R}^{\mathbb{Z}_2^n} \times \mathbb{Z}_2^n \times \mathbb{Z}_2^n \to \mathbb{R}$$

given by

$\forall (b, k, x) \in \mathbb{R}^{\mathbb{Z}_2^n} \times \mathbb{Z}_2^n \times \mathbb{Z}_2^n$,

$$t_n(b, k, x) = \prod_{r=0}^{n-1} \left(\frac{b_r^{-1} + b_r}{2} \right)^{-\frac{1}{2}} (-1)^{k_{(r)} x_{(r)}} b_r^{-\frac{1}{2}(-1)^{k_{(r)} \oplus x_{(r)}}}.$$

The orthonormality property of these functions may then be written

$\forall b \in \mathbb{R}^{\mathbb{Z}_2^n}$, $\forall (k, h) \in \mathbb{Z}_2^n \times \mathbb{Z}_2^n$,

$$\sum_{x\in \mathbb{Z}_2^n} t_n(b, k, x)\, t_n(b, h, x) = 2^n \delta_{kh}.$$

Epitome

We have sought to throw some light on the significance of the logical differential calculus by extending the concept of logical derivative in various directions. The following results appear to be new:

The classical (dichotomous) logical derivative is defined for functions on the unit interval (Appendix 1). In contrast with the ordinary derivative, the logical derivative of any positive integer order exists, provided only that the function be bounded in the interval. The Walsh-Paley functions on $[0, 1)$ satisfy the logical differential equation

$$\phi^{[2;\ 1]} = \kappa\phi.$$

The logical differential equation

$$\phi^{[n,\ 2;\ 1]} = \kappa\phi$$

satisfied by the discrete Walsh-Paley functions is looked upon as a dichotomous wave equation in a Boolean vector space (Appendix 2).

This view proves fruitful (Appendix 3): a certain polychotomous wave equation generates in a similar way the discrete Walsh-Lévy-Chrestenson functions; at the same time, we obtain the definition of a polychotomous logical derivative by writing the polychotomous wave equation in the form

$$\phi^{[n,\ p;\ 1]} = \kappa\phi.$$

The polychotomous logical derivative is defined for functions on the unit interval (Appendix 4). The Walsh-Lévy-Chrestenson functions on $[0, 1)$ satisfy the polychotomous logical differential equation

$$\phi^{[p;\ 1]} = \kappa\phi.$$

A distinct kind of generalization of the dichotomous wave equation enables us to generate as solutions the Takayasu Ito functions (Appendix 5).

References

BARRETT, R.A., and LINES, P.D., edit., 1971, Symposium on Theory and Applications of Walsh Functions, Hatfield, England (Hatfield: The Hatfield Polytechnic)

BASS, C.A., edit., 1970, Proceedings of the Symposium and Workshop on Applications of Walsh Functions, Washington, D.C. (AD 707 431) (Springfield, Va 22151: National Technical Information Service)

CHRESTENSON, H.E., 1955, Pacific J. Math., 5, 17-31. A class of generalized Walsh functions

FINE, N.J., 1949, Trans. Amer. Math. Soc., 65, 372-414. On the Walsh functions

FINE, N.J., 1950, Trans. Amer. Math. Soc., 69, 66-77. The generalized Walsh functions

GIBBS, J.E., 1967, National Physical Laboratory, Teddington, Middlesex, England: Unpublished report. Walsh spectrometry, a form of spectral analysis well suited to binary digital computation

GIBBS, J.E., 1969, NPL: DES Rept No. 3. Some properties of functions on the non-negative integers less than 2^n.

GIBBS, J.E., 1970 a, Proc. Sympos. Applic. Walsh Functions, Washington, D.C., 106-122. Discrete complex Walsh functions

GIBBS, J.E., 1970 b, Proc. Sympos. Applic. Walsh Functions, Washington, D.C., 260-274. Sine waves and Walsh waves in physics

GIBBS, J.E., 1970 c, NPL: DES Rept No. 4. Functions that are solutions of a logical differential equation

GIBBS, J.E., 1970 d, NPL: DES Rept No. 5. Digital filtering in dyadic-time and sequency

GIBBS, J.E., 1971 May 21, NPL News, 1-4. A contribution to a revolution?

GIBBS, J.E., and MILLARD, Margaret J., 1969 a, NPL: DES Rept No. 1. Walsh functions as solutions of a logical differential equation

GIBBS, J.E., and MILLARD, M.J., 1969 b, NPL: DES Rept No. 2. Some methods of solution of linear ordinary logical differential equations

HARMUTH, H.F., 1970, "Transmission of Information by Orthogonal Functions" Second printing corrected (Berlin: Springer)

ITO, Takayasu, 1968, Information and Control, 12, 206-211. A note on a general expansion of functions of binary variables

KACZMARZ, S., 1929, C.R. 1er Cong. Math. Pays Slaves (Varsovie), 189-192. Ueber ein Orthogonalsystem

LÉVY, P., 1944, Comment. Math. Helv., 16, 146-152. Sur une généralisation des fonctions orthogonales de M. Rademacher

PALEY, R.E.A.C., 1932, Proc. Lond. Math. Soc., 34, 241-279. A remarkable series of orthogonal functions

22

PICHLER, F.R., 1967, Diss. Univ. Innsbruck. Das System der sal- und cal-Funktionen und die Theorie der sal- und cal-Fouriertransformation

POLYAK, B.T., and SCHREIDER, Yu. A., 1962, Voprosy Teor. Matem. Mashin, Coll. II, 174-190. The application of Walsh functions in approximate calculations (translation by J.E. Gibbs)

SELFRIDGE, R.G., 1955, Pacific J. Math., 5, 451-480. Generalized Walsh transform

SUNOUCHI, G., 1951, Proc. Amer. Math. Soc., 2, 5-11. On the Walsh-Kaczmarz series

WALSH, J.L., 1923, Amer. J. Math., 45, 5-24. A closed set of normal orthogonal functions

ZEEK, R.W., and SHOWALTER, A.E., edit., 1971, Proceedings of the Symposium on Applications of Walsh Functions, Washington, D.C. (AD 727 000) (Springfield, Va 22151: National Technical Information Service)

<u>Distribution</u>

Authors	(250)
Deputy Director (B)	
Mr. A.E. Bailey	(2)
Mr. C.H. Dix	
Mr. J. McA. Steele	
Dr. N.W.B. Stone	
Mr. E.C. Pyatt	

Chapter 4
Early Contributions from the Aachen School to Dyadic Walsh Analysis with Applications to Dyadic PDEs and Approximation Theory

Paul L. Butzer, Heinrich Josef Wagner

This chapter presents an overview of the contributions from the Aachen School, research achieved by six senior graduate students and post-docs, beginning in 1970. First, major actors dealing with Walsh polynomials and Walsh-Fourier series world-wide are listed, together with various meetings and workshops concerned with these and with early dyadic analysis; Aachen members participated in many of them. Contrary to the original Gibbs "logical derivative", the Butzer-Wagner form—coined Gibbs- Butzer derivative by Wei-yi Su—has properties parallel to those of the classical derivative. A first application is to PDEs in the dyadic sense, namely the dyadic wave equation, in which second order dyadic derivatives occur, one with respect to the time. The second one is to best approximation of $L^p(0,1)$—functions by Walsh polynomials. Sect. 4.5 concerns dyadic Haar analysis as well as dyadic derivatives of fractional order. They demonstrate the applicability of dyadic analysis.

The authors among them, those concerned more with theory than with applications, include R.E.A.C. Paley (1932) [A-27], P. Lèvy (1944) [A-24], Vilenkin (1947,1952) [A-42], [A-43], [A-44], Fine (1949, 1950) [A-10], [A-11], S. Yano (1951) [A-58], [A-59], H.E. Chrestenson (1955) [A-8], R.G. Selfridge (1955) [A-34], Ch. Watari (1956-70) [A-51], [A-52], [A-53], [A-54], [A-55], Morgenthaler (1957) [A-25], A.A. Talaljan and F.G. Arutunjan (1964) [A-1], V.M. Kokilasvili (1965) [A-23].

It seems that the first institute to conduct a meeting concerned with Walsh functions was the *Research Institute of the Telecommunication Engineering Centre* of the Deutsche Bundespost in Darmstadt. It took place on May 20, 1968, and was conducted by Dr.H. Huebner.

The next was another one-day meeting at the *Communications Science Division* of the Naval Research Laboratory, Washington, D.C., on April 1, 1969; both had no proceedings. These two were followed up by the second meeting at the NRL, dated March 1 - April 3, 1970, discussed below, a meeting at the Departmental Auditorium, Washington, DC., on April 13 - 15, 1971, a meeting at the Hatfield

Paul L. Butzer
Lehrstuhl A für Mathematik, RWTH Aachen, Germany, e-mail: butzer@rwth-aachen.de

Heinrich Josef Wagner
Krauthausener Str. 6A, 52223 Stolberg-Dorff, Germany

R.S. Stanković et al. (eds.), *Dyadic Walsh Analysis from 1924 Onwards Walsh-Gibbs-Butzer Dyadic Differentiation in Science Volume 1 Foundations*, Atlantis Studies in Mathematics for Engineering and Science 12, DOI 10.2991/978-94-6239-160-4_4

Polytechnic, Herfordshire, UK., on June 27 - 30, 1973; at the Catholic University of America, Washington, D.C., on March 27 - 29, 1972; Regional Engineering College, Warangal, Andhra, India, on October 9 - 13, 1972 (no proceedings); National Electronic Conference, Chicago, 1972, a second symposium at Hatfield on June 28 - 29, 1973; a Special Session in Sequency Technics (in Cooperation with the first Symposium & Technical Exhibition, EMC, at Montreux on May 20 - 22, 1975. After these pioneering meetings, the study of Walsh and other non-sinusoidal functions and their applications in different areas become a standard subject of many conferences and journals. For example, the journal IEEE Transaction on Electromagnetic Compatibility, had a special session and an associate editor (Henning F. Harmuth) devoted just to Walsh functions and their applications, for several years. Few special issues of this journal devoted to Walsh functions were published.

At the Naval Research Laboratory (NRL), Washington, D.C.; there took place during March 31 - April 3, 1970, the Symposium & Workshop "Applications of Walsh Functions", the Steering Committee consisting of L.B. Wetzel and C.A. Bass, both of the NRL, as well as J.L. Walsh and H.F. Harmuth. The proceedings, of 274 pp., were edited by C.A. Bass.

Particularly in view of the fact that the Workshop could possibly bring me into contact with potential applications of our work in approximation theory, harmonic analysis and integral transform theory, I decided to accept this invitation by Prof. Harmuth and travel to the USA already during my two-months spring break. On top, it was always my [1] desire to meet Joseph L. Walsh, a student of Maxime Bôcher (and G. Birkhof), in turn a student of Felix Klein, who was a world expert in complex approximation theory.

Moreover, my first solid teacher in mathematics, Eric O' Connor S.J. (1907 - 1980), at Loyola College, Montreal, who had completed his Ph.D. in Mathematics under Walsh in the late thirties, always spoke highly of him. He was the star mathematician of the workshop, among the ca. 35 participants in the engineering and mathematical areas.

Staying at the time in the home of my brother Karl in Flossmore, IL, USA, always the home base during my yearly 4-6 week lecture tours, covering a number of universities in the USA and Canada at a time, usually during the periods September-October, the huge spring snow fall of that year forced Chicago Airport to close up completely on April 1, and I could only leave early Friday, and arriving at the main gate of the Naval Laboratory just before noon, presented my Canadian passport, was taken by car to the building where the workshop took place, accompanied by a "driver" not only during lunch but also waiting outside the washroom, making sure I was in good hands, and there met the participants who were getting ready for their final lectures. Having most unfortunately been able to spend just a few hours due to the weather circumstances at the Workshop I asked the "driver" whether he could take me to a local mathematician with whom I had been corresponding for some time. It now being 4 p.m. on Friday, I convinced him to phone him up, we met in

[1] Paul L. Butzer

front of a building with a large hall which I was not allowed to enter although it was icy cold, with snow at least a meter high, piled up on the road sides.

Nevertheless, the "driver" allowed us to talk for some 15 minutes.

These details, were brought up to explain why Prof. Walsh invited some ten of the participants to come to his home that Saturday morning to talk with them, without the presence of guarding "drivers". What an exceptional day for me! Now one could talk freely, one topic being a possible differentiation-type concept for Walsh functions. The British electrical engineer Dr. James Edmund Gibbs from the National Physical Laboratory at Teddington, Middlesex, UK, in his very first manuscript of January 13, 1967 had expressed important first ideas leading to the concept of his "logical derivative". See [A-49] for the reprint of the corresponding page of personal notices of J. Edmund Gibbs published by the courtesy of Ms. Merion Gibbs.

Involved in the discussion also included the Austrian electrical engineer Dr. Henning F. Harmuth, the author of some 150 journal publications and eight books, stationed at the Catholic University of America, as well as the Austrian System Theorist Dr. Franz Pichler who was spending a year at the University of Maryland at the time. What an exceptional day for me. Prof. Walsh, already 73 at the time truly impressed me with his broad outlook upon mathematics but also his stately spouse for being able to prepare a tasty lunch for the participants. This day also initiated a fundamental research direction at the Lehrstuhl A für Mathematik, Rheinisch-Westfaelische Technische Hochschule (RWTH) Aachen, namely to one aiming at possible application areas, of course not neglecting its usual fundamental theoretical research. As soon as I returned home I suggested to my student Heinrich Josef Wagner to tackle the problem I brought back with me.

4.1 Butzer-Wagner Dyadic Derivative

Beginning with the complete orthonormal Walsh-Paley system $w_0(x), w_1(x), \ldots$, defined on the unit interval $[0,1)$, which enjoys the relation $w_k(x \oplus y) = w_k(x) \cdot w_k(y)$, $x \oplus y$ being dyadic addition, consider the Walsh-Fourier series

$$f(x) = \sum_{k=0}^{\infty} f^{\wedge}(k) w_k(x), \qquad f^{\wedge}(k) := \int_0^1 f(u) w_k(u) du.$$

Together with Dr. H.J. Wagner we defined the dyadic differential operator $D^{[r]}$ of order $r \in N$, for $f \in L^p(0,1)$, $1 \le p < \infty$, by

$$D^{[r]} f(x) = \sum_{k=0}^{\infty} k^r f^{\wedge}(k) w_k(x)$$

for which $D^{[r]} w_k(x) = k^r w_k(x)$, $k \in N_0$. So the $w_k(x)$ are eigenfunctions of the operator $D^{[r]}$, each $r \in N_0$.

As to the definition in the original function space, for the function f defined on $[0,1)$, then the "logical", "dyadic" or also Gibbs-Butzer derivative, for $r=1$ is defined for $x \in [0,1)$ by

$$D^{[1]}f(x) = \frac{1}{4}\sum_{j=1}^{\infty} 2^j (f(x) - f(x \oplus 2^{-j})), \tag{4.1}$$

where $x = \sum_{j=1}^{\infty} x_j 2^{-j}$, $y = \sum_{j=1}^{\infty} y_j 2^{-j}$, $x_j, y_j \in [0,1)$, and $x \oplus y = \sum_{j=1}^{\infty} |x_j - y_j| 2^{-j}$.

4.2 Further Investigations

The basic paper of Dr. Gibbs with respect to Walsh functions and differentiation is his in collaboration with Dr. Bryan Ireland, University of Bath, namely [A-13], and another treatise of the same authors in 1974 [A-14], in which they took over the precise Butzer-Wagner form of the pointwise derivative for the infinite dyadic group. In his letter to me of 1972, Dr. Gibbs writes: "I am intrigued by the presence of the factor 2^j in place of the factor $2^{(-j)}$ in my definition. I have not yet got to the bottom of this discrepancy, but I associate it with the fact that you set out to define a derivative with properties parallel to those of the classical derivative - and I think you have succeeded".

This is the basic difference between the original Gibbs "logical" derivative and that of Butzer and Wagner.

Unfortunately the authors were not aware at the time of the reports in 1970 by Dr. Pichler [A-28], see also [A-29] in which he modified the Gibbs - Millard definition - their main interest was that of finite discrete structures [A-15] - to a theory of dyadic differentiation defined for real-valued functions of a continuous nonnegative real variable.

Following an invitation by Prof. J.B. Rosser, Director of the Mathematical Research Center, University of Wisconsin; Dr. Wagner presented a colloquium lecture there, namely "A new calculus for Wash functions with applications" on April 13, 1973, the 45-minute following discussion being conducted by me. That same night my dear friend Iso Schoenberg invited us two to a large dinner party in his home, present being a great number of the Madison mathematicians with their wives, including Walter and Mary Rudin, Dick Askey, Carl de Boor, T.J. Higgins.

Laurence C. Young suggested an interesting interpretation of our dyadic derivative in the realm of Wash analysis.

The main reason for this USA-trip was however the "Symposium on Applications of Wash Functions", held at the Catholic University of America that April 15-18, 1973 which was conducted by Prof. Henning F. Harmuth, who had invited us two, and who presented an interesting overview lecture of the various applications of Walsh functions, especially in sequency theory. The two of us gave a joint lecture on the same topic as at Madison; the many participants included. G. Redinbo and R.B.

Lackey of Electrical Engineering at Madison, and Ohio State University, Columbus, respectively, J.S. Lee as well as C. Boeswetter, Batelle Institute, Frankfurt, and H. Huebner, Research Institute, German Federal Mail Office. Harmuth had suggested an interpretation of dyadic differentiation in the realm of atomic physics.

Another conference at which Walsh functions were a part of the program, an extended session chaired by Robert Redinbo, was the "National Electronics Conference" held at Chicago on October 9-11, 1972. Although no one from Aachen had time to participate, our paper with Dr. Wagner, "On a Gibbs-type derivative in Walsh-Fourier analysis with applications", was nevertheless published in its proceedings (pp. 393-398).

At the Hatfield Polytechnic, Herfordshire, there took place the second colloquium in the UK in the area (the first was in 1971), namely "Theory and Applications of Walsh and other Nonsinusoidal Functions" on June 27-30, 1973, with some 30 lectures, presented by almost the whole group of those active in the field at the time. Dr. Wagner took part and lectured on "A new calculus for Walsh functions, with applications". Prof. P.D. Lines, who conducted the Symposium of 1973 also invited us to participate in his symposium of July 1-3, 1975 which combined two separate symposia which were to be held in the UK and the USA. Dr. Wagner was accompanied by Dipl. Math. Robert Weis, a further research assistant at the Lehrstuhl A für Mathematik, RWTH Aachen, (with whom P.L. Butzer later wrote a paper on the fundamental Richtmeyr-Lax equivalence theorem in numerical analysis). Although our RWTH could not support the trips due to lack of funds, our paper, "On dyadic calculus for functions defined on R_+" , was fortunately again included in its proceedings.

However, a little earlier, May 20-23, 1975, the RWTH did support the trip of Dr. Wagner as well as that of Dipl. Math. W. Splettstösser to the EMC Symposium at Montreux, Switzerland, where they talked on "Ein Infinitesimalkalkuel fuer Haarfunktionen". They were the first to define the concepts of a "HAAR" derivative and integral, and to apply them to approximation theory.

4.3 A Dyadic PDE, the Wave Equation

In this section, we present an application of the dyadic derivative in solving the wave equation in the form envisaged by H. F. Harmuth [A-19].

We may set up in dyadic sense a partial differential equation analogous to the equation

$$\frac{\delta^2}{\delta x^2} w(x,t) = \frac{1}{c^2} \frac{\delta^2}{\delta t^2} w(x,t) \tag{4.2}$$

the general solution of which, due to J.L. d'Alambert, is given by

$$w(x,t) = f(x+ct) - g(x-ct),$$

where f and g are arbitrary functions determined by the initial and boundary conditions.

Until the advent of J.E. Gibbs, an equation of this type in dyadic space and time, could not be even thought of since the classical derivative of any Walsh function is zero almost everywhere, they being step-functions. For this purpose it was desirable to find a new derivative concept with respect to which the Walsh functions themselves, at least, are non-trivially differentiable. Moreover, this new concept should play the same fundamental role as does the classical derivative in the mathematical sciences.

In this section, definition (4.1) is generalized to non-periodic functions by extending definition (4.1) to the case of functions f defined on R_+ having period k, i.e., $f(x) - f(x+k) = 0$ for $k \in P$ and all $x \in R_+$ [B-5-R], [A-47]. In that case,

$$\frac{1}{4}\sum_{j=1}^{\infty} 2^j(f(x) - f(x \oplus 2^{-j})) = \frac{1}{4}\sum_{j=-\infty}^{\infty} 2^j(f(x) - f(x \oplus 2^{-j})), \tag{4.3}$$

where now j runs from $-\infty$ to ∞.

We first fix the notation that will be used.

$L_1(R_+)$ is the space of absolutely integrable functions on the positive part of the real line R_+ with the norm $\|f\| = \int_0^\infty |f(u)|du$. For functions $w(x,t)$, $x,t \in R_+$, we set

$$\|w\|^* = \int_0^\infty \int_0^\infty |w(x,t)|dxdt,$$

and denote the set of all functions $w(x,t)$ for which $\|w\|^* < \infty$ by $L_1^* \equiv L_1(R_+ \times R_+)$.

If for $f(x,t) \in L_1^*$ there exists $g(x,t) \in L_1^*$ such that

$$\lim_{m_1,m_2\to\infty} \|\frac{1}{4}\sum_{j=-m_1}^{m_2} (f(\cdot,t) - f(\cdot \oplus 2^{-j},t))2^j - g(\cdot,t)\| = 0,$$

then $g(x,t)$ is called the first strong dyadic partial derivative of $f(x,t)$ with respect to x, denoted by $D_x^{\{1\}} f$. The second such derivative is defined by $D_{xx}^{\{2\}} f = D_x^{\{1\}}(D_x^{\{1\}} f)$. Correspondingly one defines $D_t^{\{1\}} f$.

We now set up the partial differential equation in terms of the Butzer-Wagner derivative in the space L_1^* as

$$c^2 D_{xx}^{\{2\}} w - D_{tt}^{\{2\}} w = 0, \tag{4.4}$$

with initial condition

$$\lim_{t\to 0_+} \|w(\cdot,t) - f(\cdot)\| = 0, \quad f \in L_1(R_+). \tag{4.5}$$

In order to find a solution in L_1^*, we proceed formally as follows: Apply the Walsh-Fourier transform with respect to x, $w^\wedge(v,t) = \int_0^\infty w(x,t)\psi(v,t)dx$, to equation (4.4). Since, for $f, D^{\{r\}} f \in L_1(R_+)$, it holds

$$[D^{\{r\}} f]^\wedge(v) = v^r f^\wedge(v), \quad v \in R_+, \tag{4.6}$$

where $v = \sum_{j=-N}^{\infty} v_j 2^{-j}$, $v_j \in [0,1)$, $N \in P$, it holds analogusly to this result that

$$(cv)^2 w^\wedge(v,t) = [D_{tt}^{\{2\}} w(\cdot,t)]^\wedge(v). \tag{4.7}$$

Now, would

$$[D_{tt}^{\{2\}} w(\cdot,t)]^\wedge(v) = [w^\wedge(v,t)]_{tt}^{\{2\}}, \tag{4.8}$$

then the transformed equation would have the form

$$(cv)^2 w^\wedge(v,t) = [w^\wedge(v,t)]_{tt}^{\{2\}}. \tag{4.9}$$

For each $v \in R_+$, this is an "ordinary" dyadic differential equation of second order with respect to t. Since $\psi^{\{r\}}(v,x) = v^r \psi(v,x)$ for $v,x \in R_+$, where the Walsh-Paley functions are given by $\psi(v,x) = \exp(\pi i \sum_{j=-N}^{M+1} v_j x_{1-j})$, $N, M \in P$, a solution of (4.9) is given by

$$w^\wedge(v,t) = A(v)\psi(cv,t), \tag{4.10}$$

$A(v)$ being independent of t. The initial condition (4.5) yields $A(v) = f^\wedge(v)$.

Since for all c

$$\psi(cv,t) = \psi(v,ct), \quad (t, v \in R_+) \tag{4.11}$$

one therefore has

$$w^\wedge(v,t) = f^\wedge(v)\psi(v,ct) \tag{4.12}$$

as a solution of (4.9). In order to represent this solution in terms of the original functions, i.e., to find a solution of the original equation (4.4), we must apply a suitable inversion theorem. Indeed, the uniqueness theorem says if $f^\wedge(v) = 0$ a.e., then $f = 0$ a.e. Now if $f \in L_1(R_+)$, whence since $[f(\cdot + ct)]^\wedge(v) = f^\wedge(v)\psi$ (v, ct) for all $t, v \in R_+$, one has $[f(., +ct)]^\wedge(v) = w^\wedge(v, t)$. This yields analogously that

$$w(x,t) = f(x \oplus ct)$$

is a solution of (4.4).

There are two remarks. Whether $f(x \oplus ct)$ is the most general form of the solution of (4.4) depend upon whether (4.10) is the general solution of (4.4). An answer is possible since the inverse operation to dyadic derivative is available. Secondly, equation (4.11) holds for all $c = 2^n$, $n \in Z$, since if $n \in Z = \{0, \pm 1, \pm 2, \ldots\}$, then $\psi(2^n v, x) = \psi(v, 2^n x)$. For other c values the solution $w(x,t)$ is for almost all $t \in R_+$ equal to the limit for $n \to \infty$ in the $L_1(R_+)$-norm (taken with respect to x) of

$$\int_0^{2^n} (f^\wedge(v)\psi(cv,t))\psi(v,x)dv.$$

This follows since $w \in L_1^*$ implies that $w(x,t)$ belongs to $L_1^*(R_+)$ with respect to x [B-5-R], [B-6-R]. The dyadic versions of the classical versions of the heat conduction equation with initial condition f, or of the equation of motion of a vibrating string with displacement f and velocity g, could probably be considered just as well. As a model one could take the corresponding classical PDEs, treated explicitly in P.L. Butzer and R.J. Nessel, *Fourier Analysis and Approximation*, Vol. 1, Birkhäuser Verlag and Academic Press, 1971, pp. 278–304.

4.4 Applications to Approximation by Walsh Polynomials and Walsh-Fourier Series

Each $k \in P = \{0,1,2,\ldots\}$ has a unique dyadic expansion

$$k = \sum_{j=0}^{K} k_j 2^j, \quad 2^K \le k \le 2^{K+1}, \quad K \in P,$$

with $k_j \in [0,1)$, $j \in P$. If $k = 0$, all $k_j = 0$.

Likewise, each $x \in [0,1)$ has a unique expansion

$$x = \sum_{j=1}^{\infty} x_j 2^{-j}, \quad x_j \in [0,1), \quad j \in N = \{1,2,\ldots\},$$

the finite expansion being taken, holding for all x outside a certain denumerable set.

This enables one to define the Walsh-Paley functions $\{\psi_k(x)\}_{k=0}^{\infty}$ on $[0,1)$ by

$$\psi_k(x) = \exp\{\pi i \sum_{j=0}^{k} k_j x_{j+1}\},$$

and on R by periodic expansion.

These functions form an orthonormal system. which has the property $\psi_k(x \oplus y) = \psi_k(x)\psi_k(y)$ for all $x \in R$ and a.e. $y \in R$.

A finite linear combination

$$p_n(x) = \sum_{k=0}^{n-1} c_k \psi_k(x),$$

c_k being real numbers, is called a Walsh polynomial of degree smaller or equal to n, the set of all such combinations being denoted by $\mathscr{P}_n$.

After these preliminary results, we consider the approximation of dyadically differentiable functions by Walsh polynomials.

In the notation of G.W. Morgenthaler [A-25], define for $f \in L^p(0,1)$, $1 \le p < \infty$, $\alpha > 0$, the W-modulus of continuity by

$$\omega_w(f,\delta) = \sup_{0 \le h < \delta} \|f(\cdot) - f(\cdot \oplus h)\|, \quad \delta > 0,$$

the associated Lipschitz class $Lip_w(\alpha)$ by

$$Lip_w(\alpha) := \{f; \|f(\cdot) - f(\cdot \oplus h)\| = O(h^\alpha)\} \quad (h \to 0),$$

and the best approximation of $f \in L^p(0,1)$ by $p_n \in \mathscr{P}_n$ by

$$E_n(f,L^p) = \inf_{p_n \in \mathscr{P}_n} \|f - p_n\|.$$

For each $f \in L^p(0,1)$, there exists a $p_n^* = p_n^*(f,L^p) \in \mathscr{P}_n$ such that

$$E_n(f,L^p) = \|f - p_n^*\|.$$

Of fundamental importance in classical approximation theory are the Berstein- and Jackson-type inequalities (see, e.g. [A-6]). In the setting of Walsh approximation theory, they read

(a) For $p_n \in \mathscr{P}_n$ one has

$$\|D^{[r]} p_n\| \le A n^r \|p_n\|, \quad n \in P, \quad r \in N.$$

(b) If $D^{[r]} f$ exists and belongs to $L^p(0,1)$, then

$$E_n(f,L^p) \le B n^{-r} \|D^{[r]} f\|.$$

Here A and B are constants independent of n, p_n, and f.

Then, the fundamental theorem of best approximation by Walsh polynomials reads

Theorem 4.1 *The following five assertions are equivalent for $D^{[r]} f \in L^p(0,1)$:*

(i) $D^{[r]} f \in Lip_w(\alpha)$, $\alpha > 0$,
(ii) $\omega_w(D^{[r]} f, \frac{1}{n}) = O(n^{-\alpha})$,
(iii) $E_n(f,L^p) = O(n^{-\alpha-r})$,
(iv) $D^{[s]} f \in L^p(0,1)$, $0 \le s \le r$, *and* $\|D^{[s]} f - D^{[s]} p_n^*\| = O(n^{-\alpha-r+s})$,
(v) $\|D^{[1]} p_n^*\| = O(n^{-\alpha-r+1})$, $\quad (0 < \alpha + r < 1)$.

It was C. Watari [A-53] who established the first three equivalencies in the particular case of $r = 1$. In classical approximation this theorem is associated with the names of S.N. Bernstein, D. Jackson, M. Zamansky, S.B. Stečkin, G.I. Sunouchi , P.L. Butzer, and K. Scherer (see [A-6].

C. Watari was a student of the renowned Gen-Ichirô Sunouchi (see [A-5]) who participated in our first Oberwolfach Conference "On Approximation Theory" in 1963.

But if one considers the basic problem of the rate of approximation of $f \in L^p(0,1)$ by $p_n \in \mathscr{P}_n$, one cannot get along without the existence of derivatives, in the present case the dyadic derivative. This enables us to handle (i), (ii), and (iii) for all $r > 0$ plus the important additional assertions (iv) and (v), the first being on simultaneous approximation of the derivatives of f by those of p_n^* understood of course in the dyadic sense.

Similar results apply to the rate of approximation by partial sums of Walsh-Fourier series.

In many problems of communication engineering, systems theory, and digital signal processing, various types of errors occur in the procedures involved. In order to achieve mathematical precision in these procedures, the concept of a derivative is necessary. This is precisely our dyadic derivative.

4.5 Later Contributions from Aachen

Having achieved respectable results in dyadic Walsh analysis, the next open problem was to treat a differential and integral calculus in the Haar setting. Recall that Alfred Haar introduced his functions already in 1910 [A-18]. H.J. Wagner introduced Wolfgang Splettstoesser [2] into this field (see [B-8-R], [B-21], [B-24]).

Splettstoesser, whose main research fields were sampling analysis in signal processing and linear prediction theory, treated the well-known Shannon sampling theorem, functions sequency limited to $[0,2^n]$, in the setting of dyadic Walsh analysis in [B-10], as well as for duration-limited functions, both results together with sharp error-estimates; see also [B-11], [B-13]. For Walsh differentiable functions in dyadic stationary processes, see [B-12], [B-14].

A calculus for classical fractional order derivatives considered already by Leibniz in the seventeenth century, had a revival from ca. 1970 onwards. There are now even three journals devoted to this growing subject. Our work with Ursula Westphal set in 1974; see [A-7]. Engels introduced U. Wipperfürth into the field; see [B-19], [B-20], [B-25], [B-26].

Our final paper with Engels [B-22] was in the conference in Kupari-Dubrovnik, former Yugoslavia, in 1989, (see [B-28]). There was extended the standard dyadic derivative which is not restricted to piecewise constant functions, but also covers piecewise polynomial functions, several classically smooth functions, even such unsmooth functions as $x^n d(x)$, ($d(x)$ being Dirichlet's function) which is a "polynomial" of order n possessing infinitely many discontinuities which even lie dense in $[0,1)$.

[2] For a paper dedicated to W. Splettstoesser, together with a short biography, see Butzer et al. [A-4].

The extension results by first equipping the sum defining the standard dyadic derivative with the multiplicative factor ± 1 and then applying Euler's summation process to it.

One co-author of this book suggested that this paper be also reprinted in our joint monograph, but we feel that adding 36 further pages would blow up the list of reprinted papers.

4.6 Decimal and Dyadic Systems

The following anecdote recorded by Prof. Butzer illustrates well the relationships between the decimal and dyadic systems, and also the simplicity of the dyadic system making it easily acceptable even by young generations that in their education are not yet suppressed by habits of dealing with classical numerical systems and, therefore, are more open for accepting alternatives.

At a Christmas party in 2014, I (Prof. Butzer) met Mattias Hempel, a ten year old youngster attending grade 5 at the "Stiftisches Gymnasium" in Düren (the city where Lejeune Dirichlet was born). He asked me for my age and whether I would like to see my age 86, written in the dual system. I was unbelievably surprised and said *Go ahead.* In less than one minute he completed his evaluation written with his pencil, using the following procedure

1	0	1	0	1	1	0
2^6	2^5	2^4	2^3	2^2	2^1	2^0
64	32	16	8	4	2	1

$$86 = 2^6 + 2^4 + 2^2 + 2^1$$

$$86_{decimal} = 1010110_{binary}$$

I ought to add that Mattias, whose mathematics teacher is Dr.J. Schulte, is a son of Gero and Susanne Hempel, Susanne being a daughter of the late Prof. Dr. Ing. Hans -Dieter Lüke (1935 - 2005) and his spouse Bernhardine. Dieter was Professor of Communications Engineering and one of my best friends at RWTH-AACHEN.

4.7 Further Suggestions for Studying Our Monograph

When studying the present volumes, it would not be a bad idea to follow the order of authors and their reprinted papers in the order found in the Table of Contents.

The papers [B-3-R] and [B-6-R] are the first papers of Butzer and Wagner in dyadic analysis which appeared in international journals; the proofs of the results presented there are complete; the papers in the various engineering proceedings usually contain no proofs. The reprinted article by Schipp [A-31] is the first and very fitting continuation of [B-3-R] and [B-1-R].

At Aachen, we followed up the work of Joseph Walsh [A-48], (the earlier work of Alfred Haar [A-18] somewhat later), that of Henning Harmuth [A-19], [A-20], [A-21], that of Franz Pichler, e.g. [A-28], [A-29], and of Edmund Gibbs [A-13], [A-14], [A-15]. It was a fortune for me that I had met these colleagues already in 1970 at the Naval research Laboratory (NRL) at Washington, D.C.

Then, Ferenc Schipp and his great school in Budapest, see e.g. [A-31], [A-32], [A-33] followed up essentially the work of Heinrich Josef Wagner and myself starting from [B-1-R] to [B-8-R]. Then, came almost simultaneously Kees Onneweer from New Mexico (e.g. [A-26]), Bill Wade from Tennessee, who came to Aachen after spending a year with his family in Moscow (see e.g. [A-38], [A-46]) and the Patriach Zheng Wei-Xing from Nanjing who come to Aachen in the seventies, followed by his student Su Weiyi, who visited Aachen in little later, with their papers [A-39], [A-40], (see especially the papers [A-12], [A-56], [A-57] translated and discussed in this monograph), as well as that of his student Zelin He (who passed away too soon) [A-60].

It is ought to be mentioned that Su and He had translated my book with Rolf Nessel *Fourier Analysis and Approximation*, Birkhäusser and Academic Press, 1971, into Chinese in the seventies.

The work of Boris Golubov at Moscow whom I met at the workshop honouring my 85-th birthday, *SAMPTA 13* in Bremen in July 2013, followed from 2002 onwards (e.g. [A-16], [A-17]).

Yasushi Endow whom I also met at the Kupari conference received his master and doctoral degree at Keio University, Japan. There he surely must have heard the basic lectures of the Fourier and stochastic analyst Tatsuo Kawata with whom we corresponded and exchanged various reprints and books in the early seventies. The private Keio University is a partner university of our RWTH Aachen.

As for the applications of Section 4.3 and 4.4, we have concentrated on the dyadic wave equation (see [B-5-R] and [B-6-R]) and error estimates in approximation theory (see e.g. [B-7-R]). The authors do not recall seeing similar results for Neumann's problem for the unit disc nor for the heat conduction equation.

The Nanjing group examined different problems in dyadic approximation theory, Wei-xing Zheng being an expert in approximation.

Acknowledgment The authors would like to express their deepest thanks to Prof. R.L. Stens (Aachen) for invaluable help in preparing this survey over some two years. Without him it would not have been possible to complete it.

It was Prof. R.S. Stanković who most kindly typed in LaTex our handwritten material in expert form, never complaining, always offering to help further. He coordinated the ideas and work of the nine authors in a harmonious way - my personal thanks are due to him.

Our sincere thanks are also due to Nick Krasnikoff (Aachen) and Prof. Wolfgang Engels (Gelsenkirchen) for their kind assistance.

References

[A-1] Arutunjan, F. G., Talaljan, A. A., "On uniqueness of Haar and Walsh series", *Izv. Akad. Nauk. SSSR Ser. Mat.*, Vol. 28, 1964, 1391-1408, (in Russian).

[A-2] Beauchamp, K. G., *A Classified Bilography for Walsh and Related Functions*, Cranfield Computing Center, Cranfield, Bedford, England, 1972.

[A-3] Bramhall, J. N., *An Annotated Bibliography on Walsh and Walsh Related Functions*, John Hopkins University, Applied Physics Lab, Silver Spring, Maryland, 1972, 1973, 1974. Reprinted in Schreiber, H., Sandy, G. F., (eds.) *Applications of Walsh Functions and Sequency Theory*, IEEE Press, 1974, *Proc. 1974 Symp. on Applications of Walsh Functions*, held at The Catholic University of America, Washington, D. C., 1974.

[A-4] Butzer, P. L., Dodson, M. M., Ferreira, P. J. S. G., Higgins, J. R., Schmeisser, G., Stens, R. L., "Seven pivotal theorems of Fourier analysis, signal analysis, numerical analysis and number theory: their interconnections", *Bull. Math. Sci.*, Vol. 4, No. 3, 2014, 481 - 525.

[A-5] Butzer, P. L., Ferreira, P. J. S. G., Higgins, J. R., Saitoh, S., Schmeisser, G., Stens, R. L., "Interpolation and sampling: E. T. Whittaker, K. Ogura and their followers", *J. Fourier Anal. Appl.*, Vol. 17, No. 2, 2011, 320 - 354.

[A-6] Butzer, P.L., Scherer, K., "On the fundamental approximation theorems of D. Jackson, S. Bernstein and theorems of M. Zamansky and S.B. Stechkin", *Aequationes Math.*, 3, 1969, 170-185.

[A-7] Butzer, P.L., Westphal, U., "An access to fractional differentiation via fractional difference quotients", in *Proc. Int. Conf. "Fractional Calculus and Its Applications to the Mathematical Sciences"*, New Haven, June 13-16, 1974, 116-145.

[A-8] Chrestenson, H.E., "A class of generalized Walsh functions", *Pacific J. Math.*, Vol. 5, 1955, 17-31.

[A-9] Efimov, A.V., "On some approximation properties of periodic multiplicative orthonormal systems", *Mat. Sb.*, Vol. 63, No. 3, 1966, 354-370.

[A-10] Fine, N.J., "On the Walsh functions", *Trans. Amer. Math. Soc.*, No. 3, 1949, 372-414.

[A-11] Fine, N. J., "The Generalized Walsh functions", *Transactions of the American Mathematical Society*, Vol. 69, No. 1, 1950, 66-77.

[A-12] Fuxian, R., Su, W., Weixing, Z., "The generalized logical derivative and its applications", *J. of Nanjing Univ.*, Vol. 3, 1978, 1-8 (in Chinese).

[A-13] Gibbs, J.E., Ireland, B., "Some generalizations of the logical derivative", *DES Report No. 8, Nat. Phys. Lab.*, August 1971, 22+ii

[A-14] Gibbs, J.E.. Ireland, B., "Walsh functions and differentiation", *Proc. Int. Conf. Applications of Walsh Functions and Sequency Theory*, 1974, 147-176.

[A-15] Gibbs, J.E., Millard, Margaret J., Walsh functions as solutions of a logical differential equation, NPL DES Rept., No. l, 1969, 9 pp. 10.

[A-16] Golubov, B.I., "On a modified strong dyadic integral and derivative", *Sb. Math.*, Vol., 193, No. 3-4, 2002, 507-529, translation from *Mat. Sbornik*, Vol. 193, No. 4, 2002, 37-60.

[A-17] Golubov, B.I., "Modified dyadic integral and derivative of fractional order on R_+", (in Russian), *Math. Notes*, Vol. 79, No. 1-2, 2006, 196-214, translation from *Mat. Zametki*, Vol. 79, No. 2, 2006, 213-233.

[A-18] Haar, A., "Zur theorie der orthogonalen Funktionsysteme", *Math. Annal.*, 69, 1910, 331-371.
[A-19] Harmuth, H.F., *Transmission of Information by Orthogonal Functions*, Springer, 1969.
[A-20] Harmuth, H. F., *Transmission of Information by Orthogonal Functions*, 2nd ed. Springer-Verlag, New York and Berlin, 1972. Honorable (nonremunerative) Russian translation of the first edition by N. G. Djadjunova and A. I. Senina, Svas, Moscow, 1975.
[A-21] Harmuth, H. F., *Sequency Theory - Foundations and Applications*, Academic Press, 1977.
[A-22] Kaczmarz, S., "Uber ein Orthogonalsystem", *Comptes rendus du premier congres des mathematiciens des pays slaves*, (Varsovie, 1929), (1930), 189-192.
[A-23] Kokilasvili, V. M., "On best approximations by Walsh polynomials and the Walsh-Fourier coefficients", *Bull. Acad. Polon. Sci. Ser. Math. Astronom. Phys.*, Vol. 13, 1965, 405-410, (in Russian).
[A-24] Levy, P., "Sur une generalization des fonctions orthogonales de M. Rademacher", *Comment. Math. Helv.*, Vol. 16, 1944, 146-152.
[A-25] Morgenthaler, G. W., "On Walsh-Fourier series", *Trans. Amer. Math. Soc.*, Vol. 84, 1957, 472-507.
[A-26] Onneweer, C.W., "On the definition of dyadic differentiation", *Applicable Anal.*, 9, 1979, 267-278.
[A-27] Paley, R.E.A.C., "A remarkable series of orthogonal functions", *Proc. Lond. Math. Soc.*, Vol. 34, 1932, 241-279.
[A-28] Pichler, F., *Walsh Functions and Linear System Theory*, Technical Research Report, T-70-05, Dept. of Electrical Engineering, University of Maryland, College Park, Maryland 20742, April 1970, ii+46.
[A-29] Pichler, F., *Some Aspects of Theory of Correlation with Respect to Walsh Harmonic Analysis*, Technical Research Report, T-70-11, Dept. of Electrical Engineering, University of Maryland, College Park, Maryland 20742, August 1970, ii+72.
[A-30] Rademacher, H., "Einige Sätze von allgemeinen Orthogonalfunktionen", *Math. Annalen*, Vol. 87, 1922, 122-138.
[A-31] Schipp, F., Über einen Ableitungsbegriff von P.L. Butzer und H.J. Wagner", *Mathematica Balkanica*, 4, 103, 1974, 541-546.
[A-32] Schipp, F., "On the dyadic derivative", *Acta Math. Acad. Scient. Hungar.*, Vol. 28, No. 1-2, 1976, 145-152.
[A-33] Schipp, F., Wade, W.R., Simon, P., with Pál, J., *Walsh Series: An Introduction to Dyadic Harmonic Analysis*, Adam Hilger, Bristol, 1990, x+560.
[A-34] Selfridge, R. G., "Generalized Walsh transform", *Pacific J. Math.*, Vol. 5, 1955, 451-480.
[A-35] Shneider, A.A., "On series in terms of Walsh functions with monotonic coefficients", *Izv. Akad. Nauk SSSR, Ser. Mat.*, Vol. 12, 1948, 179-192.
[A-36] Shneider, A.A., "On the uniqueness of series in terms of Walsh functions", *Mat. Sb.*, 1949, Vol. 24, 279-300.
[A-37] Shneider, A.A., "On the convergence of Fourier series in terms of Walsh functions", *Mat. Sb.*, Vol. 34, No. 2, 1954, 441-472.
[A-38] Skvorcov, V.A., Wade, W.R., "Generalization of some results concerning Walsh series and the dyadic derivative", *Anal. Math.*, Vol. 5, 1979, 249-255.
[A-39] Su, Weiyi, "The derivatives and integrals on local fields", *J. Nanjing Univ., Math. Biquarterly*, No.1, 1985, 32-40.
[A-40] Su Weiyi, "Gibbs-Butzer derivatives and their applications", *Numer. Funct. Anal. and Optimiz.*, Vol. 16, No. 5-6, 1995, 805-824.
[A-41] Uljanov, P.L, "On Haar series", *Mat. Sb., (N.S.)*, Vol. 63, No. 105, 1964, 356-391.
[A-42] Vilenkin, N.Ya., "On a class of complete orthogonal systems", *Izv. Akad. Nauk, Ser. Math.*, Vol. 11, 1947, 363-400.
[A-43] Vilenkin, N.Ya., "On the theory of Foruier integrals on topological groups", *Math. Sbornik, (N.S.)*, Vol. 30, No. 72, 1952, 233-244.
[A-44] Vilenkin, N.Ya., "On the theory of Fourier integrals on toplogical groups", *Math. Sbornik (N.S.)*, Vol. 30, No. 72, 1952, 363-400.

[A-45] Vilenkin, N.Ya., "Suplement to theory of orthogonal series", *Amer. Math. Soc. Transl.*, Vol. 17, 1961, 219-250.
[A-46] Wade, W.R., "Recent developments in the theory of Walsh series", *Internal. J. Math. Math. Sci.*, Vol. 5, No. 4, 1982, 625-673.
[A-47] Wagner, H.J., "On dyadic calculus for functions defined on R_+", *Arbeitsber., Lehrstuhl A für Math., RWTH Aachen*, ii+27 pp. 1974.
[A-48] Walsh, J.L., "A closed set of normal orthogonal functions", *Amer. Math.*, Vol. 45, 1923, 5-24.
[A-49] *Walsh and Dyadic Analysis*, Proceedings of Workshop, R.S. Stanković, (ed.), Niš, Serbia, October 18-19, 2007, Elektronski fakultet, Niš, Serbia, ISBN 978-86-85195-47-1.
[A-50] Watari, C., "On generalized Walsh-Fourier series", *Tohoku Math. J.*, Vol. 2, No. 10, 1948, 211-241.
[A-51] Watari, C., "On generalized Walsh-Fourier series, I", *Proc. Japan Acad.*, Vol. 73, No. 8, 1957, 435-438.
[A-52] Watari, Ch., "On generalized Walsh-Fourier series", *Tohoku Math. J.*, Vol. 10, No. 2, 1958, 211-241.
[A-53] Watari, C., "Best approximation by Walsh polynomials", *Tohoku Math. J.*, Vol. 15, No. 1, 1963, 1-5.
[A-54] Watari, C., "Mean convergence of Walsh-Fourier series", *Tohoku Math. J.*, Vol. 16, No. 2, 1964, 183-188.
[A-55] Watari, Ch., "Multipliers for Walsh-Fourier series", *Tohoku Math. J.*, Vol. 16, 1964, 239-251.
[A-56] Weixing, Z., "The generalized Walsh transform and an extremum problem", *Acta Math. Sinica*, Vol. 22, No. 3, 1979, 362-374 (in Chinese).
[A-57] Weixing, Z., Su, W., "The logical derivatives and integrals", *J. Math. Res. & Exposition*, Vol. 1, 1981, 79-90 (in Chinese).
[A-58] Yano, S., "On Walsh-Fourier series", *Tohoku Math. J. Ser. 2*, Vol. 3, 1951, 223-242.
[A-59] Yano, S., "On approximation by Walsh functions", *Tohoku Math. J. Ser. 2*, Vol. 15, 1951, 962-967.
[A-60] Zelin, H., "The derivatives and integral of fractional order in Walsh-Fourier analysis with applications to approximation theory", *Journal of Approximation Theory*, Vol. 39, 1983, 361-373.

4.8 Publications by Members of the Aachen School of Dyadic Analysis

In this list, the label R means that the publication has been reprinted in the present book.

B-1-R Butzer, P.L., Wagner, H.J., "Approximation by Walsh polynomials and the concept of a derivative", in *Applic. of Walsh Functions, Proc. Sympos., Catholic Univ. America*, Washington, D.C., 27.-29.3.1972, R.W. Zeek, A.E. Showalter (Eds.), 388- 392. Nat. Techn. Inform. Service, Springfield, Va., 1972; xi + 401 pp.
B-2-R Butzer, P.L., Wagner, H.J., "On a Gibbs-type derivative in Walsh-Fourier analysis with applications", in *Proceedings of the National Electronics Conference*, Vol. XXVII (Chicago, 111., 9.-11.10.1972; R.E. Horton (Ed.) 393-398, National Electronic Conference, Oak Brook, III, 1972; xxvi+457 pp.
B-3-R Butzer, P.L., Wagner, H.J., "Walsh-Fourier series and the concept of a derivative", *Applicable Anal.*, 3, 1973, 29-46.

B-4-R Butzer, P.L., Wagner, H.J., "A new calculus for Walsh functions with applications", in *Theory and Applications of Walsh Functions and Other Nonsinusoidal Functions* (Coll., Hatfield Polytechnic, 28.-29.6.1973; Eds. P.D. Lines - R. Barrett), 16 pp. Hatfield Polytechnic, Hatfield, 1973.

B-5-R Butzer, P.L., Wagner, H.J., "A calculus for Walsh functions defined on R_+", in *Applications of Walsh Functions*, (Proc. Sympos., Catholic Univ. America, Washington, D.C., 16.-18.4.1973; Eds. R.W. Zeek - A.E. Showalter), 75-81, Nat. Techn. Inform. Service, Springfield, Va., 1973; xi + 298 pp.

B-6-R Wagner, H.J., "On dyadic calculus for functions defined on R_+", *Arbeitsber., Lehrstuhl A für Math., RWTH Aachen*, ii+27 pp. 1974, also in *Proc. Sympos. Theory & Applic. Walsh & Other Non-sinus. Functions*, Hatfield, UK, July 1-3, 1975.

B-7-R Butzer, P.L., Wagner, H.J., "On dyadic analysis based on the pointwise dyadic derivative", *Anal. Math.*, Vol. 1, 1975, 171-196.

B-8-R Butzer, P.L., Splettstösser, W., Wagner, H.J., "On the role of Walsh and Haar functions in dyadic analysis", in *Proceedings of the special Session on Sequency Techniques 1975* (in Cooperation with the First Symposium and Technical Exhibition on Electromagnetic Compatibility, Montreux, Switzerland, 20.-22. 5. 1975; Ed. H. Hübner), 1-6, Fachbereich Elektrische Nachrichtentechnik, TH Darmstadt, Darmstadt, 1975, iv + 78 pp.

B-9 Splettstösser, W., Wagner, H.J., "Eine dyadische Infinitesimalrechnung für Haarfunktionen", *Z. Angew. Math. Mech.*, Vol. 57, 1977, 527-541.

B-10 Butzer, P.L., Splettstösser, W., "Sampling principle for duration-limited signals and daydic Walsh analysis", *Information Sci.*, Vol. 14, 1978, 93-106.

B-11 Splettstösser, W., "Ein Sampling Theorem in der Dyadischen Analysis mit Fehlerabschätzungen", in *Theorie und Anwendung diskreter Signale*, (Kurzfassung, 2. Aachener Kolloquium, 7.-9.10.1976; Ed. H.J. Tafel) Aachen 1976, xxi + 258 pp.; 128-131.

B-12 Engels, W., "Über Walsh-differenzierbare dyadische stationare stochastische Prozesse", in *Stochastische Signale* (Kurzfassungen, 3. Aachener Kolloquium, 4.-6.10.1979, Ed. H. Meyr) Aachen 1979, 251 pp.; pp. 53-57.

B-13 Splettstösser, W., "Der Abtastsatz in der Walsh-Analysis", in *Kolloquium DFG-Schwerpunktprogramm Digitale Signalverarbeitung*, Göttingen 20. - 21.2.1980. (Ed. H.-G. Zimmer - V. Neuhoff) Gottingen 1980, vii + 123 pp., 57-60.

B-14 Engels, W., Splettstösser, W., "On Walsh differentiable dyadic stationary random processes", *IEEE Trans. Information Theory*, Vol. IT - 28, 1982, 612-619.

B-15 Splettstösser, W., "Error analysis in the Walsh sampling theorem", in *1980 IEEE Int. Symp. on Electromagnetic Compatibility*, Baltimore, The Institute of Electrical and Electronic Engineers, Piscataway, N.J., 1980, 409 pp., 366-370.

B-16 Butzer, P.L., Engels, W., "Dyadic calculus and sampling theorems for functions with multi-dimensional domain I. General theory", *Information and Control*, Vol. 52, 1982, 333-351.

B-17 Butzer, P.L., Engels, W., "Dyadic calculus and sampling theorems for functions with multi-dimensional domain II. Applications to dyadic sampling representations", *Information and Control*, Vol. 52, 1982, 352-363.

B-18-R Engels, W., "On the characterization of the dyadic derivative", *Act. Math. Humgar.*, Vol. 46, 1985, 47-56.

B-19 Butzer, P.L., Engels, W., Wipperfürth, U., "An extension of the dyadic calculus with fractional order derivatives. General theory", *Computers and Mathematics with Applications*, Vol. 12 B, 1986, 1073-1090.

B-20 Butzer, P.L., Engels, W., Wipperfürth, U., "An extension of the dyadic calculus with fractional order derivatives. Further theory and applications", *Computers and Mathematics with Applications*, Vol. 12 A, 1986, 921-943.

B-21 Splettstösser, W., Ziegler, W., "The generalized Haar functions, the Haar transform on R_{+}, and the Haar sampling theorem", in *Colloquia Mathematica Societatis Janos Bolyai, 49. A. Haar Memorial Conference*, Budapest, Aug. 12-16, 1985 (Eds. J. Szabados, K. Tandori) North Holland 1987. 2 Vols., Vol. II, 873-896.

B-22 Butzer, P.L., Engels, W., "Background to an extension of Gibbs differentiation in Walsh analysis", in *Theory and Applications of Gibbs Derivatives, Proc. Workshop*, Kupari - Dubrovnik, Sept. 1989, P.L. Butzer, R.S. Stanković, Math. Inst., Belgrade, Yugoslavia, 1990, 312 pp., 19-57.

Proceedings, Doctoral and Diploma Theses

B-23 Wagner, H.J., *Ein Differential - und Integralkalkül in der Walsh-Fourier-Analysis mit Anwendungen*, Forschungsberichte des Landes Nordrhein-Westfalen, No. 2334, Westdeutscher Verlag, Köln - Opladen, 1972, 71 pp, Dissertation, 1972.

B-24 Splettstösser, W., *Ein Haarscher Infinitesimalkalkül mit Anwendungen*, Diplomarbeit, Lehrstuhl A für Mathematik, Aachen 1975, 75 pp.

B-25 Engels, W., *Ein neuer Differentialkalkül in der dyadischen Walsh-Analysis*, Dissertation, Aachen 1982, 93 pp.

B-26 Wipperfürth, U., *Verschiedene Differential- und Intgralkalküle in der Walsh-Analysis*, Diplomarbeit, Lehrstuhl A für Mathematik, Aachen 1984, 98 pp.

B-27 Butzer, P.L., Splettstösser, W., Stens, R.L., "The sampling theorem and linear prediction in signal analysis", *Jber. d. Dt. Math.-Verein.* 90, 1988, 1-70.

B-28 Butzer, P.L., Stanković, R.S. (Editors), *Theory and Applications of Gibbs Derivatives*, Proc. Workshop, Kupari Dubrovnik, Sept. 1989, Math. Inst. Belgrade, Yugoslavia, 1990, 312 pp.

Applicable Analysis, 1973, Vol. 3, pp. 29-46

Printed in Great Britain

Walsh–Fourier Series and the Concept of a Derivative†

P. L. BUTZER and H. J. WAGNER‡

Technological University of Aachen, Aachen, Germany

Communicated by R. P. GILBERT

In the theory of Walsh–Fourier series on the dyadic group G developed so far, such essential concepts as continuity, Lipschitz conditions, modulus of continuity, etc., have been employed. However, with the exception of several articles by J. E. Gibbs, no attempt has been made to define a derivative for functions f given on G. In this paper such a derivative $D^{[1]}f$ is defined which has most properties in common with the ordinary derivative. $D^{[1]}$ is a linear, closed operator, the inverse operator or integral is defined, the fundamental theorem of the calculus is valid for these two notions. The derivative enables one to estimate the order of magnitude of the Walsh-Fourier coefficients, the degree of approximation of f by the partial sums of the Walsh-Fourier series of f, etc. Moreover the Walsh functions are the non-trivial eigensolutions of a first-order linear differential equation. The proofs depend upon a Walsh–Fourier "coefficient method".

1. INTRODUCTION

The system of Walsh functions [20] has been the subject of a great deal of study, particularly in the past few years. Thus, in 1970, there was held a symposium on "Applications of Walsh Functions" in Washington, D.C., which was followed up by a yearly symposium§ stressing their applications,

† Dedicated to Eberhard Hopf on the occasion of his 70th birthday.

‡ The contribution of this named author was supported by grant No. B/1–4858 from the Minister für Forschung des Landes Nordrhein-Westfalen—Landesamt für Forschung. The authors wish to thank the Landesamt for permission to publish the results here.

§ The first named author was fortunate to be able to attend the conference in March 30 to April 3, 1970 at the invitation of Dr. H. F. Harmuth, University of Maryland. At a stately dinner at the home of Professor J. L. Walsh he had the opportunity to listen in on an interesting conversation between Dr. J. E. Gibbs (National Physical Laboratory, Teddington, England), Dr. F. Pichler (Linz, Austria) and others on the fundamental importance of a suitable derivative concept for Walsh functions.

especially to problems of computer science and electrical engineering. There are two ways of considering the Walsh functions $\{\psi_k(x)\}_{k=0}^{\infty}$; either they may be defined on the unit interval [0, 1], or, following N. J. Fine [4], they may be identified with the full set of characters of the dyadic group G. G is the set of all sequences $x = \{x_s\}$, $s = 1, 2, \ldots$ such that $x_s = 0$ or 1, with termwise addition modulo 2, this operation being denoted by $\dot{+}$. Our considerations are confined to the dyadic group.

The Walsh system $\{\psi_k(x)\}_{k=0}^{\infty}$ has the essential property that it is orthonormal as well as complete with respect to the space $C(G)$ of continuous functions on G, or to the space $L^p(G)$, $1 \leqq p < \infty$, of pth power integrable functions.

In analogy with the theory of trigonometric Fourier series, there has been built up an extensive† parallel theory of Walsh–Fourier series. In its development, such essential concepts as continuity, Lipschitz conditions, modulus of continuity, etc., have been employed. However, hardly any papers, with the exception‡ of several articles by J. E. Gibbs [5, 6, 7], make an attempt to define a derivative for functions f defined on G, although the concept of a derivative plays a decisive role in all of analysis. This is perhaps due to the fact that G is not a Lie group, and so not a differentiable manifold.

In the case of functions defined on the discrete set $B_n = \{0, 1, \ldots, 2^n - 1\}$, Gibbs [9, 10] proceeded as follows: Each $x \in B_n$ has the unique representation

$$x = \sum_{s=0}^{n-1} x_s 2^s, \quad x_s \in \{0, 1\}$$

and B_n becomes a discrete abelian group with respect to term-wise addition modulo 2. Then Gibbs defined his logical derivative by

$$f^{[1]}(x) = \tfrac{1}{2} \sum_{j=0}^{n-1} 2^j [f(x) - f(x \dot{+} 2^j)] \quad (x \in B_n)$$

and in recent paper [8] extended this concept also to functions defined on the interval [0, 1].

The purpose of the present paper is to define a derivative on our non-discrete group G. This will lead to a modified form of Gibbs definition, given by an infinite series ($\mathbb{N} = \{1, 2, \ldots\}$)

$$(D^{[1]}f)(x) = \tfrac{1}{2} \sum_{j=0}^{\infty} 2^j [f(x) + f(x \dot{+} e_{j+1})], \tag{1.1}$$

† A list of papers on Walsh functions and their applications is to be found in [1]; it comprises more than 200 articles written up to 1970. See also the list in H. F. Harmuth [12].

‡ G. Pál–F. Schipp [16] define the concept of a derivative for the Haar-system. See also V. M. Kokilasvili [14].

where for each fixed $j \in \mathbb{N}$, $e_j = \{x_s^j\}$ with $x_s^j = 0$ for $j \neq s$, $= 1$ for $j = s$. More precisely, definition (1.1) is to be understood in the norm of the space $C(G)$ or $L^p(G)$, and not in the pointwise sense (see (3.1) below). Our definition of a derivative has practically all properties in common with the ordinary concept of a derivative.† For example, $D^{[1]}$ is a linear, closed operator. If f, $D^{[1]}f \in C(G)$ or $L^p(G)$ and $D^{[1]}f = 0$, then $f =$ const. However, this definition would only prove to be of actual value if it can be shown that the parallelism which is known to exist between the properties of trigonometric Fourier-series and those of Walsh–Fourier series can actually be enlarged by bringing this modified derivative into play. And this is indeed the case.

For example, denoting $\mathbb{Z} = \{0, \pm 1, \pm 2, \ldots\}$, and the Fourier coefficients of $f \in C_{2\pi}$ (space of 2π-periodic continuous functions on the real axis) by

$$f_F^\wedge(k) = (1/2\pi) \int_{-\pi}^{\pi} f(x) e^{-ikx}\, dx \quad (k \in \mathbb{Z}),$$

then it is a fundamental result of classical Fourier analysis (compare [2, p. 172]) that the class ($r \in \mathbb{N}$)

$$\{f \in C_{2\pi} \mid f^{(r)}(x) = g(x), \quad g \in C_{2\pi}\}$$

is equivalent to

$$\{f \in C_{2\pi} \mid (ik)^r f_F^\wedge(k) = g_F^\wedge(k), \quad g \in C_{2\pi}, \quad k \in \mathbb{Z}\}.$$

Defining higher derivatives successively by $D^{[r]}f = D^{[1]}(D^{[r-1]}f)$, the Walsh analogue will be shown to be ($\mathbb{P} = \{0, 1, 2, \ldots\}$) (see Theorem 4.2)

$$\{f \in C(G) \mid D^{[r]}f = g, g \in C(G)\}$$
$$\Leftrightarrow \{f \in C(G) \mid k^r f^\wedge(k) = g^\wedge(k), g \in C(G), k \in \mathbb{P}\}, \tag{1.2}$$

where (see definition (2.14))

$$f^\wedge(k) = \int_G f(x)\, \psi_k(x)\, dx \quad (k \in \mathbb{P})$$

are the Walsh–Fourier coefficients of $f \in C(G)$. Thus the Walsh coefficients convert differentiation to multiplication by k. These results are also valid in $L^p(G)$, $1 \leq p < \infty$.

Now Walsh has shown that the Walsh–Fourier coefficients of any integrable function converge to zero with $k \to \infty$. By introducing the derivative

† In marked contrast, any bounded function on [0, 1] possesses derivatives in the sense of Gibbs of all positive integer orders.

it is possible to estimate the order of magnitude of these coefficients. Indeed, if $D^{[r]}f$ exists and belongs to $C(G)$ or $L^p(G)$, then (see Theorem 5.1)

$$|f^\wedge(k)| = O\,(k^{-r}). \tag{1.3}$$

The analogue for trigonometric Fourier series is well-known, compare [11, p. 25ff].

It is generally true in this theory that in order to establish straight convergence results one does not need the concept of a derivative. However, as soon as one asks for the order of magnitude or degree of convergence, derivates usually enter the scene. This is again revealed by the following result of Lebesgue. Denoting the nth partial sum of the Fourier series $\sum_{k=-\infty}^{\infty} f_F^\wedge(k)e^{ikx}$ by $s_{n,F}(f, x)$, he showed that (compare [2, p. 105])

$$\sup_{x\in[-\pi,\pi]} |s_{n,F}(f, x)-f(x)| = O\left(\frac{\log n}{n^{r+\alpha}}\right) \tag{1.4}$$

provided the rth derivative $f^{(r)}$ belongs to the class Lip $(\alpha, C_{2\pi})$, $0 < \alpha \leqq 1$.

Its analogue will turn out to be (see Theorem 5.2)

$$D^{[r]}f \in \text{Lip}\,(\alpha, C(G)) \Rightarrow \sup_{x\in G} |s_n(f, x)-f(x)| = O\left(\frac{\log n}{n^{r+\alpha}}\right), \tag{1.5}$$

where $s_n(f, x)$ is the nth partial sum of the Walsh–Fourier series of f (see (2.16)), and Lip $(\alpha, C(G))$ is defined suitably (see (5.2)). The situation is similar for $L^p(G)$, $1 \leqq p < \infty$.

During the course of the work it will also be seen that once the derivative is defined, the way is clear to a fitting definition of the corresponding inverse operation or integral, including the fundamental theorem of the calculus for these two notions.

The authors are indebted to Dr. H. Johnen for his constructive criticism and ever helpful advice during the entire preparation of this paper. The proof of the vital Theorem 4.1, for example, is due to him. The authors also wish to thank Drs. J. E. Gibbs (letters of 6 and 28 Aug. 71) and F. Pichler (letter of 9 Aug. 71) for their encouragement and valuable suggestions.

2. NOTATIONS AND PRELIMINARY RESULTS

The group G already being defined in the introduction, taking as neighbourhoods of its elements $x = \{x_s\}$, $s \in \mathbb{N}$, the set of points $\{x_1, x_2, \ldots, x_n, z_{n+1}, \ldots\}$ in which $x_1, x_2, \ldots, x_n$ are fixed and $z_{n+1}, \ldots$ vary independently, then these neighbourhoods induce a topology on G with respect to which G forms a compact Abelian group. The mapping $\lambda: G \to [0, 1]$.

$$\lambda(x) = \sum_{s=1}^{\infty} 2^{-s} x_s,$$

defines a metric on G by

$$d(x, y) = \lambda(x \dot{+} y) = \sum_{s=1}^{\infty} 2^{-s}[(x_s + y_s) \text{ mod. } 2]. \tag{2.1}$$

This metric induces a topology on G which is equivalent to the original topology (compare N. J. Fine [4], G. W. Morgenthaler [15]). For purposes of a short notation in the following, let us denote the neighbourhoods of the identity $O = (0, 0, \ldots)$ on G by (see [21])

$$V_0 = G, \quad V_n = \{x \in G; x_1 = \ldots = x_n = 0\} \quad (n \in \mathbb{N}). \tag{2.2}$$

Since G is a compact group, the Haar measure dx on G may be normalized to $\int_G dx = 1$. A basic property of the Haar integral is its translation-invariance, namely, that for each fixed $y \in G$

$$\int_G f(x \dot{+} y)\, dx = \int_G f(x)\, dx. \tag{2.3}$$

The character group $\hat{G}$ of G is by definition the set of continuous functions ψ on G satisfying the functional equation

$$\psi(x \dot{+} y) = \psi(x)\, \psi(y) \quad (x, y \in G), \tag{2.4}$$

endowed with the compact open topology. Fine has shown that these functions are given by

$$\psi_0(x) = 1, \quad \psi_n(x) = \varphi_{n_1}(x)\, \varphi_{n_2}(x) \ldots \varphi_{n_i}(x) \tag{2.5}$$

and called the Walsh functions, where the Rademacher functions

$$\varphi_n(x) = (-1)^{x_{n+1}} \quad (n \in \mathbb{P}) \tag{2.6}$$

and the number $n \in \mathbb{N}$ has the unique representation

$$n = 2^{n_1} + 2^{n_2} + \ldots + 2^{n_i} \quad (n_1 > n_2 > \ldots > n_i \geqq 0,\ n_i \in \mathbb{P}). \tag{2.7}$$

A finite linear combination of the $\psi_j(x)$, namely ($\mathbb{C}$ the set of complex numbers)

$$p_n(x) = \sum_{j=0}^{n-1} a_j\, \psi_j(x) \quad (a_j \in \mathbb{C}), \tag{2.8}$$

is called a Walsh polynomial of degree n. The set of Walsh polynomials of degree $\leqq n$ is denoted by $\mathscr{P}_n$. These polynomials have the typical property that (see [23])

$$p_m(x \dot{+} h) = p_m(x) \quad (p_m \in \mathscr{P}_{2^n},\ h \in V_n). \tag{2.9}$$

A particular Walsh polynomial of degree m is the so-called Walsh-Dirichlet kernel

$$D_m(x) = \sum_{j=0}^{m-1} \psi_j(x), \tag{2.10}$$

which in the case $m = 2^n$ has the form

$$D_{2^n}(x) = \begin{cases} 2^n, & x \in V_n \\ 0, & x \notin V_n \end{cases}. \tag{2.11}$$

Let $X(G)$ denote one of the spaces $C(G)$, $L^p(G)$, $1 \leq p < \infty$, of continuous functions of f on G or integrable to the pth power on G, respectively, with norms

$$\|f\|_{C(G)} = \sup_{x \in G} |f(x)|,$$

$$\|f\|_{L^p(G)} = \{\textstyle\int_G |f(x)|^p \, dx\}^{1/p} \quad (1 \leq p < \infty).$$

The orthonormal system of Walsh functions $\{\psi_n(x)\}_{n=0}^{\infty}$ is complete in $X(G)$; in other words: The Weierstrass approximation-theorem is valid in $X(G)$ (see [23]); to each $f \in X(G)$ and $\varepsilon > 0$ there exists a Walsh polynomial $p_n(x)$ of degree $n = n_\varepsilon$ such that

$$\|f - p_n\|_{X(G)} < \varepsilon. \tag{2.12}$$

If $f \in X(G)$, $g \in L^1(G)$, the function $f * g$ defined by (compare W. Rudin [19, p. 3], E. Hewitt-K.A. Ross [13, p. 262])

$$(f*g)(x) = \textstyle\int_G f(x \dot{+} u)\, g(u)\, du = \int_G f(u)\, g(x \dot{+} u)\, du$$

and called the resultant of f and g exists for all $x \in G$ in case $X(G) = C(G)$, and for almost all $x \in G$ with respect to Haar measure on G in case $X(G) = L^p(G)$, $1 \leq p < \infty$. Moreover, $f*g \in X(G)$ and

$$\|f*g\|_{X(G)} \leq \|f\|_{X(G)} \|g\|_{L^1(G)}. \tag{2.13}$$

Denoting the Walsh-Fourier coefficients of $f \in X(G)$ by

$$f^\wedge(k) = \textstyle\int_G f(u)\, \psi_k(u)\, du \quad (k \in \mathbb{P}), \tag{2.14}$$

then for each $y \in G$

$$\textstyle\int_G f(y \dot{+} u)\, \psi_k(u)\, du = \psi_k(y) f^\wedge(k) \quad (k \in \mathbb{P}). \tag{2.15}$$

$\sum_{k=0}^{\infty} f^\wedge(k)\, \psi_k(x)$ is called the Walsh-Fourier series of f, and for its mth partial sum

$$s_m(f, x) = (f*D_m)(x) = \sum_{k=0}^{m-1} f^\wedge(k)\, \psi_k(x) \tag{2.16}$$

it is known that (see Ch. Watari [23]) for $m = 2^n$

$$\lim_{n\to\infty} \|s_{2^n}(f, \circ)-f(\circ)\|_{X(G)} = 0. \tag{2.17}$$

On account of (2.16), (2.17) there follows the uniqueness theorem: For $f \in X(G)$

$$f^\wedge(k) = 0 \quad (k \in \mathbb{P}) \Rightarrow f = 0 \text{ on } G.\dagger \tag{2.18}$$

Moreover, it is obvious that for $f \in X(G)$

$$|f^\wedge(k)| \leqq \|f\|_{X(G)} \quad (k \in \mathbb{P}), \tag{2.19}$$

and the convolution theorem states that for $f, g \in L^1(G)$

$$(f*g)^\wedge(k) = f^\wedge(k)\, g^\wedge(k) \quad (k \in \mathbb{P}). \tag{2.20}$$

Finally, we need the fact that if $f, f_n \in X(G)$, $n \in \mathbb{N}$, then for $k \in \mathbb{P}$

$$\lim_{n\to\infty} \|f_n-f\|_{X(G)} = 0 \Rightarrow \lim_{n\to\infty} f_n^\wedge(k) = f^\wedge(k). \tag{2.21}$$

3. THE DERIVATIVE AND ITS SIMPLER PROPERTIES

DEFINITION 3.1 *If for $f \in X(G)$ there exists $g \in X(G)$ such that*

$$\lim_{m\to\infty} \Big\|\tfrac{1}{2} \sum_{j=0}^{m} 2^j[f(\circ)-f(\circ \dotplus e_{j+1})]-g(\circ)\Big\|_{X(G)} = 0, \tag{3.1}$$

where for each fixed $j \in \mathbb{N}$, $e_j = \{x_s^j\}$ with $x_s^j = 0$ for $j \neq s$, $= 1$ for $j = s$, then g is called the strong derivative of f and is denoted by $D^{[1]}f$. For any $r \in \mathbb{N}$, the rth strong derivative of $f \in X(G)$ is defined successively by

$$D^{[r]}f = D^{[1]}(D^{[r-1]}f).$$

Let U be the set of all strongly differentiable functions belonging to $X(G)$. The operator $D^{[r]}: U \subset X(G) \to X(G)$ is linear. For, if $f_1, f_2 \in U$ and $c_1, c_2 \in \mathbb{C}$, then $c_1f_1+c_2f_2 \in U$ and

$$D^{[r]}(c_1f_1+c_2f_2) = c_1D^{[r]}f_1+c_2D^{[r]}f_2.$$

The algebraic or trigonometric polynomials are arbitrary often differentiable in the ordinary sense. An analogous result holds for the Walsh polynomials (2.8) with respect to the derivatives (3.1).

† More precisely, $f(x) = 0$ for all $x \in G$ if $X(G) = C(G)$, and $f(x) = 0$ a.e. if $X(G) = L^p(G)$, $1 \leqq p < \infty$. More generally, we write $f = g$ provided $\|f-g\|_{X(G)} = 0$.

PROPOSITION 3.1 *The Walsh functions $\psi_n(x)$ of (2.5) have the property*

$$D^{[r]}\,\psi_n = n^r\,\psi_n \quad (n \in \mathbb{P},\ r \in \mathbb{N}). \tag{3.2}$$

Proof To prove (3.2) we need the following identity

$$\sum_{j=0}^{\infty} 2^j[1-\psi_n(e_{j+1})] = 2n \quad (n \in \mathbb{P}). \tag{3.3}$$

Indeed, by (2.5) $\psi_n(e_{j+1}) = \varphi_{n_1}(e_{j+1}) \ldots \varphi_{n_l}(e_{j+1})$ where n is given by (2.7). By definition of e_{j+1} in (3.1) and (2.6) one has $\varphi_{n_k}(e_{j+1}) = -1$ for $j = n_k$, $= +1$ for $j \neq n_k$. Therefore $\psi_n(e_{j+1}) = 1$ for $j > n_1$. This yields

$$\sum_{j=0}^{\infty} 2^j[1-\psi_n(e_{j+1})] = \sum_{j=0}^{n_1} 2^j[1-\varphi_{n_1}(e_{j+1}) \ldots \varphi_{n_l}(e_{j+1})]$$

$$= 2(2^{n_1}+2^{n_2}+\ldots+2^{n_l}) = 2n.$$

To establish (3.2) itself for $r = 1$, note that (2.4) and (3.3) imply

$$\lim_{m\to\infty} \Big\|\tfrac{1}{2}\sum_{j=0}^{m} 2^j[\psi_n(\circ)-\psi_n(\circ\dot{+}e_{j+1})]-n\,\psi_n(\circ)\Big\|_{X(G)}$$

$$= \lim_{m\to\infty} \Big\|\psi_n(\circ)\,\tfrac{1}{2}\sum_{j=0}^{m} 2^j[1-\psi_n(e_{j+1})]-n\,\psi_n(\circ)\Big\|_{X(G)} = 0.$$

The proof for the higher strong derivatives follows by induction.

From the proof it is clear that the strong derivative of Walsh functions, and thus of Walsh polynomials, is equal to its derivative defined in the pointwise sense.

Definition 3.1 implies immediately that the strong derivative of a constant function vanishes. Conversely we have

PROPOSITION 3.2 *If f, $D^{[1]}f \in X(G)$ and $D^{[1]}f = 0$; then $f = const.$*

Proof By hypothesis

$$\lim_{m\to\infty} \Big\|\tfrac{1}{2}\sum_{j=0}^{m} 2^j[f(\circ)-f(\circ\dot{+}e_{j+1})]\Big\|_{X(G)} = 0.$$

Now for the Walsh–Fourier coefficients of the expression within the norms we have by (2.15) and (3.3)

$$[\tfrac{1}{2}\sum_{j=0}^{m} 2^j(f(x)-f(x\dot{+}e_{j+1}))]^\wedge(k)$$

$$= \tfrac{1}{2}\sum_{j=0}^{m} 2^j[1-\psi_k(e_{j+1})]\,f^\wedge(k) = kf^\wedge(k) \quad (k \in \mathbb{P}) \tag{3.4}$$

provided m is sufficiently large. It therefore follows in view of (2.21) that $kf^\wedge(k) = 0$, $k \in \mathbb{P}$. This proof is complete by (2.18).

We are now ready to establish one direction of the result (1.2) announced in the introduction.

PROPOSITION 3.3 *If for $f \in X(G)$ there exists $D^{[r]}f$ for some $r \in \mathbb{N}$, then*

$$[D^{[r]}f]^\wedge(k) = k^r f^\wedge(k) \quad (k \in \mathbb{P}). \tag{3.5}$$

Proof In the instance $r = 1$ (3.5) follows readily by the known result (2.21) by making use of definition (3.1) and identity (3.4). The result for general r follows by induction.

In the next section we consider the converse to this assertion.

4. FURTHER PROPERTIES OF THE DERIVATIVE, ITS INVERSE OPERATOR

A fundamental rôle in this section will be played by the functions $W_r(x)$ which are defined by means of their Walsh–Fourier coefficients

$$W_r^\wedge(k) = \begin{cases} 1, & k = 0 \\ k^{-r}, & k \in \mathbb{N} \end{cases} \quad (r \in \mathbb{N}). \tag{4.1}$$

The Walsh–Fourier series of $W_r(x)$ therefore has the form

$$1 + \sum_{k=1}^{\infty} k^{-r}\, \psi_k(x). \tag{4.2}$$

Functions of this type were apparently first investigated by Ch. Watari [22]. However, instead of working with k^{-r}, he used $2^{-k_1 r}$, k_1 being the largest exponent in the expansion $k = 2^{k_1} + \ldots + 2^{k_t}$, where $k_1 > \ldots > k_t \geqq 0$. For this purpose we give complete proofs of the following two propositions, essentially due to Watari [22].

PROPOSITION 4.1 $W_r(x) \in L^1(G)$ for $r \in \mathbb{N}$.

Proof If $r \geqq 2$ the series (4.2) is uniformly convergent and represents a function $g_r \in C(G)$ with coefficients equal to $g_r^\wedge(0) = 1$, $g_r^\wedge(k) = k^{-r}$ for $k \in \mathbb{N}$. By the uniqueness result (2.18) this implies $g_r(x) = W_r(x)$ for all $x \in G$.

In the case $r = 1$ we set

$$W_{1,m}(x) = (W_1 * D_{2^m})(x) = 1 + \sum_{k=1}^{2^m - 1} \frac{\psi_k(x)}{k}.$$

Employing Abel's formula for partial summation with $n > m$ one obtains

$$W_{1,n}(x) - W_{1,m}(x) = \sum_{k=2m}^{2n-1} \frac{\psi_k(x)}{k}$$

$$= \sum_{k=2m}^{2n-2} \frac{1}{k(k+1)} D_{k+1}(x) - \frac{1}{2^m} D_{2m}(x) + \frac{1}{2^n-1} D_{2n}(x). \tag{4.3}$$

Making use of the results that $\|D_2^j(\circ)\|_{L^1(G)} = 1$ (see (2.11)) and $\|D_k(\circ)\|_{L^1(G)} = O(\log k)$ (see Fine [4]), then (4.3) may be estimated by

$$\|W_{1,n} - W_{1,m}\|_{L^1(G)} \leqq \text{const.} \sum_{k=2m}^{2n-2} \frac{\log k}{k^2} + \frac{1}{2^m} + \frac{1}{2^n-1}.$$

Now the right side tends to zero for $m, n \to \infty$ and so $W_{1,m}$ converges in $L^1(G)$-norm to an element $g_1 \in L^1(G)$ since $L^1(G)$ is complete. Since $W_{1,m}^\wedge(k) = 1$ for $k = 0$, $= k^{-1}$ for $k \in \mathbb{N}$, m being sufficiently large, it follows by (2.21) that $g_1(k) = 1$ for $k = 0$, $= k^{-1}$ for $k \in \mathbb{N}$. This delivers $g_1 = W_1$ by (2.18), and the proof is complete.

Now let $W_r^{(m)}(x)$ be the function whose coefficients are of the form

$$[W_r^{(m)}(x)]^\wedge(k) = \begin{cases} 0, & 0 \leqq k < 2^m \\ k^{-r}, & 2^m \leqq k, \end{cases}$$

so having Walsh–Fourier series $\sum_{k=2m}^{\infty} k^{-r} \psi_k(x)$.

PROPOSITION 4.2 *One has for* $m \to \infty$

(a) $\|W_r^{(m)}\|_{L^1(G)} = O(2^{-mr})$,

(b) $\|W_r(\circ) - W_r(\circ \dotplus h)\|_{L^1(G)} = O(2^{-mr}) \quad (h \in V_m)$.

Proof Concerning part (a), setting $K_k(x) = k^{-1} \sum_{j=1}^{k} D_j(x)$, this being the Fejer kernel, it follows from (4.3) by a second application of Abel's formula that

$$\sum_{k=2m}^{2n-1} \frac{\psi_k(x)}{k} = \sum_{k=2m}^{2n-3} \left(\frac{1}{k} - \frac{1}{k+2}\right) K_{k+1}(x) - \frac{K_{2m}(x)}{(2^m+1)}$$

$$+ \frac{K_{2n-1}(x)}{(2^n-2)} - \frac{D_{2m}(x)}{2^m} + \frac{D_{2n}(x)}{2^n-1}.$$

Since $\|K_k(\circ)\|_{L^1(G)} \leqq 2$ (see [4]), this yields for $2^n - 3 > 2^m$

$$\left\| \sum_{k=2m}^{2n-1} \frac{\psi_k(\circ)}{k} \right\|_{L^1(G)} \leqq \sum_{k=2m}^{2n-3} \frac{2}{k^2} \|K_{k+1}(\circ)\|_{L^1(G)} + O\left(\frac{1}{2^m}\right) = O(2^{-m}).$$

Letting $n \to \infty$ the proof of (a) follows for $r = 1$.

If $r = 2$, note that the coefficients of $W_1^{(m)} * W_1^{(m)}$ are equal to 0 for $0 \leq k < 2^m$, and to k^{-2} for $2^m \leq k$. According to (2.18) this gives $W_2^{(m)} = W_1^{(m)} * W_1^{(m)}$ and

$$\|W_2^{(m)}\|_{L^1(G)} < (\|W_1^{(m)}\|_{L^1(G)})^2 = O(2^{-2m}).$$

The proof for $r \geq 3$ follows by induction.

Concerning part (b), the Walsh polynomial $W_{r,m}(x) \in \mathscr{P}_{2^m}$, and so $W_{r,m}(x) = W_{r,m}(x \dotplus h)$ for $h \in V_m$ by (2.9). This delivers

$$\begin{aligned}\|W_r(\circ) - W_r(\circ \dotplus h)\|_{L^1(G)} &= \|W_r(\circ) - W_{r,m}(\circ) - W_r(\circ \dotplus h) - W_{r,m}(\circ \dotplus h))\|_{L^1(G)} \\ &\leq 2\|W_r - W_{r,m}\|_{L^1(G)} = 2\|W_r^{(m)}\|_{L^1(G)} \\ &= O(2^{-mr}).\end{aligned}$$

These results will enable us to obtain a partial converse, to begin with, of Proposition 3.3. This is given by

PROPOSITION 4.3 *If for $f \in X(G)$ there exists $g \in X(G)$ such that*

$$k^r f^\wedge(k) = g^\wedge(k) \quad (k \in \mathbb{P}), \tag{4.4}$$

$r \in \mathbb{N}$ *being fixed, then*

$$f = W_r * g + f^\wedge(0). \tag{4.5}$$

Proof According to (2.20) one has

$$(W_r * g)^\wedge(k) = \begin{cases} k^{-r} k^r f^\wedge(k), & k \in \mathbb{N}, \\ 0, & k = 0. \end{cases}$$

This implies (4.5) in view of (2.18).

The next theorem is actually the fundamental theorem of the calculus for our concept of a derivative, at the same time it solves the problem: given $D^{[r]} f \in X(G)$, to find the original function f. This is the question of integration.

THEOREM 4.1 *Let $f \in X(G)$ such that $f^\wedge(0) = \int_G f(x)\,dx = 0$.*

(a) *If there exists $D^{[r]} f \in X(G)$ for some $r \in \mathbb{N}$, then*

$$W_r * D^{[r]} f = f. \tag{4.6}$$

(b) *One has*

$$D^{[r]}(W_r * f) = f. \tag{4.7}$$

Proof If $D^{[r]}f \in X(G)$, then $[D^{[r]}f]^\wedge(k) = k^r f^\wedge(k)$ by Proposition 3.3. Part (a) now follows by Proposition 4.3.

Concerning part (b) we first consider the case $r = 1$ and for this purpose establish the identity

$$\tfrac{1}{2}\sum_{j=0}^{m-1} 2^j[(W_1*f)(x)-(W_1*f)(x\dot{+}e_{j+1})]-f(x)$$
$$= s_{2^m}(f, x)-f(x)+(f*F_m)(x), \tag{4.8}$$

where

$$F_m(x) = \tfrac{1}{2}\sum_{j=0}^{m-1} 2^j[W_1^{(m)}(x)-W_1^{(m)}(x\dot{+}e_{j+1})].$$

Indeed, in view of the convolution formula (2.20) and identity (3.3), the Walsh–Fourier coefficients of the left side of (4.8) equal

$$\frac{1}{k}f^\wedge(k)\,\tfrac{1}{2}\sum_{j=0}^{m-1} 2^j[1-\psi_k(e_{j+1})]-f^\wedge(k) \quad (k \in \mathbb{N}), \tag{4.9}$$

they vanish for $0 \leqq k < 2^m$. Concerning the right side,

$$(f*F_m)^\wedge(k) = \begin{cases} 0, & 0 \leqq k < 2^m \\ (2k)^{-1} f^\wedge(k)\sum_{j=0}^{m-1} 2^j[1-\psi_k(e_{j+1})], & 2^m \leqq k; \end{cases}$$

and

$$[s_{2^m}(f, x)]^\wedge(k) = \begin{cases} f^\wedge(k), & 0 \leqq k < 2^m, \\ 0, & 2^m \leqq k. \end{cases}$$

Therefore the Walsh coefficients of both sides of (4.8) are equal and so the identity follows by (2.18).

To prove assertion (4.7), in view of the triangle inequality applied to (4.8) it just remains to show that

$$\lim_{m\to\infty} \|(f*F_m)\|_{X(G)} = 0 \tag{4.10}$$

on account of (2.17).

For this purpose, first note that $F_m \in L^1(G)$ since by Proposition 4.2

$$\|F_m\|_{L^1(G)} \leqq \sum_{j=0}^{m-1} 2^j \|W_1^{(m)}\|_{L^1(G)} \leqq \sum_{j=0}^{m-1} 2^{j-m} = \text{const.}$$

We now show that (4.10) holds for f replaced by any Walsh polynomial $p_n(x)$. Indeed, the convolution $(F_m{*}p_n)(x)$ is a Walsh polynomial of degree n since

$$\int_G F_m(u\dot{+}x) \sum_{j=0}^{n-1} a_j\,\psi_j(u)\,du = \sum_{j=0}^{n-1} a_j\,\psi_j(x) \int_G F_m(u\dot{+}x)\,\psi_j(u\dot{+}x)\,du$$
$$= \sum_{j=0}^{n-1} a_j\,F_m^{\wedge}(j)\,\psi_j(x).$$

This implies that $(F_m{*}p_n)(x) = 0$ for $2^m > n$. But, according to Weierstrass' theorem, to each $f \in X(G)$ and each $\varepsilon > 0$ there exists $p_n \in \mathscr{P}_n$ with $n = n_\varepsilon$ such that (2.12) holds. Hence

$$\|(F_m{*}f)(\circ)\|_{X(G)} \leqq \|F_m\|_{L^1(G)}\,\|f-p_n\|_{X(G)} + \|F_m{*}p_n\|_{X(G)}$$
$$\leqq \text{const.}\,\|f-p_n\|_{X(G)} < \varepsilon$$

by (2.13) for m sufficiently large.

The case for general $r \geqq 2$ follows by induction. Indeed, since $W_{r+1} = W_r{*}W_1$ by (2.20), (2.18), and as the operation is associative, it follows from the validity of (4.7) that $D^{[r+1]}(W_{r+1}{*}f) = D^{[1]}D^{[r]}(W_r{*}(W_1{*}f)) = D^{[1]}(W_1{*}f) = f$. This completes the proof of the theorem.

The above proof reveals that the resultant of any function $f \in X(G)$ with W_r is r-times differentiable.

Observe that Theorem 4.1 reveals that the operator $I = (W_1{*}): X(G) \to X(G)$ defined by

$$If(x) = (W_1{*}f)(x) = \int_G f(x\dot{+}u)\,W_1(u)\,du \tag{4.11}$$

is actually our operator of integration. It is obvious that I is linear and continuous.

With Theorem 4.1 we have also shown that the operator $D^{[1]}$ of differentiation is closed. Indeed,

PROPOSITION 4.4 *The operator* $D^{[1]}: U \subset X(G) \to X(G)$ is closed.

Proof One must show that relations

$$\lim_{n\to\infty} \|f_n - f\|_{X(G)} = 0, \tag{4.12}$$

$$\lim_{n\to\infty} \|D^{[1]}f_n - g\|_{X(G)} = 0 \tag{4.13}$$

imply that $f \in U$ and $g = D^{[1]}f$. Indeed, (4.13) implies by Proposition 3.3 and (2.21) that

$$\lim_{n\to\infty} [D^{[1]}f_n]^{\wedge}(k) = \lim_{n\to\infty} k\,f_n^{\wedge}(k) = g^{\wedge}(k) \quad (k \in \mathbb{P}).$$

By (4.12) and (2.21) again, this yields $g^\wedge(k) = kf^\wedge(k)$ for $k \in \mathbb{P}$. By Proposition 4.3 it therefore follows that (4.5) is valid for $r = 1$. Hence $f \in U$ and $D^{[1]}f = g$ by part (b) of Theorem 4.1.

We may at last formulate the full converse to Proposition 3.3. This yields the complete characterization of the class of differentiable functions in the sense of Definition 3.1 mentioned in the introduction.

THEOREM 4.2 *If $f \in X(G)$ the following are equivalent for $r \in \mathbb{N}$:*

(*i*) *$D^{[r]}f = g$ exists with $g \in X(G)$,*

(*ii*) *there exists $g \in X(G)$ such that $k^r f^\wedge(k) = g^\wedge(k)$, $k \in \mathbb{P}$,*

(*iii*) *there exists $g \in X(G)$ such that $f = W_r * g + f^\wedge(0)$.*

Proof The implication (i) $\Rightarrow$ (ii) follows by Proposition 3.3, (ii) $\Rightarrow$ (iii) by Proposition 4.3, and (iii) $\Rightarrow$ (i) by Theorem 4.1.

G being a locally compact Abelian group, it would be possible to define derivatives of functions on G by means of the Fourier transform and its inverse operator, as, for example, in M. Engert [3]. However, we prefer to define the derivative directly as a limit of elements belonging to the original space, and not involving transforms.

5. APPLICATIONS TO WALSH-FOURIER ANALYSIS

One of the reasons why Gibbs introduced his logical derivative was to show that the Walsh functions are eigensolutions of a differential equation. As our first application we consider a first-order linear differential equation that has the Walsh functions $\psi_n(x)$ as non-trivial eigensolutions, namely

$$D^{[1]}f(x) - \lambda f(x) = 0 \quad (\lambda \text{ real}) \tag{5.1}$$

with initial value $f(0) = 1$.

Since the solution must be differentiable, let us apply the Walsh–Fourier coefficients to this equation, obtaining by Theorem 4.2

$$(k-\lambda) f^\wedge(k) = 0 \quad (k \in \mathbb{P}).$$

This implies that for arbitrary fixed $\lambda \in \mathbb{P}$, $f(x) = \text{const.}\ \psi_\lambda(x)$. Since $f(0) = 1$ the solution is $f(x) = \psi_\lambda(x)$. In case $\lambda \notin \mathbb{P}$ there only exists the trivial solution $f = 0$.

As our second application we wish to consider the order of magnitude of the Walsh coefficients. For this purpose we need some further definitions.

If $f \in X(G)$, we write $f \in \text{Lip}\,(\alpha, X(G))$ provided (recall 2.1))

$$\|f(\circ) - f(\circ \dot{+} h)\|_{X(G)} = O[(\lambda(h))^\alpha] \quad (\alpha > 0, \text{ real}). \tag{5.2}$$

This concept is connected with the modulus of continuity of $f \in X(G)$, defined by

$$\omega(f, V_n) = \sup_{h \in V_n} \|f(\circ) - f(\circ \dotplus h)\|_{X(G)} \tag{5.3}$$

by the following result due to Morgenthaler [15] (in the formulation of Watari [23])

$$f \in \text{Lip}\,(\alpha, X(G)) \Leftrightarrow \omega(f, V_n) = O[2^{-n\alpha}] \quad (n \to \infty). \tag{5.4}$$

This enables us to establish the following important inequality.

PROPOSITION 5.1 *If $D^{[r]}f$ exists and belongs to $X(G)$, then*

$$\omega(f, V_n) = O[2^{-nr}\, \omega(D^{[r]}f, V_n)]. \tag{5.5}$$

Proof We need an identity for $h \in V_n$, namely

$$f(x) - f(x \dotplus h) = W_r^{(n)} * [D^{[r]}f(x) - D^{[r]}f(x \dotplus h)]. \tag{5.6}$$

To establish it we again apply the Walsh–Fourier coefficient method. The coefficients of the left side are by (2.15) and (2.9) $[1 - \psi_k(h)]f^\wedge(k)$ for $k \geqq 2^n$, and zero for $0 \leqq k < 2^n$. The coefficients of the right side are $[W_r^{(n)}]^\wedge(k) = 0$ for $0 \leqq k < 2^n$, and $k^{-r}k^r[1 - \psi_k(h)]f^\wedge(k)$ for $2^n \leqq k$. This gives (5.6).

In view of (5.6)

$$\omega(f, V_n) \leqq \|W_r^{(n)}\|_{X(G)}\, \omega(D^{[r]}f, V_n),$$

and this inequality implies (5.5) by Proposition 4.2.

COROLLARY 5.1 *If $D^{[r]}f \in X(G)$, then*

$$\omega(f, V_n) = O[2^{-nr}\, \|D^{[r]}f\|_{X(G)}].$$

Our application now depends upon a result due to Fine [4], namely, if $f \in X(G)$, then for $k \in \mathbb{N}$

$$|f^\wedge(k)| \leqq \omega(f, V_n) \quad (2^n \leqq k < 2^{n+1}).$$

This immediately yields by Proposition 5.1.

THEOREM 5.1 *If $D^{[r]}f \in X(G)$, then for $2^n \leqq k < 2^{n+1}$, $k \in \mathbb{N}$,*

$$|f^\wedge(k)| = O[2^{-nr}\, \omega(D^{[r]}f, V_n)].$$

This gives the inequality (1.3) of the introduction, more precisely

COROLLARY 5.2 *If $D^{[r]}f \in \text{Lip}\,(\alpha, X(G))$, then*

$$|f^\wedge(k)| = O[k^{-r-\alpha}].$$

Our third application is concerned with the order of approximation of the partial sums $s_n(f; x)$ of (2.16) to $f(x)$. This being connected with the concept of best approximation we need some auxiliary results.

If $f \in X(G)$, then

$$E_n(f, X(G)) = \inf_{p_n \in \mathscr{P}_n} \|f - p_n\|_{X(G)}$$

is called best approximation of f by Walsh polynomials of degree n in $X(G)$-space. There exists a Walsh polynomial $p_n^* = p_n^*(f, X(G))$ of best approximation to $f \in X(G)$ for which $E_n(f, X(G)) = \|f - p_n^*\|_{X(G)}$.

In view of the fact that $E_{2^n}(f, X(G)) \leqq \omega(f, V_n)$ (compare [23]) we have directly by Proposition 5.1 the following Jackson-type inequality

PROPOSITION 5.2 *If $D^{[r]}f \in X(G)$, then*

$$E_{2^n}(f, X(G)) = O[2^{-nr}\,\omega(D^{[r]}f, V_n)].$$

The third application, the analogue of the theorem of Lebesgue in the theory of Fourier series (see, e.g., [2, p. 105]), now reads

THEOREM 5.2 *If $D^{[r]}f \in X(G)$, then for $2^m \leqq n < 2^{m+1}$*

$$\|s_n(f, \circ) - f(\circ)\|_{X(G)} = O[n^{-r} \log n\, \omega(D^{[r]}f, V_m)].$$

Proof Indeed, by the standard method of proof one has since $s_n(p_n^*, x) = p_n^*(x)$

$$\|s_n(f, \circ) - f(\circ)\|_{X(G)} \leqq (1 + \|D_n\|_{L^1(G)})\, \|f - p_n^*\|_{X(G)} = O[\log n\, E_n(f, X(G))].$$

The theorem now follows by Proposition 5.2.

As a fourth application (recall definition of U in Section 3) we have

THEOREM 5.3 *For any $\alpha > 1$ and $1 \geqq \beta > 0$*

$$\mathrm{Lip}\,(\alpha, X(G)) \subset U \subset \mathrm{Lip}\,(\beta, X(G)).$$

Proof If $f \in \mathrm{Lip}\,(\alpha, X(G))$ for $\alpha > 1$, then $f \in U$ by Definition 3.1. If $f \in U$, then one has by Theorem 4.2 and Proposition 4.2 for $n \to \infty$

$$\|f(\circ) - f(\circ \dotplus h)\|_{X(G)} = O[2^{-n}\|D^{[1]}f\|_{X(G)}]$$

with $h \in V_n$. This gives the result.

The foregoing theorems are just a sample of possible applications of our derivative in Walsh–Fourier analysis.

Concerning our concept of the integral, it can easily be shown that any Walsh–Fourier series of $f \in C(G)$, whether convergent or not, may be integrated (in the sense of (4.11)) term by term, and the integrated series is uniformly convergent.

In a second paper under preparation the second named author will present the definition of the derivative, this time in its setting on the unit interval

[0, 1], also in the pointwise sense. Further applications will be given to Walsh–Fourier series, especially to the theory of approximation connected with Walsh polynomials.

References

[1] C. A. Bass, edit., Proceedings of the Symposium and Workshop on Applications of Walsh functions. Washington, D.C. 1970, viii+274 pp.

[2] P. L. Butzer and R. J. Nessel, Fourier Analysis and Approximation. Vol. 1: One-dimensional Theory, Academic Press, New York 1971, xvi+553 pp.

[3] M. Engert, A characterisation of differential operators on locally compact Abelian groups, Doctoral Dissertation. Stanford University, 1965, iv+55 pp.

[4] N. J. Fine, On the Walsh functions. *Trans. Amer. Math. Soc.* **65** (1949), 372–414.

[5] J. E. Gibbs, Some properties of functions on the non-negative integers less than 2^n, NPL (National Physical Laboratory), Middlesex, England. DES Report No. 3, 1969.

[6] J. E. Gibbs, Functions that are solutions of a logical differential equation. NPL DES Report No. 4, 1970.

[7] J. E. Gibbs, Digital filtering in dyadic-time and sequency. NPL DES Report No. 5, 1970.

[8] J. E. Gibbs and B. Ireland, Some generalizations of the logical derivative. NPL DES Report No. 8, 1971.

[9] J. E. Gibbs and M. J. Millard, Walsh functions as solutions of a logical differential equation. NPL DES Report No. 1, 1969.

[10] J. E. Gibbs and M. J. Millard, Some methods of solution of linear ordinary differential equations, NPL DES Report No. 2, 1969.

[11] G. H. Hardy and W. W. Rogosinski, Fourier Series, Cambridge Univ. Press (1944) 1962, 100 pp.

[12] H. F. Harmuth, Transmission of Information by Orthogonal Functions. Springer, Berlin and New York, 1969, xii+322 pp.

[13] E. Hewitt and K. A. Ross, Abstract Harmonic Analysis. Vol. I: Structure of Topological Groups, Integrations Theory, Group Representation. Springer 1963, viii+519 pp.

[14] V. M. Kokilasvili, On the best approximation by Walsh polynomials and Walsh–Fourier coefficients. *Bull. Acad. Polon. Sci. Sér. Sci. Math. Astronom, Phys.* **13** (1965), 405–410.

[15] G. W. Morgenthaler, On Walsh-Fourier series. *Trans. Am. Math. Soc.* **84** (1957), 472–507.

[16] L. G. Pál-F. Schipp, On Haar and Schauder series. *Acta Sci. Math. (Szegzed)* **31** (1970), 53–58.

[17] R. E. A. C. Paley, A remarkable series of orthogonal functions. *Proc. London Math. Soc.* **34** (1932), 241–279.

[18] F. Pichler, On state-space description of linear dyadic-invariant systems. Proc. Sympos. Applic. Walsh Functions, Washington, D.C. 1971, 166–170.

[19] W. Rudin, Fourier Analysis on Groups, Interscience Publ. 1962, ix+285 pp.

[20] J. L. Walsh, A closed set of orthogonal functions. *Am. J. Math.* **55** (1923), 5–24.

[21] Ch. Watari, Best approximation by Walsh polynomials. *Tôhoku Math. J.* **15** (1963), 1–5.

[22] Ch. Watari, Multipliers for Walsh Fourier series, *Tohoku Math. J.* (2) **16** (1964), 239–251.

[23] Ch. Watari, Approximation of functions by a Walsh-Fourier series [in C. A. Bass, Reference [1], 166–169].

[24] S. Yano, On Walsh-Fourier series, *Tôhoku Math. J.*, (2) 3 (1951), 223–242.

[25] P. L. Butzer and H. J. Wagner, Approximation by Walsh polynomials and the concept, of a derivative. In: Applications of Walsh functions (Proc. Symp. Naval Res. Lab. Washington, D.C., 27-29 March 1972; Ed. R. W. Zeek—A. E. Showalter) Washington, D.C. 1972, xi+401 pp.; pp. 388–392.

[26] P. L. Butzer and H. J. Wagner, On a Gibbs-type derivative in Walsh-Fourier analysis with applications. In: Proceedings of the National Electronics Conference, Chicago, 9–11 Oct. 1972; 27 (1972), xviii+457 pp.; pp. 393–398.

[27] H. J. Wagner, Ein Differential—und Integralkalkül in der Walsh-Fourier-Analysis mit Anwendungen. Dissertation, Rheinisch-West-fälische Technische Hochschule Aachen, 71 pp.; 1972 = (Forschungsberichte ges Landes Nordrhein-Westfalen) Köln Opladen (in print).

Date communicated: January 6, 1972.

APPROXIMATION BY WALSH POLYNOMIALS AND THE CONCEPT OF A DERIVATIVE

P.L. Butzer and H.J. Wagner
Technological University of Aachen
Aachen, Germany

1. Introduction

In a recent paper the authors [3] defined a derivative for functions f given on the dyadic group G. This derivative turned out to be a linear, closed operator, its inverse operator or integral was introduced, and the fundamental theorem of the calculus was found to hold for these two concepts. This enabled one to obtain first results on the approximation of f by the partial sums of the Walsh-Fourier series of f. The purpose of this paper is to present the definition of the derivative, this time in its setting on the unit interval [0,1], as well as to give a new application, namely to establish the fundamental theorem on best approximation (in the version of [1,2] for functions f defined on [0,1] by Walsh polynomials.

2. Preliminary results

Defining the Rademacher functions by

$$\phi_0(x) = \begin{cases} 1, & 0\leq x<1/2 \\ -1, & 1/2\leq x<1 \end{cases}, \quad \phi_0(x+1) = \phi_0(x),$$

$$\phi_n(x) = \phi_0(2^n x) \qquad (n\in\mathbb{N} = \{1,2,\ldots\}),$$

then the Walsh functions are given by

(2.1) $\Psi_0(x)=1, \Psi_n(x)=\phi_{n_1}(x)\cdots\phi_{n_i}(x)$

$n=2^{n_1}+\ldots+2^{n_i}, n_1>\ldots>n_i\geq 0$, n_j being integers. Let G be the dyadic group consisting of all sequences $\bar{x}=\{x_n\}_{n=1}^{\infty}$ such that $x_n=0$ or 1, the operation of G being addition modulo 2 (notation $\dotplus$). G is related to the unit interval by the mapping $\lambda: G\to[0,1]$

$$\lambda(\bar{x}) = \sum_{n=1}^{\infty} 2^{-n}x_n .$$

λ does not have a single-valued inverse since the dyadic rationals (=D.R.) have two representations in G. We shall agree to take the finite expansion in that case. If μ is the inverse of λ, then, according to N.J. Fine [4], one has for all real x

$$\lambda(\mu(x)) = x - [x],$$

$[x]$ denoting the greatest integer $\leq x$. Moreover, $\mu(\lambda(\bar{x})) = \bar{x}$ for all $\bar{x}\in G$ provided $\lambda(\bar{x}) \notin$ D.R. Denoting $\lambda(\mu(x)\dotplus\mu(y))$ in short by $x\oplus y$, then Fine [4] showed that

$$\int_0^1 f(x\oplus y)dx = \int_0^1 f(x)dx$$

for every fixed y provided that f is Lebesgue integrable on [0,1]. Furthermore, for each fixed y and for all x outside a certain denumerable set (depending on y), one has.

(2.2) $\Psi_n(x\oplus y) = \Psi_n(x)\Psi_n(y)$.

More generally, (2.2) is also valid for all $(x,y)\in[0,1]\times[0,1]$ outside a denumerable set.

A finite linear combination $\sum_{k=0}^{n-1} c_k\Psi_k(x)$ (c_k being complex numbers) is called a Walsh polynomial of degree n, the set of such polynomials of degree $<n$ being denoted by P_n. If $p_m\in P_{2^n}$, $h\in[0,2^{-n})$, then, outside the above denumerable set,

(2.3) $p_m(x\oplus h) = p_m(x)$.

Denoting by $L^p(0,1)$, $1\leq p<\infty$, the set of all functions f of period 1 which are pth power integrable with norm $\|f\| = \|f\|_p = \{\int_0^1|f(x)|^p dx\}^{1/p}$, it is known that the Walsh-system $\{\Psi_k(x)\}_{k=1}^{\infty}$ is closed with respect to the space $L^p(0,1)$, i.e.,

(2.4) $f^\wedge(k)=0,\ k\in \mathbb{P}, \Longrightarrow f(x)=0$ a.e.,

the Walsh-Fourier coefficients of f being

$$f^\wedge(k) = \int_0^1 f(u)\Psi_k(u)du \qquad (k\in\mathbb{P}=\{0,1,2,...\}).$$

By (2.2) for each real y and $k \in \mathbb{P}$

(2.5) $\int_0^1 f(y \oplus u)\Psi_k(u)du = \Psi_k(y)f^\wedge(k)$.

Defining the convolution of $f\in L^p(0,1)$ and $g\in L^1(0,1)$ by

$$(f*g)(x) = \int_0^1 f(x\oplus u)g(u)du$$
$$= \int_0^1 f(u)g(x\oplus u)du,$$

it is also known that $f*g$ exists for almost all real x, $f*g\in L^p(0,1)$ and

(2.6) $\|f*g\|_p \leq \|f\|_p\|g\|_1$.

Moreover, it is obvious that for $f\in L^p(0,1)$

$$|f^\wedge(k)| \leq \|f\| \qquad (k\in\mathbb{P}),$$

and the convolution theorem states that for $f,g \in L^1(0,1)$

(2.7) $(f*g)^\wedge(k)=f^\wedge(k)\cdot g^\wedge(k) \quad (k\in\mathbb{P})$.

Finally, if $f_n,f\in L^p(0,1)$, $n\in\mathbb{N}$, then for all $k \in \mathbb{P}$

(2.8) $\lim_{n\to\infty}\|f_n-f\| = 0 \Longrightarrow \lim_{n\to\infty} f_n^\wedge(k) = f^\wedge(k)$

3. <u>The Derivative and its properties</u>

<u>Definition</u>

If for $f\in L^p(0,1)$ there exists $g\in L^p(0,1)$ such that

(3.1) $\lim_{m\to\infty}\|\frac{1}{2}\sum_{j=0}^m 2^j[f(\cdot)-f(\cdot\oplus 2^{-j-1})]-g(\cdot)\|=0$,

then g is called the strong derivative of f, denoted by $D^{[1]}f$. For r=2,3,... the rth strong derivative of $f\in L^p(0,1)$ is defined successively by

$$D^{[r]}f = D^{[1]}D^{[r-1]}f .$$

Note that this definition differs from that given by J.E. Gibbs [5] essentially in the sense that the factor 2^j is replaced by 2^{-j}; however, Gibbs' definition is taken in the pointwise sense.

$D^{[r]}$ is a linear operator with the property

<u>Proposition 3.1</u>

(3.2) $D^{[r]}\Psi_n = n^r\Psi_n \qquad (n\in\mathbb{P},r\in\mathbb{N})$.

Using (2.2), the proof for r=1 follows readily from the identity ($n\in\mathbb{P}$)

(3.3) $\sum_{j=0}^{\infty} 2^j[1-\Psi_n(2^{-j-1})] = 2n$.

To prove (3.3), note that for $m,j\in\mathbb{P}$ $\phi_m(2^{-j-1}) = \phi_0(2^m 2^{-j-1}) = -1$ for m=j, =1 for $m\neq j$. Then apply (2.1).

Hence Walsh polynomials are arbitrarily often differentiable. Furthermore, (3.3) together with (2.8) implies that

$$[D^{[r]}f]^\wedge(k) = k^r f^\wedge(k) \qquad (k\in\mathbb{P})$$

under the existence of $D^{[r]}f\in L^p(0,1)$. It is obvious that the strong derivative of a constant vanishes and, conversely, $D^{[1]}f = 0$ implies f = const. in view of (2.5), (3.3), (2.8) and (2.4).

Of importance is the function $W_r(x)$ defined by (see Watari [7])

$$W_r^{\wedge}(k) = \begin{cases} 1, & k = 0 \\ k^{-r}, & k \in \mathbb{N} \end{cases} \qquad (r \in \mathbb{N}).$$

Having the theory developed in Butzer-Wagner [3] at our disposal, it can be shown by parallel methods that $W_r(x) \in L^1(0,1)$. Moreover, if $f,g \in L^p(0,1)$ such that $k^r f^{\wedge}(k) = g^{\wedge}(k)$, $k \in \mathbb{P}$, $r \in \mathbb{N}$, then

$$f = W_r * g + f^{\wedge}(0).$$

Theorem 3.1

Let $f \in L^p(0,1)$ such that $f^{\wedge}(0) = \int_0^1 f(x)dy = 0$.

a) If there exists $D^{[r]} f \in L^p(0,1)$ for some $r \in \mathbb{N}$, then

$$(W_r * D^{[r]} f) = f$$

b) $D^{[r]}(W_r * f) = f$

This theorem may be regarded as the fundamental theorem of the calculus for our concept of a derivative. In this sense, our operator of integration I is given by

$$If \equiv (W_1 * f) := \int_0^1 f(x \oplus u) W_1(u) du .$$

It is obvious that I is linear and continuous. On the other hand, it can be shown that the operator $D^{[r]}$ of differentiation is closed on $L^p(0,1)$.

All of the results established in Butzer-Wagner [3], including the applications, carry over from the dyadic group G to the interval [0,1]. The following new applications are proved in detail.

4. Applications to Approximation by Walsh Polynomials

In the notation of G.W. Morgenthaler [6] define for $f \in L^p(0,1)$, $\alpha > 0$

$$\text{Lip}\ \alpha(W) = \{f; \| f(\cdot) - f(\cdot \oplus h)\| = O(h^{\alpha}), h \to 0\}$$

$$\omega_W(f;\delta) = \sup_{0<h<\delta} \| f(\cdot) - f(\cdot \oplus h)\| ,$$

$$E_n(f) = \inf_{p_n \in P_n} \| f - p_n \| ,$$

the latter quantity being called best approximation of f by Walsh polynomials of degree n in $L^p(0,1)$-space. It is known that there exists a Walsh polynomial p_n^* $(=p_n^*(f))$ of best approximation to $f \in L^p(0,1)$ for which $E_n(f) = \| f - p_n^* \|$.

According to Morgenthaler and Watari [8] it is known that for $\delta \to 0$

(4.1) $f \in \text{Lip}\ \alpha(W) \Leftrightarrow \omega_W(f;\delta) = O(\delta^{\alpha})$,

(4.2) $E_{2^n}(f) \leq \omega_W(f;\frac{1}{2^n}) \leq 2\, E_{2^n}(f)$,

respectively. Analogous to [3] we have

Proposition 4.1

If $D^{[r]} f$ exists and belongs to $L^p(0,1)$, then

$$\omega_W(f;\frac{1}{2^n}) = O[\frac{1}{2^{nr}}\, \omega_W(D^{[r]} f;\frac{1}{2^n})] .$$

Of fundamental importance are the Bernstein and Jackson-type inequalities given by

Proposition 4.2

a) For $p_n \in P_n$ one has

$$\| D^{[r]} p_n \| \leq A n^r \| p_n \| \qquad (n \in \mathbb{P}, r \in \mathbb{N}).$$

b) If $D^{[r]} f$ exists and belongs to $L^p(0,1)$, then

$$E_n(f) \leq B n^{-r} \| D^{[r]} f \| .$$

Here A and B are constants independent of n, p_n, f.

390

Proof

If n has the representation (2.1), then $p_n \in P_{2^{n_1+1}}$. By (2.3) one has

$$\frac{1}{n}\|D^{[1]}p_n\|$$

$$= \frac{1}{n}\|\frac{1}{2}\sum_{j=0}^{\infty} 2^j [p_n(\cdot)-p_n(\cdot \oplus 2^{-j-1})]\|$$

$$< \frac{1}{n}\sum_{j=0}^{n_1} 2^j \|p_n\| < \|p_n\| \sum_{j=0}^{n_1} 2^{j-n_1} < 2\|p_n\|.$$

The proof of a) now follows by induction.

Under the hypotheses of b) the assertion of b) follows immediately by (4.2) and Prop. 4.1.

The following, which is the fundamental theorem on best approximation by Walsh polynomials, shows the usefulness of our derivative concept.

Theorem 4.1

If $f \in L^p(0,1)$, the following statements are equivalent:

(i) $D^{[r]}f \in \text{Lip}\,\alpha(W)$ $(\alpha > 0)$,

(ii) $\omega_W(D^{[r]}f; \frac{1}{n}) = O(n^{-\alpha})$,

(iii) $E_n(f) = O(n^{-\alpha-r})$,

(iv) $D^{[\nu]} f \in L^p(0,1)$, $0 < \nu < r$,

$\|D^{[\nu]}f - D^{[\nu]} p_n^*\| = O(n^{+\nu-\alpha-r})$,

(v) $\|D^{[1]}p_n^*\| = O(n^{1-\alpha-r})$ $(0<\alpha+r<1)$.

Proof

The equivalence (i)$\Leftrightarrow$(ii) follows by Morgenthaler [6] (who considered case $r = 0$; see (4.1)).The implication (ii) $\Rightarrow$(iii) follows by (4.2) and Prop. 4.1.

To prove (iii)$\Rightarrow$(iv) first note that (iii) implies $\|U_k^*\| = O[2^{-k(\alpha+r)}]$,

$$\text{where } U_k^* = \begin{cases} p_{2^k}^* - p_{2^{k-1}}^*, & k=2,3,\ldots, \\ p_1^*, & k=1. \end{cases}$$

Therefore by Prop. 4.2 a) for $k \in \mathbb{N}$

$$(4.3)\quad \|D^{[\nu]}U_k^*\| = O[2^{k(\nu-\alpha-r)}].$$

Now the sum $\sum_{k=1}^{m} U_k^*$ converges in $L^p(0,1)$-norm to f. Since vor $0 < \nu < r$

$$\|\sum_{k=1}^{m} D^{[\nu]}U_k^*\| < C\sum_{k=1}^{m} 2^{k(\nu-\alpha-r)} < \infty,$$

there exist functions $g_\nu \in L^p(0,1)$ with

$$\lim_{m\to\infty} \|\sum_{k=1}^{m} D^{[\nu]}U_k^* - g_\nu\| = 0.$$

Since the operators $D^{[\nu]}$ are closed, $D^{[\nu]}f = g_\nu$ for $0 < \nu < r$. By (4.3) we have

$$\|D^{[\nu]}(f-p_{2^m}^*)\| = \|D^{[\nu]}f - \sum_{k=1}^{m} D^{[\nu]}U_k^*\|$$

$$< \|\sum_{k=m+1}^{\infty} D^{[\nu]}U_k^*\| = O[2^{m(\nu-\alpha-r)}].$$

Now for $2^m < n < 2^{m+1}$ one has by (4.3) $\|D^{[\nu]}(p_n^*-p_{2^m}^*)\| = O(n^{\nu-\alpha-r})$. Combining the two inequalities yields (iv).

(iv)$\Rightarrow$(ii) by (4.2). It remains to show that (iii)$\Leftrightarrow$(v). This follows along standard lines, compare Butzer-Scherer [1,p.118 f, or 2].

In contrast to the situation for classical Fourier series, the assertion (ii) need only be formulated for the modulus of continuity of the first difference.

The results of this paper may also be established for the space $L^\infty(0,1)$, taking into account slight modifications.

The contribution of the second-named author was supported by the Minister für Forschung des Landes Nordrhein-Westfallen - Landesamt für Forschung.

References

[1] P.L. Butzer - K. Scherer: Über die Fundamentalsätze der klassischen Approximationstheorie in abstrakten Räumen, in: "Abstract Spaces and Approximation" (Proceedings of the Oberwolfach Conference 1968, P.L. Butzer and B. Sz.-Nagy, Eds.), ISNM, Vol. 10. (Basel 1969), 113 - 125.

[2] P.L. Butzer - K. Scherer: Jackson and Bernstein-type inequalities for families of commutative operators in Banach spaces, J. Approximation Theory 5 (1972) (in print).

[3] P.L. Butzer - H.J. Wagner: Walsh-Fourier series and the concept of a derivative, Applicable Analysis (in print).

[4] N.J. Fine: On the Walsh functions, Trans. Amer. Math. Soc., 65 (1949), 372 - 414.

[5] J.E. Gibbs - B. Ireland: Some generalizations of the logical derivative, NPL : DES Report No. 8, 1971.

[6] G.W. Morgenthaler: On Walsh-Fourier series, Trans. Amer. Math. Soc., 84 (1957), 472 - 507.

[7] C. Watari: Multipliers for Walsh-Fourier series, Tôhoku Math. J. (2) 16 (1964), 239 - 251.

[8] C. Watari: Approximation of functions by a Walsh-Fourier series, Proceedings of the Symposium and Workshop on Applications of Walsh Functions, Naval Research Laboratory, Washington D. C., 1970, 166 - 169.

ON A GIBBS-TYPE DERIVATIVE IN WALSH-FOURIER ANALYSIS WITH APPLICATIONS

P.L. Butzer and H.J. Wagner

Technological University of Aachen

Aachen, West Germany

I. INTRODUCTION

The idea of representing numerical data in binary notation is of current interest in communication engineering, information science, system theory, etc. The equivalent mathematical idea is that of a dyadic expansion, which, in turn, is associated with the dyadic group G (= set of all infinite sequences $x = \{x_s\}$ with $x_s = 0$ or 1, $s \in N = \{1,2,\ldots\}$, with termwise addition modulo 2, in symbols $\dotplus$) and its characters the Walsh functions

$$\{\phi_k(x)\}_{k=0}^{\infty}.$$

In the analysis of functions f defined on G developed so far, such essential concepts as continuity, Lipschitz conditions, etc, have been employed (see Fine [4], Morgenthaler [8]). However, with the exception of several articles by J.E. Gibbs [5,6] as well as by F. Pichler [10] and recent ones by the authors [2,3], no attempt has been made to define the concept of a derivative, one which is fundamental in all of analysis.

It is our purpose to present a modified Gibbs derivative in the norm-sense (compare [2,3]) as well as in the more-classical point-wise sense, this time with respect to the Walsh-Kaczmarz system $\{\phi_k\}_{k=0}^{\infty}$ ($\phi_k(x) = \mathrm{wal}\,(k,x)$) - a system which is sequency ordered in the sense of H.F. Harmuth [7]. It is shown that the norm-derivative is equal to the point-wise derivative on G, an operator inverse to the derivative, namely the integral operator is introduced, the counterpart of the fundamental theorem of the calculus is valid for these two concepts. Applications are given to speed of convergence of Walsh-Fourier series as well as to a wave-type "differential" equation.

II. DEFINITIONS AND PRELIMINARY RESULTS

If $k \in P = \{0,1,2,\ldots\}$ and $2^K \le k < 2^{K+1}$, $K \in P$, then (for $k = 0$: $k_s = 0$, $s \in P$)

$$k = \sum_{s=0}^{K} k_s 2^s \quad (k_s \in \{0,1\};\ s \in P)$$

is the unique dyadic expansion of k. The Walsh-Kaczmarz [11] system on G is defined as ($x_0 = 0$)

$$\phi_k(x) = \exp\left[\pi i \sum_{s=0}^{K} k_s (x_s + x_{s+1})\right]. \quad (1)$$

A second definition of the Walsh functions, actually a different ordering, is the Walsh-Paley system (see R.E.A.C. Paley [9]) defined by

$$\psi_k(x) = \exp\left[\pi i \sum_{s=0}^{K} k_s x_{s+1}\right]. \quad (2)$$

The relationship between the two systems (see Pichler [10]) is

$$\phi_k(x) = \begin{cases} \psi_{k/2}(x)\, \psi_k(x), & k = 0,2,4,\ldots \\ \psi_{(k-1)/2}(x)\, \psi_k(x), & k = 1,3,5,\ldots \end{cases}$$

Moreover, the set $\{\phi_o,\ldots,\phi_{2^n-1}\}$ contains the same functions as $\{\psi_o,\ldots,\psi_{2^n-1}\}$. Therefore the Walsh-Dirichlet kernel

$$D_{2^n}(x) \equiv \sum_{k=o}^{2^n-1} \phi_k(x) = \sum_{k=o}^{2^n-1} \psi_k(x). \quad (3)$$

If $\lambda(x) = \sum_{s=1}^{\infty} 2^{-s} x_s$, then

$$\lambda(x\dot{+}y) = \sum_{s=1}^{\infty} 2^{-s}|x_s - y_s| \quad (4)$$

defines a metric on G. f is said to be continuous at $x\varepsilon G$ if to each $\varepsilon > 0$ there is $\delta = \delta(\varepsilon, x) > 0$ such that

$$|f(x)-f(y)|<\varepsilon \quad (y\varepsilon N_\delta(y) = \{y|\lambda(x\dot{+}y)<\delta\}).$$

Denoting by $X(G)$ one of the spaces $C(G)$, $L^p(G)$, $1 \leq p < \infty$, of continuous functions f on G or integrable to the pth power on G, with norms

$$\|f\|_{C(G)} = \sup_{x\varepsilon G}|f(x)|,$$

$$\|f\|_{L^p(G)} = \{\int_G |f(x)|^p dx\}^{1/p},$$

respectively, it is known that

$$\lim_{n\to\infty} \|s_{2^n}(f;x) - f(x)\|_{X(G)} = o, \quad (5)$$

where

$$s_m(f;x) = \sum_{k=o}^{m-1} f^\wedge(k)\, \phi_k(x),$$

the Walsh-Fourier coefficients of $f\varepsilon X(G)$ being

$$f^\wedge(k) = \int_G f(u)\, \phi_k(u)du \quad (k\varepsilon P).$$

III. DEFINITION OF NORM-DERIVATIVE

For each fixed $j\varepsilon P$ let $h_j=\{h_{j,s}\}$, $s\varepsilon N$, where

$$h_{j,s} = \begin{cases} 0 \text{ for } 1 \leq s \leq j \\ 1 \text{ " } j < s. \end{cases}$$

DEFINITION 1

If for $f\varepsilon X(G)$ there exists $g\varepsilon X(G)$ such that

$$\lim_{m\to\infty}\|\frac{1}{2}\sum_{j=o}^{m} 2^j[f(x)-f(x\dot{+}h_j)]-g(x)\|_{X(G)}=o,$$

then g is called the norm (or strong) derivative of f, denoted by $D^{[1]}f$. For any $r = 2,3,\ldots$ the rth strong derivative of $f\varepsilon X(G)$ is defined successively by

$$D^{[r]}f = D^{[1]}(D^{[r-1]}f).$$

$D^{[r]}$ is a linear, closed operator with the property

PROPOSITION 1

If for $f\varepsilon X(G)$ there exists $D^{[r]}f\varepsilon X(G)$ for some $r\varepsilon N$, then

$$[D^{[r]}f]^\wedge(k) = k^r f^\wedge(k) \quad (k\varepsilon P).$$

Moreover, the Walsh-Kaczmarz functions are arbitrary often differentiable with

$$D^{[r]}\phi_n = n^r \phi_n \quad (r\varepsilon N,\ n\varepsilon P) \quad (6)$$

The proofs of Prop. 1 and (6) follow readily (compare [2,3]) from the identity $(n\varepsilon P)$

$$\sum_{j=o}^{\infty} 2^j [1-\phi_n(h_j)] = 2n. \quad (7)$$

To establish (7), note that

$$\phi_n(h_j)=\exp[\pi i\sum_{s=o}^{k} n_s(h_{j,s}+h_{j,s+1})]$$

$$= -1 \text{ for } n_j=1, =+1 \text{ for } n_j=o.$$

Let $W_r(x)$ be the function defined for $r\varepsilon N$ by $W_r^\wedge(k) = 1$ for $k = o$, $=k^{-r}$ for $k\varepsilon N$; its Walsh-Fourier series has the form $1 + \sum_{k=1}^{\infty} k^{-r}\phi_k(x)$.

This enables one to introduce an operator inverse to $D^{\{r\}}$, actually the integral of $f\epsilon X(G)$ of order r, defined by

$$(I^r f)(x) = (W_r * f)(x) = \int_G f(x \dotplus u) W_r(u)du; \quad (8)$$

this integral, the resultant of f and W_r, is meaningful since $W_r(x)\epsilon L^1(G)$.

It is obvious that I^r is linear and continuous; the fact that it plays the role of an integral follows by

THEOREM 1

Let $f\epsilon X(G)$ such that $f^{\wedge}(o)=\int_G f(x)dx=o$.

a) $D^{\{r\}}(I^r f) = f$,

b) If $D^{\{r\}} f\epsilon X(G)$, then

$$I^r(D^{\{r\}}f) = f.$$

This theorem, which is really the counterpart of the fundamental theorem of the differential and integral calculus for our definitions of a derivative and integral, may be established along the lines of Butzer-Wagner [2]. It will enable us to set up a calculus parallel to the Leibniz-Newton calculus but which fits the analysis concerned with Walsh functions.

As a first application, by Prop. 1 the Walsh-Kaczmarz functions $\phi_n(x)$ are the non-trivial proper solutions of the first order linear "differential" equation

$$D^{\{1\}} - \lambda f = o \qquad (\lambda \text{ real})$$

with initial value $f(o) = 1$.

IV. THE POINT-WISE DERIVATIVE ON C(G)

Although our strong derivative (Def. 1) possesses properties as well applications which are analogous to those for the ordinary derivative, the situation is very different for the derivative considered in the point-wise sense.

If f is defined on G, then

$$f^{[1]}(x) = \frac{1}{2}\sum_{j=o}^{\infty} 2^j [f(x) - f(x \dotplus h_j)] \qquad (9)$$

is called its first (point-wise) derivative at $x\epsilon G$. The rth point-wise derivative of f is defined as usual.

If to $f\epsilon C(G)$ there exists $D^{\{r\}}f\epsilon C(G)$ for some $r\epsilon N$, then $f^{[r]}(x)$ exists at each $x\epsilon G$, $f^{[r]}(x)\epsilon C(G)$ and $f^{[r]}(x)=(D^{\{r\}}f)(x)$ for all $x\epsilon G$, i.e. both derivatives are equal. This follows since $|g(x)| \leq \|g(x)\|_{C(G)}$.

The converse is also true. Indeed

THEOREM 2

If $f(x)$, $f^{[r]}(x)$ exist for each $x\epsilon G$ and belong to $C(G)$, then there exists $D^{\{r\}}f\epsilon C(G)$ with $(D^{\{r\}}f)(x) = f^{[r]}(x)$ on G.

To prove Theorem 2 we need three propositions which also reveal the structure of the point-wise derivative.

PROPOSITION 2

If f defined on G possesses an absolute maximum (or minimum) at x (or y) and $f^{[1]}(x)$ (or $f^{[1]}(y)$) exist, then

$$f^{[1]}(x) \geq o \qquad (\text{or } f^{[1]}(y) \leq o).$$

The proof follows directly by (9).

PROPOSITION 3

If $f\epsilon C(G)$ and $f^{[1]}(x) = c$ on G, then $c = o$.

Since G is compact, for each continuous f defined on G there exist $u,v\epsilon G$ such that

$$\left.\begin{matrix} M \\ m \end{matrix}\right\} = \left\{ \begin{matrix} \sup \\ \inf \\ x\epsilon G \end{matrix} f(x) \right\} = \left\{ \begin{matrix} f(u) \\ f(v) \end{matrix} \right. .$$

Assume $f^{[1]}(x) = c \neq o$. By Prop. 2 $f^{[1]}(u) \geq o$, $f^{[1]}(v) \leq o$, thus $c = o$.

The derivative of a constant is zero. Conversely,

PROPOSITION 4

If $f \in C(G)$ and $f^{[1]}(x) = o$ for all $x \in G$, then $f(x) = $ const.

Following the first steps of the proof of Prop. 3, if $M = m$, then $f = $ const.

If $M > m$ then $f(u) - f(v) = M - m > o$. Since f is uniformly continuous on the compact G, one has (recall (4))

$$|f(v)-f(y)|<\varepsilon<M-m \quad (y \in N_{\delta(\varepsilon)}(v)).$$

We construct a point for which f takes on its absolute maximum and which lies in $N_{\delta(\varepsilon)}(v)$ - this contradicts the assumption. Now there exists $w \in G$ with $u \dotplus w = v$. So for $w' = \{w_1, \ldots, w_n, o, \ldots\}$ one has $u \dotplus w' = \{v_1, \ldots, v_n, u_{n+1}, \ldots\} \in N_{\delta(\varepsilon)}(v)$ if $2^{-n} < \delta(\varepsilon)$ by (4).

It remains to show $f(u \dotplus w') = M$. Since $f^{[1]}(u) = o$, $f(u) = f(u \dotplus h_j) = M$ for all $j \in P$ by (9). More generally, one has for

$$t = \delta_o h_o \dotplus \ldots \dotplus \delta_m h_m \quad (\delta_j \in \{o,1\}) \quad (10)$$

with $\delta_j h_j = \{\delta_j h_{j,s}\}$, $s \in N$, that $f(u)=f(u \dotplus t)=M$. But w' is representable as

$$w'=w_1 h_o \dotplus w_1 h_1 \dotplus \ldots \dotplus w_n h_{n-1} \dotplus w_n h_n.$$

With (10) this implies $f(u)=f(u \dotplus w')=M$.

PROOF OF THEOREM 2

Let $g = I^1(f^{[1]} - [f^{[1]}]^\wedge(o))$. Then g is strongly differentiable on $C(G)$ by Theorem 1 and so also point-wise, i.e.,

$$(D^{[1]}g)(x)=g^{[1]}(x)=f^{[1]}(x)-[f^{[1]}]^\wedge(o).$$

By Prop.3 $[f^{[1]}]^\wedge(o) = o$ and so $g(x) = f(x) + $ const. by Prop. 4. Since g is strongly differentiable so is f with $(D^{[1]}f)(x) = f^{[1]}(x)$.

Therefore the strong derivative is equal to the point-wise derivative for $C(G)$. The corresponding result holds for the derivatives $D^{[1]}f$, $f^{[1]}(x)$ in [2,3].

V. PECULIARITIES OF THE POINT-WISE DERIVATIVE

The fundamental difference between the ordinary concept of a derivative and ours in the sense of (9) is that the former is a <u>local</u> property (the function need only be defined in a small neighborhood about the point when setting up the classical differential quotient), whereas the latter depends upon elements x, $x \dotplus h_j$ which are at a definite distance apart.

This is one intuitive reason why such peculiarities as Props. 2 and 3 as well as the following dissimilarities arise:

a. In contrast with the ordinary derivative, differentiability of a function f at a point $x \in G$ in the sense of (9) does not imply continuity of f at x. This follows by the example

$$f_1(x) = \begin{cases} o & \text{for } x = o,\ h_j (j \in P) \\ 1 & \text{otherwise.} \end{cases}$$

$f_1(x)$ is differentiable at $x=o=\{o,o,\ldots\}$ with $f_1^{[1]}(o) = o$, but $f_1(x)$ is not continuous at $x = o$.

On the other hand, just as in the classical instance, there exist continuous functions which are not differentiable at some point. For example,

$$f_2(x) = \lambda(x) = \sum_{j=1}^{\infty} x_j 2^{-j}$$

is continuous on G but $f_2^{\{1\}}(o)$ does not exist.

b. In contrast with the ordinary integral, the integral $(I^1f)(x) = (W_1 * f)(x)$ of a positive function need not be positive; for $W_1(x)$ is not positive. For example, take (see (3))

$$D_4(x) = \begin{cases} 4 \text{ for } x_1 = x_2 = o, x_3, .. \text{arbitrary} \\ o \text{ otherwise.} \end{cases}$$

Here $(I^1D_4)(x) = 1 + \sum_{k=1}^{3} k^{-1}\phi_k(x)$; $(I^1D_4)(x) < o$ at all $x \in G$ with $x_1=1, x_2=o$ and x_s, $s > 3$ arbitrary.

These counterexamples reveal that classical concepts such as continuity, positivity (or monotonicity) do not fit in with our definitions of a derivative or integral. The classical derivative is associated with the slope of a tangent, the integral with area; both concepts are associated with geometrical notions. An open problem is an interpretation of our derivative, not via geometrical or physical notions, but perhaps in the language of information science (compare Gibbs [6] and literature cited).

VI. SOME APPLICATIONS

In connection with filtering a function f (signal), Walsh-Fourier series turn out to be an important tool. In this respect, our derivative enables one to estimate the degree of approximation of a function f by the partial sums of the Walsh-Fourier series (recall (5)). Indeed, by methods similar to those of Butzer-Wagner [2, Thm. 5.2] it can be shown that:

If $f^{\{r\}}(x) \in C(G)$, then

$$\| s_n(f;x) - f(x) \|_{C(G)} \leq M \frac{\log n}{n^r} \| f^{\{r\}}(x) \|_{C(G)}$$

for some constant $M > o$.

In general, the results of this lecture, in particular Def. 1, may be restated for functions defined on the interval [0,1] (instead of on G) provided $f \in L^p(0,1), 1 < p < \infty$ (compare [3]).

In this setting one can solve a partial "differential" equation, namely the wave equation for $o < x,t < 1$:

$$\frac{\partial^{\{2\}}}{\partial t^{\{2\}}} w(x,t) = \frac{\partial^{\{2\}}}{\partial x^{\{2\}}} w(x,t) \qquad (11)$$

under the conditions

$$w(x,o) = f(x), \quad w(o,t) = w(1,t),$$

$$\frac{\partial^{\{1\}}}{\partial t^{\{1\}}} w(x,o) = \frac{\partial^{\{1\}}}{\partial t^{\{1\}}} w(x,1).$$

Using the Walsh-Fourier coefficient method of [2], which is parallel to that applied in Butzer-Nessel [1, Chap.7], the solution is obtained formally as follows: Applying Walsh-Fourier coefficients $w^\wedge(k,t) = \int_0^1 w(x,t)\phi_k(x)dx$ to both sides of (11), one has (by Prop. 1)

$$\frac{\partial^{\{2\}}}{\partial t^{\{2\}}} w^\wedge(k,t) = k^2\, w^\wedge(k,t) \qquad (k \in P).$$

This ordinary second-order differential equation has solution

$$w^\wedge(k,t) = A(k)\phi_k(t) \qquad (o < t < 1)$$

with constant $A(k)$ independent of t. Since $w(x,o) = f(x)$, one has $A(k) = f^\wedge(k)$, $k \in P$. The formal solution of (11) is now readily seen to be

$$w(x,t) = \sum_{k=o}^{\infty} f^\wedge(k)\phi_k(t)\phi_k(x).$$

It is actually the Walsh-Fourier expansion of $f(x \oplus t)$, $\oplus$ being the analog of the operation $\dot{+}$ for [0,1].

Walsh functions have recently been widely applied in communication engineering including coding theory, band-width compression, digital and analog multiplexing, voice coding, and analog and digital filtering. It would be of great interest to find an appropriate meaning for the above differential equation, e.g., in terms of these sciences. Much work is still to be done.

ACKNOWLEDGMENT

The contribution of the second-named author was supported by the Minister für Forschung des Landes Nordrhein-Westfalen-Landesamt für Forschung.

REFERENCES

[1] P.L. Butzer - R.J. Nessel: Fourier Analysis and Approximation. Vol.I: One-dimensional Theory, Academic Press, New York 1971, xvi + 553 pp.

[2] P.L. Butzer - H.J. Wagner: Walsh-Fourier series and the concept of a derivative, Applicable Analysis 3 (1973) (in print).

[3] P.L. Butzer - H.J. Wagner: Approximation by Walsh polynomials and the concept of a derivative, Proc.Symp. Applications of Walsh Functions, Catholic University of America, Washington D.C., March 27-29, 1972.

[4] N.J. Fine: On the Walsh functions, Trans. Amer.Math.Soc., 65 (1949), 372-414.

[5] J.E. Gibbs - M.J. Millard: Walsh functions as solutions of a logical differential equation, National Physical Laboratory, DES Report No. 2, (1969).

[6] J.E. Gibbs - B. Ireland - D.R. Perkins: A generalization of the dyadic expansion, NPL, DES Report No.10, (1971).

[7] H.F. Harmuth: Transmission of Information by Orthogonal Functions. Second Printing, Springer Verlag, Berlin 1970, xi + 322 pp.

[8] G.W. Morgenthaler: On Walsh-Fourier series, Trans.Amer.Math.Soc., 84 (1957), 472-507.

[9] R.E.A.C. Paley: A remarkable series of orthogonal functions, Proc.London Math.Soc., 34 (1932), 241-279.

[10] F.R. Pichler: Walsh functions and linear system theory, In "Proc. Symp. and Workshop Applications of Walsh Functions" (C.A.Bass edit.), Wahington, DC 1970,175-182.

[11] J.L. Walsh: A closed set of normal orthogonal functions, Amer.J.Math., 45 (1923), 5-24.

Chapter 5
Dyadic Derivative, Summation, Approximation *

S. Fridli, F. Schipp

Abstract The "Hungarian school" has played an active role in the development of the theory of dyadic derivative, in particular in exploring its properties related to the fundamental theorem, right from the beginning. New concepts and techniques have emerged on which summaries can be found in the relevant chapters of the monographes [SchWadSim90], [Wei02]. In this survey paper we present the results on the fundamental theory of dyadic derivative, and their effect on the solutions of problems regarding to summation, approximation of Walsh–Fourier series, Cesáro operators etc., by focusing on the involvement of the Hungarian mathematicians. We note, that in the last ten years intensive research have been carried on in Russia (see e.g. [Gol13a*], [GolVol13a*]) by taking the generalized concept, due to Onneweer [Onn13*], instead of the original concept of the dyadic derivative. We think that these results augment each other well by showing the similarities and the differences between the two approaches. The latest results for the multidimension case are summed up in the paper of Weisz [Wei13*] in this volume. In general the * sign in the citations indicate that the paper, or its copy can be found in this volume.

5.1 Introduction

The era of modern theory of real analysis began by the introduction of Lebesgue's integration. The new concept not only opened the door for the extension of the

Sándor Fridli
Department of Numerical Analysis, Eötvös L. University, Budapest, Pázmány P. sétány 1\C, H-1117 Hungary, e-mail: fridli@inf.elte.hu

Ferenc Schipp
Department of Numerical Analysis, Eötvös L. University, Budapest, Pázmány P. sétány 1\C, H-1117 Hungary,e-mail: schipp@numanal.inf.elte.hu

* The Project is supported by the European Union (grant agreement no. TÁMOP 4.2.2/A-11/1/KONV-2012-0051).

R.S. Stanković et al. (eds.), *Dyadic Walsh Analysis from 1924 Onwards Walsh-Gibbs-Butzer Dyadic Differentiation in Science Volume 1 Foundations*, Atlantis Studies in Mathematics for Engineering and Science 12, DOI 10.2991/978-94-6239-160-4_5

Newton–Leibniz calculus but had a substantial influence on the development of several branches of mathematics as well. The new concept of integration played a vital role in the fundamental results of analysis such as the theory of L^p spaces and the Riesz–Fischer theorem. Even Lebesgue himself considered his theorems on the almost everywhere differentiability of the integral function, i.e.

$$\lim_{h\to 0}\frac{1}{h}\int_x^{x+h} f(t)\,dt - f(x) = \lim_{h\to 0}\frac{1}{h}\int_x^{x+h} (f(t)-f(x))\,dt = 0 \tag{5.1}$$

holds almost everywhere, and the a.e. differentiability of monotonic functions as the summits of the new theory. The new approach appeared in the theory of Fourier series very soon. In 1905 Lebesgue [Leb05] generalized Fejér's [Fej04] result concerning the $(C,1)$ summability of continuous functions. Namely, he proved that the arithmetic means of the Fourier series of a 2π periodic Lebesgue–integrable function converge to the function a.e.. In connection with that he introduced a descriptive condition that holds for integrable functions at a.e. point. Namely, Lebesgue showed that the condition

$$\lim_{h\to 0}\frac{1}{h}\int_x^{x+h} |f(t)-f(x)|\,dt = 0 \tag{5.2}$$

which is stronger than (5.1) holds at a.e. point. Since then points satisfying (5.2) are called the *Lebesgue* points of the locally integrable function f. Lebesgue proved that at every such point the Fejér–means $F_n f$ of the integrable function f converge to the function, i.e.

$$\lim_{n\to\infty} (F_n f)(x) = f(x).$$

In 1930 Hardy and Littlewood [HarLit30] introduced the concept of the maximal function f^* of a function $f \in L^1[0,1]$

$$f^*(x) := \sup_h \frac{1}{h}\int_x^{x+h} |f(t)|\,dt \qquad (x \in [0,1]) \tag{5.3}$$

which later became an effective tool in analysis. Hardy and Littlewood proved that the maximal operator $Mf := f^*$ is bounded on L^p $(p>1)$ and is of weak type $(1,1)$, i.e. for every $f \in L_{2\pi}$ there exists a $K>0$ such that

$$\operatorname{mes}\{x \in [0,1] : f^*(x) > y\} \le K\|f\|_1/y \qquad (y>0). \tag{5.4}$$

Such type of inequalities play an important role in the investigations on a.e. convergence. It turned out [Ste61] for example that (5.4) is necessary and sufficient for the existence and the values of the limits in (5.1) a.e. for every $f \in L^1_{2\pi}$.

It is the rule

$$\frac{d}{dx}\varepsilon_n(x) = ni\varepsilon_n(x) \qquad (x \in \mathbb{R}, n \in \mathbb{Z}) \tag{5.5}$$

that connects the complex trigonometric system $\varepsilon_n(x) := e^{inx}$ $(x \in \mathbb{R}, n \in \mathbb{Z})$ with the differentiation. In other words the trigonometric functions are eigenfunctions of the differential operator.

In 1967 Gibbs [Gib67*] introduced the operator of dyadic differentiation by means of the translations of finite dyadic groups. It is related to the Walsh system w_n $(n \in \mathbb{N})$ in a similar way as the trigonometric system in (5.5):

$$dw_n = nw_n \qquad (n \in \mathbb{N}).$$

Therefore the dyadic differentiation can be interpreted as a special Walsh multiplier operator. The dyadic derivative due to Gibbs provides a link between dyadic translations and multipliers, which proved to be very effective in handling various problems.

5.2 Dyadic Derivative

For notations and concepts in dyadic analysis we refer the reader to the monograph [SchWadSim90]. Instead of the dyadic group we will take the unit interval $\mathbb{I} := [0,1)$. Let $x \dot{+} t$ stand for the dyadic sum

$$x \dot{+} t := \sum_{k=0}^{\infty} |x_k - t_k| 2^{-k-1} \qquad (x, t \in \mathbb{I})$$

of

$$x = \sum_{k=0}^{\infty} x_k 2^{-k-1}, t = \sum_{k=0}^{\infty} t_k 2^{-k-1} \qquad (x_k, t_k \in \{0,1\}).$$

Then the members of the Walsh system in the Paley enumeration can be expressed by the binary digits

$$w_n(x) = (-1)^{\sum_{k=0}^{\infty} n_k x_k} \qquad (x \in \mathbb{I}, n = \sum_{k=0}^{\infty} n_k 2^k \in \mathbb{N}, n_k \in \{0,1\}). \tag{5.6}$$

We note that the original Walsh system, defined by Walsh [Wal23] in 1923, can be received from the $(w_n,\ n \in \mathbb{N})$ system by transforming the bits of the variable ([SchWadSim90] p. 19). This enables the transmission of the results on the Walsh–Paley system for the original Walsh–system.

5.2.1 Dyadic derivative of functions of one variable

Gibbs [Gib67*] introduced the operators

$$(d_n f)(x) := \sum_{k=0}^{n-1} 2^{k-1} (f(x) - f(x \dot{+} e_k)) \qquad (x \in \mathbb{I}, e_k := 2^{-k-1}, k, n \in \mathbb{N}) \tag{5.7}$$

as the dyadic analogues of the difference operators $(\Delta_h f)(x) := (f(x+h) - f(x))/h$. It follows from the decomposition in (5.6) that

$$d_n w_m = m w_m \qquad (m < 2^n, n \in \mathbb{N}). \tag{5.8}$$

Based on this relation Butzer and Wagner [ButWag72*], [ButWag73a*] extended the concept of dyadic derivative for functions defined on $\mathbb{I}$ in 1972. They introduced the notion of strong derivative with respect to the norm of the space X, where $X := L^p := L^p(\mathbb{I})$ $(1 \le p < \infty)$ or X the closure of the linear hull of the Walsh functions in the $L^\infty(\mathbb{I})$ norm. A function $f \in X$ is said to be strongly differentiable in X if there exists a $g \in X$ with

$$\lim_{n\to\infty} \|d_n f - g\|_X = 0.$$

The function $df := g$ is called the *strong dyadic derivative* of f. They showed that the operator d plays the same role in dyadic analysis as the classical derivative in the theory of Fourier series. By (5.8) it is obvious that the Walsh–functions $(w_n,\ n \in \mathbb{N})$ are differentiable also by the new definition and $dw_m = m w_m$ $(m \in \mathbb{N})$. They also showed that the operator d is closed and the dyadic derivative of f can be characterized by the Walsh–Fourier coefficients

$$\widehat{f}(n) := \int_0^1 f(t) w_n(t)\, dt \qquad (n \in \mathbb{N}).$$

Namely, $df = g$ if and only if $\widehat{g}(n) = n\widehat{f}(n)$ $(n \in \mathbb{N})$. The inverse of the operator d, i.e. the dyadic integral (antiderivative) I was given as well. They proved that the antiderivative can be described by the function $W \in L^1$ the Walsh–Fourier coefficients of which are

$$\widehat{W}(0) := 1\,, \quad \widehat{W}(n) := \frac{1}{n} \qquad (n \in \mathbb{N}^* := \mathbb{N} \setminus \{0\}).$$

Then the dyadic antiderivative, i.e. the inverse of d can be given as dyadic convolution operator generated by the kernel function W:

$$(Ig)(x) := (g * W)(x) := \int_0^1 W(x \dotplus t) g(t)\, dt \qquad (g \in L^1).$$

Indeed, the following statement holds true: If $g = df$ is the dyadic derivative of f then $Ig = I(df) = f$. This can be verified directly by comparing the Walsh–Fourier coefficients.

In the other direction the relation $d(If) = f$ was proved: For any $f \in X$ the function $If = f * W \in X$ is strongly differentiable in X, i.e.

$$\|d_n(If) - f\|_X \to 0 \qquad (n \to \infty). \tag{5.9}$$

Taking the operator sequence

$$d_n(If) = f * d_n W \qquad (f \in X, n \in \mathbb{N}) \tag{5.10}$$

we have [ButWag72*] that (5.9) is equivalent to

$$\sup_{n\in\mathbb{N}} \|d_n W\|_{L^1} < \infty.$$

Consequently, *the analogue of the Newton–Leibniz fromula holds for the dyadic derivative if it is taken in the strong sense.*

In connection with this Wagner raised the question in his dissertation [Wag73] whether the convergence in norm could be replaced by a.e. convergence. In other words is the theorem on the a.e. differentiability of the integral function true in the dyadic setting? If yes then it means that for any $f \in L^1$ the sequence in (5.10) converges to f a.e.. This implied the investigation of the maximal operator

$$(d^* f)(x) := \sup_{n\in\mathbb{N}} |(d_n If)(x)| \qquad (x \in \mathbb{I}, f \in L^1),$$

which is the dyadic version of the Hardy–Littlewood maximal operator (5.3). Schipp [Sch74*] proved that similarly to the classical Hardy-Littlewood maximal operator *the dyadic maximal operator d^* is of weak type (1,1), and of (∞,∞). Consequently, d^* is bounded on L^p if $1 < p \leq \infty$.* Hence the analogue of Lebesgue's theorem (5.9) follows by standard arguments [Sch74*], [Sch86]: For any $f \in L^1$ we have

$$\lim_{n\to\infty} d_n(If)(x) = f(x) \qquad (a.e.\, x). \tag{5.11}$$

Based on this result Butzer and Wagner [ButWag75*] introduced *the dyadic version of absolutely continuous functions* and discussed its role in dyadic analysis. The convergence result in (5.11) can be extended to Lebesgue–Stieltjes measure δF generated by a function F of bounded variation:

$$I(\delta F)(x) := \int_0^1 W(x \dot{+} t)\, \delta F(t) \qquad (x \in [0,1)).$$

Obviously, if F is absolutely continuous with respect to the Lebesgue measure then we get the original definition of If. As it was shown in [Sch76b] also in this case $\lim_{n\to\infty}(d_n I(\delta F))(x)$ *exists for a.e. $x \in \mathbb{I}$, and it is equal a.e. to the absolutely continuous component of F. In particular, if F is singular, i.e. $dF(x)/dx = 0$ for a.e. $x \in [0,1)$, then $dI(\delta F) = 0$ a.e..* This is the dyadic analogue of Lebesgue's theorem on the differentiability of monotonic functions. A detailed discussion of this topic can be found in Sections 1.7, 6.2 and 8.2 of the monograph [SchWadSim90]. For the history of dyadic derivative see [ButSta89*].

In the following sections we are concerned with a few results on the various generalizations of the fundamental theorem.

5.2.2 Dyadic derivative in the multivariable case

It is known that the following analogue of Lebesgue's theorem on the differentiability of the integral function holds true for the two dimensional case: Taking the integral means with respect to squares we have that for every $g \in L^1(\mathbb{I}^2)$ $(\mathbb{I}^2 := \mathbb{I} \times \mathbb{I})$

$$\lim_{h\to 0} \frac{1}{h^2} \int_x^{x+h} \int_y^{y+h} g(s,t)\,ds\,dt = g(x,y)$$

for a.e. $(x,y) \in \mathbb{I}^2$. Replacing the squares by rectangles the statement holds only for functions belonging to $(L^1 \log_+ L^1)(\mathbb{I}^2)$.

The study of the dyadic derivative in the multivariable case has started by the papers of Butzer and Wagner [ButWag75*], and of Butzer and Engels [ButEng82]. In connection with it the dyadic analogues of the above theorems were proved in [SchWad89], and also the two dimensional maximal operator was defined. Moreover, weak type (1,1) inequalities were given by means of the hybrid Hardy norms introduced there. A summary on how this research program has started can be found in [Sch90].

The results with respect to the dyadic maximal operator were generalized for the spaces $H^p(\mathbb{I})$ $(p > 1/2)$ by Weisz by enhancing the technique of atomic decomposition of dyadic Hardy spaces. These imply not only the former results on functions in $L^1(\mathbb{I})$ [Wei98] but he was able to apply this technique also in the multivariable case [Wei96a], [Wei96b], [Wei00]. These are detailed in the papers [GatTol13*] and [Wei13*] published in this volume.

5.2.3 Dyadic derivative on the real line

Butzer and Wagner [ButWag73b*] introduced the notion of dyadic derivative for functions defined on $\mathbb{R}_+ := [0,\infty)$ by modifying the dyadic difference operator (5.7) as follows

$$(d_n f)(x) := \sum_{k=-n}^{n} 2^{k-1}(f(x) - f(x \dotplus e_k)) \qquad (x \in \mathbb{R}_+, n \in \mathbb{N}, f \in L^1(\mathbb{R}_+)). \quad (5.12)$$

Then the dyadic derivative operator satisfies $dw_x = xw_x$ $(x \in \mathbb{R}_+)$, where the w_x's are the Walsh functions on $\mathbb{R}_+$, pointwise and also in the strong sense. Wagner [Wag75] characterized the dyadic integral by a sequence of dyadic convolutions generated by the functions $W_n \in L^1(\mathbb{R}_+)$ $(n \in \mathbb{N})$, which were determined by their Walsh–Fourier transforms:

$$\widehat{W}_n(y) = \begin{cases} 1/y, & y \geq 2^{-n}; \\ 0, & 0 \leq y < 2^{-n}. \end{cases}$$

Following the definition of Wagner a function $f \in L^1(\mathbb{R}_+)$ is dyadically integrable if there exists a $g \in L^1(\mathbb{R}_+)$ such that

$$\lim_{n\to\infty} \|W_n * f - g\|_{L^1(\mathbb{R}_+)} = 0.$$

$If := g$ is called the strong dyadic integral function of f. Wagner showed that if $f, g \in L^1(\mathbb{R}_+)$ then $g = If$ if and only if $\widehat{g}(y) = \widehat{f}(y)/y$ $(y > 0)$, and $\widehat{g}(0) = 0$ hold for the Walsh–Fourier transform of g. We note that unlike the dyadic group it can happen that a function in $L^1(\mathbb{R}_+)$ does not have dyadic integral. One such example is the characteristic function of the unit interval $[0,1)$.

Pál [Pal75], [Pal77], [Pal78] proved the pointwise version of the fundamental theorem for special dyadically integrable functions. Later in [Pal85], [PalSch87*], [PalSch90] it was shown that the existence of If implies $d_n(If) = d_n W_n * f$. Moreover the maximal operator

$$T^* f := \sup_{n\in\mathbb{N}} |d_n W_n * f| \qquad (f \in L^1(\mathbb{R}_+))$$

is of weak type (1,1), and of (∞,∞). Hence the fundamental theorem corresponding to the pointwise derivative follows: *If $f \in L^1(\mathbb{R}_+)$ is dyadically integrable then If is dyadically differentiable a.e. and the equality $d(If) = f$ holds a.e.*. For details we refer the reader to [SchWadSim90], pp. 435–445.

Golubov modified (5.12) to obtain a Gibbs-type derivative on $\mathbb{R}_+$ for which the derivative of the Walsh function w_x $(x \in \mathbb{R}_+)$ is $\|x\| w_x$ $(x \in \mathbb{R}_+)$, where $\|x\| := 2^n$ $(2^n \le x < 2^{n+1})$ is the non-archimedean norm of x. This definition complies with the concept of derivative introduced by Onneweer on locally compact fields. The theory with respect to this concept, which is simpler than the original one, was developed by Golubov. For a review on this see the survey paper of Golubov and Volosivets [GolVol13a*] and the list of references in it.

5.2.4 Dyadic derivative on groups

The basic idea that led to the definition of dyadic derivative was employed by Gibbs and his collaborators [Gib90*], [GibIre74*], [GibIreMar73] for other groups as well. Namely, using the translations of the group they defined a linear operator of which the characters of the group are eigenfunctions. Similar program has been carried out among others for finite cyclic groups and for their finite direct products. Onneweer [Onn77a], [Onn78] constructed a Gibbs–type derivative on local fields. His definition was adapted by Pál and Simon for Vilenkin groups. *For the so called bounded case they have identified the inverse of the differential operator and showed the corresponding fundamental theorem in both pointwise, and strong senses [PalSim77a], [PalSim77b].*

Changing his original definition Onneweer [Onn79*] introduced a different derivative for which the additive characters of the local field are eigenfunctions, and the eigenvalues are equal to the norm of the character. In case of the dyadic group the derivative of w_k $(2^n \le k < 2^{n+1})$ is $2^n w_k$ $(2^n \le k < 2^{n+1})$ by this concept.

This corresponds to a special martingale transform. This concept was used also by Golubov [Gol13a*], [Gol13b*], [Gol92].

For finite non-commutative groups and for their direct products a Gibbs–type derivative was constructed by Stanković [Sta86], [Sta89*] (see also [GatTol13*]). In connection with dyadic derivatives of fractional orders we mention the papers of Butzer and Engels [ButEng89*], Golubov [Gol13a*], Onneweer [Onn77a], [Onn80], and He Zelin [HeZ83]. For dyadic derivatives of the Wash–Kaczmarz system see [GatTol13*].

5.3 Summation, Strong summation

The technique developed for the study of dyadic derivative can be applied successfully in the investigations related to summation of Walsh–Fourier series. Let

$$\begin{aligned} S_n f &:= \sum_{k=0}^{n-1} \widehat{f}(k) w_k = f * D_n, \\ \sigma_n f &:= \frac{1}{n} \sum_{k=1}^{n} S_k f = f * K_n \qquad (n \in \mathbb{N}^*) \end{aligned}$$

be the partial sums and the (C,1) means of the Walsh–Fourier a function $f \in L^1$, where

$$\begin{aligned} D_n &:= \sum_{k=0}^{n-1} w_k, \\ K_n &:= \frac{1}{n} \sum_{k=1}^{n} D_k = \sum_{k=0}^{n-1} (1 - k/n) w_k \qquad (n \in \mathbb{N}^*) \end{aligned}$$

are the Walsh–Dirichlet, and the Walsh–Fejér kernels. By the Paley formula we have that the Walsh–Dirichlet kernel with index 2^n is of the form

$$D_{2^n}(t) = \begin{cases} 2^n, \ 0 \le t < 2^{-n}; \\ 0, \ \ 2^{-n} \le t < 1. \end{cases}$$

5.3.1 $(C,1)$ *summability of Walsh–Fourier series*

The analogue of Lebesgue's theorem on the Walsh system proved by Fine [Fin95]: *For every* $f \in L^1$ *the convergence* $\lim_{n\to\infty}(\sigma_n f)(x) = f(x)$ *holds for a.e.* $x \in [0,1)$.

By means of the dyadic difference operator d_n the kernel function K_{2^n} can be written as an expression of D_{2^n} :

$$
\begin{aligned}
K_{2^n}(x) &= \sum_{k=0}^{2^n-1}(1-k/2^n)w_k(x) = D_{2^n}(x) - 2^{-n}(d_n D_{2^n})(x) = \\
&= \left(\frac{1}{2}+\frac{1}{2^{n+1}}\right)D_{2^n}(x)) + \frac{1}{2^{n+1}}\sum_{k=0}^{n-1}2^k D_{2^n}(x \dot{+} e_k) \qquad (n \in \mathbb{N}).
\end{aligned}
\tag{5.13}
$$

Consequently,

$$K_{2^n}(x) \geq 0, \int_0^1 K_{2^n}(t)\,dt = 1 \qquad (x \in [0,1), n \in \mathbb{N}).$$

It can be seen from this decomposition that the necessary conditions of the theorem of Fadeev (see e.g. [Ale61], Theorem 4.4.2. pp. 272) on the convergence of singular integrals in the Lebesgue points are not satisfied for the kernel functions K_{2^n}. Therefore, there exists an $f \in L^1$ and a Lebesgue point x of f such that $(\sigma_{2^n} f)(x)$ diverges as $n \to \infty$.

Let $I_n(x)$ denote the dyadic interval of length 2^{-n} that contains $x \in [0,1)$. Then by the decomposition (5.13) of K_{2^n} we have

$$(\sigma_{2^n} f)(x) - f(x) = \frac{2^n}{2}\int_{I_n(x)}(f(t)-f(x))\,dt + \frac{1}{2}\sum_{k=0}^{n}2^k\int_{I_n(x\dot{+}e_k)}(f(t)-f(x))\,dt.$$

Based on this we introduce the following concept: An $x \in [0,1)$ is called a *Walsh–Lebesgue (WL) point* of $f \in L^1$ if

$$\sum_{k=0}^{n}2^k\int_{I_n(x\dot{+}e_k)}|f(s)-f(x)|\,ds \to 0 \qquad (n \to \infty). \tag{5.14}$$

It is clear by the definition that $\lim_{n\to\infty}(\sigma_{2^n} f)(x) = f(x)$ holds at the WL points. By proving the analogue of the Fadeev theorem Weisz [Wei89] gave a necessary and sufficient condition for the convergence in the Walsh–Lebesgue points.

By means of the class of operators introduced in [Sch86] it was possible to handle not only the dyadic maximal operator but also the maximal operator $\sigma^* f = \sup_m |\sigma_m f|$ of the $(C,1)$ means. This way the extension of the theorem of Fine [Fin95] was proved in [Sch86]: σ^* *is of weak type* $(1,1)$, *which implies that* $\lim_{n\to\infty}\sigma_n f = f$ *holds a.e. for every* $f \in L^1$. In the same way one can show that for any $f \in L^1$ a.e. $x \in [0,1)$ is a WL point.

The special relationship between the dyadic derivative and the $(C,1)$ means of Walsh–Fourier series became transparent not only in the study of the pointwise convergence but also in the study of the rate of norm convergence, in particular uniform convergence, of these means. The following result proved by Fridli in [Fri94b] is an example for demonstrating this relationship. Let X denote the space of dyadically continuous functions or one of the spaces $L^p[0,1)$ $(1 \leq p < \infty)$. *If* (α_k) *is a sequence of nonnegative numbers tending monotonically to* 0 *then* $\|f - \sigma_k f\|_X = O(\alpha_k)$ *as* $k \to \infty$ *if and only if* $\|dS_{2^n} f\|_X = O(2^n \alpha_{2^n})$ *and* $E_{2^n}(f,X) = O(\alpha_{2^n})$ *as* $n \to \infty$ *hold.* Here $E_n(f,X)$ denotes the best approximation of $f \in X$ by Walsh polynomials of order at most n. In other cases the dyadic derivative does not appear in the result itself but plays essential role in the proofs. This is for instance the situation on the connection between the rate of convergence of $(C,1)$ means and the modulus of continuity. By taking the dyadic translation the dyadic modulus of continuity $\omega(\delta, f, X)$ can be defined similarly to the classical case. There are however some properties in which the two cases differ essentially. For instance the dyadic modulus of continuity can be completely characterized by sequences (see [Rub83], [Fri85]). The connection between the Hölder classes and the convergence classes of $(C,1)$ means was studied by Fridli [Fri94a]. The result in the spaces of dyadic continuous functions and in L^1 reads as follows. Let X denote one of these spaces. *If* $\sum_{k=1}^{n} \omega(1/k, f, X) = O(n\alpha_n)$ *then* $\|f - \sigma_n f\|_X = O(\alpha_n)$ $(n \to \infty)$. *On the other hand if* $\|f - \sigma_n f\|_X = O(\alpha_n)$ *then* $\max_{0<k<n} k\,\omega(1/k, f, X) = O(n\alpha_n)$ $(n \to \infty)$. *It was shown that these results are best possible.* The corresponding L^p versions were also given and special cases like Lipschitz classes were studied as well. For instance $\|f - \sigma_n f\|_X = O((\log n)^{X_p}/n)$ $(n \to \infty, f \in Lip(1,X))$, *where* $X_p = 1/p$ *if* $X = L^p$ *with* $1 \le p \le 2$, $X_p = 1/2$ *if* $X = L^p$ *with* $2 < p < \infty$, *and* $X_p = 1$ *if* X *is the space of diadically continuous functions, and the estimates are best possible.*

The investigations that led to these results were inspired by problems posed by Móricz and Siddiqi in their paper [MorSid92] on Walsh–Nörlund means. Their results on the approximation properties of Walsh–Nörlund means in L^p norm were later generalized by Fridli, Manchanda and Siddiqi [FriManSid08] for homogeneous Banach spaces. The two-dimensional version is due Nagy [Nag10]. Concerning Nörlund means, in particular logarithmic means, the authors Gát, Goginava and Nagy have made intensive research (see e.g. [GatGog06], [GogNag09]). We note that the rate of convergence of (C,β) means in Lipschitz classes were studied earlier by Yano [Yan51], Jastrebova [Jas66], and Skvorcov [Skv81].

5.3.2 Strong summability

In Fourier analysis not only the arithmetic means but also the strong

$$h_n^p := \left(\frac{1}{n} \sum_{k=1}^{n} |s - s_k|^p \right)^{1/p} \qquad (n \in \mathbb{N}^*, p > 0)$$

means are investigated frequently. We say that the sequence $(s_n, n \in \mathbb{N}^*)$ is of strong h_p summable if there exists a number s such that $h_{n,p} \to 0$ if $n \to \infty$. Then s is called the h_p sum of the series. It is clear by the inequality

$$\left| \frac{1}{n} \sum_{k=1}^{n} s_k - s \right| \le h_{n,p} \qquad (p \ge 1, n \in \mathbb{N}^*)$$

that every h_p $(p \geq 1)$ summable series is $(C,1)$ summable, and that the $(C,1)$ and the h_p sums are the same. It turned out that many of the results on the convergence and approximation properties of the arithmetic means remain valid also for the strong sums. One of the first results of this type is due to Hardy and Littlewood [HarLit13]. In 1913 they proved that the trigonometric series of a function $f \in L^r_{2\pi}$ $(r > 1)$ is strongly h_p $(p < \infty)$ summable. Generalizing Lebesgue's theorem Marcinkiewicz [Mar39] showed that the trigonometric series of every $f \in L^1_{2\pi}$ is strongly h_2 summable. This result was extended from h_2 to h_p $(0 < p < \infty)$ by Zygmund [Zyg41] in 1941, see also [Zyg59].

The Walsh–Fourier analogue of Marcinkiewicz's result was proved for $p = 2$ in [Sch69] in 1969. The general case involving strong BMO means is due to Rodin [Rod89] in the trigonometric case, and Rodin [Rod91] and Schipp [Sch98] in the Walsh case.

The following question has arisen even in connection with Marcinkiewicz's result. Is it possible to give a simple condition, similar to the definition of Lebesgue point, that guarantees the pointwise strong summability of the trigonometric Fourier series? The problem remained open for a long time. Finally, Gabisonia [Gab73] gave an answer to that in 1973 by introducing the concept of strong Lebesgue points. In order to define the concept of strong Lebesgue point let $J_n(x) := (x - \pi 2^{-n}, x + \pi 2^{-n})$ $(n \in \mathbb{N})$ denote the interval of length $2\pi 2^{-n}$, and centered at x. It is easy to see that

$$\lim_{n \to \infty} \frac{2^n}{2\pi} \int_{J_n(x)} |f(t) - f(x)\, dt = 0$$

is equivalent to x being a Lebesgue point of f. Instead of the integral means Gabisonia considered the weighted strong means of their translates

$$(\Lambda_n^{(p)} f)(x) := \left(\sum_{t \in T_n} \left| \frac{1}{t} \int_{J_n(x+t)} |f(s) - f(x)|\, ds \right|^p \right)^{1/p}, \tag{5.15}$$

where

$$T_n := \{2\pi(k + 1/2)2^{-n} : -2^{n-1} \leq k < 2^{n-1}\}.$$

Then $x \in \mathbb{R}$ is called a *strong p-Lebesgue point* of the function f if

$$\lim_{n \to \infty} (\Lambda_n^{(p)} f)(x) = 0.$$

Gabisonia [Gab73] proved that *for any $f \in L^1_{2\pi}$ a.e. $x \in \mathbb{R}$ is a strong p-Lebesgue point. Moreover in every such point the trigonometric Fourier series of f is strongly h_p summable.*

Starting from the definition of the WL points (5.14) and following the scheme in (5.15) let us introduce the weighted strong means of dyadic translates of the sums in (5.14):

$$(W_n^{(p)}f)(x) := \left(\sum_{t\in Q_n} \left| 2^k \chi_{[0,2^{-k})}(t) \int_{I_n(x\dot{+}e_k\dot{+}t)} |f(x)-f(s)|\,ds \right|^p \right)^{1/p}, \qquad (5.16)$$

where $Q_n := \{\ell 2^{-n} : 0 \le \ell < 2^n\}, n \in \mathbb{N}, p > 0$. Then $x \in [0,1)$ is called a strong WL point of $f \in L^1$ if

$$\lim_{n\to\infty} (W_n^{(p)}f)(x) = 0.$$

Schipp proved in [Sch98] that *for every $f \in L^1$, $p > 0$, a.e. $x \in [0,1)$ is a strong p-WL point. Moreover the Walsh–Fourier series of f is strongly h_p summable in these points.*

Concerning the convergence and approximation properties of strong Walsh–Fourier means of dyadically continuous functions we refer to the work of Fridli and Schipp [FriSch98]. Their results are based on a duality relation that they proved for strong convergence, approximation and the so called Sidon type inequalities. We note that these results hold in a more general setting in terms of orthogonal systems, and also the strong means were defined in a wider sense. For instance they proved the following strong convergence result. If f is a dyadically continuous function and ψ is a monotonically increasing function defined on $[0,\infty)$ for which $\lim_{u\to 0+} \psi(u) = 0$ then

$$\lim_{n\to\infty} \frac{1}{n} \sum_{k=1}^{n} \psi\big(|S_k f(x) - f(x)|\big) = 0 \qquad (0 \le x < 1)$$

if and only if there exists $A > 0$ such that $\psi(t) \le \exp(At)$ $(0 \le t < \infty)$. Moreover, the convergence is uniform in x. The corresponding result for the trigonometric case was proved earlier by Totik [Tot80].

A similar theorem on strong approximation [FriSch98] reads as follows. Let $\varphi : [0,\infty) \mapsto \mathbb{R}$ be a monotonically increasing function with $\lim_{u\to 0+} \varphi(u) = 0$ for which there exists A such that $\varphi(u) \le \exp(Au)$ $(u \ge 0)$. Then

$$\frac{1}{r} \sum_{k=1}^{r} \varphi\big(|S_{k+\ell} f(x) - V_{r,\ell} f(x)|\big) \le C E_\ell f$$

$(0 \le x < 1, r, \ell \in \mathbb{N}, \|f\|_\infty \le 1)$, where $V_{r,\ell} f$ is the generalized de la Vallée Poussin mean defined as

$$V_{r,\ell} f = \frac{1}{r} \sum_{k=1}^{r} S_{k+\ell} f \qquad (r, \ell \in \mathbb{N}).$$

5.3.3 Summability of multivariable Walsh–Fourier series

The first result on almost everywhere convergence of multivariable Walsh (C,1) means was published by Móricz, Schipp and Wade [MorSchWad92] in 1992 (see also [Sch90]). Since then a new theory has emerged due to Gát, Simon and Weisz mainly. In this theory the one and the multi dimensional dyadic Hardy spaces play a central role. They have investigated not only the multivariate (C,α) and θ means but also the Riesz, the Marcinkiewicz means by taking several types of convergence of multidimensional series. A great deal of the results were generalized to Vilenkin systems. In connection with this we mention the papers of Gát G. [Gat98], [Gat99], [Gat00], [Gat02], [Gat03], and the survey paper [GatTol13*]. A good summary on these results and methods can be found in Chapter 3. and in the bibliography of the monograph [Wei02].

5.3.4 Dyadic Cesáro and Copson operators

Instead of $(C,1)$ summability, i.e. averaging partial sums of the Fourier series, Hardy [Har29] initiated the research of the effect of averaging the Fourier–coefficients. He showed that the $L^p_{2\pi}$ $(1 \le p < \infty)$ spaces are invariant with respect to this operation. Since then many mathematicians have investigated the corresponding *Cesáro operators* and their duals the *Copson operators* related to this problem. We note that different terms exist for Cesáro operators in the literature. Some people call them Hardy or Bellmann operators. In connection with results on the trigonometric system we refer the reader to the papers of Móricz [Mor99], [Mor02].

The dyadic version of this topic was investigated in two different ways by Eisner and Schipp in Hungary, and by Golubov [Gol94] and Volosivets [Vol08] in Russia. The dyadic derivative played important role in these investigations. For a function $f \in L^1(\mathbb{I})$ there is one and only one such function $g \in L^1(\mathbb{I})$ for the Walsh-Fourier coefficients of which we have $\widehat{g}(n-1) := (\sum_{k=0}^{n-1} \widehat{f}(k))/n$ $(n \in \mathbb{N}^*)$. The operator $\mathscr{C}f := g$ is called dyadic Cesáro operator. In the papers [Eis98a],[Eis98b], [Eis99], [Eis02] the operator $\mathscr{C}$ and its dual were studied in one and two dimensional L^p, diadic Hardy, and BMO spaces. In [EisSch00] the research was extended to functions defined on the dyadic field.

There is a special relation between the Walsh–Dirichlet kernels and the modified dyadic difference operators that played an important role in these investigations. Let us modify the definition of the operator d_n (5.7):

$$\begin{aligned}(d_n^- f)(x) :&= \sum_{k=0}^{n-1} 2^{k-1}(f(x) - f(x \dotplus e_k) - 2^{n-1}(f(x) - f(x \dotplus e_n)) = \\ &= (d_n f)(x) - 2^{n-1}(f(x) - f(x \dotplus e_n)).\end{aligned} \tag{5.17}$$

Then the following equality holds

$$D_n(x) = d_{s-1}^- w_n(x) \qquad (x \in [2^{-s}, 2^{-s+1}),\ s,n \in \mathbb{N}).$$

Hence the Cesáro operator can expressed as an integral operator

$$(\mathscr{C}f)(x) = \int_0^1 f(t)K(x,t)\,dt \qquad (f \in L^1(\mathbb{I}),$$

the kernel function of which is

$$K(x,t) = \sum_{s\in\mathbb{N}^*} \chi_s(t)(d_{s-1}^- W)(x \dotplus t) \qquad (x,t \in \mathbb{I})$$

where χ_s stands for the characteristic function of the interval $[2^{-s}, 2^{-s+1})$. These formulas, that have versions in $\mathbb{I}^2$ and in $\mathbb{R}_+$, play a key role in the theory of dyadic Cesáro and Copson operators.

5.3.5 Multipliers

The dyadic derivative of a function is defined as limit of the dyadic difference operators (5.7). On the other hand the Walsh functions are eigenfunction of dyadic differentiation, i.e. $dw_n = nw_n$ $(n \in \mathbb{N})$. Therefore in the Walsh frequency space the dyadic differentiation can be interpreted as a multiplier operator. This is, however, a simplification of the situation for the relation between the two approaches in terms of pointwise dyadic differentiation is more delicate. In connection with this issue we refer to the work of Butzer and Wagner [ButWag75*], Onnewer [Onn77b], Powell and Wade [PowWad91], Schipp [Sch76a], Fridli and Wade [FriWad94]. This section is about giving examples for this dual feature of dyadic differentiation, and is by no means intended to be a thorough summary on multiplier results in Walsh analysis. A simple but very transparent example for this duality is the expression (5.13) given for the Fejér kernel, where both interpretations appear. As a result the kernel function is associated with dyadic translations of the Dirichlet kernels which explains why the dyadic derivative is so effective in investigations with respect to $(C,1)$ means. On the other hand the multiplier aspect of the dyadic derivative became especially important in the further generalizations of the concept including the concept of fractional derivatives, extended dyadic derivative or dyadic derivatives on p-adic groups. Here we only mention the works of Butzer, Engels and Wipperfürth [ButEngWip86], Onneweer [Onn77a], [Onn78], and He Zelin [HeZ83] as examples. The transition between the two models, namely the kernel function and the multiplier is sometimes challenging. Such an example is the case of boundedness of Hörmander multipliers in dyadic Hardy spaces H^p $(0 < p \leq 1)$. Let (ϕ_k) be a real sequence. The Walsh multiplier operator T_ϕ on H^p corresponding to ϕ is given by

$$\widehat{T_\phi f}(k) = \phi_k \hat{f}(k) \qquad (k \in \mathbb{N}).$$

Then T_ϕ is said to be bounded on H^p if there exists an absolute constant C such that $\|T_\phi f\|_{H^p} \leq \|f\|_{H^p}$. There is a rich history of multiplier theorems for Hardy spaces in which conditions are made on the dyadic blocks of the Walsh series of the kernel $\boldsymbol{\Phi}$ corresponding to ϕ. The relevant results can be found in Daly and Fridli

[DalFri04], Daly and Phillips [DalPhi98a], [DalPhi98b], Kitada [Kit87], Onneweer-Quek [OnnQue89], and Simon [Sim85]. In all of these papers the conditions involve the growth of the dyadic blocks $\sum_{j=2^{n-1}}^{2^n-1} \phi_k w_k$ of the kernel Φ. In other words the conditions are given via the kernel function and not directly for the terms of the multiplier, i.e not in the frequency space. This inspired Fridli and Daly to study of Hörmander multipliers for Walsh series. Among others they proved in [DalFri03] that *if the multiplier ϕ is bounded and satisfies*

$$2^j\left(\sum_{k=2^j+1}^{2^{j+1}} \frac{|\Delta\phi_k|^r}{2^j}\right)^{1/r} \leq C \qquad (j \in \mathbb{N})$$

for some $1 < r \leq 2$ then T_ϕ is bounded on H^p with $p > r/(2r-1)$. They also showed that the condition on p can not be relaxed. The two dimensional version of this result is given in Daly and Fridli [DalFri08]. We note that the famous Marcinkiewicz multiplier condition corresponds to the special case $r = 1$.

References

[Ale61] Alexits, G. "Convergence problems of orthogonal functions", Pergamon Press, New York, 1961.

[ButEng82] Butzer P.L., Engels W. *Dyadic Calculus and sampling theorems for functions with multidimensional domain,* Information and Control **52** (1982), 33–351.

[ButEng89*] Butzer P.L., Engels W. *Background to an extension of Gibbs differentiation in Walsh analysis,* First International Workshop on Gibbs Derivatives, Kupari-Dubrovnik, Yugoslavia **52** (1989), 19–57.

[ButEngWip86] Butzer, P.L.; Engels, W.; Wipperfürth, U. *An extension of the dyadic calculus with fractional order derivatives: general theory,* Comp. Math. with Appls., **12B(5/6)**, (1986), 1073–1090.

[ButSta89*] Butzer P.L., Stankovic R. S. "Theory and Application of Gibbs Derivatives", Proc First Internat. Workshop on Gibbs Derivatives Kupari-Dubrovnik, Yugoslavia, 1989.

[ButWag72*] Butzer P.L., Wagner H.J. *Approximation by Walsh polynomials and the concept of a derivative,* Proc. Sympos. Applic. Walsh Functions, Washington D.C. **3** (1972), 388–392.

[ButWag73a*] Butzer P.L., Wagner H.J. *Walsh-Fourier series and the concept of a derivative,* Applicable Anal. **3** (1973), 29–46.

[ButWag73b*] Butzer P.L., Wagner H.J. *A calculus for Walsh functions defined on $\mathbb{R}_+$,* Proc. Symp. on Appl. of Walsh Functions, Washington D.C. **29** (1973), 75–81.

[ButWag75*] Butzer P.L., Wagner H.J. *On the dyadic analysis based on the pointwise dyadic derivative,* Analysis Mathematica **1** (1975), 171–196.

[DalFri03] Daly, J.; Fridli, S. *Walsh multipliers for dyadic Hardy spaces,* Appl. Anal. **82** (2003), no. 7, 689–700.

[DalFri04] Daly, J.; Fridli, S. *Translation invariant operators on Hardy spaces over Vilenkin groups,* Acta Math. Acad. Paedagog. Nyházi. (N.S.) **20** (2004), 131–140.

[DalFri08] Daly, J.; Fridli, S. *Hörmander multipliers on two-dimensional dyadic Hardy spaces,* J. Math. Anal. Appl. **348** (2008), no. 2, 977–989.

[DalPhi98a] Daly, J.; Phillips, K. *Walsh multipliers and square functions for the Hardy space H^1*, Acta Math. Hungar., **79** (4) (1998), 311–327.

[DalPhi98b] Daly, J., Phillips, K. *A note on H^1 multipliers for locally compact Vilenkin groups,* Canad. Math. Bull., **41** (4) (1998), 392–397.

[Eis98a] Eisner, T. *The dyadic Cesáro operators,* Acta Sci. Math. (Szeged) **64** (1998), 99–111.

[Eis98b] Eisner, T. *The two-parameter Cesáro operators,* Math. Pannon. **9** (1998), 243–258.

[Eis99] Eisner, T. *Dyadic Cesáro operators on Hölder spaces,* Functions, Series, Operators, Budapest, Hungary 1999, János Bolyai Math. Soc., Budapest, 2002, 213–223.

[Eis02] Eisner, T. *Dyadic Cesáro operators on Hardy spaces,* Acta Sci. Math. (Szeged) **68** (2002), 203–228.

[EisSch00] Eisner, T., Schipp, F. *The dyadic Cesáro operator on* $\mathbb{R}_+$, Anal. Math. **26** (2000), 263–274.

[Fej04] Fejér, L. *Untersuchungen über Fouriersche Reihen,* Math. Annalen **58** (1904), 501–569.

[Fin95] Fine, N. J. *Cesàro summability of Walsh-Fourier series,* Nat. Acad. Sci. USA **41** (1995), 588–591.

[Fri85] Fridli, S. *On the modulus of continuity with respect to functions defined on Vilenkin groups,* Acta Math. Hung. **45** (1985), no. 3-4, 393–396.

[Fri94a] Fridli, S. *On the rate of convergence of Cesáro means of Walsh-Fourier series,* J. Approximation Theory **76** (1994), no. 1, 31–53.

[Fri94b] Fridli, S. *Convergence classes of Walsh-Fejér means in homogeneous Banach spaces,* Festschrift for the 50th birthday of Karl-Heinz Indlekofer. Ann. Univ. Sci. Budap. Rolando Eötvös, Sect. Comput. **14** (1994), 47–59.

[FriManSid08] Fridli, S.; Manchanda, P.; Siddiqi, A. H. *Approximation by Walsh-Nörlund means,* Acta Sci. Math. (Szeged) **74** (2008), no. 3-4, 593–608.

[FriSch98] Fridli, S.; Schipp, F. *Strong approximation via Sidon type inequalities,* J. Approximation Theory **94** (1998), no. 2, 263–284.

[FriWad94] Fridli, S.; Wade, W.R. *On the rate of convergence and dyadic differentiability of Walsh series,* Journal d'Analyse Mathématique **62** (1994), 287–305.

[Gab73] Gabisonia O. D *On strong summability points for Fourier series,* Mat. Zametki **14(5)** (1973), 615–626.

[Gat98] Gát G. *On the two-dimensional pointwise dyadic calculus,* Approximation Theory **92(2)** (1998), 191–215.

[Gat99] Gát G. *Pointwise convergence of the Fejér means of functions on unbounded Vilenkin groups,* J. Approximation Theory **101(1)** (1999), 1–36.

[Gat00] Gát G. *On the divergence of the* $(C,1)$ *means of double Walsh-Fourier series,* Proc. Am. Math. Soc. **128(6)** (2000), 1711–1720.

[Gat02] Gát G. *On the divergence of the two-dimensional dyadic difference of dyadic integrals,* J. Approximation Theory **116(1)** (2002), 1–27.

[Gat03] Gát G. *Cesàro means of integrable functions with respect to unbounded Vilenkin systems,* J. Approximation Theory **124(1)** (2003), 25–43.

[GatGog06] Gát, G.; Goginava, U. *Uniform and Lconvergence of Logarithmic Means of Walsh–Fourier Series,* Acta Mathematica Sinica, **22(2)** (2006), 497–506.

[GatTol13*] Gát Gy. *Calculus on Walsh and Vilenkin groups,* (2013). (present volume)

[Gib67*] Gibbs J.E. *Walsh spectrometry, a form of spectral analysis well suited to binary digital computation,* NPL DES Repts., Nat.Phys. Lab.Teddington, Middx. UK **3** (1967), 24 pp.

[Gib90*] Gibbs, J.E. *Local and global veiws of differentiation,* Proc. First Internat. Workshop on Gibbs Derivatives, Kupari-Dubrovnik (1990), 1–18.

[GibIre74*] Gibbs, J.E., Ireland, B. *Walsh Function and differentiation,* Application of Walsh Functions and Sequency Theory, IEEE New York (1974), 147–176.

[GibIreMar73] Gibbs, J.E., Ireland, B., Marshall J. E. *A generalization of the Gibbs differentiator,* Theory and Applications of Walsh and other Non-Sinusoidal Functions, Coll. Hatfield Polytechnic, (1973), 32 pp.

[GogNag09] Goginava, U.; Nagy, K. *Weak type inequality for logarithmic means of Walsh-Kaczmarz-Fourier series,* Real Analysis Exchange, **35(2)** (2009), 445–462.

[GolVol13a*] Golubov B. I., Volosivets S. S. *Generalized Derivatives and Integrals on Vilenkin Groups,* (2013). (present volume)

[Gol92] Golubov B.I. "Fourier Series: Theory and Applications", Kiev (1992), 18–26. (in Russian)

[Gol94] Golubov B. I. *On a theorem of Bellmann on Fourier coefficients,* Mat. Sb. **185/II.** (1994), 31–40.

[Gol02] Golubov B. I. *On the modified strong dyadic integral and derivative,* Math. Sbornik **193(4)** (2002), 37–60. (in Russian)

[Gol13a*] Golubov B. I. *How I came to the study of Gibbs derivatives,* (2013). (present volume) (2013).

[Gol13b*] Golubov B. I. *Dyadic Derivative and Walsh-Fourier Transform,* (2013). (present volume)

[Har29] Hardy, G.H. *Notes on some points in the integral calculus,* LXVI Messenger of Math. **58** (1929), 50–52.

[HarLit13] Hardy, G. H. and Littlewood, J. E. *Sur la séries de Fourier d'une fonction à carré sommable,* Comptes Rendus (Paris) **156** (1913), 1307–1309.

[HarLit30] Hardy, G. H. and Littlewood, J. E. *A maximal theorem with function-theoretic applications,* Acta Math., Comptes Rendus (Paris) **63** (1930), 81-116.

[HeZ83] He Zelin *The derivatives and integrals of fractional order in Walsh-Fourier analysis, with applications to approximation theory,* J. Approx. Theory **39** (1983), 361–373.

[Jas66] Jastrebova, M.A. *On approximation of functions satisfying the Lipschitz condition by arithmetic means of their Walsh-Fourier series,* Mat. Sb. **71** (1966), 214–226. (in Russian)

[Kit87] Kitada, K. *H^p multiplier theorems on certain totally disconnected groups*, Sci. Rep. Hirosaki Univ., **34** (1987), 1–7.

[Leb05] Lebesgue, H. *Recherches sur la convergence des séries de Fourier,* Math. Annalen **61** (1905), 251–280.

[Mar39] Marcinkiewicz J. *Sur la sommabilité forte des series de Fourier,* J. London Math. Soc. **14** (1939), 162–168.

[Mor99] Móricz, F. *The harmonic Cesàro and Copson operators on the spaces L^p, $1 \leq p \leq \infty$, H^1 and BMO,* Acta Sci. Math. (Szeged) **65** (1999), 293–310.

[Mor02] Móricz, F. *The harmonic Cesàro and Copson operators on the spaces $L^p(\mathbb{R})$, $1 \leq p \leq 2$,* Studia Math. **149** (2002), 267–279.

[MorSchWad92] Móricz F,. Schipp F., Wade W.R. *Cesáro summability of double Walsh-Fourier series,* Trans. Amer. Math. Soc. **329** (1992), 131–140.

[MorSid92] Móricz, F.; Siddiqi, A.H. *Approximation by Nörlund means of Walsh-Fourier series,* Journal of Approximation theory **70(3)** (1992), 375–389.

[Nag10] Nagy, K. *Approximation by Nrlund means of quadratical partial sums of double Walsh-Fourier series,* Analysis Mathematica, **36(4)** (2010), 299–319.

[Onn77a] Onneweer, C.W. *Fractional differentiation on the group of integers of the p-adic or p-series field,* Analysis Math. **3** (1977), 119–130.

[Onn77b] Onneweer, C.W. *Differentiabilityfor Rademacher series on groups,* Acta Sci. Math. (Szeged) **39** (1977), 121–128.

[Onn78] Onneweer, C.W. *Differentiation on a p-adic or p-series field,* Proc. Conference *Linear Spaces and Approximation*, Birkhäuser Analysis Math. **40** (1978), 187–198.

[Onn79*] Onneweer, C.W. *On the definition of dyadic differentiation,* Applicable Anal. **9** (1979), 267–278.

[Onn80] Onneweer, C.W. *Fractional differentiation and Lipschitz spaces on local fields,* Trans. Amer. Math. Soc. **258** (1980), 155–165.

[Onn13*] Onneweer, C.W. *My Involvement with the Dyadic Derivative,* (2013). (present volume)

[OnnQue89] Onneweer, C.W.; T.S. Quek, T.S. *H^p multiplier results on locally compact Vilenkin groups*, Quart. J. Math Oxford(2), **40** (1989), 313–323.

[Pal75] Pál, J. *On the connection between the concept of a derivative defined on the dyadic field and the Walsh-Fourier transform,* Annales Univ. Sci. Budapest, Sect. Math. **18** (1975), 49–54.

[Pal77] Pál, J. *On a concept of derivative among function defined on the dyadic field,* SIAM J. Math. Anal. **8** (1977), 375–391.

[Pal78] Pál, J. *On almost everywhere differentiability of dyadic integral functions on $\mathbb{R}^+$,* Colloq. Math. Soc. J. Bolyai, Fourier Analysis and Approximation Theory,SIAM J. Math. Anal. **19** (1978), 591–601.

[Pal85] Pál, J. *The almost everywhere differentiability of Wagner's dyadic integral function on* $\mathbb{R}^+$, Colloq. Math. Soc. J. Bolyai, Alfred Haar Memorial Conference **49** (1985), 695–701.

[PalSim77a] Pál, J., Simon P. *On a generalization of the concept of derivative,* Acta Math. Acad. Sci. Hung. **29** (1977), 155–164.

[PalSim77b] Pál, J., Simon P. *On a generalized Butzer-Wagner type a.e. differentiability of integral function,* Annales Univ. Sci. Budapest, Sect. Math. **20** (1977), 155–164.

[PalSch87*] Pál, J., Schipp, F. *On the dyadic differentiability of dyadic integral function on* $\mathbb{R}_+$, Annales Univ. Sci. Budapest, Sect. Comp. **8** (1987), 91–108.

[PalSch90] Pál, J., Schipp, F. *On the a.e. dyadic differentiability of dyadic integral on* $\mathbb{R}_+$, Theory and Application of Gibbs Deivatives **8** (1990), 103–113.

[PowWad91] Powell, C.H.; Wade, W.R. *Term by term differentiation and rapidly convergent Walsh series,* Approx. Theory and Appl. **7** (1991), 20–40.

[Rod89] Rodin, V. A. *A BMO strong means of Fourier series,* Funk. Anal. i Prilozhen. **23(2)** (1989), 73–74.

[Rod91] Rodin, V. A. *The space BMO and strong means of Fourier-Walsh series,* Mat. Sbornik **182(10)** (1991), 1463–1478. (In Russian)

[Rub83] Rubinstein, A.I. *On the modulus of continuity and best approximation of functions defined by lacunary Walsh series in* L^p, Izvv. Vysš. Učeb. Zaved. **252** (1983), 61–68. (in Russian)

[Sch69] Schipp, F. *Über die starke Summation der Walsh-Fourierreihen,* Acta Sci. Math. (Szeged) (1969), 203–218.

[Sch74*] Schipp, F. *Über einen Ableitungsbegriff von P.L.Butzer und H.J.Wagner,* Math. Balkanica **4** (1974), 541–546.

[Sch76a] Schipp, F. *On term by term dyadic differentiation of Walsh series,* Analysis Math. **2** (1976), 149–154.

[Sch76b] Schipp, F. *On the dyadic derivative,* Acta Math. Acad. Sci. Hungar. **28** (1976), 145–152.

[Sch86] Schipp, F. *Über gewisse Maximaloperatoren,* Annales Univ. Sci. Budapest, Sect. Math. **28** (1986), 145–152.

[Sch90] Schipp, F. *Multiple Walsh Analysis,* Theory and Application of Gibbs Derivatives (1990), 73–90.

[Sch98] Schipp, F. *On the strong summability of Walsh series,* Publ. Math. (Debrecen) **53(3-4)** (1998), 611–633.

[SchWad89] Schipp, F., Wade, W. R. *A Fundamental Theorem of Dyadic Calculus for the Unit Square,* Applicable Analysis **34** (1989), 77–87.

[SchWadSim90] Schipp, W. R. Wade, P. Simon, J. Pál "Walsh series" Adam Hilger, Bristol New-York, 1990.

[Sim85] Simon, P. *(L^1, H)-type estimations for some operators with respect to the Walsh-Paley system*, Acta Math. Hung. **46** (1985), 307–310.

[Skv81] Skvorcov, V. A. *Certain estimates of approximation of functions by Cesaro means of Walsh-Fourier series,* Mat. Zametki **29** (1981) 539-547. (in Russian)

[Sta86] Stankovic R. S. *A note on differential operators on finite non-abelian groups,* Applicable Analysis **21** (1986), 31–41.

[Sta89*] Stankovic R. S. *Gibbs derivatives on finite non-Abelian groups,* First Internat. Workshop on Gibbs Derivatives, Kupari-Dubrovnik, Yugoslavia (1989), 269–41.

[Ste61] Stein, E. M. *On the limits of sequences of operators,* Annals Math. **74(2)** (1961), 140–170.

[Tot80] Totik, V. *On the generalization of Fejér's summation theorem,* "Functions, Series, Operators " Coll. Math. Soc. J. Bolyai (Budapest) Hungary, North Holland, Amsterdam, Oxford, New York**35** (1980), 1195–1199.

[Vol08] Volosivets, S.S. *Hardy and Bellman transformations of series in multiplicative systems,* Mat. Sb. 199 (2008), no. 8, 3–28. (in Russian)

[Wag73] Wagner, H. J. "Ein Differential- und Integralkalkül in der Walsh-Fourier-Analysis mit Anwendungen", Westdeutscher Verlag, Köln-Opladen, 1973.

[Wag75] Wagner, H. J. *On dyadic calculus for functions defined on* $\mathbb{R}_+$, Theory and Applications of Walsh Functions, Proc. Symp. Hatfield (1975), 101–129.

[Wal23] Walsh, J.L. *A closed set of orthogonal function,* Amer. J. Math. **55** (1923), 5–24.

[Wei89] Weisz, F. *Convergence of singular integrals,* Annales Univ. Sci. Budapest, Sect. Math. **32** (1989), 243–256.

[Wei96a] Weisz, F. *(H_p, L_p)-type inequalities for the two-dimensional dyadic derivative,* Studia Math. **120** (1996), 271–288.

[Wei96b] Weisz, F. *Some maximal inequalities with respect to two-parameter dyadic derivative and Cesàro summability,* Appl. Anal. **62** (1996), 223–238.

[Wei98] Weisz, F. *Martingale Hardy spaces and the dyadic derivative,* Anal. Math. **24** (1998), 59–77.

[Wei00] Weisz, F. *The two-parameter dyadic derivative and the dyadic Hardy spaces,* Anal. Math. **20** (2000), 143–160.

[Wei02] Weisz, F. "Summability of Multi-dimensional Fourier Series and Hardy Spaces", Mathematics and Its Applications, Kluwer Academic, Dordrecht, Boston, London, 2002.

[Wei13*] Weisz, F. *Hardy spaces in the theory of dyadic derivative,* (2013). (present volume)

[Yan51] Yano, Sh. *On Walsh series,* Tôhoku Math. J. **3** (1951), 223–242.

[Zyg41] Zygmund, A. *On the convergence and summability of power series on the circle of convergence II.,* Proc. London Math. Soc. **47** (1941), 326–350.

[Zyg59] Zygmund, A. "Trigonometric Series", Cambridge University Press, New York, N.Y., 1959.

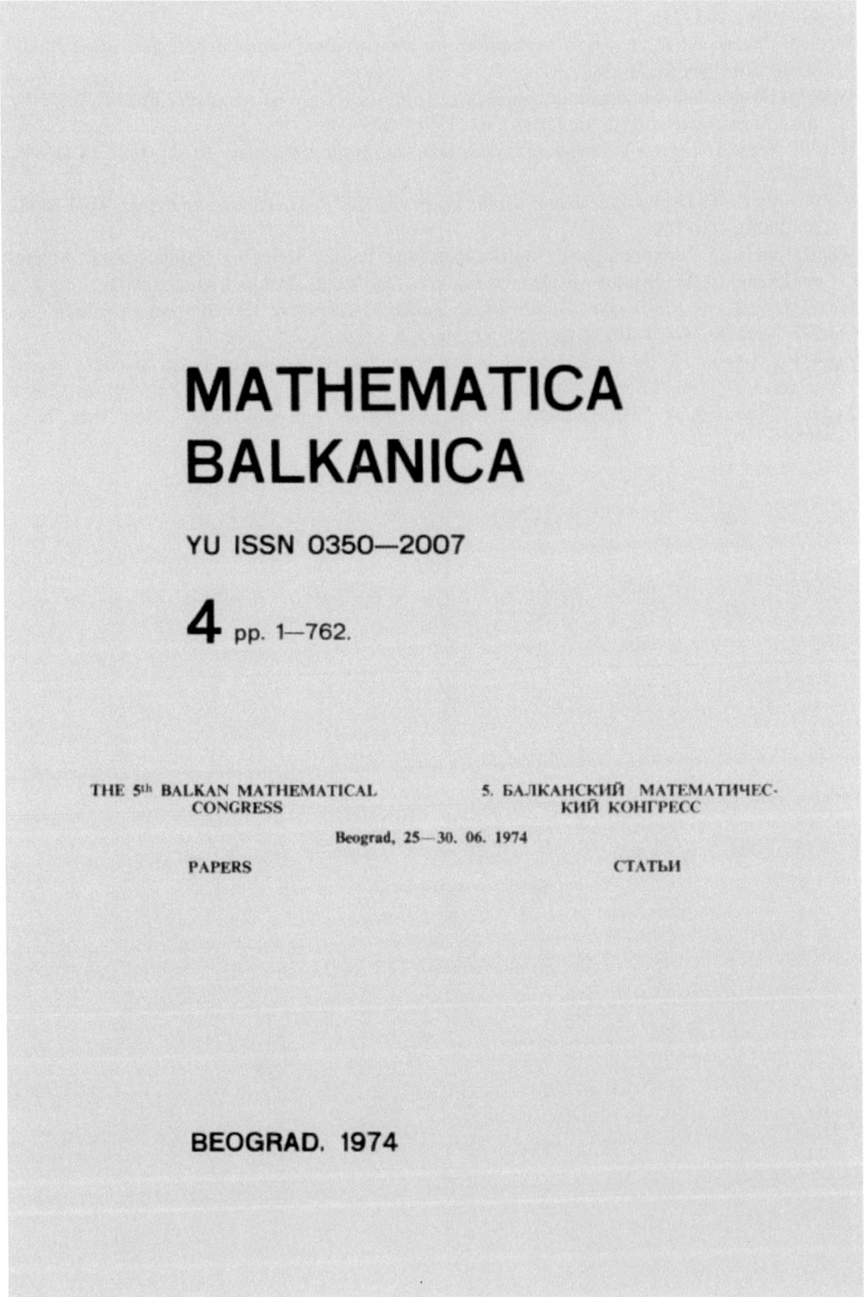

MATHEMATICA BALKANICA

YU ISSN 0350—2007

4 pp. 1—762.

THE 5th BALKAN MATHEMATICAL CONGRESS

5. БАЛКАНСКИЙ МАТЕМАТИЧЕСКИЙ КОНГРЕСС

Beograd, 25—30. 06. 1974

PAPERS

СТАТЬИ

BEOGRAD. 1974

MATEMATICA BALKANICA
4.103 (1974) 541—546

SCHIPP FERENC (Budapest, Ungarn)

UBER EINEN ABLEITUNGSBEGRIFF VON P. L. BUTZER UND H. J. WAGNER*

Einleitung. Die klassischen Orthogonalsysteme sind als Lösungen einer Differentialgleichung derstellbar. J. E. GIBBS [1], P. L. BUTZER und H. J. WAGNER [2], [3] führten Ableitungsbegriffe ein, unter denen die WALCHfunktionen sinvolle, nichttriviale „Ableitungen" besitzen. Für $f\in L^p(0,1)$ $(1\leq p\leq\infty)$ sei

$$(1)\qquad (d_nf)(x)=\sum_{k=1}^{n-1}2^{k-1}(f(x)-f(x\overset{\circ}{+}2^{-(k+1)}))\quad (n\in P,\ x\in[0,1)),^{1)}$$

wobei $\overset{\circ}{+}$ die FINE-sche Operation bezeichnet (vergl. [4]).

Definition (BUTZER-WAGNER): Existiert für $g\in L^p(0,1)$ ein $g\in L^p(0,1)$ $(1\leq p\leq\infty)$ mit $\lim_{n\to\infty}\|d_nf-g\|_p=0^{2)}$, so heißt g *die starke L^p-W-Ableitung von f* und wird mit df bezeichnet.

In [2] und [3] wurde unter anderem bewiesen, daß d ein linearer, abgeschlossener Operator ist und die Operation

$$(If)(x)=\int_0^1 f(t)\quad W(x\overset{\circ}{+}t)\,dt=(f*W)(x)\qquad (x\in[0,1))$$

ist die zu d inverse Transformation, wobei W durch die Walsh-Fourierreihe $W\sim\sum_{k=1}^{\infty}\Psi_k/k$ definiert ist.

Die m-te Walsh-Funktion Ψ_m ist durch

$$(2)\qquad \Psi_m(x)=(-1)^{\sum_{k=0}^{\infty}x_km_k}$$

$$\left(x=\sum_{k=0}^{\infty}x_k2^{-(k+1)}\in[0,1),\ m=\sum_{k=0}^{\infty}m_k2^k\in N: x_k,\ m_k\in\{0,1\}\right)^{3)}$$

---

* Dargelegt auf dem 5. Balkan-Mathematiker-Kongress (Beograd, 24—30. 06. 1974)

1) $P=\{1,2,3,\ldots\}$.

2) $\|\cdot\|_p$ bezeichnet die $L^p(0,1)$ — Norm $(1\leq p\leq\infty)$.

3) $N=\{0,1,2,\ldots\}$.

gegeben. Nach (2) ist (vergl. [5])

$$(3)\qquad (d_n\Psi_m)(x)=\sum_{k=0}^{n-1}2^{k-1}(\Psi_m(x)-\Psi_m(x\overset{\circ}{+}2^{-(k+1)})=$$
$$=\Psi_m(x)\sum_{k=0}^{n-1}2^k(1-(-1)^{m_k})/2=\Psi_m(x)\,(\sum_{k=0}^{n-1}m_k 2^k).$$

Daraus folgt, daß die Walsh-Paley-Funktionen L^p-W-differenzierbar sind, und $d\Psi_m=m\Psi_m$ $(m\in N)$ ist.

In dieser Arbeit betrachten wir die fast überall Konvergenz der Folge $d_n f$. Wir beweisen den folgenden

Satz. *Der Operator* $(Tf)(x)=\sup_n|(d_n(If))(x)|$ $(f\in L(0,1),\ x\in[0,1))$ *ist vom Typus* (∞,∞) *und von schwachem Typ* (1, 1), *dh.*

a) *für* $f\in L^\infty(0,1)$ *gilt* $\|Tf\|_\infty\leq A\|f\|_\infty$.

b) *für* $f\in L(0,1)$ *und für* $y<0$ *gilt* $\text{mes}\{x:(Tf)(x)>y\}\leq A\|f\|_1/y$, *wobei* A *eine von* f *und* y *unabhangige Konstante ist.*

Da $(I\Psi_m)(x)=(W*\Psi_m)(x)=\frac{1}{m}\Psi_m(x)$ $(m\in P)$ ist, auf Grund von (3) ergibt sich, daß für Walsh-Polynome $P=\sum_{j=0}^{m}a_j\Psi_j$ $\lim d_n(IP)(x)=P(x)$ gilt.

Daraus, mit Berücksichtigung, daß die Menge der Walsh-Polynome in $L(0,1)$ dicht liegt, ergibt sich das folgende.

Korollar 1. Für $f\in L(0,1)$ gilt $\lim_{n\to\infty}d_n(If)=f$ fast überall.

Diese Behauptung ist das Analogon des nachstehenden Fundamentalsatzes der klassischen Differential und Integralrechnung: für $g\in L(0,1)$ gilt

$$\frac{d}{dx}\left(\int_0^x g(t)\,dt\right)=g(x)\quad \text{für fast alle } x\in(0,1).$$

Korollar 1. beantwortet zugleich eine Frage von H. J. WAGNER (vergl. [5], Seite 32).

Aus dem Satz durch Anwendung eines Satzes von MARCINKIEWICZ (vergl. [6], Seite 111) ergibt sich.

Korollar 2. Für $f\in L^p(0,1)$ $(1<p<\infty)$ gilt $\|Tf\|_p\leq A_p\|f\|_p$, wobei A_p eine von f unabhangige Konstante ist.

Hilfssätze. Zum Beweis des Satzes werden wir einige Hilfssätze benützen. Wir bezeichnen mit $f*g$ die W-Faltung der Funktion von f mit g, und mit D_n bzw. mit K_n die n-te WALSH-DIRICHLETsche, bzw. die n-te WALSH-FEJÉRsche Kernfunktion, dh. es wird

$$(f*g)(x)=\int_0^1 f(t)\,g(x\overset{\circ}{+}t)\,dt,\quad D_n=\textstyle\sum_{k=0}^{n-1}\Psi_k\qquad(n\in P),$$
$$K_n=\frac{\sum_{k=1}^{n}D_k}{n},\qquad(n\in P)$$

gesetzt und sei $\qquad e_n=2^{-(n+1)}\qquad(n\in N).$

Hilfssatz 1. *Ist* $x\in(0,1)$, *so konvergiert die Reihe*

$$\sum_{k=1}^{\infty}\Psi'_k(x)/k \quad \textit{und für} \quad W(x)=\sum_{k=1}^{\infty}\Psi'_k(x)/k \quad \textit{gilt}$$

$$|W(x)|\leq\sum_{k=0}^{\infty}2^{-k+1}D_{2^k}(x). \tag{4}$$

Beweis. Hat $n\in N$ die dyadische Darstellung $n=\sum_{k=0}^{\infty}n_k 2^k$, $n_k\in\{0,1\}$, so gilt

$$D_n=\Psi'_n\sum_{k=0}^{\infty}n_k r_k D_{2^k} \quad (n\in N), \tag{5}$$

wobei $r_k\,(k\in N)$ die k-te RADEMACHERsche Funktion bedeutet (vergl. [7], Hilfs satz I.)

Daraus folgt, daß $|D_n(x)|\leq\sum_{k=0}^{\infty}D_{2^k}(x)$ $(x\in(0,1),\ n\in N)$ ist und deshalb konvergiert die Reihe $\sum_{k=0}^{\infty}\Psi'_k(x)/k$ $(x\in(0,1))$.

Mit Hilfe der ABELschen partiellen Summation erhält man

$$W(x)=\sum_{k=1}^{\infty}\frac{1}{k}\Psi'_k(x)=\sum_{k=2}\frac{1}{k(k-1)}D_{2^k}(x)-1$$

und daraus auf Grund von (5):

$$W(x)+1=\sum_{k=2}^{\infty}\frac{1}{k(k-1)}D_k(x)=\sum_{k=2}^{\infty}\frac{1}{k(k-1)}\Psi'_k(x)\sum_{i=0}^{\infty}k_i r_i(x)D_{2^i}(x)=$$

$$=\sum_{i=0}^{\infty}r_i(x)D_{2^i}(x)\sum_{\substack{k\geq 2\\ k_i=1}}\frac{1}{k(k-1)}\Psi'_k(x).$$

Da

$$\left|\sum_{\substack{k\geq 2\\ k_i=1}}\frac{1}{k(k-1)}\Psi'_k(x)\right|\leq\sum_{k=2^i}^{\infty}\frac{1}{k(k-1)}=\frac{1}{2^i-1}$$ ist, gilt $|W(x)|\leq\sum_{i=0}^{\infty}2^{-i+1}D_{2^i}(x)$.

Damit ist Hilfssatz 1 bewiesen.

Hilfssatz 2. *Es sei* $2^{n-1}\leq m<2^n$. *Dann gilt*

$$|K_m(x)|\leq\sum_{i=0}^{n-1}2^{i-n}\sum_{j=i}^{n-1}(D_{2^j}(x)+D_{2^j}(x\overset{\circ}{+}e_i)).$$

Beweis. Aus der Definition von K_m und aus (3) folgt
$mK_m=mD_m-\sum_{i=0}^{m-1}i\Psi'_i=mD_m-d_n D_m$, woraus sich auf Grund von (1) und (5)

$$mK_m(x)=\Psi'_m(x)\sum_{i=0}^{n-1}m_i 2^i\sum_{j=0}^{n-1}m_j r_j(x)D_{2^j}(x)-$$

$$-\Psi'_m(x)\sum_{i=0}^{n-1}2^i\sum_{j=0}^{n-1}\frac{1}{2}m_j r_j(x)(D_{2^j}(x)-\Psi'_m(e_i)r_j(e_i)D_{2^j}(x\overset{\circ}{+}e_i))=$$

$$=\Psi'_m(x)\sum_{i=0}^{n-1}2^i\sum_{j=0}^{n-1}c_{ij}(x)$$

ergibt.

Da $D_{2^j}(x \overset{\circ}{+} e_i) = D_{2^j}(x)$ $(i>j)$, $r_j(e_i) = (i>j)$ und $(1-\Psi_m(e_i))/2 = (1-(-1)m_i)/2 = m_i$ $(i \in N)$ sind, deshalb ist $c_{ij}(x) = 0$ $(j<i,\ x \in (0,1))$ und so gilt

$$m|K_m(x)| < \sum_{i=0}^{n-1} 2^{i-1} \sum_{j=i}^{n-1} (D_{2^j}(x) + D_{2^j}(x \overset{\circ}{+} e_i)).$$

Daraus folgt die Behauptung.

Hilfssatz 3. Für $n \in P$ ist $d_n W$ von der Gestalt $d_n W = D_{2^n} + F_n + G_n$, wobei

(6) a) $$|F_n(x)| < \sum_{k=0}^{\infty} 2^{-k} D_{2^{n+k}}(x) + \sum_{l=0}^{n-1} 2^l \sum_{k=0}^{\infty} 2^{-(n+k)} D_{2^{n+k}}(x \overset{\circ}{+} e_l),$$

b) $$|G_n(x)| < 24\{D_{2^n}(x) + K_{2^n-1}(x) + 2^{-n}(\sum_{i=1}^{2^n} |K_i(x)|)$$

sind.

Beweis. Auf Grund von (3) ergibt sich

$$d_n W = \sum_{m=1}^{\infty} \frac{1}{m} \Psi_m(\sum_{i=0}^{n-1} m_i 2^i) - \sum_{m=0}^{2^n-1} \Psi_m + \sum_{i=1}^{\infty} \sum_{k=1}^{2^n-1} \frac{k}{i 2^n + k} \Psi_{i 2^n + k}.$$

Es sei nun $a_k(i,n) = i\left(\frac{k}{2^n} - \frac{ik}{i 2^n + k}\right)$ $(i \in P,\ n, k \in N)$.

Dann hat $d_n W$ die Gestalt

$$(d_n W)(x) = D_{2^n}(x) + \sum_{i=1}^{\infty} \frac{1}{i} \Psi_i(2^n x) \sum_{k=0}^{2^n-1} \frac{ik}{i 2^n + k} \Psi_k(x) =$$

$$= D_{2^n}(x) + W(2^n x) \sum_{k=0}^{2^n-1} k 2^{-n} \Psi_k(x) - \sum_{i=1}^{\infty} \frac{1}{i^2} \Psi_i(2^n x) \sum_{k=0}^{2^n-1} a_k(i,n) \Psi_k(x) =$$

$$= D_{2^n}(x) + F_n(x) + G_n(x).$$

Für F_n auf Grund von

$$2^{-n} \sum_{k=0}^{2^n-1} k \Psi_k(x) = 2^{-n}(d_n D_{2^n})(x) = 2^{-n-1} \sum_{i=0}^{n-1} 2^i (D_{2^n}(x) - D_{2^n}(x \overset{\circ}{+} e_i)) =$$

$$= (2^n - 1) 2^{-n-1} D_{2^n}(x) - \sum_{i=0}^{n-1} 2^{i-n-1} D_{2^n}(x \overset{\circ}{+} e_i)$$

und nach Hilfssatz 1 gilt die folgende Abschätzung:

$$|F_n(x)| < (\sum_{k=0}^{\infty} 2^{-k} D_{2^k}(2^n x))\ (2^{-n} D_{2^n}(x) + \sum_{i=0}^{n-1} 2^{i-n} D_{2^n}(x \overset{\circ}{+} e_i)).$$

Da $D_{2^k}(x) = \prod_{j=0}^{k-1} (1 + r_j(x))$ $(x \in [0,1),\ k \in N)$ ist, so gilt für $i<n$

$$D_{2^k}(2^n x) D_{2^n}(x \overset{\circ}{+} e_i) = \prod_{j=0}^{k-1} (1 + r_j(2^n x)) \prod_{l=0}^{n-1} (1 + r_l(x \overset{\circ}{+} e_i)) =$$

$$= \prod_{j=0}^{k-1} (1 + r_{j+n}(x \overset{\circ}{+} e_i)) \prod_{l=0}^{n-1} (1 + r_l(x \overset{\circ}{+} e_i)) = \prod_{j=0}^{n+k-1} (1 + r_j(x \overset{\circ}{+} e_i)) = D_{2^{n+k}}(x \overset{\circ}{+} e_i) =$$

$$= D_2 n + k_{(x \overset{\circ}{+} a)}$$

und ähnlich $D_{2^k}(2^n x) D_{2^n}(x) = D_{2^{n+k}}(x)$. Daraus folgt

$$|F_n(x)| < \sum_{k=0}^{\infty} 2^{-k} D_{2^{n+k}}(x) + \sum_{i=0}^{n-1} 2^{-(n+k)} D_{2^{n+k}}(x \overset{\circ}{+} e_i),$$

womit die Behauptung für F_n bewiesen ist.

Zum Beweis von (b) führen wir die Bezeichnung

$$B_n(x) = \sup_i \left| \sum_{k=0}^{2^n-1} a_k(i, n) \Psi_k(x) \right| \text{ ein; wir werden zeigen, daß}$$

(7) $$|B_n(x)| < 12 \left(D_{2^n}(x) + |K_{2^n-1}(x)| + 2^{-n} \sum_{k=1}^{2^n-1} |K_k(x)|\right)$$

gilt. Da $|G_n(x)| < B_n(x) \sum_{k=1}^{\infty} 1/k^2 < 2 B_n(x)$ ist, folgt die Behauptung (b) unmittelbar aus (6).

Es sei $g_i(t) = \dfrac{t^2}{1+t/i}$ $(i>1,\ 0<t<1)$. Dann ist $a_k(i,n) = g_i\left(\dfrac{k}{2^n}\right)$ und durch abelsche partelle Summation erhält man

$$\sum_{k=0}^{2^n-1} a_k(i,n)\Psi_k = \sum_{k=1}^{2^n-1} \left(g_i((k-1)2^{-n}) - 2g_i(k2^{-n}) + g_i(k+1)2^{-n})\right) k K_k +$$

$$+ (2^n-1)\left(g_i((2^n-2)2^{-n}) - g_i((2^n-1)2^{-n})\right) K_{2^n-1} + g_i((2^n-1)2^{-n}) D_{2^n}.$$

Durch einfache Rechnung ergibt sich $\max\limits_{0 \le j \le 2} \max\limits_{0 \le t \le 1} |g_i^{(j)}(t)| < 6$ $(i \in P)$, woraus

$$|g_i((2^n-1)2^{-n})| < 6, \quad |g_i((2^n-2)2^{-n}) - g_i((2^n-1)2^{-n})| < 6 \ 2^{-n},$$

$|g_i((k-1)2^{-n}) - 2g_i(k2^{-n}) + g_i((k+1)2^{-n})| < 12 \ 2^{-2n}$ folgt und so erhält man

$$\left| \sum_{k=0}^{2^n-1} a_k(i,n)\Psi_k(x) \right| < 12 \left\{ D_{2^n}(x) + |K_{2^n-1}(x)| + 2^{-n} \sum_{k=1}^{2^n-1} |K_k(x)| \right\}.$$

Damit ist die Ungleichung (7) und der Hilfssatz 3 bewiesen.

Hilfssatz 4. *Es seien* $A = (a_{ij}^n)$ $(i, j, n \in N)$, $B = (b_{ij})$ $(i, j \in N)$ *Matrizen und* $(\lambda(i))$ $(i \in N)$ *eine positive Zahlenfolge mit folgenden Eigenschaften:*

1° $0 < b_{ij} < 2^{-\lambda(i)}$ $(i, j \in N)$; 2° $a_{ij}^n > 0$, $(i, j, n \in N)$;

3° $\sup\limits_n \left(\sum\limits_{i,j=0}^{\infty} a_{ij}^n\right) = K_1 < \infty$; 4° $\sup\limits_m \sum\limits_{\lambda(i) \le m} \sum\limits_{j=m}^{\infty} (\sup\limits_{n<j} a_{ij}^n) = K_2 < \infty$;

5° $\sup\limits_m \sum\limits_{\lambda(i) \le m} \sum\limits_{j=m}^{\infty} 2^{-j} \sup\limits_{n \ge j} 2^n a_{ij}^n = K_3 < \infty$.

Es sei weiterhin $f_n(x) = \sum\limits_{i,j=0}^{\infty} a_{ij}^n D_{2^j}(x \overset{\circ}{+} b_{ij})$ $(n \in N)$; *dann ist der Operator* $(T^* f)(x) = \sup\limits_n |(f_n * f)(x)|$ *von Typ* (∞, ∞) *und von schwachem Typ* $(1, 1)$.

Beweis des Satzes. Aus der Definition von F_n und G_n folgt

$$d_n(If) = d_n(W * f) = (d_n W) * f = D_{2^n} * f + F_n * f + G_n * f.$$

Zum Beweis führen wir die folgenden Bezeichnungen ein:

$$f_n^1 = D_{2^n}, \quad f_n^2 = \sum_{j=n}^{\infty} 2^{n-j} D_{2^j}, \quad f_n^3(x) = \sum_{i=0}^{n-1} 2^i \sum_{j=n}^{\infty} 2^{-j} D_{2^j}(x \overset{\circ}{+} e_i),$$

$$f_n^4 = \sum_{i=0}^{n-1} \sum_{j=i}^{n-1} 2^{i-n} D_{2^j}, \qquad f_n^5(x) = \sum_{i=0}^{n-1} 2^{i-n} \sum_{j=i}^{n-1} D_{2^j}(x \overset{\circ}{+} e_i).$$

Die Funktionen f_n^i $(i = 1, \ldots, 5)$ genügen der Bedingung des Hilfssatzes 4. Daraus folgt, daß die Operatoren

$$(T^i f)(x) = \sup (f_n^i * f)(x) \qquad (i = 1, \ldots, 5)$$

vom Typ (∞, ∞) und von schwalchem Typ $(1, 1)$ sind.

Da nach Hilfssatz 2 $|K_m(x)| \leq f_n^4(x) + f_n^5(x)$ $(2^{n-1} \leq m < 2^n)$ gilt, ist

$$(T^6 f)(x) = \sup_m |(|K_m| * f)(x)|$$

auch ein Operator vom Typ (∞, ∞) und von schwachem Typ $(1, 1)$.

Aus der Ungleichung (6) b) folgt

$$\sup_n (G_n * f)(x) \leq 24 \{\sup_n (f_n^1 * |f|)(x) + 2 \sup_n (|K_m| * |f|)(x)\} \leq$$

$$\leq 24 \{T^1(|f|)(x) + 2 T^6(|f|)(x)\}.$$

und so ist der Operator $(G^* f)(x) = \sup_n |(G_n * f)(x)|$ auch vom gewünschten Typ. Da endlich nach (6) a) $|F_n(x)| \leq f_n^2(x) + f_n^3(x)$ ist, hat man

$$(Tf)(x) \leq (T^1 f)(x) + (T^2 f)(x) + (T^3 f) + (G^* f)(x),$$

woraus sich die Behauptung ergibt.

Literaturverzeichnis

[1] J. E. Gibbs, *Some properties of functions of the non-negative integers less than 2^n*, NPL National Physical Laboratory Middlesex, England, DES Rep. No. 3 (1969).

[2] P. L. Butzer — H. J. Wagner, *Walsh-Fourier series and the concept of a derivative*, Applicable Analysis 3 (1973), 29—46.

[3] __________, *Approximation by Walsh polynomials and the concept of a derivative*, Proc. Symp. on Applications of Walsh Functions, Washington, D. C. (1972).

[4] N. J. Fine, *On the Walsh functions*, Trans. Amer. Math. Soc., 65 (1949), 372—414.

[5] H. Wagner, *Ein Differential-und Integralkalkül in der Wals-Fourier-Analysis mit Anwendungen*, Forschungsberichte des Landes Nordrhein-Westfalen (Nr 2334), Westdeutscher Varlag Opladen (1973).

[6] A. Zygmund, *Trigonometric series*, Vol. II. (Cambridge, 1959).

[7] F. Schipp, *Über die Grössenordnung der Pastialsummen der Entwicklung integrierbarer Funktionen nach W-Sytemen*, Acta Sci. Math., Szeged, 28 (1967), 123—134.

[8] ________, *Über gewisse Maximaloperatoren*, Annales Univ. Sci. Budapest (Im Druck).

(Došlo 04. 9. 1974)

Schipp Ferenc
1088 Budapest
Muzeum krt 6—8

Chapter 6
How I Started My Research in Walsh and Dyadic Analysis

Ferenc Schipp

I graduated [1] at Eötvös Lorand University and earned a diploma as a mathematics and physics teacher. After graduation in 1962 I stayed at the University and started working there as an assistant. At that time I established a closer professional connection with an older colleague of mine, László Pál, who had been an aspirant of Frigyes Riesz. Thanks to him we learned about the spiritual climate that had dominated the Mathematical Institute during the 1950's and about the stories concerning the true friendship between Frigyes Riesz and Lipót Fejér, and also the anecdotes about them. When I started my university studies in 1957, Riesz was already dead and Fejér did not give analysis lectures anymore. However, I once saw Fejér at the University.

Due to Fejér, Riesz and Alfréd Haar, there is still an important tradition of the theory of Fourier-series, approximation theory and functional analysis. This heritage was continued by György Alexits and Pál Turán, in Budapest, and Béla Szőkefalvi-Nagy and Károly Tandori, in Szeged, who established the domestic schools of approximation theory and functional analysis. Unfortunately, they are all dead by now.

It was that environment where I started my work with Fourier-series. I first heard about the Walsh system in connection with a problem of László Pál. He worked on a problem of Hardy on the convergence of products of multiplication of hyper harmonic series. The convergence with probability one of them led to the question of convergence of two-dimensional Rademacher-series, which are Walsh-series of a special type. We found Walsh's paper in the library of the University of Szeged.

Ferenc Schipp
Department of Numerical Analysis, Eötvös L. University, Budapest, Pázmány P. sétány 1\C, H-1117 Hungary,e-mail: schipp@numanal.inf.elte.hu

[1] The following description of involvement of F. Shipp in Walsh and dyadic analysis is an excerpt from *Reprints from the Early Days of Information Sciences Reminiscences of the Early Work in Walsh Functions Interviews with Franz Pichler, William R. Wade, Ferenc Schipp*, Radomir S. Stanković, Jaakko T. Astola, (eds.), TICSP Series # 58, ISBN 978-952-15-2598-8, ISSN 1456-2744.

R.S. Stanković et al. (eds.), *Dyadic Walsh Analysis from 1924 Onwards Walsh-Gibbs-Butzer Dyadic Differentiation in Science Volume 1 Foundations*, Atlantis Studies in Mathematics for Engineering and Science 12, DOI 10.2991/978-94-6239-160-4_6

The original definition of Walsh was too complicated to solve this problem. After having found the basic paper of N.J. Fine on Walsh-series, and applying the explicit form of the Lebesgue constants for the Walsh system, Pál solved the problem [1].

Fine showed that the seemingly artificial Walsh system can be considered as the trigonometric system (the character system) of the Cantor (dyadic) group. This work made a great impression on me, since this new concept allowed us to handle the problems with the tools of harmonic analysis. I did not know about the paper of N.Ja. Vilenkin at that time, in which this idea had already appeared in a much more general form. The character systems of those groups were named after him. In my first paper, which was the basis of my university doctor dissertation, I constructed a continuous function whose Walsh-series diverges at a point [2].

Later I constructed another very simple counter example, which can be considered as the dyadic analogue of the Fejér construction for divergent trigonometric Fourier series. This construction was further developed by Peter Simon by giving several examples for divergent Vilenkin series [3].

Another fundamental paper of this subject is due to R.E.A.C. Paley. He introduced a new and very useful enumeration of the Walsh system. Later this paper had a great influence on the development of martingale theory.

At that time in Hungary, thanks to mainly the works of György Alexits and Károly Tandori, many people were dealing with the convergence and summability problems of general orthogonal series. The monograph of Alexits was published at that time, and it contained the famous deep and subtle divergence constructions of Tandori. In the Soviet Union, the Russian translation of the book of Kaczmarz and Steinhaus, with an appendix on Vilenkin systems written by P.L. Uljanov, came out. These works influenced my scientific activities in a great deal.

Here I would like to mention our connection with the Polish mathematicians which was initiated by László Pál. My first official visit was to Poznań, Poland in 1964. It was a great experience to me to meet Professor W. Orlicz, the head of the Department of Mathematics and Professor Z. Ciesielski, who was then writing his important papers on Franklin and Schauder systems. This relationship deepened during mutual visits and the international conferences organized at the Banach Centre. It was then when I first heard about the massacre in Katyn. Later I learned that, J. Marczinkiewicz, whose result I respected very much had been one of the victims there. We can only guess the great loss that his death caused to the theory of Fourier series.

After having the university doctor degree I continued my research work in order to earn the so called candidate degree in mathematics. My supervisor was Károly Tandori. He continued, significantly improved and in a certain sense finalized the results of professor Menchoff, the leader of the Moscow school of real functions theory. As a consequence he reached great reputation in Moscow which reflected also on me and Ferenc Móricz, his students at that time. Alexits and Nikolskii have initiated a series of international conferences on approximation theory and Fourier series. Later, besides Hungary and the Soviet Union also Bulgaria and Poland joined the program, and colleagues over there organized successful conferences. These conferences provided good opportunities to keep in contact not only with the researchers from socialist countries but also with western researchers. An important point in the collaboration was the foundation of the journal *Analysis Mathematica* jointly by the Soviet and the Hungarian Academies.

The reputation of the journal was well established by the first chief editors, professors Nikolskii and Szőkefalvi-Nagy. Throughout the years Analysis Mathematica became a worldwide acknowledged periodical in the theory of Fourier series including dyadic analysis. I felt honored when I became a member of the editorial board. Professor Nikolskii at age 105 is still the chief editor from the Russian side.

References

1. Pál, L.G., "On general multiplication of infinite series", *Acta Sci. Math. (Szeged)*, 29, 1968, 317-330.
2. Schipp. F., "Construction of a continuous function whose Walsh series diverges at a prescribed point", *Annales Universitatis Scientiarum Budapestinensis de Roland Eötvös Nominatae, Sectio Mathematics*, Tomus IX, 1966, 103-108.
3. Simon, P., "On the divergence of Vilenkin-Fourier series", Acta Math. Hungar., 41 (1983), No. 3-4, 359-370.

Acta Mathematica Academiae Scientiarum Hungaricae
Tomus 28 (1—2), (1976), 145—152.

ON THE DYADIC DERIVATIVE

By
F. SCHIPP (Budapest)

1. Introduction

In [2], P. L. BUTZER and H. J. WAGNER introduced the following concept of derivate for a function f given on the interval $[0, 1)$.

DEFINITION. For $f\colon [0, 1)\to \mathbf{R}$ let

(1) $$(d_n f)(x) = \sum_{i=0}^{n-1} 2^{i-1}(f(x)-f(x\dotplus 2^{-i-1})) \quad (x\in[0, 1), n\in \mathbf{P}),$$

where $\dotplus$ denotes the addition introduced by N. J. FINE [4], and $\mathbf{P}=\{1, 2, 3, \ldots\}$.

(i) If for the function f there exists

$$\lim_{n\to\infty} (d_n f)(x) = c$$

at some point $x\in[0, 1)$, then c is called the pointwise dyadic derivative of f at x denoted by $f^{[1]}(x)=c$.

(ii) If for $f\in L^p(0, 1)$ $(1\leq p\leq\infty)$ there exists $g\in L^p(0, 1)$ such that

$$\lim_{n\to\infty} \|d_n f-g\|_p = 0,^1$$

then g is called the strong dyadic derivative of f in $L^p(0, 1)$ denoted by $Df=g$.

This is a modified form of GIBBS' definition, introduced in [6], [7], [8].

The strong dyadic derivative turned out to be a linear, closed operator, its inverse operator or "integral" was introduced, and the fundamental theorem of the calculus was also found to hold for these two concepts [2], [3], [12].

The most important property of the dyadic derivative is the fact that the Walsh—Paley functions ψ_k $(k\in \mathbf{N}=\mathbf{P}\cup\{0\})$ are arbitrarily often differentiable (in both sense) and one has

$$\psi_k^{[1]} = D\psi_k = k\psi_k \quad (k\in\mathbf{N}).$$

The inverse operator to the differential operator D, namely the "integral" operator I, is defined by the convolution

(2) $$(If)(x) = (f * W)(x) = \int_0^1 f(u)W(x\dotplus u)\, du,$$

[1] $\|\cdot\|_p$ denotes the $L^p(0,1)$-norm $(1\leq p\leq\infty)$.

where the function $W\in L^1(0,1)$ is given by its Walsh—Fourier coefficients

(3) $$\hat{W}(0)=1, \qquad \hat{W}(k)=k^{-1} \quad (k\in\mathbf{P}).$$

(See [2], [3].)

In [10], the following analogon of the fundamental theorem of the differential and integral calculus is proved for the pointwise dyadic derivative:

THEOREM 1. *For $f\in L^1(0,1)$ and for almost every $x\in(0,1)$ we have*

(4) $$(If)^{[1]}(x)=f(x).$$

In this paper we give the following generalization of the above theorem.

THEOREM 2. *If F is a function of bounded variation on* [0, 1], *then for almost every $x\in(0,1)$*

(5) $$(W*dF)^{[1]}(x)=F'(x),$$

where $(W*dF)(x)=\int_0^1 W(x\dot{+}t)\,dF(t)$ and the integral is taken in the Lebesgue—Stieltjes sense.

It is obvious, that Theorem 1 follows from Theorem 2, if we apply it to the absolutely continuous function $F(x)=\int_0^x f(t)\,dt$ $(x\in[0,1])$. To the proof of Theorem 2 we apply the Calderón—Zygmund decomposition lemma in a form concerning Borel measures, and use stopping time technics. (For these concepts see e.g. [13].)

An immediate consequence of Theorem 2 is

COROLLARY 1. *If*

(6) $$\sum_{n=0}^{\infty} na_n\psi_n$$

is the Walsh—Fourier—Stieltjes series of a function F of bounded variation, then the series

(7) $$\sum_{n=0}^{\infty} a_n\psi_n$$

converges a.e. to a function $f\in L^2(0,1)$, moreover f is a.e. dyadic differentiable and for almost every $x\in(0,1)$

(8) $$f^{[1]}(x)=F'(x)=\lim_{n\to\infty}\left(\sum_{k=0}^{2^n-1} ka_k\psi_k(x)\right).$$

Indeed, $na_n=\int_0^1 \psi_n\,dF=O(1)$ implies that (7) is the Walsh—Fourier series of a function $f\in L^2(0,1)$ and by a result of P. BILLARD [1] the series (7) converges a.e. to f. On the other hand by $W\in L^1(0,1)$ (see [2]) $(W*dF)\in L^1(0,1)$ and $(W*dF)\hat{}\,(n)=a_n$ we have $W*dF=f$ a.e., thus from Theorem 2 follows $f^{[1]}=F'$ a.e.

For every $n\in\mathbf{N}$ and $x\in(0,1)$ we define the pair of dyadic rationals $\alpha_n(x)$, $\beta_n(x)$ by the inequalities

$$\alpha_n(x)=k2^{-n}\leqq x<(k+1)2^{-n}=\beta_n(x)\qquad (k<2^n,\ k,n\in\mathbf{N}).$$

Then the well-known relation

$$D_{2^n}(x\dotplus y)=\sum_{k=0}^{2^n-1}\psi_k(x\dotplus y)=\begin{cases}2^n & (y\in[\alpha_n(x),\beta_n(x))),\\ 0 & (y\notin[\alpha_n(x),\beta_n(x)))\end{cases}\tag{9}$$

gives

$$\sum_{k=0}^{2^n-1}ka_k\psi_k(x)=\int_0^1 D_{2^n}(x\dotplus y)\,dF(y)=2^n(F(\beta_n(x))-F(\alpha_n(x)))\to F'(x)\tag{10}$$

a.e. if $n\to\infty$. Thus Corollary 1 is proved.

A standard argument shows (see e.g. [5], [9]), that if

$$\left\|\sum_{k=0}^{2^n-1}b_k\psi_k\right\|_1=O(1),$$

and for every dyadic national point α we have

$$\left(\sum_{k=0}^{2^n-1}b_k\psi_k\right)(\alpha-0)=O_\alpha(2^n),$$

then $\sum_{k=0}^{\infty}b_k\psi_k$ is a Walsh—Fourier—Stieltjes series. This implies

COROLLARY 2. *If*

$$\left\|\sum_{k=0}^{2^n-1}ka_k\psi_k\right\|_1=O(1),\qquad \sum_{k=0}^{2^n-1}ka_k\psi_k(\alpha-0)=O_\alpha(2^n)$$

(α dyadic rational), then (7) converges a.e. to a function $f\in L^2(0,1)$ which is a.e. dyadic differentiable and for almost every $x\in(0,1)$ we have

$$f^{[1]}(x)=\lim_{n\to\infty}\left(\sum_{k=0}^{2^n-1}ka_k\psi_k(x)\right).$$

Let

$$V_n^0=D_{2^n},\qquad V_n^1(x)=\sum_{i=0}^{n-1}2^i\sum_{k=n}^{\infty}2^{-k}D_{2^k}(x\dotplus e_i),$$

(11)

$$V_n^2(x)=\sum_{i=0}^{n-1}2^i\sum_{k=i}^{n-1}2^{-n}D_{2^k}(x\dotplus e_i)\quad (x\in[0,1],\ n\in\mathbf{P}),$$

where $e_i=2^{-i-1}$ $(i\in\mathbf{N})$ and for an arbitrary Borel measure μ we set

$$(T^i\mu)(x)=\sup_n|(V_n^i*\mu)(x)|\quad (i=0,1,2).\tag{12}$$

Here $V*\mu$ denotes the dyadic convolution of the measure μ with the function $V\in L(0, 1)$, i.e.

$$(V*\mu)(x) = \int_0^1 V(x \dotplus y)\, d\mu(y) \quad (x\in[0, 1]).$$

We shall prove, that Theorem 2 is an immediate consequence of the following

THEOREM 3. *Let μ be a Borel measure and $y>0$. Then*

$$\lambda\{x: (T^1\mu)(x) > y\} \leqq 16\|\mu\|/y, \tag{13}$$

where λ denotes the Lebesgue measure and $\|\mu\| = \sup_B |\mu(B)|$ (the sup *is taken over all Borel sets $B\subset(0, 1)$).*

Theorem 3 is a generalization of part b) of the theorem in [10].

By means of the method, used in the proof of Theorem 3, we can also show the following theorem of N. J. FINE (see [5], Theorem 8).

THEOREM 4. *Let F be a function of bounded variation and denote $\sigma_n(dF)$ the n-th $(C, 1)$ mean of the Walsh—Fourier—Stieltjes series of F. Then for almost every $x\in(0, 1)$ we have*

$$\lim_{n\to\infty} \sigma_n(dF)(x) = F'(x). \tag{14}$$

2. Proofs

To prove Theorem 3 we shall use the following decomposition lemma of Calderón—Zygmund type.

LEMMA. *Let μ be a positive Borel measure and $y>\mu([0, 1)) = \|\mu\|$. Then there exist a sequence of dyadic intervals $(I_m, m\in\mathbf{P})$ and Borel-measures μ_m $(m\in\mathbf{N})$ such that*

(15) (i) $I_n \cap I_m = \emptyset$ $(n\neq m)$ *and* $\sum_{n=1}^{\infty} \lambda(I_n) \leqq \|\mu\|/y.$

(ii) $\mu = \sum_{n=0}^{\infty} \mu_n,$

(iii) $\mu_n(H) = 0$, *if* $H\supseteq I_n$ $(n\in\mathbf{P})$, $\operatorname{supp}\mu_n \subseteq I_n$ *and* $\|\mu_n\| \leqq 2\mu(I_n)$.

(iv) $\|\mu_0 * D_{2^n}\|_\infty \leqq 3y$ $(n\in\mathbf{N})$.

PROOF. Denote $\mathcal{A}_n$ $(n\in\mathbf{N})$ the σ-algebra, generated by the dyadic intervals of length 2^{-n}, and set

$$\tau(x) = \inf\{n\in\mathbf{N}: (\mu * D_{2^n})(x) > y\},$$

where $\inf\emptyset = \infty$. Then τ is a stopping time with respect to the sequence $(\mathcal{A}_n, n\in\mathbf{N})$ and the set

$$X = \bigcup_{\tau(x)<\infty} [\alpha_{\tau(x)}(x), \beta_{\tau(x)}(x)) = \{x: \tau(x) < \infty\}$$

is the union of mutually disjoint dyadic intervals: $X=\bigcup_{m=1} I_m$, and by the definition of τ and I_m we have

$$\sum_{m=1}^{\infty} \lambda(I_m) \leqq \sum_{m=1}^{\infty} \mu(I_m)/y \leqq \|\mu\|/y.$$

For an arbitrary Borel set $B\subseteq[0, 1)$ let

$$\mu_m(B) = \mu(B\cap I_m) - \frac{\mu(I_m)}{\lambda(I_m)}\lambda(B\cap I_m) \quad (m\in\mathbf{P}),$$

$$\mu_0(B) = \mu(X'\cap B)+ \sum_{m=1}^{\infty} \frac{\mu(I_m)}{\lambda(I_m)}\lambda(B\cap I_m),$$

where $X'=[0, 1)\setminus X$. Then $\mu=\sum_{m=0} \mu_m$ and $B\supseteq I_m$, $m\in\mathbf{P}$ implies $\mu_m(B)=0$ and obviously $\|\mu_m\|\leqq 2\mu(I_m)$ and $\operatorname{supp}\mu_m\subseteq I_m$ $(m\in\mathbf{P})$. By the definition of τ we have $\tau\geqq 1$ and $(\mu * D_{2^n})(x)\leqq y$, if $n<\tau(x)$, thus

$$(\mu * D_{2^{\tau(x)}})(x) \leqq 2(\mu * D_{2^{\tau(x)-1}})(x) \leqq 2y \quad (x\in[0, 1]).$$

From this we get

$$\frac{\mu(I_m)}{\lambda(I_m)} \leqq 2y, \quad (\mu_0 * D_{2^m})(x) \leqq 3y \quad (m\in\mathbf{N}, x\in[0, 1])$$

and the lemma is proved.

Proof of Theorem 3. To prove (13) we need only to consider a positive measure. We shall apply the Lemma and show, that

(16) (i) $\|T^i\mu_0\|_\infty \leqq 6y$,

(ii) $\|\chi_{X'}T^i\mu_m\|_1 \leqq 2\mu(I_m) \quad (m\in\mathbf{P}, i = 0, 1, 2)$,

where $\chi_{X'}$ denotes the characteristic function of the set X'. It is easy to see that (16) implies (13). Indeed, from (16) (ii) for the measure $\mu'=\sum_{m=1}^{\infty} \mu_m$ by (15) (i) and (ii) we have

$$\|\chi_{X'}T^i\mu\|_1 \leqq 2\|\mu\| \quad (i=0, 1, 2),$$

thus

$$\lambda\{x: (T^i\mu)(x) > 12y\} \leqq \lambda\{x: (T^i\mu_0)(x) > 6y\}+\lambda\{x\in X': (T^i\mu')(x) > 6y\}+$$
$$+\lambda\{x\in X: (T^i\mu')(x) > 6y\} \leqq (1/3+1)\|\mu\|/y.$$

Hence (13) follows.

To see (16), we use (15) (iv). Hence $\|\mu_0 * V_n^0\|_\infty\leqq 3y$,

$$\|\mu_0 * V_n^1\|_\infty \leqq \sum_{i=0}^{n-1} 2^i \sum_{k=n}^{\infty} 2^{-k}\|\mu_0 * D_{2^k}\|_\infty \leqq 3y2\sum_{i=0}^{n-1} 2^{i-n} \leqq 6y,$$

and

$$\|\mu_0 * V_n^2\|_\infty \leqq \sum_{i=0}^{n-1} 2^i \sum_{k=i}^{n-1} 2^{-n}\|\mu_0 * D_{2^k}\|_\infty \leqq 3y\sum_{i=0}^{n-1} 2^{i-n}(n-i) \leqq 6y,$$

which gives (16) (i).

To prove (16) (ii) we note that $e_i \leqq \lambda(I_m)$ and $x \in I_m' = [0, 1) \setminus I_m$ implies $x \dotplus e_i \notin I_m$, thus by (15) (iii) we have

$$(\mu_m * D_{2^k})(x \dotplus e_i) = 0$$

if $2^{-k} \geqq \lambda(I_m)$, or $2^{-k} < \lambda(I_m)$ and $e_i \leqq \lambda(I_m)$. This gives for $x \notin I_m$ $(\mu_m * V_n^0)(x) = 0$ $(n \in \mathbf{N})$ and

$$|(\mu_m * V_n^i)(x)| \leqq \sum_{e_i \in \lambda(I_m)} 2^i \sum_{2^{-k} \geqq \lambda(I_m)} 2^{-k}(\mu_m * D_{2^k})(x \dotplus e_i) \quad (n \in \mathbf{N},\ i = 1, 2).$$

Hence by $\|\mu * D_{2^m}\|_1 \leqq \|\mu\| \, \|D_{2^m}\|_1 = \|\mu\|$ $(m \in \mathbf{N})$, we get

$$\|\chi_{X'} T^i \mu_m\|_1 \leqq \|\mu\| \sum_{e_i \leqq \lambda(I_m)} 2^i 2\lambda(I_m) \leqq 2\|\mu_m\| \quad (m \in \mathbf{P}).$$

This completes the proof of Theorem 3.

Proof of Theorems 2 and 4. To prove Theorems 2 and 4, by Theorem 1 we need only to consider an increasing singular function F. (See also [11], Korollar 2).

We shall use some facts from [10] (see Hilfssatz 2, 3). For $d_n W$ $(n \in \mathbf{P})$ the following estimates holds:

$$(17) \qquad |d_n W| \leqq A_1 D_{2^n} + A_2 \left(2^{-n} \sum_{i=1}^{2^n} |K_i| + |K_{2^n - 1}| \right) + \sum_{k=0}^{\infty} 2^{-k} D_{2^{n+k}} + V_n^1,$$

where

$$D_n = \sum_{k=0}^{n-1} \psi_k, \quad K_n = (D_1 + \ldots + D_n)/n \quad (n \in \mathbf{P})$$

and A_1, A_2 are constants. Further for $2^{n-1} \leqq m < 2^n$

$$(18) \qquad |K_m| \leqq \sum_{i=0}^{n-1} 2^{i-n} \sum_{j=i}^{n-1} D_{2^j} + V_n^2$$

holds.

We shall prove that for almost every $x \in (0, 1)$

$$(19) \qquad \lim_{n \to \infty} (V_n^i * dF)(x) = 0 \quad (i = 1, 2).$$

From this using Toeplitz summation theorem by

$$\lim_{n \to \infty} (D_{2^n} * dF)(x) = \lim_{n \to \infty} 2^n (F(\beta_n(x)) - F(\alpha_n(x))) = 0 \quad \text{a.e.},$$

and by (17) and (18) we get

$$\lim_{n \to \infty} (d_n W * dF)(x) = \lim_{n \to \infty} \sigma_n(dF)(x) = 0 \quad \text{a.e.}$$

Let

$$H_0 = \left\{ x \in [0, 1) : \lim_{n \to \infty} (D_{2^n} * dF)(x) \neq 0 \right\}, \quad H = H_0 \bigcup_{i=0}^{\infty} (H_0 \dotplus e_i).$$

Then $\lambda(H_0)=\lambda(H)=0$ and using Toeplitz summation theorem we get

$$\text{(20) (i)} \lim_{n\to\infty}\left(\sum_{k=n}^{\infty} 2^{-k+n}(D_{2^k} * dF)(x \dotplus a)\right) = 0 \quad (x\in H,\ a\in\{0, e_i : i\in \mathbf{N}\}$$

$$\text{(ii)} \lim_{n\to\infty}\left(\sum_{k=0}^{n-1} 2^{-n}(D_{2^k} * dF)(x \dotplus e_i)\right) = 0 \quad (x\in H,\ i\in \mathbf{P}),$$

$$\text{(iii)} \lim_{n\to\infty} \sum_{i=0}^{n-1} 2^i \sum_{k=i}^{n-1} 2^{-n}(D_{2^k} * dF)(x)\Big) = 0 \quad (x\in H_0).$$

Let $I\subset[0, 1)$ be a dyadic interval and introduce the Borel-measure $\mu_I(B)=\int\limits_{B\cap I} dF$. Since $x\notin I$, $e_i \leqq \lambda(I)$ and $2^{-k}\leqq\lambda(I)$ implies $(\mu_I * D_{2^k})(x \dotplus e_i)=0$, from (20) follows, that for all $x\notin I\cap H$ we have

$$\text{(21)} \qquad \lim_{n\to\infty} (\mu_I * V_n^i)(x) = 0 \quad (i = 1, 2,\ x\notin I\cap H).$$

To prove (19) it is enough to show that for every $\varepsilon>0$

$$\text{(22)} \qquad \lambda\left\{x: \limsup_{n\to\infty} (V_n^i * dF)(x) > 2\sqrt{\varepsilon}\right\} \leqq 3\varepsilon + 16\sqrt{\varepsilon}.$$

Since the Borel measure $\mu(B)=\int\limits_B dF$ is singular with respect to the Lebesgue measure λ, there exist mutually disjoint sets A and B such that $\lambda(A)=\mu(B)=0$ and $A\cup B=[0, 1]$. It is obvious, that we can cover A and B with open intervals I_n and J_n, respectively $(n\in\mathbf{P})$ such that

$$\text{(23)} \qquad \sum_{i=1}^{\infty}\lambda(I_i) < \varepsilon, \qquad \sum_{j=1}^{\infty}\mu(J_j) < \varepsilon$$

and thus by Borel theorem there exist finitely many intervals $I_{m_1}, \dots, I_{m_s}, J_{n_1}, \dots, J_{n_r}$ with the property

$$\left(\bigcup_{k=1}^{r} J_{n_k}\right)\cup\left(\bigcup_{i=1}^{s} I_{m_i}\right) \supseteqq [0, 1].$$

Let $\tilde\mu(H)=\mu(H\cap Y)$, where $Y=\bigcup\limits_{k=1}^{r} J_{n_k}$. Then by (23) $\tilde\mu(Y)<\varepsilon$. It is easy to see that for every I_{m_i} there exist at most three dyadic intervals I_i^j $(j=1, 2, 3)$ such that

$$\sum_{j=1}^{3}\lambda(I_i^j) \leqq 3\lambda(I_{m_i}) \quad\text{and}\quad \bigcup_{j=1}^{3} I_i^j \supseteqq I_{m_i}\cap[0, 1].$$

Since

$$\mu \leqq \tilde\mu + \sum_{i=1}^{s}\sum_{j=1}^{3}\mu_{I_i^j} = \tilde\mu + \tilde{\tilde\mu},$$

and by (21)

$$\lim_{n\to\infty} (\mu_{I_i^j} * V_n^k)(x) = 0, \quad \text{if} \quad x\notin I_i^j \quad (k = 1, 2),$$

we have

$$\lambda\left\{x: \limsup_{n\to\infty} (V_n^k * \bar{\bar{\mu}})(x) > \sqrt{\varepsilon}\right\} \leqq 3\varepsilon.$$

Finally by (23) and Theorem 2

$$\lambda\left\{x: \limsup_{n\to\infty} (V_n^k * \bar{\mu})(x) > \sqrt{\varepsilon}\right\} \leqq 16\|\bar{\mu}\|/\sqrt{\varepsilon} \leqq 16\sqrt{\varepsilon}$$

and this yields (22).

Theorems 2 and 4 are proved.

References

[1] P. Billard, Sur la convergence presque partout des séries de Fourier—Walsh des fonctions de l'espace $L^2(0,1)$, *Studia Math.*, **28** (1967), 263—388.

[2] P. L. Butzer—H. J. Wagner, Walsh—Fourier series and the concept of a derivative, *Applicable Anal.*, **3** (1973), 29—46.

[3] P. L. Butzer—H. J. Wagner, On dyadic analysis based on the pointwise dyadic derivative, *Analysis Math.*, **1** (1975), 171—196.

[4] N. J. Fine, On the Walsh functions, *Trans. Amer. Math. Soc.*, **65** (1949), 372—414.

[5] N. J. Fine, Fourier-Stieltjes series of Walsh function, *Trans. Amer. Math. Soc.*, **86** (1957), 246—255.

[6] J. E. Gibbs, Some properties of functions on the non-negative integers less than 2^n, *NPL* (National Physical Laboratory) Middlesex, England, DES Rept., **3** (1969).

[7] J. E. Gibbs—B. Ireland, Some generalizations of the logical derivative, *NPL*, DES Rept., **8** (1971).

[8] J. E. Gibbs—M. J. Millard, Walsh functions as a solution of a logical differential equation, *NPL*, DES Rept., **1** (1969).

[9] L. G. Pál—F. Schipp, On Haar and Schauder series, *Acta Sci. Math. Szeged*, **31** (1970), 53—58.

[10] F. Schipp, Über einen Ableitungsbegriff von P. L. Butzer und H. J. Wagner, *Matematica Balkanica*, **4** (1974), 541—546.

[11] F. Schipp, Über gewissen Maximaloperatoren, *Annales Univ. Sci. Budapest, Sectio Math.*, **18** (1975), 189—195.

[12] H. J. Wagner, *Ein differential- und Integralkalkül in der Walsh—Fourier-Analysis mit Anwendungen* (Forschungsber. des Landes Nordrhein-Westfalen Nr 2334), Westdeutscher Verlag, Köln-Opladen (1973), 71 pp.

(Received June 11, 1975)

EÖTVÖS LORÁND UNIVERSITY
DEPARTMENT II OF ANALYSIS
1088 BUDAPEST, MÚZEUM KRT. 6—8.

Chapter 7
My Involvement with the Dyadic Derivative

Kees (C.W.) Onneweer

My first encounter with the theory of compact Vilenkin groups occurred in 1968 when my Ph.D. advisor, Professor Daniel Waterman, asked me to read and study the paper [7] in which N. Ya. Vilenkin introduced these groups. This resulted in a series of papers, some of them with D. Waterman as co-author, in which various types of convergence of Vilenkin-Fourier series were discussed.

Around 1974, I first became aware of the papers by P.L. Butzer and H.J. Wagner [1] in which they introduced the concept of the dyadic derivative. They proved that the Walsh-Paley functions ϕ_n are the eigenfunctions of the dyadic differentiation operator with n as the corresponding eigenvalue. Using a modification of their original definition they also proved that the Walsh functions in the Kaczmarz ordering ψ_n are the eigenfunctions of this modified differential operator with n as the corresponding eigenvalue. This raised an obvious question: *is it necessary to define a different definition of dyadic differentiation for each different ordering of the Walsh functions*?

My own earliest contributions to the theory of dyadic differentiation were two-fold:

(i) An extension to dyadic differentiation of fractional order, and
(ii) An extension of the definition of dyadic differentiation to functions defined on a local field K [5].

The new definition was quite complicated and, as before, the characters of K were again eigenfunctions of the differentiation operator and the corresponding eigenvalues were intrinsically related to a "natural" ordering of these characters. These results were presented at a conference held at the Mathematical Research Institute in Oberwolfach, organized by Professor P.L. Butzer. For me, this conference was a truly inspiring event.

Kees (C.W.) Onneweer
Department of Mathematics and Statistics, University of New Mexico, Albuquerque, USA,
e-mail: neweer@math.unm.edu

R.S. Stanković et al. (eds.), *Dyadic Walsh Analysis from 1924 Onwards Walsh-Gibbs-Butzer Dyadic Differentiation in Science Volume 1 Foundations*, Atlantis Studies in Mathematics for Engineering and Science 12, DOI 10.2991/978-94-6239-160-4_7

Next, in a 1979 paper [6], I introduced a much simplified definition of the dyadic derivative for functions defined on a class of topological groups that included all previously considered examples. These groups were initially introduced by R. Spector and were studied extensively by R.E. Edwards and G.I. Gaudry, who called these groups "groups having a suitable family of open subgroups" [2]. In 1988, I introduced the name locally compact (L.C.) Vilenkin Groups for these groups, because of their close resemblance to the groups studied by Vilenkin.

Definition 7.1 *A locally compact Vilenkin group is a locally compact Abelian group G containing a strictly decreasing sequence of open compact subgroups $(G_n)_{-\infty}^{\infty}$ such that*

(i) $sup\{order G_n/G_{n+1} : n \in Z\} < \infty$*, and*
(ii) $\cup_{-\infty}^{\infty} G_n = G$ *and* $\cap_{-\infty}^{\infty} G_n = \{0\}$.

Let Γ denote the dual group of G and for each $n \in Z$ let $\Gamma_n = \{\gamma \in \Gamma : \gamma(x) = 1 \text{ for all } x \in G_n\}$. Choose Haar measures μ on G and λ on Γ so that $\mu(G_0) = \lambda(\Gamma_0) = 1$. Then $\mu(G_n) = (\lambda(\Gamma_n))^{-1} := (m_n)^{-1}$.

Definition 7.2 *Let* $\alpha > 0$ *and* $f \in L_{1,loc}(G)$*. For* $x \in G$ *and* $m \in Z$ *define* $E_{m,\alpha}f(x)$ *by*

$$E_{m,\alpha}f(x) = \sum_{l=-\infty}^{m-1} ((m_{l+1}) - (m_l)^{\alpha})(f - \Delta_l * f)(x),$$

where $\Delta_l * f(x) = m_n \int_{x+G_n} f(t) d\mu(t)$.

(i) If $\lim_{m\to\infty} E_{m,\alpha}f(x)$ *exists, the limit is called the pointwise dyadic derivative of order* α *of* f *at* x*, denoted by* $f^{(\alpha)}(x)$.
(ii) If $f \in L^p(G)$, $1 < p < \infty$ *and if* $\lim_{m\to\infty} E_{m,\alpha}f(x)$ *exists in* $L^p(G)$*, the limit is called the strong derivative of order* $\alpha > 0$ *of* f*, denoted by* $D_p^{\alpha} f$.

Like in the previously discussed cases the characters of the locally compact Vilenkin groups are the eigenfunctions of the dyadic differentiation operator as defined here and satisfy the equality $\gamma^{(\alpha)}(x) = (m_n)^{-\alpha}\gamma(x)$ whenever $\gamma \in \Gamma_n \setminus \Gamma_{n-1}$. Thus, the characters of G are again eigenvectors of the differential operator and the corresponding eigenvalues are directly related to the measure of the sets of constancy of these characters, i.e., the eigenvalues are related to what can be thought of as their "frequency".

The definition of dyadic derivative given above enabled me to study various function spaces like Lipschitz spaces, Bessel Potential spaces, etc. on L.C. Vilenkin groups. In around 1988, Toshiyuki Kitada found a solution to a question raised earlier by me, namely to give a suitable analogue of the Hormander Multiplier Theorem in the context of L.C. Vilenkin groups, using the definition of dyadic derivative as given in Definition 7.2. A very successful collaboration with Professor Kitada resulted in a paper [4] containing several generalized Hormander Multiplier theorems, both for power-weighted Lebesgue spaces and for power-weighted Hardy spaces on a L.C. Vilenkin group. Some additional results in this direction were obtained by Kitada, see [3].

In conclusion, my work in the subject of dyadic differentiation was initially strongly influenced by the work of Professor P.L. Butzer and Dr. H.J. Wagner. My latest and possibly most interesting work was the result of a very successful collaboration with Professor T. Kitada.

References

1. ∗Butzer, P.L., Wagner, H.J., "Walsh-Fourier series and the concept of a derivative", *Applicable Analysis*, Vol. 3, 1973, 29-46.
2. Edwards, R.E., Gaudry, G.I., *Littlewood-Paley and Multiplier Theory*, Springer Verlag, Berlin, 1977.
3. Kitada, T., "Potential operators and multipliers on locally compact Vilenkin groups", *Bull. Austral. Math. Soc.*, Vol. 54, 1996, 459-471.
4. Kitada, T., Onneweer, C.W., "Hormander-type multiplier theorems on locally compact Vilenkin groups", *Theory and Applications of Gibbs Derivatives*, 1990, 115-125.
5. Onneweer, C.W., "Differentiation on a *p*-adic or *p*-series field", *Linear Spaces and Approximation*, ISNM 40, 1978, 197-198.
6. ∗Onneweer, C.W., "On the definition of dyadic differentiation", *Applic. Analysis*, Vol. 9, 1979, 267-278.
7. Vilenkin, N. Ya., "On a class of complete orthonormal systems", *Amer. Math. Soc. Transl.*, Vol. 28, 1963, 1-35.

The ∗ sign in the citations indicates that the paper is reprinted in this book.

Applicable Analysis, 1979, Vol. 9, pp. 267–278
0003-6811/79/0904-0267 $04.50/0

On the Definition of Dyadic Differentiation

C. W. ONNEWEER

Department of Mathematics, University of New Mexico, Albuquerque, NM 87131

Communicated P. L. Butzer

AMS (MOS) Classification: 43A70, 43A75, 42A56.

(*Received April 25, 1978*)

1. INTRODUCTION

In a number of papers by Gibbs, Butzer and Wagner, Pál and the present author, various attempts have been made to define a derivative for complex-valued functions defined on the dyadic group **D**, the dyadic field **K**, or some of their natural generalizations. Since the Leibniz–Newton formula for derivatives cannot be used in this context, all attempts to obtain a so-called dyadic derivative have been based on the idea that it should be defined in such a way that the characters of **D** or **K** are eigenfunctions of the differentiation operator and that the corresponding eigenvalues are in a simple way related to these eigenfunctions. This last condition has resulted in two somewhat different approaches which we will discuss in this paper.

In Section 2 we define the groups **D** and **K** and present some notation needed later on. In Section 3 we first compare the definitions given in [1], [6], [7] and [8] for dyadic derivatives. Following that, we give a new definition which lends itself more easily to an interpretation of its meaning and to possible applications in the physical sciences. In Section 4 we prove that our new definition leads to essentially the same results as can be obtained from the earlier definitions. We conclude that section with some comments on possible generalizations of the theory presented here. We shall not be concerned with such generalizations in the main body of this paper because the dyadic case already shows all the essential features of the theory and this case is the most interesting as far as possible applications are concerned.

2. THE GROUPS D AND K AND THEIR DUAL GROUPS

Let **C** denote the complex numbers, **R** the real numbers, **Z** the integers, **N** the natural numbers and $\mathbf{P}=\mathbf{N}\cup\{0\}$. We shall use G to denote both **D** and **K**. The dyadic group **D** can be defined either on $[0,1)$, or else as the set of all sequences $x=(x_0, x_1, x_2, \ldots)$ with $x_i\in\{0,1\}$ for all $i\in\mathbf{P}$. We shall use only the second description of **D**. For a nice explanation of the relationship between these two ways of describing **D** we refer to [3, Appendix C]. Addition in **D** is defined termwise modulo 2. Thus, if $x=(x_i)_{i\in\mathbf{P}}\in\mathbf{D}$ and $y=(y_i)_{i\in\mathbf{P}}\in\mathbf{D}$ then $x\dot{+}y$ is given by $(x\dot{+}y)_i\equiv(x_i+y_i)\bmod 2$ for each $i\in\mathbf{P}$. For $n\in\mathbf{P}$ we define $e_n\in\mathbf{D}$ by $(e_n)_i=\delta_{n,i}$, where $\delta_{n,i}$ denotes the Kronecker δ symbol. Next we define a sequence of subgroups $G_n (n\in\mathbf{P})$ of **D** by $G_0=\mathbf{D}$ and if $n\geqq 1$ then

$$G_n=\{x\in\mathbf{D};\quad x_i=0\quad\text{for } i<n\}.$$

The G_n are a fundamental set of neighborhoods of $0\in\mathbf{D}$ and they determine the topology of **D**. The following properties hold.

(a) each G_n is a compact and open subgroup of **D**,

(b) $G_{n+1}\subsetneqq G_n$ for each $n\in\mathbf{P}$,

(c) order $(G_n/G_{n+1})=2$ for each $n\in\mathbf{P}$, (1)

(d) $\bigcup_{n\in\mathbf{P}} G_n=\mathbf{D}$ and $\bigcap_{n\in\mathbf{P}} G_n=\{0\}$.

Let Γ denote the dual group or character group of **D**, and for each $n\in\mathbf{Z}$ let $\Gamma_n\subset\Gamma$ denote the annihilator of G_n, that is,

$$\Gamma_n=\{\chi\in\Gamma;\quad \chi(x)=1\ \text{for } x\in G_n\}.$$

The elements of Γ are the Walsh functions. Using a method due to Paley, the Walsh-Paley (WP) functions $\{\chi_n\}_{n\in\mathbf{P}}$ can be described as follows. For each $n\geqq 0$ and $x=(x_i)\in\mathbf{D}$ let $r_n(x)=\exp(\pi i x_n)$. Thus, $r_n(x)$ is the n-th Rademacher function on **D**. Next define χ_0 by $\chi_0(x)=1$ for all $x\in\mathbf{D}$, and if $k\in\mathbf{N}$ can be expressed as $k=a_0 2^0+\cdots+a_t 2^t$, with $a_i\in\{0,1\}$ for each i and if $x=(x_i)\in\mathbf{D}$ then define $\chi_k(x)$ by

$$\chi_k(x)=(r_0(x))^{a_0}\cdot\ldots\cdot(r_l(x))^{a_l}=\exp(\pi i(a_0x_0+\cdots+a_lx_l)). \tag{2}$$

A second method of ordering the Walsh functions yields the Walsh-Kaczmarz (WK) functions $\{\psi_n\}_{n\in\mathbf{P}}$, see [2]. They are obtained by using functions $R_n(x)$ instead of $r_n(x)$ in (2), where $R_0(x)=r_0(x)$ and $R_n(x)=r_{n-1}(x)\cdot r_n(x)$ for $n\geqq 1$. The WK functions are often used in applications of Walsh functions in information theory, because they give the enumeration of the Walsh functions according to increasing sequency, see [5]. The following holds for the dual group of $\mathbf{D}$:

$$\Gamma=\{\chi_k;\ k\in\mathbf{P}\}=\{\psi_k;\ k\in\mathbf{P}\}.$$

Furthermore, $\Gamma_0=\{\chi_0\}=\{\psi_0\}$ and for each $n\geqq 1$ we have

$$\Gamma_n=\{\chi_k\in\Gamma;\ 0\leqq k<2^n\}=\{\psi_k\in\Gamma;\ 0\leqq k<2^n\}.$$

The dyadic field $\mathbf{K}$ can be defined either on $[0,\infty)$, or else as the set of all sequences $x=(x_i)_{i\in\mathbf{Z}}$ with $x_i\in\{0,1\}$ for each $i\in\mathbf{Z}$ and $x_i=0$ for $i<M$ for some $M\in\mathbf{Z}$. Addition in $\mathbf{K}$ is defined coordinate-wise modulo 2, see [8] or [11] for further details. For $n\in\mathbf{Z}$ we define $e_n\in\mathbf{K}$ by $(e_n)_i=\delta_{n,i}$. The subgroups $G_n(n\in\mathbf{Z})$ of $\mathbf{K}$ are defined by

$$G_n=\{x\in\mathbf{K};\ x_i=0 \text{ for } i<n\}.$$

The subgroups G_n are a fundamental set of neighborhoods of $0\in\mathbf{K}$ and they satisfy the properties mentioned in (1) after replacing $\mathbf{P}$ by $\mathbf{Z}$. The elements of the character group Γ of $\mathbf{K}$ are the generalized Walsh functions, which can be described as follows. Let $\chi_0(x)=1$ for all $x\in\mathbf{K}$. For $y=(y_i)$ in $\mathbf{K}$ with $y_i=0$ for $i<s$ and $y_s=1$, and $x=(x_i)$ in $\mathbf{K}$ set

$$\chi_y(x)=\exp\left(\pi i\left(\sum_{i=-s}^{-s-1} x_iy_{-1-i}\right)\right).$$

Then $\Gamma=\{\chi_y;\ y\in\mathbf{K}\}$ and, if Γ_n denotes the annihilator of G_n, we have

$$\Gamma_n=\{\chi_y\in\mathbf{K};\ y\in G_{-n}\}.$$

Generally, we use the notation G, G_n, Γ or Γ_n to refer to both $\mathbf{D}$ and $\mathbf{K}$ and the groups associated with them. If only one of the groups $\mathbf{D}$ or $\mathbf{K}$ is meant this will be clear from the context.

Let dx or m denote Haar measure on G, normalized so that $m(G_0)=1$ and let $(m_n)^{-1}$ denote the measure of G_n for each $G_n \subset G$. Thus $m_n=2^n$ for all $n \in \mathbf{Z}$ if $G=\mathbf{K}$, whereas $m_n=2^n$ for all $n \in \mathbf{P}$ if $G=\mathbf{D}$. If $G=\mathbf{D}$ we set $m_n=0$ for $n<0$. For $f \in L_1(G)$ and $\chi \in \Gamma$ the Fourier transform of f at χ is defined by

$$\hat{f}(\chi)=\int_G f(t)\overline{\chi(t)}\,dt.$$

Usually one uses $\hat{f}(n)$ instead of $\hat{f}(\chi_n)$, if $G=\mathbf{D}$, and $\hat{f}(y)$ instead of $\hat{f}(\chi_y)$, if $G=\mathbf{K}$. However, the present notation allows us to use the same formalism for both groups. Finally, $X(G)$ will denote either $C(G)$, the space of continuous functions on G with the supremum norm, or $L_p(G)$, $1 \leq p < \infty$, the space of measurable functions f on G for which

$$\|f\|_p=\left(\int_G |f(t)|^p\,dt\right)^{1/p}<\infty.$$

3. DEFINITIONS OF DYADIC DIFFERENTIATION

We begin this section by comparing the definitions of dyadic differentiation that are given in [1], [6], [7] and [8]. We shall see that in each case the characters on G are eigenfunctions of the differentiation operator. However, the choice of the corresponding eigenvalues has led to two different theories, one due to Butzer and Wagner (if $G=\mathbf{D}$) and to Pál (if $G=\mathbf{K}$), the other due to the present author (both if $G=\mathbf{D}$ and if $G=\mathbf{K}$). In [1] the following definition can be found, which is originally due to Gibbs.

DEFINITION 1 For $f:\mathbf{D}\to\mathbf{C}$ and $m \in \mathbf{N}$ let

$$d_m f(x)=\sum_{j=0}^{m-1} 2^{j-1}(f(x)-f(x\dotplus e_j)).$$

(a) If $\lim_{m\to\infty} d_m f(x)$ exists, it is called the pointwise BW-derivative of f at x, denoted by $f^{[1]}(x)$.

(b) If $f \in X(\mathbf{D})$ and if $\lim_{m\to\infty} d_m f$ exists in $X(\mathbf{D})$ it is called the strong BW-derivative of f in $X(\mathbf{D})$, denoted by $D^{[1]}_{X(\mathbf{D})}f$.

The definition of Pál for differentiation on **K**, see [8], is obtained by replacing **D** by **K** and $d_m f(x)$ by $\bar{d}_m f(x)$ in Definition 1, where

$$\bar{d}_m f(x) = \sum_{j=-m}^{m-1} 2^{j-1}(f(x) - f(x \dot{+} e_j)).$$

The following results can be found in [1] and [8].

THEOREM 1. *If* $\{\chi_k\}_{k \in \mathbf{P}}$ *denotes the WP functions and* $\{\chi_y\}_{y \in \mathbf{K}}$ *the generalized Walsh functions then*

i) *for each* $k \in \mathbf{P}$ *and* $x \in \mathbf{D}$, $\chi_k^{[1]}(x)$ *exists and* $\chi_k^{[1]}(x) = k\chi_k(x)$,
ii) *for each* $x, y \in \mathbf{K}$, $\chi_y^{[1]}(x)$ *exists and* $\chi_y^{[1]}(x) = \bar{y}\chi_y(x)$, *where* $\bar{y} = \sum_{i=-\infty}^{\infty} y_i 2^{-i-1}$.

In [6] and [7] a different choice for the eigenvalues was made. This choice was motivated by the following considerations. (i) The choice of eigenvalues made by Butzer and Wagner clearly depends on the ordering of the Walsh functions. Thus, for each ordering of the Walsh functions a corresponding definition of dyadic differentiation is needed to preserve the results of Theorem 1. For the WK functions such a definition was given in [2]. However, it seems desirable to choose the eigenvalues independent of a particular enumeration of the Walsh functions. (ii) If we consider the characters of the circle group T, that is, the functions $e_k(x) = \exp(ikx) = \cos kx + i \sin kx (k \in \mathbf{Z})$, then these functions are eigenfunctions of the classical differentiation operator with eigenvalues ik. The physical interpretation of k is that it is the frequency of the function e_k, and this frequency is a multiple of the inverse of the length of the intervals on which the functions $\cos kx$ and $\sin kx$ are positive or negative. The same is true for the functions $e_y(x) = \exp(iyx)$ $(y \in \mathbf{R})$, which are the characters of **R**. Now we note that with each character $\chi \neq \chi_0$ in Γ we can associate a similar number. Namely, if $\chi \neq \chi_0$, then $\chi \in \Gamma_k \backslash \Gamma_{k-1}$ for a unique $k \in \mathbf{Z}$, and such a character χ is constant on each of the cosets of G_k in G but not on the cosets of G_{k-1} in G. Let $\|\chi\|$ denote the inverse of the measure of G_k. Using the notation introduced in Section 2, we have $\|\chi\| = m_k$ whenever $\chi \in \Gamma_k \backslash \Gamma_{k-1}$. Furthermore, we set $\|\chi_0\| = 0$. Finally, we mention that if $\chi = \chi_y$ is a generalized Walsh function then $\|\chi\| = \|y\|$, where $\|y\|$ is the usual norm of $y \in \mathbf{K}$, see [11]. The foregoing considerations have led us to choose the quantity $\|\chi\|$ as eigenvalue corresponding to χ. In [6] and [7] the following definitions can be found. We use the notation $\sum_{(m,l)}$ to denote summation over all elements $z_{q,m} \in G_l \backslash G_{l+1}$, where $z_{q,m}$ is any element of G for which $(z_{q,m})_i = 0$ when $i \geqq m$.

DEFINITION 2 For $f: \mathbf{D} \to \mathbf{C}$ and $m \in \mathbf{N}$ let

$$D_m f(x) = \tfrac{2}{3} \sum_{l=0}^{m-1} 2^{-m}(2^{2l+1}+1) \sum_{(m,l)} (f(x) - f(x \dot{+} z_{q,m})).$$

(a) If $\lim_{m\to\infty} D_m f(x)$ exists, the limit is called the pointwise derivative of f at x, denoted by $f^{(1)}(x)$.
(b) If $f \in X(\mathbf{D})$ and if $\lim_{m\to\infty} D_m f$ exists in $X(\mathbf{D})$, the limit is called the strong derivative of f in $X(\mathbf{D})$, denoted by $D^{(1)}_{X(\mathbf{D})} f$.

The definition of differentiability for functions on $\mathbf{K}$ is obtained by replacing $\mathbf{D}$ by $\mathbf{K}$ and $D_m f(x)$ by $\bar{D}_m f(x)$ in Definition 2, where

$$\bar{D}_m f(x) = \tfrac{2}{3} \sum_{l=-m}^{m-1} 2^{-m}(2^{2l+1}+2^{-m}) \sum_{(m,l)} (f(x) - f(x \dot{+} z_{q,m})).$$

The following has been proved in [6, Thm. 1] and [7, Thm. 1].

THEOREM 2 *If $\chi \in \Gamma$ and $x \in G$, then $\chi^{(1)}(x)$ exists and $\chi^{(1)}(x) = \|\chi\| \chi(x)$.*

Though the result in Theorem 2 is simpler and more natural than the result in Theorem 1, the second definition is more complicated than the first. Thus it may be harder to apply the dyadic derivatives $f^{(1)}(x)$ or $D^{(1)} f$ than the derivatives $f^{[1]}(x)$ or $D^{[1]} f$ in any of the fields in which the (generalized) Walsh functions are used. Therefore, we will present another definition of dyadic differentiation which is much simpler than any of the previous definitions and so that Theorem 2 again holds. Also, the same definition will apply to both functions on $\mathbf{D}$ and on $\mathbf{K}$.

In [11, p. 23] it is shown that if we define $\Delta_n: G \to \mathbf{C}$ $(n \in \mathbf{Z})$ by

$$\Delta_n(x) = \begin{cases} m_n, & \text{if } x \in G_n, \\ 0, & \text{if } x \notin G_n, \end{cases} \tag{3}$$

then

$$(\Delta_n)^{\wedge}(\chi) = \begin{cases} 1, & \text{if } \chi \in \Gamma_n, \\ 0, & \text{if } \chi \notin \Gamma_n. \end{cases} \tag{4}$$

Note that if Γ is the set of WP functions or WK functions, then the functions Δ_n ($n \in \mathbf{P}$) are precisely the Dirichlet kernels of order 2^n. Also, if for $f \in L_1(G)$, $n \in \mathbf{Z}$ and $x \in G$, $\Delta_n * f$ denotes the convolution of f and Δ_n then

$$\begin{aligned}\Delta_n * f(x) &= \int_G f(x \dot{-} t)\Delta_n(t)dt \\ &= m_n \int_{x \dot{+} G_n} f(t)dt \\ &= (m(x \dot{+} G_n))^{-1} \int_{x \dot{+} G_n} f(t)dt.\end{aligned}$$

Thus, $\Delta_n * f(x)$ is the average of f over the coset $x \dot{+} G_n$ in G. Moreover, we have

$$\lim_{n \to \infty} \|\Delta_n * f - f\|_1 = 0. \tag{5}$$

In the following definition we use $L_{1,\mathrm{loc}}(G)$ to denote the set of all functions on G which are locally integrable, that is, they are integrable on each compact subset of G. Note that each $f \in X(G)$ belongs to $L_{1,\mathrm{loc}}(G)$.

DEFINITION 3 For $f \in L_{1,\mathrm{loc}}(G)$ and $m \in \mathbf{N}$ let

$$E_m f(x) = \sum_{l=-\infty}^{m-1} (m_{l+1} - m_l)(f(x) - \Delta_l * f(x)).$$

If $\lim_{m \to \infty} E_m f(x)$ exists we call this limit the pointwise D-derivative of f at x, denoted by $f^{[1]}(x)$.

Strong D-derivatives are defined similarly and will be denoted by $D^{[1]}_{X(G)} f$, whenever $f \in X(G)$. Note that if $G = \mathbf{D}$ then $m_{l+1} - m_l = 2^l$ for $l \geq 0$, $m_{l+1} - m_l = 1$ for $l = -1$ and $m_{l+1} - m_l = 0$ for $l < -1$, whereas if $G = \mathbf{K}$ then $m_{l+1} - m_l = 2^l$ for all $l \in \mathbf{Z}$. We prefer to use the notation $m_{l+1} - m_l$ because in its present form Definition 3 can be used to define differentiation on groups other than $\mathbf{D}$ or $\mathbf{K}$, see Remark 3 at the end of this paper.

4. PROPERTIES OF D-DERIVATIVES

In this section we show that the strong $L_1(G)$ differentiation theory based on Definition 3 is essentially the same as the $L_1(G)$ theory based on Definition 2. To simplify the notation we shall write $D^{(1)}$ for $D^{(1)}_{L_1(G)}$ and $D^{[1]}$ for $D^{[1]}_{L_1(G)}$, and $\mathcal{D}(D^{(1)})$ or $\mathcal{D}(D^{[1]})$ for the domain of $D^{(1)}$ and $D^{[1]}$, respectively. In [6, Cor. 1, 3] and [7, Thm. 2] it was shown that $\mathcal{D}(D^{(1)}) = W(L_1(G), \|\chi\|)$, where

$$W(L_1(G), \|\chi\|)$$
$$= \{f \in L_1(G);\ \text{there exists } g \in L_1(G) \text{ with } \hat{g}(\chi) = \|\chi\| \hat{f}(\chi),\ \chi \in \Gamma\}.$$

In this section we shall prove the same equality for $\mathcal{D}(D^{[1]})$. We first present some basic properties for D-differentiability.

THEOREM 3

(a) *If* $\chi \in \Gamma$ *and* $x \in G$ *then* $\chi^{[1]}(x)$ *exists and* $\chi^{[1]}(x) = \|\chi\| \chi(x)$.
(b) *If* $G = \mathbf{D}$ *then each* $\chi \in \Gamma$ *is strongly D-differentiable in every space* $X(\mathbf{D})$*; if* $G = \mathbf{K}$ *then each* $\chi \in \Gamma$ *is strongly D-differentiable in* $C(\mathbf{K})$.
(c) *If* $f \in L_1(G)$ *and* $n \in \mathbf{N}$ *then*

$$(E_n f)\hat{\ }(\chi) = \begin{cases} \|\chi\| \hat{f}(\chi), & \text{if } \chi \in \Gamma_n, \\ m_n \hat{f}(\chi), & \text{if } \chi \notin \Gamma_n. \end{cases}$$

Proof (a) If $\chi = \chi_0$ and $x \in G$ then $\Delta_l * \chi_0(x) = \chi_0(x)$ for each $l \in \mathbf{Z}$. Thus, $E_m \chi_0(x) = 0$ for each $m \in \mathbf{N}$, which implies that $\chi_0^{[1]}(x) = 0 = \|\chi_0\| \chi_0(x)$. Next, let $\chi \in \Gamma \setminus \{\chi_0\}$. Then $\chi \in \Gamma_k \setminus \Gamma_{k-1}$ for some $k \in \mathbf{Z}$. Thus χ is constant on the cosets of G_k, which implies that $\Delta_l * \chi(x) = \chi(x)$ for $l \geq k$ and $x \in G$. Also, if $l < k$ then

$$\Delta_l * \chi(x) = m_l \int_{x \dot{+} G_l} \chi(t)\,dt = 0.$$

Consequently, for $m \geq k$ we have

$$E_m \chi(x) = \sum_{l=-\infty}^{k-1} (m_{l+1} - m_l) \chi(x)$$
$$= m_k \chi(x) = \|\chi\| \chi(x),$$

and hence, $\chi^{[1]}(x) = \|\chi\| \chi(x)$.

(b) If $G=\mathbf{D}$ then every $\chi\in\Gamma$ belongs to each of the spaces $X(\mathbf{D})$, whereas if $G=\mathbf{K}$ then $\chi\in\Gamma$ belongs to $C(\mathbf{K})$ but not to $L_p(\mathbf{K})$ when $1\leqq p<\infty$. It follows immediately from the representation of $E_m\chi(x)$ obtained in the proof of (a) that χ is strongly D-differentiable in each of the indicated spaces.

(c) Let $f\in L_1(G)$ and fix $n\in\mathbf{N}$. It follows from (4) that for $\chi\in\Gamma$ and $l\in\mathbf{Z}$ we have

$$(\Delta_l * f)\hat{\ }(\chi)=(\Delta_l)\hat{\ }(\chi)\hat{f}(\chi)=\begin{cases}\hat{f}(\chi), & \text{if } \chi\in\Gamma_l,\\ 0, & \text{if } \chi\notin\Gamma_l.\end{cases}$$

Therefore, if $\chi\in\Gamma_n\backslash\{\chi_0\}$ then $\chi\in\Gamma_k\backslash\Gamma_{k-1}$ for some $k\leqq n$ and $\chi\notin\Gamma_l$ for all $l\leqq k-1$. Consequently,

$$(E_nf)\hat{\ }(\chi)=\sum_{l=-\infty}^{k-1}(m_{l+1}-m_l)\hat{f}(\chi)=m_k\hat{f}(\chi)=\|\chi\|\hat{f}(\chi).$$

If $\chi\notin\Gamma_n$ then $\chi\notin\Gamma_l$ for all $l\leqq n$. Therefore,

$$(E_nf)\hat{\ }(\chi)=\sum_{l=-\infty}^{n-1}(m_{l+1}-m_l)\hat{f}(\chi)=m_n\hat{f}(\chi).$$

Finally, if $\chi=\chi_0$ then we clearly have $(E_nf)\hat{\ }(\chi)=0$. This completes the proof of Theorem 3.

COROLLARY 1 *If* $f\in\mathscr{D}(D^{[1]})$ *then* $(D^{[1]}f)\hat{\ }(\chi)=\|\chi\|\hat{f}(\chi)$ *for all* $\chi\in\Gamma$.

Proof Since E_mf converges to $D^{[1]}f$ in $L_1(G)$, it follows from Theorem 3(c) that for each $\chi\in\Gamma$

$$(D^{[1]}f)\hat{\ }(\chi)=\lim_{n\to\infty}(E_nf)\hat{\ }(\chi)=\|\chi\|\hat{f}(\chi).$$

In [6, Lemma 3] and [7, Lemma 2] a proof of the following lemma was given.

LEMMA 1 *For each* $k\in\mathbf{Z}$ *there exists* $V_k\in L_1(G)$ *such that*

$$(V_k)\hat{\ }(\chi)=\begin{cases}0, & \text{if } \chi\in\Gamma_{-k},\\ \|\chi\|^{-1}, & \text{if } \chi\notin\Gamma_{-k}.\end{cases}$$

Moreover, $\|V_k\|_1=O(m_k)$.

THEOREM 4 *If* $f \in W(L_1(G), \|\chi\|)$ *then* $f \in \mathscr{D}(D^{[1]})$.

Proof Let $g \in L_1(G)$ satisfy $\hat{g}(\chi) = \|\chi\| \hat{f}(\chi)$ for all $\chi \in \Gamma$. According to Theorem 3(c) we have for each $n \in \mathbf{N}$

$$(E_n f)\hat{\ }(\chi) = \begin{cases} \hat{g}(\chi), & \text{if } \chi \in \Gamma_n, \\ m_n(V_{-n})\hat{\ }(\chi)\hat{g}(\chi), & \text{if } \chi \notin \Gamma_n. \end{cases}$$

Since $(V_{-n})\hat{\ }(\chi) = 0$ when $\chi \in \Gamma_n$ and $(\Delta_n)\hat{\ }(\chi) = 0$ when $\chi \notin \Gamma_n$ we have $V_{-n} * \Delta_n = 0$ in $L_1(G)$. Therefore, it follows from the Uniqueness Theorem for Fourier transforms that

$$E_n f - g = \Delta_n * g - g + m_n V_{-n} * g$$
$$= \Delta_n * g - g + m_n V_{-n} * (g - \Delta_n * g).$$

Consequently,

$$\|E_n f - g\|_1 \leqq (1 + m_n \|V_{-n}\|_1) \|\Delta_n * g - g\|_1.$$

From (5) and the estimate for $\|V_{-n}\|_1$ in Lemma 1 we can conclude that

$$\lim_{n \to \infty} \|E_n f - g\|_1 = 0,$$

that is, $f \in \mathscr{D}(D^{[1]})$ and $D^{[1]} f = g$.

Combining the results of Corollary 1 and Theorem 4 we obtain:

COROLLARY 2 $\mathscr{D}(D^{[1]}) = W(L_1(G), \|\chi\|) = \mathscr{D}(D^{(1)})$.

We conclude this paper with some additional comments.

Remark 1 Once Theorems 3 and 4 have been proved we can further develop the theory of D-differentiation along the lines of [6] or [7]. We only mention the following results. If $G = \mathbf{D}$ there exists a continuous inverse for the operator $D^{[1]}$ and we can prove a fundamental theorem of calculus for $D^{[1]}$ and its inverse as in [6, Thm. 3]. Also, the operator $D^{[1]}$ is a closed operator, both when $G = \mathbf{D}$ and when $G = \mathbf{K}$.

Remark 2 In [10] Splettstösser and Wagner define a so-called Haar derivative on [0, 1). Identifying [0, 1) with $\mathbf{D}$ and using the notation of this paper they define for $f \in L_p(\mathbf{D})$, $1 \leqq p < \infty$, and $m \in \mathbf{N}$ the expression $H_m f(x)$ by

$$H_m f(x) = \tfrac{1}{2} \left\{ \sum_{j=0}^{m} 2^j (f(x) - f * (2\Delta_j - \Delta_{j+1})(x)) + f(x) - \int_D f(t)\,dt \right\}.$$

If $\lim_{m\to\infty} H_m f$ exists in $L_p(\mathbf{D})$ the limit is called the strong H-derivative of f in $L_p(\mathbf{D})$. It is shown that the Haar functions φ_{2^n+j}, $0 \leq j < 2^n$, are eigenfunctions of the H-differentiation operator with corresponding eigenvalues equal to 2^n. For additional details and properties of H-derivatives, see [10]. It is easy to show that the Haar functions are also eigenfunctions of the operator $D^{[1]}$ with corresponding eigenvalues equal to 2^{n+1} and one could develop a theory for the derivatives of Haar functions along the lines of [10] more easily by basing such a theory on Definition 3.

Remark 3 Definition 3 can be applied to functions defined on any locally compact group G having the following property (compare with (1)). There exists a sequence of compact, open subgroups $\{G_n\}_{n\in I}$ of G, where $I = \mathbf{P}$, $-\mathbf{P}$ or $\mathbf{Z}$, such that

i) $G_{n+1} \subsetneqq G_n$ for each $n \in I$,
ii) order $(G_n/G_{n+1}) = k_n \in \mathbf{N}$ for each $n \in I$,
iii) $\bigcup_{n\in I} G_n = G$ and $\bigcap_{n\in I} G_n = \{0\}$.

Examples of such groups are the p-adic numbers $\mathbf{Q}_p$, the p-series numbers $\mathbf{K}_p$, the integers in $\mathbf{Q}_p$ or $\mathbf{K}_p$ and the so-called Vilenkin groups. For some general properties of such groups, see [9, Ch. V]. As in the case when $G = \mathbf{D}$ or $G = \mathbf{K}$ we normalize the Haar measure on G so that $m(G_0) = 1$ and we define m_k by $m_k = (m(G_k))^{-1}$ if $k \in I$ and $m_k = 0$ if $k \in \mathbf{Z}\setminus I$. Now we can use Definition 3 to define a derivative for complex-valued functions on G and we can again obtain the results stated in Theorems 3 and 4. Thus, the present theory unifies a number of differentiation theories on various groups which, until now, seemed to be without a common ground.

References

[1] P. L. Butzer and H. J. Wagner, Walsh-Fourier series and the concept of a derivative, *Applicable Anal.* 3 (1973), pp. 29–46.

[2] P. L. Butzer and H. J. Wagner, On a Gibbs-type derivative in Walsh-Fourier Analysis with applications, *Proc. National Electronics Conference*, Vol. 27, Chicago 1972, pp. 393-398.

[3] R. E. Edwards and G. I. Gaudry, *Littlewood–Paley and Multiplier Theory*, Springer Verlag, Berlin, 1977.

[4] J. E. Gibbs and B. Ireland, Walsh functions and differentiation, Applications of Walsh functions, *Proc. Sympos., Naval Research Lab.*, Washington, D.C., 1974, pp. 147-176.

[5] H. F. Harmuth, *Transmission of Information by Orthogonal Functions*, Springer Verlag, New York, 1969.

[6] C. W. Onneweer, Fractional differentiation on the group of integers of a p-adic or p-series field, *Anal. Math.* 3 (1977), pp. 119–130.

[7] C. W. Onneweer, Differentiation on a *p*-adic or *p*-series field, in *Linear Spaces and Approximation*, edited by P. L. Butzer and B. Sz.-Nagy, ISNM 40, Birkhäuser Verlag, Basel, 1978, pp. 187–198.

[8] J. Pál, On the concept of a derivative among functions defined on the dyadic field, *SIAM J. Math. Anal.* **8** (1977), pp. 375–391.

[9] R. Spector, Sur la structure locale des groupes abéliens localement compacts, *Bull. Soc. Math. France*, Mémoire 24, 1970.

[10] W. Splettstösser and H. J. Wagner, Eine dyadische Infinitesimalrechnung für Haar-Funktionen, *Z. Angew. Math. Mech.* **57** (1977), 527–544.

[11] M. H. Taibleson, *Fourier Analysis on Local Fields, Mathematical Notes*, Princeton University Press, Princeton, 1975.

Analysis Mathematica, 1 (1975), 171—196

On dyadic analysis based on the pointwise dyadic derivative

P. L. BUTZER and H. J. WAGNER

With admiration to Professor S. M. Nikol'skiĭ on the occasion of his seventieth birthday

1. Introduction

One essential open problem in dyadic analysis as initiated by J. E. GIBBS (and his collaborators M. J. MILLARD and B. IRELAND) [12, 13, 15] and further developped by P. L. BUTZER and H. J. WAGNER [6, 9, 10] is an actual interpretation of the dyadic derivative. Recently GIBBS and IRELAND [14] gave an abstract interpretation in the realms of locally compact abelian groups. (The paper [14] also gives an interesting historical survey of the field together with an extensive list of references.) However, just as the classical derivative may be associated with the slope of a tangent to a curve, or with the rate of speed of an object, thus associated with geometrical or physical notions, the problem here is to find an appropriate intuitive interpretation of the dyadic derivative in terms of one or more of the modern sciences which make use of Walsh analysis.

Although it is also our final aim to give a practical interpretation, it seems that one needs to have more detailed knowledge of the dyadic derivative. One aim of this paper is to examine, for example, those function spaces for which dyadic analysis is useful as well as to determine the limits of its ranges of applicability. For example, it will turn out that the space $L^p(0, 1)$, $1 \leqq p < \infty$ (of functions which are pth power Lebesgue integrable) and not the more classical $C[0, 1]$ is the more natural space on which to carry out dyadic analysis on $[0, 1)$. In this respect it follows from a result of S. V. BOČKAREV [3] that if the second dyadic derivative belongs to $C[0, 1]$ then $f = \text{const}$. We shall improve this result by showing that if the first dyadic derivative belongs to $C[0, 1]$ then already $f = \text{const}$. On the other hand, if $f \in C[0, 1]$ then the associated dyadic integral does not belong to $C[0, 1]$. Therefore our aim will be to study in greater detail than in [8, 10, 25] not only the parallelism between the classical derivative and the dyadic derivative but especially the differences that exist between classical and dyadic analysis as a whole. Rather the latter aspect may be of help in possible later interpretations of the dyadic derivative.

Received October 24, 1974.

The contribution of the second author was carried out as an associate of the research group "Informatik Nr. 14" at the Technological University of Aachen.

In their previous work the authors have been mainly concerned with the dyadic derivative considered in the *strong* sense for functions defined on the dyadic group G or for functions of period 1 defined on $[0, 1)$. Exceptions were [8, 25] dealing with the derivative taken in the *pointwise* sense for functions belonging to $C(G)$. For this reason this paper will stress the more difficult but classical pointwise approach for functions belonging to $L^p(0, 1)$, $1 \leqq p < \infty$ (the interval $[0, 1)$ replacing the dyadic group G). It will however be seen that the pointwise and strong dyadic derivative are indeed equal to another (except for a set of measure zero) for functions belonging to a subclass of $L^p(0, 1)$, namely $W^1_{L^p(0,1)}$.

Thus one of the essential goals of this paper will be to build up a differential and integral calculus for Walsh functions, i.e., to develop a dyadic analysis, the foundation of which is the dyadic derivative taken in the pointwise sense.

The body of this paper is divided into six sections. § 2 on preliminary results summarizes the fundamental properties of the strong dyadic derivative, including the fundamental theorem of the dyadic differential and integral calculus, also expressed in terms of Walsh – Fourier coefficients. § 3 is concerned with the connections between the strong and pointwise dyadic derivatives announced above, including the counterpart of the concept of absolutely continuous functions in dyadic analysis. A recent basic result by F. Schipp [23], a solution of a conjecture by Wagner [25, p. 32], which leads us to the fundamental theorem of dyadic analysis in the setting of the pointwise approach is also presented here. § 4 deals with a number of applications. In particular, § 4.2 is reserved to the W-modulus of continuity together with estimates of the order of magnitude of the Walsh-Fourier coefficients. § 4.3 handles results on best approximation by Walsh polynomials together with the Walsh function analog of the fundamental theorem of best approximation. § 4.4 is concerned with a particular partial "differential" equation, namely the dyadic wave equation. § 5 treats elementary results on termwise dyadic differentiation and integration of Walsh series. Finally, § 6 is concerned with those features that distinguish the dyadic derivative or integral from their classical counterparts, thus emphasizing the differences between dyadic and classical analysis.

2. Preliminary results; fundamentals

Each $k \in P = \{0, 1, 2, \dots\}$ has a unique dyadic expansion

$$k = \sum_{j=0}^{K} k_j 2^j \qquad (2^K \leqq k < 2^{K+1},\ K \in P)$$

with $k_j \in \{0, 1\}$, $j \in P$. In case $k=0$ all $k_j=0$. Likewise each $x \in [0, 1]$ has a unique expansion

$$x = \sum_{j=1}^{\infty} x_j 2^{-j} \qquad (x_j \in \{0, 1\},\ j \in N = \{1, 2, 3, \ldots\}),$$

the finite expansion being chosen in case x belongs to the set of all numbers of form $x = p2^{-q} \in [0, 1)$, $p \in P$, $q \in N$. We also need the concept of termwise addition modulo 2 of $x = \sum_{j=1}^{\infty} x_j 2^{-j}$ and $y = \sum_{j=1}^{\infty} y_j 2^{-j}$, defined by

$$x \oplus y = \sum_{j=1}^{\infty} |x_j - y_j| 2^{-j}.$$

If $x, y \in R = (-\infty, \infty)$, then $x \oplus y = (x - [x]) \oplus (y - [y])$, $[x]$ being the largest integer $\leq x$.

This enables one to define the Walsh(-Paley) functions $\{\psi_k(x)\}_{k=0}^{\infty}$ on $[0, 1)$ by

$$\psi_k(x) = \exp\left\{\pi i \sum_{j=0}^{K} k_j x_{j+1}\right\},$$

and on R by periodic extension. These functions form an orthonormal system, i.e.,

$$\int_0^1 \psi_k(x) \psi_l(x)\, dx = \delta_{k,l} \qquad (k, l \in P), \tag{2.1}$$

and possess the important property

$$\psi_k(x \oplus y) = \psi_k(x) \psi_k(y) \tag{2.2}$$

for fixed $x \in R$ and almost all $y \in R$.

Denote by $L^p(0, 1)$, $1 \leq p < \infty$, the set of all functions f of period 1 which are pth power Lebesgue integrable with norm

$$\|f\|_p = \left\{\int_0^1 |f(x)|^p dx\right\}^{1/p},$$

and $L^\infty(0, 1)$ that of all essentially bounded functions f of period 1 with norm

$$\|f\|_\infty = \operatorname*{ess\,sup}_{x \in [0,1)} |f(x)|.$$

Also denote the closure of the subspace of $L^\infty(0, 1)$ spanned by the Walsh functions by $L_W^\infty(0, 1)$ (Weierstrass approximation property!); $X = X(0, 1)$ always stands for one of the spaces $L^p(0, 1)$, $1 \leq p < \infty$, or $L_W^\infty(0, 1)$ with norm $\|f\|_X$. Writing the Walsh-Fourier coefficients of $f \in L^1(0, 1)$ by

$$f^\wedge(k) = \int_0^1 f(u) \psi_k(u)\, du \qquad (k \in P),$$

then

$$\sum_{k=0}^{\infty} f^{\wedge}(k)\psi_k(x)$$

is called the Walsh-Fourier series of f.

The uniqueness theorem here states that

(2.3) $$f^{\wedge}(k) = 0, \quad k \in P \Rightarrow f(x) = 0 \text{ a.e.},$$

and the translation rule that

(2.4) $$\int_0^1 f(u \oplus y)\psi_k(u)\,du = \psi_k(y) f^{\wedge}(k)$$

for each real y and $k \in P$.

We also need to define the dyadic convolution: if $f, g \in L^1(0, 1)$ then

$$(f * g)(x) = \int_0^1 f(x \oplus u) g(u)\,du,$$

and there holds the convolution theorem

(2.5) $$(f * g)^{\wedge}(k) = f^{\wedge}(k) g^{\wedge}(k) \qquad (k \in P).$$

We may now consider dyadic differentiation and integration. Stimulated by work of Gibbs [12, 15], the authors [6, 10, 25] introduced the following concept of a derivative.

Definition 2.1. If for $f \in X(0, 1)$ there exists $g \in X(0, 1)$ such that

$$\lim_{m \to \infty} \left\| \frac{1}{2} \sum_{j=0}^{m} 2^j [f(x) - f(x \oplus 2^{-j-1})] - g(x) \right\|_X = 0,$$

then g is called the strong dyadic derivative of f in $X(0, 1)$ denoted by $D^{[1]}f = g$. The higher order strong derivatives of $f \in X(0, 1)$ are defined successively by

$$D^{[r]}f = D^{[1]}(D^{[r-1]}f) \qquad (r = 2, 3, \ldots).$$

For this paper the pointwise dyadic derivative plays the basic role.

Definition 2.2. If for the function f of period 1 there exists

$$\lim_{m \to \infty} \frac{1}{2} \sum_{j=0}^{m} 2^j [f(x) - f(x \oplus 2^{-j-1})] = c$$

at some point $x \in R$, then c is called the pointwise dyadic derivative of f at x, denoted by $f^{[1]}(x) = c$. More generally, the rth derivative $f^{[r]}(x)$ is defined by induction.

At this point let us list some of the most important properties of the strong derivative. For the proofs see [6, 7, 25].

a) $D^{[r]}$ is a linear, closed operator in $X(0, 1)$;

b) The Walsh functions $\psi_k(x)$ are arbitrarily often dyadic differentiable and one has

$$D^{[r]}\psi_k(x) = \psi_k^{[r]}(x) = k^r\psi_k(x) \qquad (r\in N,\ k\in P); \tag{2.6}$$

c) If f, $D^{[r]}f\in X(0, 1)$ for some $r\in N$ then

$$[D^{[r]}f]^\wedge(k) = k^r f^\wedge(k) \qquad (k\in P). \tag{2.7}$$

The proof of the latter identity is based upon:

$$f_n,\ f\in X(0, 1),\ \lim_{n\to\infty}\|f_n - f\|_X = 0 \Rightarrow \lim_{n\to\infty} f_n^\wedge(k) = f^\wedge(k) \qquad (k\in P). \tag{2.8}$$

The counterpart of (2.7) for the pointwise derivative follows in (3.5).

The inverse operator to the differential operator $D^{[r]}$, namely the anti-differentiation operator $I_{[r]}$, is defined by the convolution

$$(I_{[r]}f)(x) = (f * W_r)(x) = \int_0^1 f(u)W_r(x\oplus u)\,du, \tag{2.9}$$

where $W_r(x)$ is the function given by

$$[W_r]^\wedge(k) = \begin{cases} 1 & \text{for } k = 0, \\ k^{-r} & \text{for } k\in N, \end{cases} \tag{2.10}$$

and for which $W_r\in L^1(0, 1)$.

For the operators $D^{[r]}$, $I_{[r]}$ there holds the following analog of the fundamental theorem of the differential and integral calculus — for the proofs see [6, 7, 8], [25, p. 26].

Theorem 2.3. *Let* $f\in X(0, 1)$ *with* $\int_0^1 f(u)\,du = f^\wedge(0) = 0$. *One has:*

a) *If* $D^{[r]}f\in X(0, 1)$ *then* $I_{[r]}(D^{[r]}f) = f$;

b) $D^{[r]}(I_{[r]}f) = f$.

3. On the equivalence of the pointwise and strong dyadic derivative concepts in $X(0, 1)$

The first question that arises is whether the properties listed above for the strong dyadic derivative are also valid for the pointwise dyadic derivative. Already the attempt to prove property (2.7) for $f^{[r]}\in X(0, 1)$ fails for there does not hold a result of type (2.8) here.

In this respect SCHIPP [23] solved a conjecture of WAGNER [25, p. 32]. Beginning with and sharpening considerably the proof used to establish Theorem 2.3 b), he was indeed able to show

Theorem 3.1. *For $f\in L^1(0,1)$ with $f^{\wedge}(0)=0$ there holds for almost all x*

(3.1) $$(I_{[r]}f)^{[r]}(x)=f(x).$$

This result will enable us to introduce classes of functions in $X(0,1)$ for which the pointwise and strong dyadic derivative are equal almost everywhere to another. Indeed,

Definition 3.2. Set

$AW^0=\{f\colon$ there exists $g\in L^1(0,1)$ with $f(x)-f^{\wedge}(0)=(I_{[1]}g)(x)$ a.e.$\}$,

$AW^{r-1}=\{f\colon f, f^{[1]},\dots,f^{[r-1]}\in AW^0\}$,

$W^r_{X(0,1)}=\{f\colon f\in AW^{r-1}, f^{[r]}\in X(0,1)\}$ (fixed $r\in N$).

Our first result for the first two classes is

Lemma 3.3. a) *If $f\in AW^0$, i.e., if*

(3.2) $$f(x)-f^{\wedge}(0)=(I_{[1]}g)(x) \text{ a.e. with } g\in L^1(0,1),$$

then $f^{[1]}(x)=g(x)$ a.e..

b) *If $f\in AW^{r-1}$ then f is dyadic differentiable of order r and*

(3.3) $$f(x)-f^{\wedge}(0)=(I_{[r]}f^{[r]})(x) \text{ a.e..}$$

Proof. a) Denoting by E_0 the set of all $x\in[0,1)$ for which (3.2) does not hold, i.e., mes $E_0=0$, then

$$f(x)-f^{\wedge}(0)=(I_{[1]}g)(x) \qquad (\text{all } x\in[0,1)-E_0)$$

by assumption. Setting

$$(I_{[1]}g)^{[1]}_m(x)=\frac{1}{2}\sum_{j=0}^{m-1}2^j[(I_{[1]}g)(x)-(I_{[1]}g)(x\oplus 2^{-j-1})],$$

$$f^{[1]}_m(x)=\frac{1}{2}\sum_{j=0}^{m-1}2^j[f(x)-f(x\oplus 2^{-j-1})],$$

then by induction there exists to each $m\in N$ a set E_m, mes $E_m=0$, such that

$$f^{[1]}_m(x)=(I_{[1]}g)^{[1]}_m(x) \qquad (\text{all } x\in[0,1)-E_m).$$

Setting $E=\bigcup_{l=0}^{\infty}E_l$, then mes $E=0$ and

$$f^{[1]}_m(x)=(I_{[1]}g)^{[1]}_m(x) \qquad (\text{all } x\in[0,1)-E, \text{ all } m\in N).$$

Letting $m\to\infty$, this gives

$$\lim_{m\to\infty}f^{[1]}_m(x)=\lim_{m\to\infty}(I_{[1]}g)^{[1]}_m(x)=(I_{[1]}g)^{[1]}(x)$$

for all $x \in [0, 1) - E$. On the other hand, $(I_{[1]}g)^{[1]}(x) = g(x)$ a.e. by Theorem 3.1 since $I_{[1]}g \in L^1(0, 1)$. Therefore $f^{[1]}(x)$ exists and $f^{[1]}(x) = g(x)$ a.e..

Part b) follows by a repeated application of part a).

Note that the counterpart of Lemma 3.3 in classical analysis states that if $f \in AC^0$, i.e., if

$$f(x) - f(0) = \int_0^x g(u)\,du \qquad (\text{all } x \in [0, 1]) \tag{3.4}$$

with $g \in L^1(0, 1)$, then the ordinary derivative $f'(x)$ exists and equals $g(x)$ a.e.. Thus the class AW^{r-1} plays the same role in Walsh analysis as does the class AC^{r-1} of all functions which are $(r-1)$-times absolutely continuous in the sense of classical analysis. In general, the condition "for all $x \in [0, 1]$" in (3.4) cannot be replaced by "for almost all $x \in [0, 1]$". The difference between (3.4) and (3.2) is due to the definitions of the two derivatives: in contrast to the classical derivative, the dyadic derivative for fixed x is the limit of a discrete sequence of numbers.

As a corollary to Lemma 3.3 a) one has

Corollary 3.4. *If $f \in AW^0$ and $f^{[1]}(x) = 0$ for almost all $x \in [0, 1]$, then $f(x) = $ const a.e..*

Indeed, $f \in AW^0$ means that (3.2) holds a.e. with $g \in L^1(0, 1)$. This implies that $f^{[1]}(x) = (I_{[1]}g)^{[1]}(x) = g(x)$ a.e. by Theorem 3.1. Together with $f^{[1]}(x) = 0$ a.e. this gives $f(x) = $ const a.e..

The analogue of (2.7) for the pointwise derivative now reads

Lemma 3.5. *If $f \in AW^{r-1}$ then*

$$[f^{[r]}]^{\wedge}(k) = k^r f^{\wedge}(k) \qquad (k \in P). \tag{3.5}$$

Proof. By Lemma 3.3 there holds relation (3.3) with $f^{[r]} \in L^1(0, 1)$. Taking Walsh-Fourier coefficients of both sides of (3.3) yields

$$f^{\wedge}(k) - f^{\wedge}(0) \int_0^1 \psi_k(x)\,dx = [f^{[r]} * W_r]^{\wedge}(k) \qquad (k \in P)$$

by (2.1) and (2.9). By the convolution result (2.5) and (2.10) one therefore has the desired (3.5).

We can now bring the class $W^r_{X(0,1)}$ into play for which we have a converse to Theorem 3.1. Indeed:

Theorem 3.6. *The following two assertions are equivalent for $f \in X(0, 1)$:*

(i) $f \in W^r_{X(0,1)}$, (ii) *$D^{[r]}f$ exists and belongs to $X(0, 1)$.*

In this event $f^{[r]}(x) = (D^{[r]}f)(x)$ a.e..

Proof. (i)⇒(ii): Assumption (i) implies relation (3.3) by Lemma 3.3. b). By Theorem 2.3. b) this implies that $D^{[r]}f(x)=f^{[r]}(x)$ a.e.. To show that (ii)⇒(i), since $D^{[r]}f\in X(0,1)$ one has by Theorem 2.3. a) that $f(x)-f^{\wedge}(0)=(I_{[r]}D^{[r]}f)(x)$ a.e.. By Theorem 3.1 this yields $f^{[r]}(x)=(D^{[r]}f)(x)$ a.e..

As was our aim here, for functions f belonging to the subclass $W^r_{X(0,1)}$ of $X(0,1)$ the existence of the pointwise dyadic derivative $f^{[r]}(x)$ as an element of $X(0,1)$ is equivalent to the existence of the strong derivative $D^{[r]}f$ as an element of $X(0,1)$. This tells us that one should be able to obtain all the results deduced in [6, 25] for the strong dyadic derivative also for the pointwise dyadic derivative provided one restricts oneself to the space $W^r_{X(0,1)}$.

Let us compare the above results, in particular Theorem 3.6, with the corresponding known ones for the strong and pointwise classical derivative, denoted by $(D^{(r)}f)(x)$, $f^{(r)}(x)$, in $L^p(0,1)$ (see Butzer and Nessel [4, p. 33 and Ch. 10]). If one sets

$$\overline{W}^r_{L^p(0,1)}=\{f\in L^p(0,1)\colon f(x)=\varphi(x) \text{ a. e., } \varphi\in AC^{r-1},\ \varphi^{(r)}\in L^p(0,1\},$$

there holds:

A function $f\in L^p(0,1)$, $1\leqq p<\infty$, belongs to the class $\overline{W}^r_{L^p(0,1)}$, i.e., $f=\varphi$ a.e. with $\varphi\in AC^{r-1}$, $\varphi^{(r)}\in L^p(0,1)$, if and only if the strong derivative $D^{(r)}f$ exists as an element of $L^p(0,1)$. In this event $(D^{(r)}f)(x)=f^{(r)}(x)$ a.e..

Thus our class $W^r_{L^p(0,1)}$ plays the same role as does the basic class $\overline{W}^r_{L^p(0,1)}$ in classical analysis.

Let us finally put together some of the results established so far. The pointwise analog of the fundamental theorem of dyadic analysis, namely Theorem 2.3, is given by Lemma 3.3 b) together with Theorem 3.1, stating that $f\in W^r_{X(0,1)}$ with $f^{\wedge}(0)=0$ implies that

$$(I_{[r]}f^{[r]})(x)=f(x) \quad \text{a.e.,}$$

and that $f\in X(0,1)$ with $f^{\wedge}(0)=0$ implies

$$(I_{[r]}f)^{[r]}(x)=f(x) \quad \text{a.e..}$$

Finally, the following characterizations of the class $W^r_{X(0,1)}$ will be of importance below.

Lemma 3.7. *The following assertions are equivalent for $f\in X(0,1)$, fixed $r\in N$:*
(i) $f\in W^r_{X(0,1)}$,
(ii) *there exists $g\in X(0,1)$ such that $f(x)-f^{\wedge}(0)=(I_{[r]}g)(x)$ a.e.,*
(iii) *there exists $g\in X(0,1)$ such that $k^r f^{\wedge}(k)=g^{\wedge}(k) \qquad (k\in P)$.*

Concerning the proof, note that assertion (i) is equivalent by Theorem 3.6 to the fact that $D^{[r]}f$ exists and belongs to $X(0,1)$, which in turn is equivalent to assertions (ii) and (iii), as established in [6, 25].

4. The role of the dyadic derivative and integral in Walsh-Fourier analysis

In this section we shall at first examine some of the applications studied by us [6, 7, 8, 25] using the strong derivative, but this time in terms of the pointwise derivative.

4.1. *A first order ordinary dyadic differential equation*

The Walsh functions in AW^0 are eigensolutions of the dyadic differential equation

(4.1) $$f^{[1]}(x) - \lambda f(x) = 0 \qquad (\lambda \in R),$$

$$\lim_{x \to +0} f(x) = 1.$$

Indeed, the Walsh coefficients applied to (4.1) yield $kf^{\wedge}(k) - \lambda f^{\wedge}(k) = 0$, $k \in P$, by Lemma 3.5 and (2.6). If $\lambda \in P$, the uniqueness theorem (2.3) together with the initial condition gives $f(x) = \psi_\lambda(x)$ a.e.. If $\lambda \notin P$ the trivial solution $f = 0$ is the only one.

4.2. *Modulus of continuity; Walsh-Fourier coefficients*

If one defines the Lipschitz class $\mathrm{Lip}_W(\alpha, X)$, $\alpha > 0$, as the set of functions $f \in X(0, 1)$ for which

(4.2) $$\|f(\cdot) - f(\cdot \oplus h\|_X = O(h^\alpha) \qquad (h \to +0),$$

and the W-modulus of continuity by

(4.3) $$\omega_W(f; \delta; X) = \sup_{0 \leq h < \delta} \|f(\cdot) - f(\cdot \oplus h)\|_X,$$

then, according to G. W. Morgenthaler [20],

$$f \in \mathrm{Lip}_W(\alpha, X) \Leftrightarrow \omega_W(f; \delta; X) = O(\delta^\alpha).$$

The fact that the role of the W-modulus of continuity in Walsh-Fourier analysis is different in a number of points to that of the classical modulus was already pointed out in [10, 25]. Although the following result follows from the fact that the strong derivative equals the pointwise derivative for functions in $W^r_{X(0,1)}$, and was established for the former in [6, 25], we sketch the proof.

Lemma 4.1. *If* $f \in W^r_{X(0,1)}$ *then*

(4.4) $$\omega_W(f; \delta; X) = O[\delta^r \omega_W(f^{[r]}; \delta; X)].$$

Proof. $f\in W^r_{X(0,1)}$ implies that $f(x)-f^\wedge(0)=(W_r*f^{[r]})(x)$ a.e. by Lemma 3.3. b). This yields for $h\in[0,2^{-n})$, as was shown in [25, p. 50], that $f(x)-f(x\oplus h)=$ $=(W_r^{(n)}*[f^{[r]}(\cdot)-f^{[r]}(\cdot\oplus h)])(x)$ a.e., where $W_r^{(n)}(x)$ is defined via

$$[W_r^{(n)}]^\wedge(k)=\begin{cases}0, & 0\leq k<2^n;\\ k^{-r}, & 2^n\leq k.\end{cases}$$

It was also shown in [25, p. 23] that

$$\|W_r^{(n)}\|_1=O(2^{-nr})\qquad (n\to\infty).$$

This implies that

$$\|f(\cdot)-f(\cdot\oplus h)\|_X\leq\|W_r^{(n)}\|_1\|f^{[r]}(\cdot)-f^{[r]}(\cdot\oplus h)\|_X,$$

from which there follows (4.4).

This result has a number of consequences, including one upon the order of magnitude of the Walsh coefficients. Indeed, it is well-known (see N. J. Fine [11]) that for $f\in X(0,1)$

$$|f^\wedge(k)|=O[\omega_W(f;k^{-1};X)]\qquad (k\to\infty).$$

Our implication is

Lemma 4.2. *If $f\in W^r_{X(0,1)}$ then*

$$|f^\wedge(k)|=O[k^{-r}\omega_W(f^{[r]};k^{-1};X)]\qquad (k\to\infty).$$

In particular, if in addition $f^{[r]}\in\mathrm{Lip}_W(\alpha,X)$ then

$$|f^\wedge(k)|=O(k^{-r-\alpha})\qquad (k\to\infty).$$

4.3. Best approximation

Another sample of possible applications of the pointwise dyadic derivative is best approximation by Walsh polynomials, a topic considered by the authors [7] for strong derivatives.

Denoting by P_n the set of all Walsh polynomials of degree $\leq n$, i.e., of all

$$p_n(x)=\sum_{k=0}^{n-1}c_k\psi_k(x),$$

c_k being real numbers, then the best approximation of $f\in X(0,1)$ by $p_n\in P_n$ is defined by

$$E_n(f;X)=\inf_{p_n\in P_n}\|f-p_n\|_X. \tag{4.5}$$

It is known that there exists a Walsh polynomial $p_n^*(=p_n^*(f))$ of best approximation to $f\in X(0,1)$ for which $E_n(f;X)=\|f-p_n^*\|_X$. Of fundamental importance are the Bernstein and Jackson-type inequalities given by (see also [7], [25, p. 53])

Lemma 4.3. a) *For* $p_n\in P_n$ *one has*

$$\|p_n^{[r]}\|_X \leqq An^r\|p_n\|_X \qquad (n\in P,\ r\in N).$$

b) *If* $f\in W^r_{X(0,1)}$ *then*

$$E_n(f;X)\leqq Bn^{-r}\|f^{[r]}\|_X.$$

Here A and B are constants independent of n, p_n*, and f.*

Together with Lemma 4.1 these enable one to prove (see [7], [25, p. 54], where analogs for the strong dyadic derivative are established)

Theorem 4.4. *The following statements are equivalent for* $f\in W^r_{X(0,1)}$, $n\to\infty$:

(i) $f^{[r]}\in \mathrm{Lip}_W(\alpha, X) \qquad (\alpha>0)$,

(ii) $\omega_W(f^{[r]};n^{-1};X)=O(n^{-\alpha})$,

(iii) $E_n(f;X)=O(n^{-\alpha-r})$,

(iv) $f^{[v]}\in X(0,1),\quad 0\leqq v\leqq r,$ *and* $\|f^{[v]}-p_n^{*[v]}\|_X=O(n^{-\alpha-r+v})$,

(v) $\|p_n^{*[l]}\|_X=O(n^{-\alpha-r+l}) \qquad (0<\alpha+r<l)$.

Note that C. Watari [26, 27] established the first three equivalences of Theorem 4.4 in the particular case $r=0$ for the space $X(0,1)$.

In classical analysis this theorem is associated with the names S. N. Bernstein, D. Jackson, M. Zamansky, S. B. Stečkin, G. I. Sunouchi, P. L. Butzer and K. Scherer (see [5]).

Finally, by methods similar to those of Wagner [25, p. 62] one can establish the following result on the degree of approximation of $f\in X(0,1)$ by the nth partial sum of the Walsh-Fourier series

$$S_n(f;x)=\sum_{k=0}^{n-1} f^\wedge(k)\psi_k(x), \tag{4.6}$$

namely,

Lemma 4.5. *If* $f\in W^r_{X(0,1)}$ *then*

$$\|S_n(f;\cdot)-f(\cdot)\|_X\leqq M\frac{\log n}{n^r}\|f^{[r]}\|_X,$$

M being some positive constant.

4.4. *The dyadic wave equation*

In order to obtain a possible interpretation of the dyadic derivative it may be useful to set up a partial differential equation in dyadic space and time and solve it. A differential equation of wave equation type is easy to handle.

Let us first introduce a pointwise partial dyadic derivative with respect to x of the function $f(x, t)$ defined on $R\times R$ which is of period 1 in both variables:

$$f_x^{[1]}(x, t) = \frac{1}{2}\sum_{j=0}^{\infty} 2^j [f(x, t) - f(x\oplus 2^{-j-1}, t)].$$

Partial dyadic derivatives with respect to t or those of higher order are defined correspondingly.

The dyadic wave equation now has the form

$$\text{(4.7)} \qquad w_{xx}^{[2]}(x, t) - w_{tt}^{[2]}(x, t) = 0 \qquad \text{(a.e. in } x, t).$$

Under the initial condition

$$\text{(4.8)} \qquad \lim_{t\to +0} \|w(\cdot, t) - f(\cdot)\|_X = 0$$

for any prescribed function $f\in W_{X(0,1)}^2$, we shall now show that the unique solution of (4.7) is given by

$$\text{(4.9)} \qquad w(x, t) = f(x\oplus t) \qquad \text{(a.e. in } x, t).$$

For this purpose we proceed as follows. Apply the Walsh-coefficients of $w(x, t)$ with respect to x, namely

$$w^\wedge(k, t) = \int_0^1 w(u, t)\psi_k(u)\, du \qquad (k\in P),$$

to equation (4.7). For this application one needs to know that

$$\text{(4.10)} \qquad \begin{aligned} &w(x, t)\in W_{X(0,1)}^2 \text{ as a function of } x \text{ for almost all } t,\\ &w(x, t)\in W_{X(0,1)}^2 \text{ as a function of } t \text{ for almost all } x. \end{aligned}$$

Under the first of these conditions there follows by Lemma 3.5 for almost all t

$$[w_{xx}^{[2]}(\cdot, t)]^\wedge(k) = k^2 w^\wedge(k, t) \qquad (k\in P).$$

The transformed equation therefore has to begin with the form

$$\text{(4.11)} \qquad k^2 w^\wedge(k, t) - [w_{tt}^{[2]}(\cdot, t)]^\wedge(k) = 0 \qquad (k\in P),$$

the second term of which must still be investigated. For this purpose one must assume further that for almost all t, $w_{tt}^{[2]}(x,t)\in X(0,1)$ as a function of x and

$$(4.12)\quad \begin{aligned}&\lim_{m\to\infty}\left\|\frac{1}{2}\sum_{j=0}^{m}2^{j}[w(\cdot,t)-w(\cdot,t\oplus 2^{-j-1})]-w_t^{[1]}(\cdot,t)\right\|_X=0,\\ &\lim_{m\to\infty}\left\|\frac{1}{2}\sum_{j=0}^{m}2^{j}[w_t^{[1]}(\cdot,t)-w_t^{[1]}(\cdot,t\oplus 2^{-j-1})]-w_{tt}^{[2]}(\cdot,t)\right\|_X=0.\end{aligned}$$

This together with (2.8) implies that for almost all t

$$(4.13)\quad \begin{aligned}&w^{\wedge}(k,t)^{[1]}=[w_t^{[1]}(\cdot,t)]^{\wedge}(k)\qquad (k\in P),\\ &w^{\wedge}(k,t)^{[2]}=[w_{tt}^{[2]}(\cdot,t)]^{\wedge}(k)\qquad (k\in P).\end{aligned}$$

Therefore the transformed equation of the original (4.7) takes on the final form

$$(4.14)\quad k^2w^{\wedge}(k,t)-w^{\wedge}(k,t)^{[2]}=0\qquad (k\in P;\ \text{a.e. in } t).$$

To solve these ordinary dyadic differential equations of second order with respect to t, first note that by the second assumption of (4.10) and Lemma 3.3. b)

$$w(x,t)=\big(I_{[2]}w_{tt}^{[2]}(x,\cdot)\big)(t)+w^{\wedge}(x,0).$$

This implies that $w^{\wedge}(k,t)\in W^2_{X(0,1)}$ by (4.13), and therefore (4.14) may be solved by a second application of the Walsh coefficients. Setting $w^{\wedge}(k,t)=h_k(t)$, this yields for each fixed $k\in P$

$$k^2h_k^{\wedge}(j)-j^2h_k^{\wedge}(j)=0\qquad (j\in P).$$

Its solution is given by

$$h_k(t)=A(k)\psi_k(t)$$

with constant $A(k)$ independent of t. In view of the initial condition (4.8) and (2.8) one has $A(k)=f^{\wedge}(k)$, $k\in P$, and so

$$w^{\wedge}(k,t)=h_k(t)=f^{\wedge}(k)\psi_k(t)\qquad (k\in P;\ \text{a.e. in } t)$$

is the solution of (4.14). Therefore by the uniqueness result (2.3) and the translation rule (2.4) one obtains the desired solution (4.9) provided it exists.

To show that $f(x\oplus t)$ actually is a solution of (4.7) satisfying the assumptions (4.8) together with (4.10) and (4.12), it remains to show that the function $w(x,t)$ given by (4.9) satisfies the assumptions stated, which is clear.

It should be mentioned that H. F. Harmuth [18] gave a first speculative interpretation of our solution in particle physics.

5. Dyadic differentiation and integration of Walsh series

5.1. Differentiation of Walsh series

One of the aims of this subsection is the interchange of summation and dyadic differentiation of Walsh series. At first a few known lemmas concerning Walsh series

$$\sum_{j=0}^{\infty} a_j \psi_j(x), \tag{5.1}$$

the $\{a_j\}_{j=0}^{\infty}$ satisfying suitable conditions. A. A. ŠNEĬDER [24] (in this respect see also the paper by L. A. BALAŠOV and A. I. RUBINŠTEIN [1] and the references cited there) showed in analogy with trigonometric series

Lemma 5.1. *If $\{a_j\}_{j=0}^{\infty}$ is a positive monotonely decreasing sequence of reals with $\lim_{j\to\infty} a_j=0$, then the series* (5.1) *converges uniformly on each interval* $(\delta, 1)$, $\delta>0$.

The question under what conditions (5.1) converges to an $L^1(0, 1)$-function was answered by S. YANO [28], namely

Lemma 5.2. *If $\{a_j\}_{j=0}^{\infty}$ is quasi-convex and $\lim_{j\to\infty} a_j=0$, then the series converges pointwise on* $(0, 1)$ *to a function $f\in L^1(0, 1)$, and* (5.1) *is the Walsh-Fourier series of $f(x)$. This convergence is uniform on each* $(\delta, 1)$, $\delta>0$.

Recall that a sequence $\{a_j\}_{j=0}^{\infty}$ is called quasi-convex if

$$\sum_{j=0}^{\infty} (j+1)|a_{j+2}-2a_{j+1}+a_j| <\infty.$$

The proofs are analogous to the trigonometric case (see BUTZER and NESSEL [4, p. 249]) and are based upon the estimate of Dirichlet's kernel

$$|D_k(x)| = \left|\sum_{j=0}^{k-1} \psi_j(x)\right| \leqq \frac{\text{const}}{x} \qquad (0<x<1).$$

This will enable us to establish several elementary results on interchange of summation and differentiation.

Theorem 5.3. *Given a series*

$$\sum_{j=0}^{\infty} a_j \psi_j(x). \tag{5.2}$$

If the sequences $\{a_j\}_0^{\infty}$, $\{ja_j\}_0^{\infty}$ are quasi-convex and $\lim_{j\to\infty} ja_j=0$, then the series (5.2) *converges pointwise towards a function* $f \in L^1(0,1)$, *which is dyadic differentiable a.e.. Moreover,*

$$f^{[1]}(x) = \sum_{j=1}^{\infty} j a_j \psi_j(x) \quad a.e.,$$

i.e., the series (5.2) *is dyadic differentiable term-by-term.*

Proof. According to Lemma 5.2 there exist two functions f, $g \in L^1(0,1)$ such that for $x \in (0,1)$

$$f(x) = \sum_{j=0}^{\infty} a_j \psi_j(x), \quad a_j = f^\wedge(j) \qquad (j \in P),$$

$$g(x) = \sum_{j=0}^{\infty} j a_j \psi_j(x), \quad j a_j = g^\wedge(j) \qquad (j \in P).$$

This implies that $jf^\wedge(j) = g^\wedge(j)$, $j \in P$, and so $f \in W^1_{L^1(0,1)}$ by Lemma 3.7. Therefore $f^{[1]}(x) = g(x)$ a.e. by Lemma 3.3. a).

Restricting the matter to Walsh series with monotone coefficients one has

Lemma 5.4. *If the sequence* $\{ja_j\}_0^\infty$ *is quasi-convex and monotonely decreasing with* $\lim_{j\to\infty} ja_j = 0$, *then the series* (5.2) *converges pointwise to a function* $f \in L^1(0,1)$, *and f is a.e. dyadic differentiable with*

$$f^{[1]}(x) = \sum_{j=1}^{\infty} j a_j \psi_j(x) \quad a.e..$$

Proof. Since $\{ja_j\}$ is monotonely decreasing to zero, so is $\{a_j\}$. Setting $a_j = \varphi(j)/j$, $\{\varphi(j)\}$ is a monotonely decreasing sequence with $\lim_{j\to\infty} \varphi(j) = 0$. This yields $\sum_{j=1}^{\infty} [\varphi(j)]^2/j^2 < \infty$, and by the Riesz-Fischer theorem

$$a_0 + \sum_{j=1}^{\infty} a_j \psi_j(x)$$

is the Walsh-Fourier series of some function h in $L^2(0,1)$, and so in $L^1(0,1)$. But for each $h \in L^1(0,1)$ it is well known that

$$\lim_{n\to\infty} S_{2^n}(h; x) = h(x) \quad \text{a.e..} \tag{5.3}$$

Since the series (5.2) converges for each $x \in (0,1)$ by Lemma 5.1, i.e.,

$$a_0 + \sum_{j=1}^{\infty} a_j \psi_j(x) = f(x),$$

it follows that $f(x)=h(x)$ a.e. by (5.3). Hence

$$f(x) = a_0 + \sum_{j=1}^{\infty} a_j \psi_j(x) \quad \text{a.e.},$$

where $a_j = f^\wedge(j)$, $j \in P$. The rest of the proof follows as in the proof of Theorem 5.3.

Let us apply these results to some examples.

a) Consider the series

$$1 + \sum_{j=1}^{\infty} \frac{\psi_j(x)}{j^\alpha} \qquad (\alpha > 1). \tag{5.4}$$

The sequence $\{j^{-\alpha}\}_1^\infty$ is monotonely decreasing, and $\{j^{-\alpha+1}\}_1^\infty$ is known to be quasi-convex (see YANO [28], compare A. ZYGMUND [29, pp. 93—94]). So by Lemma 5.4 there exists $f \in L^1(0, 1)$ such that for almost all $x \in (0, 1)$

$$f(x) = 1 + \sum_{j=1}^{\infty} \frac{\psi_j(x)}{j^\alpha}$$

with $f^\wedge(j) = j^{-\alpha}$, $j \in N$, $f^\wedge(0) = 1$. Moreover, the series (5.4) is dyadic differentiable term-by-term and

$$f^{[1]}(x) = \sum_{j=1}^{\infty} \frac{\psi_j(x)}{j^{\alpha-1}} \quad \text{a.e.}.$$

b) Now consider

$$\sum_{j=2}^{\infty} \frac{\psi_j(x)}{j \log j}.$$

The sequence $\{1/\log j\}_2^\infty$ is monotonely decreasing and quasi-convex with $\lim_{j\to\infty} 1/\log j = 0$. So according to Lemma 5.4 there exists $f \in L^1(0,1)$ such that

$$f(x) = \sum_{j=2}^{\infty} \frac{\psi_j(x)}{j \log j}$$

for almost all $x \in (0, 1)$, and f is differentiable with

$$f^{[1]}(x) = \sum_{j=2}^{\infty} \frac{\psi_j(x)}{\log j} \quad \text{a.e.}.$$

The criteria established above concerning interchange of summation and dyadic differentiation, particularly Theorem 5.3 and Lemma 5.4, are principally based upon the fact that one must show that the original series as well as the differentiated series are Walsh-Fourier series of $L^1(0, 1)$-functions. These being essentially global criteria, the question arises whether one can find *local* criteria, e.g., for a closed interval $[a, b] \subset (0, 1)$. Note that for the corresponding results for trigonometric series under the classical derivative, *uniform* convergence of the series plays the essential role. Whether this is also so in Walsh analysis seems a little unlikely, but the matter is unsolved. In this connection we have the following

Conjecture 5.5. If $\{ja_j\}_1^\infty$ is a positive monotonely decreasing sequence with $\lim_{j\to\infty} ja_j=0$, then the series

$$\sum_{j=0}^{\infty} a_j\psi_j(x) \tag{5.5}$$

converges for each x to some function $f(x)$ which is dyadic differentiable a.e. with

$$f^{[1]}(x) = \sum_{j=0}^{\infty} ja_j\psi_j(x) \quad \text{a.e..}$$

We shall show below that this conjecture would indeed hold provided one could show that

$$\Psi_m(x) = \sum_{k=2^m}^{\infty} a_k\psi_k(x)\frac{1}{2}\sum_{j=0}^{m-1} 2^j[1-\psi_k(2^{-j-1})]$$

tends a.e. to zero for $m\to\infty$.

Now to this goal: Since $\{ja_j\}$ is monotonely decreasing to zero, so is $\{a_j\}$. Lemma 5.4 then implies that there exist functions f and g with

$$f(x) = \sum_{j=0}^{\infty} a_j\psi_j(x), \tag{5.6}$$

$$g(x) = \sum_{j=1}^{\infty} ja_j\psi_j(x), \tag{5.7}$$

both valid for all $x\in(0, 1)$. The convergence here is uniform in each $(\delta, 1)$, $\delta>0$. Since $ja_j\to 0$ for $j\to\infty$, setting $a_j=\varphi(j)/j$ implies that $\varphi(j)$ is bounded and $\lim_{j\to\infty}\varphi(j)=0$. Therefore $\sum_{j=1}^{\infty} [\varphi(j)]^2/j^2<\infty$, and by the Riesz-Fischer theorem there exists some function in $L^1(0, 1)$ whose Walsh-Fourier series is the given series (5.5). This function is equal a.e. to $f(x)$ on account of (5.6), i.e., $f^\wedge(j)=a_j$, $j\in P$.

It remains to show that $f^{[1]}(x)$ exists a.e. and $f^{[1]}(x)=g(x)$ a.e.. One has by (5.6) and (2.2)

$$\left|\frac{1}{2}\sum_{j=0}^{m-1} 2^j[f(x)-f(x\oplus 2^{-j-1})]-g(x)\right| =$$

$$=\left|\frac{1}{2}\sum_{j=0}^{m-1} 2^j\sum_{k=0}^{\infty} a_k\psi_k(x)[1-\psi_k(2^{-j-1})]-g(x)\right| =$$

$$=\left|\sum_{k=0}^{\infty} a_k\psi_k(x)\frac{1}{2}\sum_{j=0}^{m-1} 2^j[1-\psi_k(2^{-j-1})]-g(x)\right| \leqq$$

$$\leqq\left|\sum_{k=0}^{2^m-1} ka_k\psi_k(x)-g(x)\right|+|\Psi_m(x)|,$$

the first term in the above estimate following in view of the identity

$$(5.8)\qquad \frac{1}{2}\sum_{j=0}^{m-1} 2^j[1-\psi_k(2^{-j-1})] = k \qquad (0 \leqq k < 2^m).$$

But this first term tends to zero for $m\to\infty$ by (5.7).

Returning to the conjecture that $\Psi_m(x)$ tends a.e. to zero for $m\to\infty$, note that provided $g\in L^1(0, 1)$, then

$$\Psi_m(x) = \sum_{k=2^m}^{\infty} a_k\psi_k(x)\frac{1}{2}\sum_{j=0}^{m-1} 2^j[1-\psi_k(2^{-j-1})] = (g*F_m)(x),$$

where $F_m(x)$ is a certain $L^1(0, 1)$-function which is of importance in dyadic analysis (see [25, p. 26]). SCHIPP [23] showed implicitly that for $g\in L^1(0, 1)$

$$\lim_{m\to\infty} (g*F_m)(x) = 0 \quad \text{a.e.}.$$

Therefore our conjecture sounds reasonable.*

Let us close off the problem of interchange of dyadic differentiation and summation by presenting a criterium the assumptions of which, however, are unfortunately so strong so that the series

$$(5.9)\qquad \sum_{j=0}^{\infty} a_j\psi_j(x)$$

represents an $L^1(0, 1)$-function which is dyadic differentiable for *all* x.

Theorem 5.6. *Under the assumption*

$$\sum_{j=0}^{\infty} j|a_j| < \infty$$

the series (5.9) *is absolutely and uniformly convergent on* [0, 1] *to a function* f *which is dyadic differentiable for all* $x\in[0, 1]$ *and for which*

$$f^{[1]}(x) = \sum_{j=1}^{\infty} ja_j\psi_j(x) \qquad (\textit{all}\ x\in[0, 1]).$$

Proof. Clearly the series (5.9) converges absolutely (and uniformly) on [0, 1] to some function f since $|\psi_j(x)|=1$, $j\in N$. We have

$$(5.10)\quad \frac{1}{2}\sum_{j=0}^{\infty} 2^j[f(x)-f(x\oplus 2^{-j-1})] = \frac{1}{2}\sum_{j=0}^{\infty} 2^j\left\{\sum_{k=0}^{\infty} a_k\psi_k(x)[1-\psi_k(2^{-j-1})]\right\}.$$

* Remark (added at proof-reading). A proof of this conjecture was given by F. SCHIPP, On term by-term-dyadic differentiability of Walsh series, *Anal. Math.* (to appear).

But according to the assumption and (5.8)

$$\sum_{k=0}^{\infty}\sum_{j=0}^{\infty}\left|\frac{1}{2}2^j a_k\psi_k(x)[1-\psi_k(2^{-j-1})]\right| \leqq \sum_{k=0}^{\infty} k|a_k| < \infty.$$

Therefore one may interchange the order of summation in (5.10) to obtain

$$f^{[1]}(x) = \frac{1}{2}\sum_{j=0}^{\infty}2^j\left\{\sum_{k=0}^{\infty}a_k\psi_k(x) - \sum_{k=0}^{\infty}a_k\psi_k(x\oplus 2^{-j-1})\right\} =$$

$$= \sum_{k=0}^{\infty}a_k\psi_k(x)\frac{1}{2}\sum_{j=0}^{\infty}2^j[1-\psi_k(2^{-j-1})] = \sum_{k=0}^{\infty}ka_k\psi_k(x).$$

As an application consider again the series

$$f(x) = 1+\sum_{j=1}^{\infty}\frac{\psi_j(x)}{j^\alpha} \qquad (\alpha > 2).$$

By Theorem 5.6 one has for all $x\in[0, 1]$

$$f^{[1]}(x) = \sum_{j=1}^{\infty}\frac{\psi_j(x)}{j^{\alpha-1}}.$$

5.2. Dyadic integration of Walsh series

Let us now consider the problem of interchange of summation and dyadic integration. First note that in contrast to Lebesgue integration, dyadic integration is only defined over the interval [0, 1), and not over a subinterval $[a, b]\subset(0, 1)$. The reason is that the dyadic integral is defined via the convolution (2.9).

Consider the Walsh-Fourier series of $f\in L^1(0, 1)$

$$\sum_{j=0}^{\infty} f^{\wedge}(j)\psi_j(x). \tag{5.11}$$

Already Watari [27] showed that the term-by-term integrated series

$$f^{\wedge}(0)+\sum_{j=1}^{\infty}\frac{f^{\wedge}(j)}{j}\psi_j(x)$$

need not necessarily be uniformly convergent to some $f(x)$, as is the case for Fourier series. For example, $\sum_{j=2}^{\infty}\psi_j(x)/\log j$ is the Walsh-Fourier series of an L^1-function, but the dyadic integrated series $\sum_{j=2}^{\infty}\psi_j(x)/j\log j$ is not uniformly convergent.

On the other hand, the integrated Walsh-Fourier series of an $L^1(0, 1)$-function $f(x)$ converges for almost all x to the dyadic integral of f, namely $I_{[1]}f$. Indeed,

Lemma 5.7. *If* $f \in L^1(0,1)$ *then*

$$f^\wedge(0)+\sum_{j=1}^{\infty}\frac{f^\wedge(j)}{j}\psi_j(x) = (I_{[1]}f)(x) \quad a.e..$$

Proof. Since

$$[f^\wedge(0)]^2+\sum_{j=1}^{\infty}\left[\frac{f^\wedge(j)}{j}\right]^2<\infty,$$

by the Riesz-Fischer theorem there exists $g \in L^2(0,1)$ such that $g^\wedge(0)=f^\wedge(0)$ and $g^\wedge(j)=f^\wedge(j)/j$ for $j\in N$. Since $g\in L^2(0,1)$, one has by a result of P. Billard [2], R. A. Hunt [19] that

$$g(x)=f^\wedge(0)+\sum_{j=1}^{\infty}\frac{f^\wedge(j)}{j}\psi_j(x) \quad \text{a.e..}$$

By (2.10), (2.5), and (2.3) one therefore has

$$g(x) = (I_{[1]}f)(x) \quad \text{a.e.,}$$

which implies our lemma.

This lemma may also be formulated for pure Walsh series. Indeed, given the Walsh series

$$\sum_{j=0}^{\infty} a_j\psi_j(x) \qquad (a_j \to 0,\ j\to\infty),$$

there is $g\in L^1(0,1)$ such that

$$a_0+\sum_{j=1}^{\infty}\frac{a_j}{j}\psi_j(x) = g(x) \quad \text{a.e.,}$$

this series also being the Walsh-Fourier series of g.

6. Peculiarities of the dyadic derivative and integral

So far we have seen that there is a very close parallelism between dyadic and classical analysis, as well as between their applications for functions in $X(0,1)$. However, there are also differences (see also [8,10]). These are rather marked in the case of dyadic analysis on the space $C[0,1]$. As a matter of fact, this space seems to be rather unnatural in dyadic analysis. As we shall see, for example, if $f, f^{[1]}\in C[0,1]$, then f=const; if $f\in C[0,1]$, then the dyadic integral $I_{[1]}f\notin C[0,1]$. That there are marked differences does not stand in contradiction to the result of [14] that dyadic and classical differentiation are specific cases of the same generic concept, both being differential operators on specific locally compact abelian groups. To establish the above assertions we need the following results.

The partial sums of order 2^n (see (4.6)) may be written as

$$S_{2^n}(f;x)=\int_0^1 f(u)D_{2^n}(x\oplus u)\,du,$$

$$D_{2^n}(x)=\begin{cases}2^n, & x\in[0,2^{-n});\\ 0, & x\in[2^{-n},1).\end{cases}$$

Since for each $x\in[0,1)$ there is an i and n such that

(6.1) $$0\le\frac{i-1}{2^n}\le x<\frac{i}{2^n}\le 1,$$

$S_{2^n}(f;x)$ may be rewritten as (see Fine [11])

(6.2) $$S_{2^n}(f;x)=2^n\int_{(i-1)/2^n}^{i/2^n} f(u)\,du,$$

a form which plays an essential role in the proofs of the following results. It is concerned with the connections between the ordinary modulus of continuity, defined by

(6.3) $$\omega(f;\delta;C)=\sup_{0\le|h|<\delta}\|f(\cdot)-f(\cdot+h)\|_C,$$

and the W-modulus of (4.3), as well as between $\|f(\cdot)-S_{2^n}(f;\cdot)\|_C$ and the best approximation $E_{2^n}(f;C)$ of (4.5).

Lemma 6.1. *If* $f\in C[0,1]$ *then*

(6.4) $$\frac{1}{2}\omega_W\left(f;\frac{1}{2^n};C\right)\le\left\{\begin{matrix}\|f(\cdot)-S_{2^n}(f;\cdot)\|_C\\ E_{2^n}(f;C)\end{matrix}\right\}\le\omega_W\left(f;\frac{1}{2^n};C\right).$$

The proof follows exactly along the lines of Watari [26] who established the corresponding result for functions defined on the dyadic group G.

Theorem 6.3 to follow will give estimates corresponding to (6.4) by means of the ordinary modulus of continuity (6.3).

For the proof of this theorem we need a further result. For $1\le i\le 2^n$ let

$$I_n^{(i)}=\left[\frac{i-1}{2^n},\frac{i}{2^n}\right),$$

$$M_n^{(i)}(f)=\sup_{x\in I_n^{(i)}}|f(x)|,\qquad m_n^{(i)}(f)=\inf_{x\in I_n^{(i)}}|f(x)|,\qquad D_n^{(i)}(f)=M_n^{(i)}(f)-m_n^{(i)}(f).$$

Then there holds

Lemma 6.2. *If* $f\in C[0,1]$ *then*

(6.5) $$\frac{1}{2}\max_{1\le i\le 2^n}D_n^{(i)}(f)\le\|f(\cdot)-S_{2^n}(f,\cdot)\|_C\le\max_{1\le i\le 2^n}D_n^{(i)}(f).$$

Proof. Each Walsh polynomial $p_{2^n}(x)$ of order 2^n, as well as $S_{2^n}(f; x)$, are constant in each interval $I_n^{(i)}$, $1 \leqq i \leqq 2^n$. So on account of (6.2)

$$m_n^{(i)}(f) \leqq S_{2^n}(f; x) \leqq M_n^{(i)}(f) \qquad (x \in I_n^{(i)}),$$

which yields

$$\frac{1}{2} D_n^{(i)}(f) \leqq \sup_{x \in I_n^{(i)}} |f(x) - S_{2^n}(f; x)| \leqq D_n^{(i)}(f).$$

This gives (6.5) by taking the maximum over all $I_n^{(i)}$, $1 \leqq i \leqq 2^n$.

Between the modulus of continuity (6.3) and $\|f(\cdot) - S_{2^n}(f; \cdot)\|_C$ there exists the estimate:

Theorem 6.3. *If* $f \in C[0, 1]$ *then*

$$\omega\left(f; \frac{1}{2^n}; C\right) \leqq 4\|f(\cdot) - S_{2^n}(f; \cdot)\|_C \leqq 4\omega\left(f; \frac{1}{2^n}; C\right). \tag{6.6}$$

Proof. For the inequality on the right note that

$$\|f(\cdot) - S_{2^n}(f; \cdot)\|_C \leqq \omega_W\left(f; \frac{1}{2^n}; C\right)$$

by Lemma 6.1. For the inequality

$$\omega_W\left(f; \frac{1}{2^n}; C\right) \leqq \omega\left(f; \frac{1}{2^n}; C\right)$$

see Fine [11] or Morgenthaler [20]. The proof uses the fact that for each $h \in [0, 1)$ one has $x \oplus h - x| \leqq h$ for all $x \in [0, 1]$.

Concerning the left hand inequality, first let $x \in I_n^{(i)}$, $h \in [0, 2^{-n})$. Then

$$|f(x) - f(x+h)| \leqq \max_{x, y \in I_n^{(i)} \cup I_n^{(i+1)}} |f(x) - f(y)| \leqq$$

$$\leqq 2 \max\left\{\max_{x \in I_n^{(i)}} \left|f(x) - f\left(\frac{i}{2^n}\right)\right|, \max_{x \in I_n^{(i+1)}} \left|f(x) - f\left(\frac{i}{2^n}\right)\right|\right\} \leqq$$

$$\leqq 2 \max\{D_n^{(i)}(f), D_n^{(i+1)}(f)\} \leqq 4\|f(\cdot) - S_{2^n}(f; \cdot)\|_C$$

by Lemma 6.2. Taking the maximum over all $x \in [0, 1)$ of the above inequalities yields (6.6).

It should be mentioned that it would be possible to replace Lemma 6.2 by the following: If $f \in C[0, 1]$ then

$$E_{2^n}(f; C) = \frac{1}{2} \max_{1 \leqq i \leqq 2^n} D_n^{(i)}(f), \tag{6.7}$$

the proof of which is a little more difficult. A related result for Haar functions is due to B. I. Golubov [16, 17]. (6.7) would enable us to replace inequality (6.6) by

(6.8) $$\omega\left(f; \frac{1}{2^n}; C\right) \leqq 2E_{2^n}(f; C) \leqq 2\omega\left(f; \frac{1}{2^n}; C\right).$$

Theorem 6.3 and Lemma 6.1 readily imply

Corollary 6.4. *If* $f \in C[0, 1]$ *then*

(6.9) $$\omega\left(f; \frac{1}{2^n}; C\right) \leqq 12\omega_W\left(f; \frac{1}{n}; C\right) \leqq 12\omega\left(f; \frac{1}{n}; C\right).$$

Had one used (6.8) instead of (6.6), the constant 12 in (6.9) could have been improved to 6.

Note that if

$$\text{Lip}\,(\alpha, C) = \{f \in C[0, 1]: \omega(f; \delta; C) = O(\delta^\alpha)\},$$

then this corollary implies that $\text{Lip}_W(\alpha, C) = \text{Lip}\,(\alpha, C)$. On the other hand, results of this type are generally not valid with the space $C[0, 1]$ replaced by $L^p(0, 1)$, $1 \leqq p < \infty$.

We now apply Theorem 6.3 and Lemma 6.1 to best approximation by Walsh polynomials, namely to $E_n(f; C)$ as defined by (4.5).

Lemma 6.5. *$E_n(f; C)$ is saturated with order $O(n^{-1})$, and the corresponding saturation class is given by* $\text{Lip}\,(1, C) = \text{Lip}_W(1, C)$.

Proof. First note that the set of functions $f \in C[0, 1]$ for which $E_n(f; C) = O(n^{-1})$ holds is not empty. Indeed, the function $f_1(x) = \sin 2\pi x$ belongs to Lip (1, *C*) and $E_n(f_1; C) = O(n^{-1})$ by (6.4) and (6.9). On the other hand, if $E_n(f; C) = o(n^{-1})$, then $f(x) = \text{const}$ by (6.4) and (6.9), noting that $\omega(f; n^{-1}; C) = o(n^{-1})$ implies that $f = \text{const}$. This means that $E_n(f; C)$ is saturated with order $O(n^{-1})$.

That the saturation class, namely the set $\{f \in C[0, 1]: E_n(f; C) = O(n^{-1})\}$, is given by Lip (1, *C*) also follows by (6.4) and (6.9).

Let us remark here that K. Scherer [22] established a result corresponding to Theorem 6.7 for best approximation by splines.

Let us summarize the equivalences established above in the following

Lemma 6.6. *The following assertions are equivalent for* $f \in C[0, 1]$, $0 < \alpha \leqq 1$:

(i) $f \in \text{Lip}_W(\alpha, C)$,
(ii) $f \in \text{Lip}\,(\alpha, C)$,
(iii) $\|f - S_{2^n}(f)\|_C = O(2^{-n\alpha})$,
(iv) $E_n(f; C) = O(n^{-\alpha})$.

The results of this lemma are "best" possible as the following theorems show.

We may now establish the results announced in the beginning of this section.

For this purpose first observe that one can show analogously to [8], [25, p. 30] that $f^{[1]} \in C[0, 1]$ implies that $D^{[1]}f$ exists together with $f^{[1]}(x) = D^{[1]}f(x)$ for all x. The converse assertion is trivial, in other words, the pointwise and strong dyadic derivatives are equal to another for $f \in C[0, 1]$.

Theorem 6.7. *If $f, f^{[1]} \in C[0, 1]$ then $f(x) = \text{const}$.*

Proof. If $f^{[1]} \in C[0, 1]$, one deduces as in [25, p. 26], recalling the remark preceding this theorem, that

$$[f^{[1]}]^{\wedge}(k) = k f^{\wedge}(k) \qquad (k \in P).$$

By the uniqueness and convolution theorem (2.3), (2.5) this implies

$$f(x) - f^{\wedge}(0) = (I_{[1]} f^{[1]})(x) \quad \text{a.e.}.$$

Just as in the proof of Lemma 4.1 (see also [25, p. 50]) one therefore has for $h \in [0, 2^{-n})$

$$|f(x) - f(x \oplus h)| = \left|(W_1^{(n)} * [f^{[1]}(\cdot) - f^{[1]}(\cdot \oplus h)])(x)\right|,$$

which leads to the estimate

$$|f(x) - f(x \oplus h)| = O[2^{-n} \omega_W(f^{[1]}; 2^{-n}; C)]$$

for $h \in [0, 2^{-n})$. Hence one has for $f \in C[0, 1]$

$$\omega_W(f; 2^{-n}; C) = o(2^{-n}) \qquad (n \to \infty),$$

implying $f(x) = \text{const}$ by (6.9).

For the dyadic integral we have

Theorem 6.8. *If $f \in C[0, 1]$ then $I_{[1]} f \notin C[0, 1]$.*

Proof. If $I_{[1]} f = W_1 * f$ would belong to $C[0, 1]$ then one can show as in the proof of Theorem 2.3. b) (see [6], [25, p. 26]) that the strong and also the pointwise derivative of $W_1 * f$ in the C-norm exists, and that

$$(W_1 * f)^{[1]}(x) = f(x) - f^{\wedge}(0) \in C[0, 1].$$

This would imply that $W_1 * f$ would be a continuous, non-constant function with continuous derivative $f - f^{\wedge}(0)$. This is a contradiction to Theorem 6.6.

This is different to the classical integration concept. The indefinite Lebesgue integral $F(x) = \int_0^x f(u)\,du$ of a function $f \in L^1(0, 1)$ is always continuous. On top of this R. Penny [21] showed by means of an example that the dyadic integral of an $L^1(0, 1)$ function need not even be Walsh-continuous.

In any case one can say that the space $L^p(0, 1)$ and not $C[0, 1]$ is the more natural one on which to carry out dyadic analysis on $[0, 1)$.

Let us mention that all of the results established in this paper on the basis of the Walsh-Paley system could also be considered for the Walsh-Kaczmarz system (which is a rearrangement of the Paley system). In this respect see also BUTZER and WAGNER [8, 10].

Finally our results could also be established for functions belonging to $L^p(G)$, $1 \leq p < \infty$ or to $X(G)$, G being the dyadic group (see also [6, 25]).

References

[1] L. A. BALAŠOV and A. I. RUBINŠTEIN, Series with respect to the Walsh system and generalizations, *J. Soviet Math.*, 1 (1973), 727—763.

[2] P. BILLARD, Sur la convergence presque partout des séries de Fourier-Walsh des fonctions de l'espace $L^2(0, 1)$, *Studia Math.*, 28 (1967), 363—388.

[3] S. V. BOČKAREV, On the Fourier-Walsh coefficients, *Izv. Akad. Nauk SSSR*, ser. mat., 34 (1970), 203—208 = *Math. USSR-Izv.*, 4 (1970), 209—214.

[4] P. L. BUTZER and R. J. NESSEL, *Fourier Analysis and Approximation*. I, Academic Press (New York, 1971).

[5] P. L. BUTZER and K. SCHERER, Jackson and Bernstein-type inequalities for families of commutative operators in Banach spaces, *J. Approximation Theory*, 5 (1972), 308—342.

[6] P. L. BUTZER and H. J. WAGNER, Walsh-Fourier series and the concept of a derivative, *Applicable Anal.*, 3 (1973), 29—46.

[7] P. L. BUTZER and H. J. WAGNER, Approximation by Walsh polynomials and the concept of a derivative, *Proc. Sympos. Naval Res. Lab., Washington, D. C., 1972;* 388—392 (Washington, D. C., 1972).

[8] P. L. BUTZER and H. J. WAGNER, On a Gibbs-type derivative in Walsh-Fourier analysis with applications, *Proc. of the Electronics Conference, Chicago 1972;* 393—398 (Oak Brook, Illinois, 1972).

[9] P. L. BUTZER and H. J. WAGNER, A calculus for Walsh functions defined on R_+. *Proc. Sympos. Naval Res. Lab., Washington, D. C., 1973;* 75—81 (Washington, D. C., 1973).

[10] P. L. BUTZER and H. J. WAGNER, A new calculus for Walsh functions with applications, *Proc. Sympos. Hatfield Polytechnic, 1973* (Hatfield, England, 1973).

[11] N. J. FINE, On the Walsh functions, *Trans. Amer. Math. Soc.*, 65 (1949), 372—414.

[12] J. E. GIBBS, *Some properties of functions on the nonnegative integers less than* 2^n, NPL (National Physical Laboratory), Middlessex, England, DES Rept. no. 3 (1969).

[13] J. E. GIBBS and B. IRELAND, *Some generalizations of the logical derivative*, NPL, DES Rept. no. 8 (1971).

[14] J. E. GIBBS and B. IRELAND, Walsh functions and differentiation, *Proc. Sympos. Naval Res. Lab., Washington, D. C., 1974;* 147—176 (Washington, D. C., 1974).

[15] J. E. GIBBS and M. J. MILLARD, *Walsh functions as solutions of a logical differential equation*, NPL, DES Rept. no. 1 (1969).

[16] B. I. GOLUBOV, Fourier series of continuous functions with respect to the Haar system, *Dokl. Akad. Nauk SSSR*, 156 (1964), 247—250 = *Soviet Math. Dokl.*, 5 (1964), 620—623.

[17] B. I. Golubov, Fourier series of continuous functions with respect to the Haar system, *Izv. Akad. Nauk SSSR*, ser. mat., 28 (1964), 1271—1296.
[18] H. F. Harmuth, Real numbers versus dyadic group as basis for models of time and space (preprint).
[19] R. A. Hunt, Almost everywhere convergence of Walsh-Fourier series of L^2 functions, *Proc. Internat. Congress Math., Nice, 1970.* II, 655—661; Gauthier-Villars (Paris, 1971).
[20] G. W. Morgenthaler, On Walsh-Fourier series, *Trans. Amer. Math. Soc.*, 84 (1957), 472—507.
[21] R. Penney, On the rate of growth of the Walsh anti-differentiation operator (to appear).
[22] K. Scherer, On the best approximation of continuous functions by splines, *SIAM J. Numer. Anal.*, 7 (1970), 418—423.
[23] F. Schipp Über einen Ableitungsbegriff von P. L. Butzer und H. J. Wagner, *Proc. 5. Balkan Math. Congress, Belgrad, 1974.*
[24] A. A. Šneider, On series with respect to Walsh functions with monotone coefficients, *Izv. Akad. Nauk SSSR*, ser. mat., 12 (1948), 179—192.
[25] H. J. Wagner, *Ein Differential- und Integralkalkül in der Walsh-Fourier Analysis mit Anwendungen*, Westdeutscher Verlag (Köln—Opladen, 1973).
[26] C. Watari, Best approximation by Walsh polynomials, *Tôhoku Math. J.*, 15 (1963), 1—5.
[27] C. Watari, Multipliers for Walsh-Fourier series, *Tôhoku Math. J.*, 16 (1964), 239—251.
[28] S. Yano, On Walsh-Fourier series, *Tôhoku Math. J.*, 3 (1951), 223—242.
[29] A. Zygmund, *Trigonometric Series*. I (Cambridge, 1959).

О двоичном анализе, основанном на понятии двоичной производной в точке

П. Л. БУТЦЕР и Х. Й. ВАГНЕР

В предыдущих работах авторы в основном развивали двоичный анализ, основанный на понятии сильной двоичной производной для функций, определенных на диадической группе или на $[0, 1)$ с периодом 1. Целью настоящей работы является построение двоичного дифференциального и интегрального исчислений на основе более сложного, но зато и более классического понятия двоичной производной в точке. Исследуются те пространства функций, для которых применим двоичный анализ, а также определяются границы его применимости. Так оказалось, что пространство $L^p(0, 1)$, $1 \leq p < \infty$, является более естественным простран ством для построения двоичного анализа, чем классическое пространство $C[0, 1]$. Например, если первая двоичная производная принадлежит $C[0, 1]$, то $f=\text{const}$. С другой стороны, если $f \in C[0, 1]$, то двоичный интеграл, построенный для f, не принадлежит $C[0, 1]$. Установлено также, что сильная двоичная производная и двоичная производная в точке совпадают почти всюду для функций, принадлежащих определенному подклассу $L^p[0, 1]$.

Полученные результаты применяются к почленному дифференцированию и интегрированию рядов по системе Уолша, к оценкам величин коэффициентов Фурье-Уолша, к доказательству аналога основной теоремы о наилучшем приближении для полиномов по системе Уолша, а также к решению двоичного волнового уравнения.

P. L. BUTZER
LEHRSTUHL A FÜR MATHEMATIK
TECHNOLOGICAL UNIVERSITY OF AACHEN
51 AACHEN, WESTERN GERMANY

A NEW CALCULUS FOR WALSH FUNCTIONS WITH APPLICATIONS

P.L. Butzer, B.Sc., M.A., Ph. D.

H.J. Wagner. Dr. rer. nat.

Lehrstuhl A für Mathematik
Technological University of Aachen, Germany

"Theory and Applications of Walsh and other non-sinusoidal Functions"

June 28th and 29th, 1973

Abstract

In a series of papers the authors modified the concept of a dyadic derivative, first introduced by J.E. Gibbs in 1969, and built up a basic differential and integral calculus for Walsh analysis. The object of this paper is to present a survey of this so-called dyadic analysis. There are applications to the theory of Walsh-Fourier series, to best approximation by Walsh polynomials, to the dyadic wave equations.

- 1 -

1. Introduction

As mentioned in [2], one of the authors was first confronted with the problem of finding a suitable derivative concept for Walsh functions on the occasion of the Walsh Functions Sympos. in Washington,D.C.,1970. In a series of papers [2,3,4,5,15], the first two of which were completed in preprint form in fall 1971, both authors, stimulated by work of J.E. Gibbs [7,8,9,10], have built up a mathematical theory comprising a notion of dyadic derivative as well as its inverse operation, the dyadic integral. This led to the fundamental theorem of the calculus for these two new concepts, thus to a socalled "dyadic analysis". Both concepts can indeed be regarded as counterparts of the classical derivative or integral since their properties and many applications to Walsh-Fourier analysis and approximation theory, etc. are practically entirely parallel to those in the classical situation. It is the object of this lecture to present a survey of the results mentioned above.

2. Walsh Functions and their Derivatives

Each $x \in (0,1]$ has a dyadic expansion

$$x = \sum_{j=1}^{\infty} x_j \, 2^{-j} \quad , \; x_j \in \{0,1\}$$

which is unique if $x \notin$ D.R. (= set of all numbers of form $q \cdot 2^{-p}$, $p, q \in N = \{1,2,...\}$). If $x \in$ D.R. there are two such expansions, a finite and infinite one, of which we choose the latter. Each $k \in P = \{0,1,2,...\}$ has the unique dyadic expansion

$$k = \sum_{j=-K}^{0} k_j \, 2^{-j} \qquad (k_j \in \{0,1\},\; 2^K \leq k < 2^{K+1}).$$

This enables one to define the Walsh (-Kaczmarz) functions

$$(2.1) \qquad \phi(k,x) = (-1)^{\sum_{j=-K}^{0} [k_s + k_{s-1}] x_{1-s}}$$

which are extended periodically th the whole R. They form an orthonormal system, namely,

$$\int_0^1 \phi(k,u)\phi(l,u)du = \delta_{k,l} \qquad (k,l \varepsilon P).$$

These functions may, in some sense, be regarded as "step-function" counterparts of the classical sine and cosine functions. Indeed, the graphs of the first four Walsh functions are

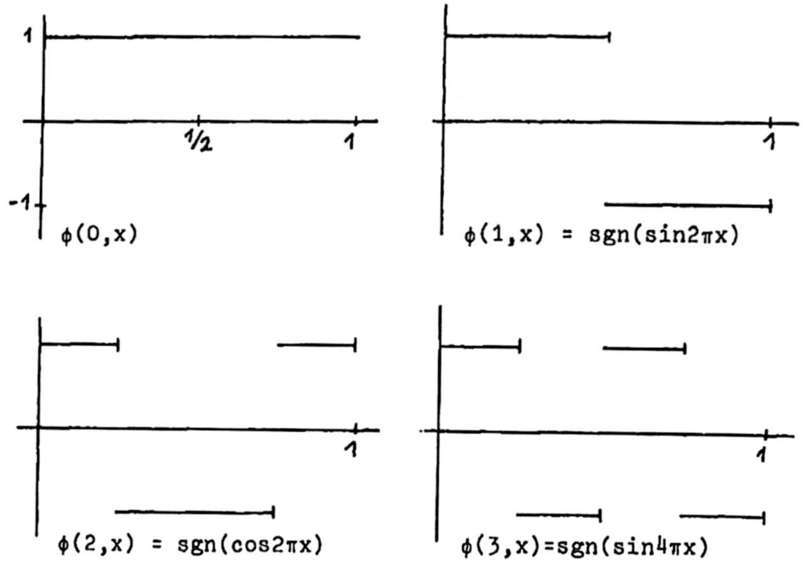

Various different notations for the Walsh functions are in common use, namely

$$\phi(2k,x) = \mathrm{wal}(2k,x) = \mathrm{cal}(k,x) \qquad (k \varepsilon P),$$

$$\mathrm{cal}(2^{n-1},x) = \mathrm{sgn}(\cos 2^n\pi x) \qquad (n \varepsilon N),$$

$$\phi(2k+1,x) = \mathrm{sal}(k+1,x) \qquad (k \varepsilon P),$$

$$\mathrm{sal}(2^{n-1},x) = \mathrm{sgn}(\sin 2^n\pi x) \qquad (n \varepsilon N).$$

To define a concept of derivative here we need the operation of addition modulo 2. If $x,y \varepsilon (0,1]$, let

- 3 -

$$x \oplus y = \sum_{j=1}^{\infty} |x_j - y_j| 2^{-j};$$

if x, y ε R (=(-∞,+∞)), then $x \oplus y = (x-[x]) \oplus (y-[y])$, [x] being the largest integer ≤ x. If f is defined on R with period 1, then the first (strong) dyadic derivative of f is defined by

$$(2.2) \qquad (D^{\{1\}}f)(x) = \frac{1}{2} \sum_{j=0}^{\infty} \frac{f(x) - f(x \oplus 2^{-j})}{2^{-j}}.$$

The convergence of the infinite sum is to be understood in the norm of the space $L_p(0,1)$, of functions integrable to the pth power (with norm $\|f\|_p = \{\int_0^1 |f(u)|^p du\}^{1/p}$, $1 \leq p < \infty$). However, if the sum is pointwise convergent, then one speaks of a pointwise dyadic derivative, denoted by $f^{\{1\}}(x)$. The derivatives of higher order are defined successively by $D^{\{r\}}f = D^{\{1\}}(D^{\{r-1\}}f)$. The curly bracket { } in $D^{\{1\}}f$, $f^{\{1\}}$ is used to distinguish the dyadic first derivative $D^{\{1\}}f$, $f^{\{1\}}$ of f with respect to the Walsh (-Kaczmarz) functions from the classical first derivative $f^{(1)}(x)$.

3. Simple Properties of the Dyadic Derivative

a) $D^{\{1\}}$ is a linear, closed operator,

b) The Walsh functions are arbitrarily often dyadic differentiable,i.e.

$$D^{\{r\}}\phi(k,x) = k^r \phi(k,x) \qquad (r \in N),$$

c) The dyadic derivative of a constant function vanishes. Conversely, $D^{\{1\}}f = 0$ implies f = const.

d) Defining the discrete finite Walsh transform (=Walsh-Fourier coefficient) of $f \in L_p(0,1)$, $1 \leq p < \infty$ by

$$f^{\wedge}(k) = \int_0^1 f(u)\phi(k,u)du \qquad (k \in P),$$

then

(i) $|f^\wedge(k)| \leq \|f\|_p$,

(ii) $f_n, f \in L_p(0,1)$, $\lim_{n\to\infty} \|f_n - f\|_p = 0 \Rightarrow \lim_{n\to\infty} f_n^\wedge(k) = f^\wedge(k)$ $(k \in P)$.

(iii) (Translation rule)

$$[f(\cdot \oplus h)]^\wedge(k) = \phi(k,h) f^\wedge(k) \qquad (k \in P,\ h \in R),$$

(iv) (Uniqueness theorem)

If $f \in L_p(0,1)$, $1 \leq p < \infty$ and $f^\wedge(k) = 0$, $k \in P$, then $f = 0$ a.e.

This leads to the fundamental

Theorem 1

If $f, D^{\{r\}}f \in L_p(0,1)$, then

$$[D^{\{r\}}f]^\wedge(k) = k^r f^\wedge(k) \qquad (k \in P).$$

The converse to this theorem is valid. This gives that a necessary and sufficient condition for the existence of $D^{\{r\}}f \in L_p(0,1)$ is that there exists some $g \in L_p(0,1)$ such that $k^r f^\wedge(k) = g^\wedge(k)$, $k \in P$.

Sketch of Proof

The proof depends essentially upon the identity

$$(3.1) \qquad \frac{1}{2}\sum_{j=0}^{\infty} 2^j [1-\phi(k,2^{-j})] = k \qquad (k \in P).$$

In case $r = 1$ it follows by 3. d) (ii), (iii) and (3.1) that if $D^{\{1\}}f \in L_p(0,1)$, then

$$\lim_{m\to\infty} \frac{1}{2}\sum_{j=0}^{m} 2^j [1-\phi(k,2^{-j})] f^\wedge(k) = k \cdot f^\wedge(k) = [D^{\{1\}}f]^\wedge(k) \quad (k \in P).$$

The result for $r > 1$ follows by induction.

- 5 -

4. Dyadic Integration Operator

Defining the convolution of $f \in L_p(0,1)$ and $g \in L_1(0,1)$ by

(4.1) $$(f * g)(x) = \int_0^1 f(x \oplus u)g(u)du \quad ,$$

then $f * g$ exists for almost all x, $f * g \in L_p(0,1)$ and $\|f * g\|_p \leq \|f\|_p \|g\|_1$. This enables us to define an operator $I_{\{1\}}$ by

(4.2) $$(I_{\{1\}}f)(x) = (f * W_1)(x) = \int_0^1 f(x \oplus u)W_1(u)du,$$

where W_1 is an $L_1(0,1)$-function given by $W_1^{\wedge}(0) = 1$, $W_1^{\wedge}(k) = k^{-1}$, $k \in N$. The operator $I_{\{1\}}$ is linear, continuous. It is indeed our operator of dyadic integration-the inverse operator to the dyadic differential operator $D^{\{1\}}$ – since the fundamental theorem of the calculus holds for these two concepts in the form:

Theorem 2

If $f \in L_p(0,1)$, $1 \leq p < \infty$ and $\int_0^1 f(u)du = f^{\wedge}(0) = 0$, then $D^{\{1\}}(I_{\{1\}}f) = f$, $I_{\{1\}}(D^{\{1\}}f) = f$.

These two concepts enable one to built up a dyadic analysis in which they play the same role as does the classical derivative and integral in ordinary analysis.

Let us consider the analogies between classical Fourier series and Walsh-Fourier series in the form of a table.

$\hat{f}_F(k)=\frac{1}{2\pi}\int_{-\pi}^{\pi} f(u)e^{-iku}du \quad (k=0,\pm1,\pm2,\ldots)$	$f^\wedge(k)=\int_0^1 f(u)\phi(k,u)du \quad (k\varepsilon P)$
$f(x)\sim\sum_{k=-\infty}^{\infty} \hat{f}_F(k)e^{ikx}$	$f(x)\sim\sum_{k=0}^{\infty} f^\wedge(k)\phi(k,x)$
$f^{(1)}(x)\sim\sum_{k=-\infty}^{\infty} (ik)\hat{f}_F(k)e^{ikx}$	$(D^{\{1\}}f)(x)\sim\sum_{k=0}^{\infty} kf^\wedge(k)\phi(k,x)$
$\int_{-\pi}^{x} f(u)du\sim\sum_{\substack{k=-\infty \\ k\neq 0}}^{\infty} \frac{\hat{f}_F(k)}{ik}\, e^{ikx} \quad (\hat{f}_F(0)=0)$	$(I_{\{1\}}f)(x) \sim \sum_{k=1}^{\infty} \frac{f^\wedge(k)}{k}\phi(k,x) \quad (f^\wedge(0)=0)$

Note that in the transformed state the integral is given by

$$[I_{\{1\}}f]^\wedge(k) = \frac{1}{k}\, f^\wedge(k) \qquad (k \ \varepsilon\ N).$$

In this form, the integration operator introduced by Butzer-Wagner [2, 3] in 1971, is an extension of that in Harmuth [11] (second edition) who considers the simpler case for the finite Walsh-Fourier series

$$\sum_{k=0}^{2^n-1} f^\wedge(k)\phi(k,x).$$

5. Basic Applications of the Concepts $D^{\{1\}}$, $I_{\{1\}}$

a) The Walsh functions may be regarded as eigensolutions of the first order ordinary dyadic differential equation

$$(5.1) \qquad \begin{aligned} &D^{\{1\}}f - \lambda f = 0 \qquad (\lambda\varepsilon R) \\ &\lim_{x\to 0+} f(x) = 1. \end{aligned}$$

Indeed, the Walsh transform applied to equation (5.1) yields $kf^\wedge(k) - \lambda f^\wedge(k)=0$ by Theorem 1. If $\lambda\varepsilon P$, the uniqueness theorem together with the initial condition gives $f(x) = \phi(\lambda,x)$. If

$\lambda \notin P$ the trivial solution is the only one.

b) In analogy to the situation for classical Fourier series, the order of magnitude of the Walsh-Fourier coefficients is given by:

If $D^{\{r\}}f \in L_p(0,1)$, $1<p<\infty$, then
$$f^\wedge(k) = \mathcal{O}(k^{-r}) \qquad (k\to\infty).$$

(Note that if the classical derivative $f^{(r)}$ exists for an $r \in N$ and belongs to $L_p(0,1)$, then $f^\wedge(k) = \mathcal{O}[k^{-1}]$. Moreover, this order cannot be improved; see Fine [6]).

c) Denoting the nth partial sum of the Walsh-Fourier series by

$$S_n(f;x) = \sum_{k=0}^{n-1} f^\wedge(k)\phi(k,x),$$

the degree of approximation of f by S_n is given by:

If $D^{\{r\}}f \in L_p(0,1)$, $1<p<\infty$, then

$$\|S_n(f;\cdot) - f(\cdot)\|_p = \mathcal{O}\left(\frac{\log n}{n^r}\right) \qquad (n\to\infty).$$

6. The Dyadic Wave Equation

Let $L_p^* = L_p((0,1)\times(0,1))$ be the set of all functions $w(x,t)$ which are of period 1 in both variables and for which

$$\|w\|_p^* = \left\{\int_0^1 \int_0^1 |w(x,t)|^p dx\, dt\right\}^{1/p} < \infty \qquad (1<p<\infty).$$

If one defines strong partial dyadic derivatives with respect to x or t analogously to (2.2) (denoting them by $D_x^{\{1\}}w(x,t)$, $D_t^{\{1\}}w(x,t)$, $D_{xx}^{\{2\}}w(x,t)$, etc.), then one can set up the dyadic wave equation

$$D_{xx}^{\{2\}}w - D_{tt}^{\{2\}}w = 0 \qquad (6.1)$$

with initial condition

(6.2) $\lim_{t \to 0+} \|w(\cdot,t)-f(\cdot)\|_p = 0$ $(f \in L^p(0,1))$.

To seek for a solution in L_p^* one applies the discrete Walsh transform $w^\wedge(k,t) = \int_0^1 w(u,t)\phi(k,u)du$ with respect to x to (6.1). The method may be explained by the diagram

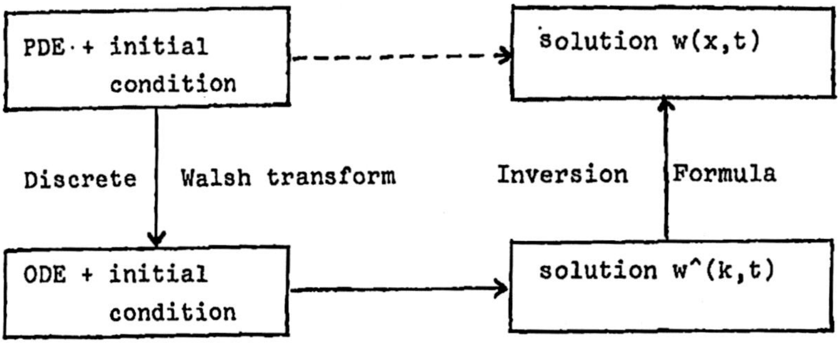

Instead of solving the PDE directly in the original function space one takes the indirect way indicated by the arrows (It is the analog of the usual transform technique in solving PDE).

The transformed equation, a second order "ordinary" dyadic differential equation, has the form

$$k^2\, w^\wedge(k,t) - D_{tt}^{\{2\}} w^\wedge(k,t) = 0 \qquad (k \in P).$$

A second application of the transform technique, this time with respect to t, making use of the initial condition (6.2) together with the uniqueness theorem yields

$$w^\wedge(k,t) = f^\wedge(k)\phi(k,t) \qquad (k \in P).$$

The uniqueness theorem again and the translation rule gives the solution of the original problem (6.1), (6.2), namely

$$w(x,t) = f(x \oplus t).$$

- 9 -

In distinction to d'Alembert's solution of the classical wave equation, the above solution of the dyadic equation (6.1) is explained by the fact that dyadic addition and substraction yield the same result; thus the solution $f(x \oplus t)$ is a degeneration of two solutions into one. H.F. Harmuth [12] gave a first physical interpretation of this solution.

7. Extension of Derivative Concept to Non-Periodic Functions defined on R_+

For this extension the continuous Walsh-Kaczmarz functions must first be defined. If $R_+ = [0,\infty)$ and D_+ is the set of dyadic rationals of R_+, every $v \in R_+$, $v \notin D_+$ has the unique dyadic representation

$$v = \sum_{j=-N}^{\infty} v_j 2^{-j} \qquad (v_j \in \{0,1\}).$$

For $v \in D_+$ one chooses the finite expansion, and for $t = \sum_{j=-M}^{\infty} t_j 2^{-j}$ and $t \in D_+$ the infinite expansion. The continuous Walsh functions are then defined by

$$\phi(v,t) = (-1)^{\sum_{j=-N}^{M+1}(v_j+v_{j-1})t_{1-j}} \qquad (v,\ t \in R_+).$$

Setting $v \oplus t = \sum_{j=-L}^{\infty} |v_j - t_j| 2^{-j}$, $L = \max(M,N)$, the strong dyadic derivative of $f \in L_1(R_+)$ is defined by (see also F.Pichler [14])

$$(7.1) \qquad (D^{\{1\}}f)(x) = \frac{1}{2}\sum_{j=-\infty}^{\infty} \frac{f(x) - f(x \oplus 2^{-j})}{2^{-j}},$$

the convergence of the sum to be understood in $L_1(R_+)$-norm. It is obvious that this definition coincides with that in (2.2) in case f has period 1; so (7.1) is an appropriate extension.

If L_1^{**} is the set of all functions $w(x,t)$ on $R_+ \times R_+$ such that $\int_0^\infty \int_0^\infty |w(x,t)|\, dx\, dt < \infty$, it is possible to set up a dyadic wave equation:

- 10 -

$$(7.2)\qquad D_{xx}^{\{2\}}w - D_{tt}^{\{2\}}w = 0$$

$$\lim_{t\to 0+} \|w(\cdot,t) - f(\cdot)\|_1 = 0 \qquad (f \in L_1(R_+)).$$

To seek for a solution in L_1^{**} which is not periodic in either variable one applies the Walsh-Fourier transform to (7.2), defined for $w(\cdot,t) \in L_1(R_+)$ by

$$w^{\wedge}(v,t) = \int_0^\infty w(u.t)\phi(v,u)du \qquad (v \in R_+).$$

The solution again turns out to be $w(x,t) = f(x \oplus t)$.

8. Differences between the Concepts $D^{\{1\}}$, $I_{\{1\}}$ and the classical Derivative and Integral

As noted, there is a very close parallelism between the dyadic and classical concepts of a derivative, their properties and applications included. However, there are also differences. These are rather marked if one considers the pointwise instead of the strong dyadic derivative for comparison purposes.

(i) The classical derivative $f'(x)=\lim_{h\to 0} \frac{f(x+h)-f(x)}{h}$ is a local property in sense that differential quotient need only be defined in neighbourhood of x.	The dyadic derivative $f^{\{1\}}(x)=\frac{1}{2}\sum_{j=0}^{\infty} \frac{f(x)-f(x \oplus 2^{-j})}{2^{-j}}$ is a global property since difference in sum depends upon points such as $x \oplus 2^{-j}$ at a distance 2^{-j} from x.
(ii) Existence of the derivative $f'(x)$ at a point x implies the continuity of f at x.	Existence of $f^{\{1\}}(x)$ at x need not necessarily imply the continuity of f at x. Consider $f(x) = 0$ for $x=0, 1/2, 1/4,..,$ $=1$ otherwise. Then $f^{\{1\}}(0) = 0$, but f is not continuous at $x=0$.

- 11 -

(iii) A formally integrated Fourier series is always uniformly convergent.	The dyadic integral of a Walsh-Fourier series need not necessarily be uniformly convergent. Consider $\sum_{k=1}^{\infty}[\log(k+1)]^{-1}\phi(k,x)$, the Walsh-Fourier series of an $L_1(0,1)$-function.
(iv) In classical analysis the product law reads: $\frac{d}{dx}[f(x)g(x)] = (f.g)^{(1)}(x) = f^{(1)}(x)g(x) + f(x)g^{(1)}(x)$	The product law in this form does not hold in dyadic analysis. Consider the dyadic derivative of $\phi(1,x)=\phi(2,x)\phi(3,x)$.

Note that the existence of the classical derivative need not imply that of the dyadic derivative (consider $f(x) = x$), and conversely.

Concerning the fundamental mean value and Rolle's theorem of the differential calculus one may establish analogs, however in a somewhat different form, in the analysis for continuous functions defined on the dyadic group G. Compare Butzer-Wagner [2], Wagner [15]. Note that for continuous functions defined on the interval [0.1] (or R_+) our concept of a dyadic derivative is not useful. Indeed, making use of a result of Bočkarev [1], one can show that if $f^{\{2\}} \in C[0,1]$, then f=const. For this reason the space $L_p(0,1)$ is very natural for dyadic analysis on (0,1].

9. Best Approximation by Walsh Polynomials

One of the basic fields of computer science is approximation theory. Dyadic analysis yields definite improvements here.

- 12 -

For this purpose we need to define the modulus of continuity in Walsh analysis, namely

(9.1) $$\omega_W(f;\delta) = \sup_{0<h<\delta} \|f(\cdot) - f(\cdot\oplus h)\|_p ,$$

and the Lipschitz classes by

$$\mathrm{Lip}_W\alpha = \{f\in L_p(0,1);\|f(\cdot) - f(\cdot\oplus h)\|_p \leq Mh^{\alpha}\} ,$$

$M>0$ being a constant.

(i) A sufficient condition such that $D^{\{1\}}f$ exists as an element of $L_p(0,1)$ is that $f \in \mathrm{Lip}_W\alpha$, $\alpha>1$. Indeed formally

$$\lim_{m\to\infty} \|\frac{1}{2}\sum_{j=0}^{m} 2^j[f(\cdot)-f(\cdot\oplus 2^{-j})] - (D^{\{1\}}f)(\cdot)\|_p$$

$$= \lim_{m\to\infty}\|\frac{1}{2}\sum_{j=m+1}^{\infty} 2^j[f(\cdot)-f(\cdot\oplus 2^{-j})]\|_p$$

$$\leq \lim_{m\to\infty} \frac{1}{2}\|\sum_{j=m+1}^{\infty} 2^j f(\cdot)-f(\cdot\oplus 2^{-j})\|_p$$

$$= \lim_{m\to\infty} \frac{1}{2}\sum_{j=m+1}^{\infty} M\, 2^{-j(\alpha-1)}=0.$$

(ii) If $f \in \mathrm{Lip}\alpha$, $0<\alpha\leq 1$, in the classical sense, then $f \in \mathrm{Lip}_W\alpha$ (see [13]).

(iii) $f \in \mathrm{Lip}\alpha, \alpha>1$, in the classical sense implies that f=const. However, this need not neccessarily be so in Walsh analysis. Indeed, the function $\phi(1,x) \in \mathrm{Lip}_W\alpha$, each $\alpha\in P$ with $|\phi(1,x)-\phi(1,x\oplus h)| \leq 2^{(\alpha+1)}h^{\alpha}$. However, $\phi(1,x) \neq$ const.

A further change with respect to classical analysis is (see [3]):

If f, $D^{\{r\}}f \in L_p(0,1)$, $1\leq p<\infty$, then

$$\omega_W(f;\delta) = \mathcal{O}[\delta^r\omega_W(D^{\{r\}}f;\delta)] \quad (\delta\to 0+).$$

- 13 -

Indeed, in the classical situation one can only establish a corresponding result by bringing in moduli of continuity of higher order.

Now, letting P_n be the set of all Walsh polynomials $p_n(x) := \sum_{k=0}^{n-1} a_k\phi(k,x)$, $a_k \in R$, of degree n, then

$$(9.2) \qquad E_n(f) = \inf_{p_n \in P_n} \|f-p_n\|_p$$

is the best approximation of f by Walsh polynomials in $L_p(\overline{0,1})$-space. There exists a polynomial $p_n^*(=p_n^*(f)) \in P_n$ of best approximation to f, i.e. $E_n(f) = \|f-p_n^*\|_p$.

Already C. Watari [16] established a Jackson and Bernstein type result in Walsh analysis, namely, for any fixed $\alpha>0$,

$$(9.3) \qquad E_n(f) = \mathcal{O}(n^{-\alpha}) \, (n\to\infty)$$

$$\Longleftrightarrow \quad \omega_W(f;\delta) = \mathcal{O}(\delta^{\alpha}) \qquad (\delta\to 0).$$

With our dyadic derivative concept it is possible to add two equivalent assertions to the latter pair, the first one being a result on simultaneous approximation (Garkavi-Stečkin type), the second on the order of magnitude of p_n^* itself (Zamansky-type). Indeed

<u>Theorem 3</u>

If $f \in L_p(0,1)$ the following statements are equivalent for the Walsh polynomial of best approximation to f for any $\alpha>0$:

(i) $\quad \omega_W(D^{\{r\}}f;\delta) = \mathcal{O}(\delta^{\alpha})$,

(ii) $\quad E_n(f) = \|f - p_n^*\|_p = \mathcal{O}(n^{-\alpha-r})$,

(iii) $\quad \|D^{\{m\}}f - D^{\{m\}}p_n^*\|_p = \mathcal{O}(n^{-\alpha-r+m})$, $0\leq m\leq r$,

(iv) $\quad \|D^{\{1\}}p_n^*\|_p = \mathcal{O}(n^{1-\alpha-r})$, $0<\alpha+r<1$.

As above, note that only the first modulus of continuity need be used.

Concluding remarks

As noted. there is a great parallelism between the classical and dyadic derivative and their properties. However, as far as the authors are aware, an actual interpretation of the dyadic derivative is still missing. The classical derivative may be associated with the slope of a tangent to a curve, or with the rate of speed of an object, and the integral with the area under a curve; they may thus be associated with geometrical or physical notions. The problem is to find an appropriate interpretation of the dyadic derivative in terms of one of the many modern sciences which make use of Walsh analysis. In the words of J.E. Gibbs [8] it is to be hoped that "our calculus will serve information science in the universal way that Newton's differential calculus has served physical science". If it is not information science, it may be one of the other sciences using Walsh functions.

The contribution by Dr. Wagner was carried out as an associat of the research group "Informatik Nr. 14" at the Technological University of Aachen.

References

[1] S.V. Bočkarev: On the Fourier-Walsh coefficients. (Russian); Izv. Akad. Nauk SSSR Ser. Mat. 34 (1970) 203-208 = Transl. Math. USSR-Izv. 4(1970) 209-214.

[2] P.L. Butzer-H.J. Wagner: Walsh-Fourier series and the concept of a derivative. Applicable Anal. 3 (1973) 29-46.

[3] P.L. Butzer-H.J. Wagner: Approximation by Walsh polynomials and the concept of a derivative. IN: Applications of Walsh Functions (Proc. Sympos. Naval Res. Lab., Washington, D.C., 27.-29. March 1972; Ed. R.W. Zeek - A.E. Showalter) Washington, D.C. 1972, xi + 401 pp.; pp. 388 - 392.

- 15 -

[4] P.L. Butzer - H.J. Wagner: On a Gibbs-type derivative in Walsh-Fourier analysis with applications. In: Proceedings of the 1972 National Electronics Conference (Chicago, 9. - 10. Oct. 1972; Ed. R.E. Horton) Oak Brook, Illinois, 1972, xxvi + 457pp.; pp. 393-398.

[5] P.L. Butzer-H.J. Wagner: A calculus for Walsh functions defined on R_+. In: Applications of Walsh Functions (Proc. Sympos. Naval Res. Lab., Washington D.C., 18.-20. April 1973; in print.)

[6] N.J. Fine: On the Walsh functions. Trans. Amer. Math. Soc. 65 (1949) 372-414.

[7] J.E. Gibbs: Functions that are solutions of a logical differential equation. NPL (National Physical Laboratory), Middlesex, England, DES Rept. No. 4 (1970).

[8] J.E. Gibbs: The dyadic group and the second industrial revolution. In : Theory and Applications of Walsh Functions. (Proc. Sympos. Hatfield Polytechnic 29.-30. June 1971).

[9] J.E. Gibbs - B.Ireland: Some generalizations of the logical derivative. NPL: DES Rept. No. 8 (1971).

[10] J.E. Gibbs - M.J. Millard: Walsh functions as solutions of a logical differential equation . NPL: DES Rept. No. 1 (1969).

[11] H.F. Harmuth: Transmission of Information by Orthogonal Functions. Second edition, Springer Berlin, New York 1972.

[12] H.F. Harmuth: Real numbers versus dyadic group as basis for models of time and space. (To appear.)

[13] G.W. Morgenthaler: On Walsh-Fourier series. Trans. Amer. Math. Soc. 84 (1967) 472-507.

[14] F. Pichler: Walsh functions and linear system theory. In:Applications of Walsh Functions (Proc. Sympos. Naval Res. Lab., Washington, D.C. 31. March-3. April 1970;

Ed. C.A. Bass) Washington, D.C. 1970, viii + 274pp; pp. 175-182.

[15] H.J. Wagner: Ein Differential- und Integralkalkül in der Walsh-Fourier-Analysis mit Anwendungen. (Forschungsber. des Landes Nordrhein-Westfalen), Westdeutscher Verlag, Köln-Opladen;(to appear).

[16] C. Watari: Best approximation by Walsh polynomials. Tohoku Math. J. 15 (1963) 1-5.

A CALCULUS FOR WALSH FUNCTIONS DEFINED ON R_+

P.L. Butzer and H.J. Wagner
Technological University of Aachen
Aachen, Western Germany

1. Introduction

Stimulated by work of J.E. Gibbs [6,7,8], the authors [1,2,3,14] defined the concept of a derivative for functions defined on the infinite dyadic group G and then built up a differential and integral "Walsh" calculus for such functions as well as for periodic functions defined on the whole real axis. Using this calculus they solved a series of problems in Walsh-Fourier analysis, including a partial "differential" equation of wave equation type, the solution of which was periodic in both variables.

The purpose of the present lecture is to generalize the derivative concept in [1,14] to non-periodic functions, as first considered by F. Pichler [11], and then to solve the wave equation in a form envisaged by H.F. Harmuth [9,p.86].

For purposes of orientation, let us begin with the basic classical one-dimensional wave equation

$$(1.1)\quad \frac{\partial^2}{\partial x^2} w(x,t) = \frac{1}{c^2}\frac{\partial^2}{\partial t^2} w(x,t),$$

the general solution of which, due to J.L. d'Alembert, is given by

$$w(x,t) = f(x+ct) + g(x-ct),$$

where f and g are arbitrary functions determined by the initial and boundary conditions. Until the advent of J.E. Gibbs, an equation of this type in dyadic space and time, could not be even thought of since the classical derivative of any Walsh-function is zero almost everywhere, they being step-functions. For this purpose it was desirable to find a new derivative concept with respect to which the Walsh functions themselves, at least, are non-trivially differentiable. Moreover, this new concept should play the same fundamental role as does the classical derivative in the mathematical sciences.

If the function f is defined on the interval [0,1), such a derivative, called the "logical", "dyadic" or also Gibbs derivative, is defined by ($x\epsilon[0,1)$),

$$(1.2)\quad f^{[1]}(x) = \frac{1}{4}\sum_{j=1}^{\infty} \frac{f(x) - f(x \oplus 2^{-j})}{2^{-j}},$$

where $x = \sum_{j=1}^{\infty} x_j 2^{-j}$, $y = \sum_{j=1}^{\infty} y_j 2^{-j}$, $x_j, y_j \epsilon \{0,1\}$, and

$$x \oplus y = \sum_{j=1}^{\infty} |x_j - y_j| 2^{-j}$$

being termwise addition modulo 2. (The square bracket [ ] is used in $f^{[1]}(x)$ to distinguish the dyadic first derivative from the classical first derivative $f^{(1)}(x)$.)

As a first result, the rth dyadic derivative of the Walsh functions $\psi(k,x)$ is given by

$$\psi^{[r]}(k,x) = k^r \psi(k,x) \quad (k \in P=\{0,1,2,\dots\}).$$

Moreover, with this new concept the authors [3,14] could for the first time set up a dyadic wave equation, namely

$$(1.3) \quad \frac{\partial^{[2]}}{\partial x^{[2]}} w(x,t) = \frac{\partial^{[2]}}{\partial t^{[2]}} w(x,t)$$

$$\lim_{t \to 0+} w(x,t) = f(x),$$

and show that its general solution has the form $w(x,t) = f(x \oplus t)$.

In distinction to d'Alembert's solution of equation (1.1), the solution of the dyadic equation (1.3) is explained by the fact that dyadic addition and subtraction yield the same result; thus the solution of (1.3) is a degeneration of two solutions into one. Then H.F. Harmuth [10] gave a first physical interpretation of this solution, at the same time giving an answer to a fundamental problem raised by Erwin Schroedinger.

The general solution of (1.3) was found under the assumption that it be periodic in the space variable x as well as in the time variable t. This being a somewhat restrictive assumption, particularly the periodicity with respect to time, we wish to solve the dyadic wave equation for functions which need be periodic in at most one variable, or in neither variable.

2. Walsh Functions and Walsh-Fourier Transforms

Let R_+ be the set of all non-negative real numbers and D_+ the set of all non-negative dyadic rationals. Each $v \in R_+$ has the dyadic representation

$$(2.1) \quad v = \sum_{j=-N}^{\infty} v_j 2^{-j} \quad (v_j \in \{0,1\}, N \in P).$$

If $v \notin D_+$ this representation is unique, otherwise there are two expansions and one customarily chooses the finite one for $v \in D_+$. If $x = \sum_{j=-M}^{\infty} x_j 2^{-j}$, then one defines

$$(2.2) \quad v \oplus x = \sum_{j=-L}^{\infty} |v_j - x_j| 2^{-j}, \quad L = \max(N,M).$$

The generalized Walsh(-Paley) functions introduced by N.J. Fine [5] then have the form $(v,x \in R_+)$

$$(2.3) \quad \psi(v,x) = \exp\{\pi i \sum_{j=-N}^{M+1} v_j x_{1-j}\};$$

the $\psi(v,x)$ coincide with the functions $\psi(k,x)$ introduced by J.L. Walsh [15] for $v \in P$.

Their basic properties are given by (see N.J. Fine [5], R.G. Selfridge [13])

Lemma 2.1

(i) $\psi(v,x) = \psi(x,v) \quad (v,x \in R_+)$,

(ii) $\psi(v,x) = \psi([v],x)\psi(v,[x])$ for all $v,x \in R_+$, $[v]$ being the greatest integer $\leq v$,

(iii) For each $y \in R_+$ and $v \in R_+$ and almost all $x \in R_+$
$\psi(v,x \oplus y) = \psi(v,x)\psi(v,y)$,

(iv) For $h \in [0,2^{-n})$ and $v \in [0,2^n)$
$\psi(v,x) = \psi(v,x \oplus h) \quad (x \in R_+)$,

(v) If $n \in Z = \{0,\pm 1,\pm 2,\dots\}$
$\psi(2^n v,x) = \psi(v,2^n x) \quad (v,x \in R_+)$.

In order to solve the desired wave equation one needs a further tool, namely the Walsh-Fourier transform. If $L_1(R_+)$ is the space of absolutely integrable functions with norm $\|f\| = \int_0^\infty |f(u)|du$, then the Walsh-Fourier transform $f^\wedge$ of f is defined by

$$(2.4) \quad f^\wedge(v) = \int_0^\infty f(u)\,\psi(v,u)du \quad (v \in R_+).$$

It exists for all $v \in R_+$ as a bounded

function satisfying

(2.5) $|f^{\wedge}(v)| \leq \|f\|$ $(v\epsilon R_+)$.

Defining the convolution of $f,g\epsilon L_1(R_+)$ by

(2.6) $(f*g)(x) = \int_0^\infty f(u)g(x \oplus u)du$,

it is also known that $f*g$ exists for almost all $x\epsilon R_+$, $f*g\epsilon L_1(R_+)$, and

$$\|f*g\| \leq \|f\| \, \|g\|.$$

The principal properties of $f^{\wedge}$ are

Lemma 2.2

(i) If $f\epsilon L_1(R_+)$, then for all $h,v\epsilon R_+$
$[f(\cdot \oplus h)]^{\wedge}(v) = \psi(v,h)f^{\wedge}(v)$,

(ii) If f, $g\epsilon L_1(R_+)$, then
$[f*g]^{\wedge}(v) = f^{\wedge}(v) \cdot g^{\wedge}(v)$ $(v\epsilon R_+)$,

(iii) If f, $f_n\epsilon L_1(R_+)$, then for all $v\epsilon R_+$
$\lim_{n\to\infty} \|f_n - f\| = 0 \Rightarrow \lim_{n\to\infty} f_n^{\wedge}(v) = f^{\wedge}(v)$.

As a further preliminary result we need the uniqueness theorem for this transform. We can derive this easily by making use of the Dirichlet kernel, introduced by Fine [5], namely

$$J(\omega,x) = \int_0^\omega \psi(v,x)dv \qquad (x,\omega\epsilon R_+),$$

and the Dirichlet partial sums

$$S(f;\omega,x) = \int_0^\omega f^{\wedge}(v)\psi(v,x)dv,$$

which may be rewritten as (Fubini !)

$$S(f;\omega,x) = \int_0^\infty f(v)J(\omega,x \oplus v)dv = (f*J(\omega,\cdot))(x).$$

This enables one to prove, first of all,

Lemma 2.3

If $f\epsilon L_1(R_+)$, then $\lim_{n\to\infty} \|S(f;2^n,\cdot)-f(\cdot)\|=0$.

Proof

We first show that for $\omega = 2^n$ $\|J(2^n,\cdot)\| = 1$. Indeed, by Lemma 2.1 (ii) and since the functions $\psi(v,[x])$ are orthogonal on each unit interval

$$J(2^n,x) = \int_0^{2^n} \psi(v,x)dv$$
$$= \int_0^{2^n} \psi([v],x)\psi(v,[x])dv$$
$$= \sum_{j=0}^{2^n-1}\psi(j,x) \int_j^{j+1}\psi(v,[x])dv$$
$$=\begin{cases} D(2^n,x), & x\epsilon[0,1) \\ 0, & x\epsilon[1,\infty)\end{cases} = \begin{cases} 2^n, & x\epsilon[0,2^{-n}) \\ 0, & x\epsilon[2^{-n},\infty),\end{cases}$$

where

$$D(m,x) = \sum_{j=0}^{m-1} \psi(j,x).$$

This yields $\|J(2^n,\cdot)\| = \int_0^{2^{-n}} 2^n du = 1$.

Concerning the result itself, noting that $f(x) = \int_0^\infty f(x)J(2^n,u)du$, it follows by Fubini that

$$\|f(\cdot) - S(f;2^n,\cdot)\|$$
$$\leq \int_0^\infty \int_0^\infty |f(x) - f(x\oplus u)|J(2^n,u)dudx$$
$$= \int_0^\infty \int_0^\infty |f(x) - f(x\oplus u)|J(2^n,u)dxdu$$
$$= \int_0^{2^{-n}} 2^n \|f(\cdot) - f(\cdot\oplus u)\| du$$
$$\leq \sup_{0<u<2^{-n}} \|f(\cdot) - f(\cdot\oplus u)\|,$$

the latter term tending to zero for $n\to\infty$ in view of the classical continuity in the mean property of $L_1(R_+)$-functions.

This yields the desired uniqueness result.

Corollary 2.1

If $f^{\wedge}(v) = 0$ a.e., then $f = 0$ a.e.

3. The Derivative for Functions on R_+

Now to the extension of definition (1.2) to non-periodic functions given on R_+.

Definition 3.1

(i) If for $f\epsilon L_1(R_+)$ there exists $g\epsilon L_1(R_+)$ such that for m_1, $m_2\epsilon P$

$$\lim_{m_1,m_2\to\infty} \left\|\frac{1}{4} \sum_{j=-m_1}^{m_2} \frac{f(\cdot) - f(\cdot\oplus 2^{-j})}{2^{-j}} - g(\cdot)\right\|=0,$$

then g is called the strong dyadic derivative of f, denoted by $D^{[1]}f$.

(ii) If for $x\epsilon R_+$

$$\frac{1}{4}\sum_{j=-\infty}^{+\infty} 2^j[f(x) - f(x\oplus 2^{-j})] = c,$$

then c is called the pointwise dyadic derivative of f at x and is denoted by $f^{[1]}(x)$.

The derivatives of order $r\varepsilon P$ are defined successively as usual. It is obvious that the strong as well as pointwise derivatives define linear operations.

First we prove that Def.3.1(ii) is an actual extension of that in (1.2) in case f has period 1, by showing that for $x\varepsilon R_+$

$$\frac{1}{4}\sum_{j=-\infty}^{\infty} 2^j[f(x) - f(x\oplus 2^{-j})]$$

$$= \frac{1}{4}\sum_{j=1}^{\infty} 2^j[f(x) - f(x\oplus 2^{-j})].$$

Indeed, f being defined on R_+ and having period 1, $f(x) - f(x\pm k) = 0$, $k\varepsilon P$, for all $x\varepsilon R_+$. Now if $x = \sum_{j=-M}^{\infty} x_j 2^{-j}$ with $x_{-k} = 0$, arbitrary fixed $k\varepsilon P$, then $x\oplus 2^k = x+2^k$. If $x_{-k} = 1$, then $x\oplus 2^k = x-2^k$. The periodicity then delivers the result.

Since the Walsh functions $\psi(v,x)$ do not belong to the space $L_1(R_+)$ they are not strongly dyadic differentiable. However, they are arbitrarily often pointwise dyadic differentiable. Indeed,

<u>Theorem 3.1</u>

$$\psi^{[r]}(v,x) = v^r\psi(v,x) \qquad (v,x\varepsilon R_+).$$

For the proof see F. Pichler [11]. It depends essentially upon the identity

$$(3.1)\quad \frac{1}{4}\sum_{j=-\infty}^{+\infty} 2^j[1-\psi(v,2^{-j})] = v \qquad (v\varepsilon R_+)$$

as well as on Lemma 2.1 (iii).

Concerning Walsh-Fourier Transforms of strong dyadic derivative we have

<u>Theorem 3.2</u>

If f, $D^{[r]}f\varepsilon L_1(R_+)$, then

$$[D^{[r]}f]^\wedge(v) = v^r f^\wedge(v) \qquad (v\varepsilon R_+).$$

<u>Proof</u>

In case r = 1 it follows by Lemma 2.2 (iii) and (3.1) that if $D^{[r]}f\varepsilon L_1(R_+)$, then

$$\lim_{m_1,m_2\to\infty} \frac{1}{4}\sum_{j=-m_1}^{m_2} 2^j[1-\psi(v,2^{-j})]f^\wedge(v)$$

$$= vf^\wedge(v) = [D^{[1]}f]^\wedge(v) \qquad (v\varepsilon R_+).$$

The result for r>1 follows by induction.

It is probably true that Theorem 3.2 also holds provided the pointwise dyadic derivative exists for all $x\varepsilon R_+$ and belongs to $L_1(R_+)$.

4. <u>Comparison with the discrete Walsh Transform</u>

In the foregoing we have set up the Walsh-Fourier transform in connection with the dyadic derivative. At this stage it pays to recall the corresponding material when the discrete Walsh functions $\psi(k,x)$ are used, namely the situation in case of the discrete finite Walsh transform considered in [2,3,14].

Denoting the space of absolutely integrable functions having period 1 by $L_1(0,1)$ (with norm $\|f\|_{L_1(0,1)} = \int_0^1|f(u)|du$), the discrete finite Walsh transform (=kth Walsh-Fourier coefficient) of $f\varepsilon L_1(0,1)$

is defined by

$$f_F^\wedge(k) = \int_0^1 f(u)\psi(k,u)du \qquad (k\varepsilon P).$$

The counterpart of Lemma 2.2 is valid here, as well as the uniqueness theorem:

$$(4.1)\quad f_F^\wedge(k) = 0, k\varepsilon P, \Rightarrow f = 0 \text{ a.e.}$$

The analog of Theorem 3.2, the derivative here being defined by (1.2) (in the strong sense), reads

(4.2) $f,\ D^{[r]}f \in L_1(0,1),$

$$\Rightarrow [D^{[r]}f]^{\wedge}_{F}(k) = k^r f^{\wedge}_{F}(k) \quad (k \in P).$$

To set forth the analogies between the discrete finite Walsh transform and Walsh-Fourier transform and the corresponding transform methods we arrange both side by side below. The signs $\sum$ and $\int$, the integer k and the continuous variable v, the intervals (0,1) and $(0,\infty)$ correspond.

$f^{\wedge}_{F}(k) = \int_0^1 f(u)\psi(k,u)du \quad (k \in P)$	$f^{\wedge}(v) = \int_0^{\infty} f(u)\psi(v,u)du \quad (v \in R_+)$
$D(m,x) = \sum_{k=0}^{m-1} \psi(k,x) \quad (m \in P, x \in [0,1))$	$J(\omega,x) = \int_0^{\omega} \psi(v,x)dv \quad (\omega, x \in R_+)$
$(f * g)(x) = \int_0^1 f(u)g(x \oplus u)du$	$(f * g)(x) = \int_0^{\infty} f(u)g(x \oplus u)du$
$S(f;m,x) = \sum_{k=0}^{m-1} f^{\wedge}_{F}(k)\psi(k,x)$	$S(f;\omega,x) = \int_0^{\omega} f^{\wedge}(v)\psi(v,x)dv$
$= (f * D(m,\cdot))(x) \quad (x \in [0,1), m \in P)$	$= (f * J(\omega,\cdot))(x) \quad (\omega, x \in R_+)$
$f^{[1]}(x) = \frac{1}{4} \sum_{j=1}^{\infty} 2^j [f(x) - f(x \oplus 2^{-j})]$	$f^{[1]}(x) = \frac{1}{4} \sum_{j=-\infty}^{+\infty} 2^j [f(x) - f(x \oplus 2^{-j})]$
$\psi^{[r]}(k,x) = k^r \psi(k,x) \quad (k \in P, x \in [0,1))$	$\psi^{[r]}(v,x) = v^r \psi(v,x) \quad (v, x \in R_+)$

In the next paragraph we shall use the discrete transform to solve the dyadic wave equation in case one of the two variables x,t is periodic (say the space variable x) and, preceding this, use the continuous Fourier-Walsh transform in case neither variable x,t is periodic.

5. Wave Equation-Type Partial Differential Equations

In order to set up partial differential equations for which the derivatives are understood in the dyadic sense (Def. 3.1), we need the following definitions. We set

$$\|w\|^* = \int_0^{\infty} \int_0^{\infty} |w(x,t)|\,dxdt,$$

and denote the set of all functions $w(x,t)$ for which $\|w\|^* < \infty$ by $L_1^* \equiv L_1(R_+ \times R_+)$. This enables us to introduce strong dyadic partial derivatives for functions $f \in L_1^*$.

Definition 5.1

If for $f \in L_1^*$ there exists $g \in L_1^*$ such that

$$\lim_{m_1,m_2 \to \infty} \Big\| \frac{1}{4} \sum_{j=-m_1}^{m_2} \frac{[f(\cdot,t) - f(\cdot \oplus 2^{-j},t)]}{2^{-j}} - g(\cdot,t) \Big\| = 0$$

for almost all $t \in R_+$, then $g(x,t)$ is called the strong dyadic partial derivative of $f(x,t)$ with respect to x, denoted by $D_x^{[1]}f$. The second such derivative is defined by $D_{xx}^{[2]}f = D_x^{1}(D_x^{[1]}f)$.

Correspondingly one may define $D_t^{[1]}f$, etc. The pointwise first dyadic partial derivative of $f(x,t)$ with respect to t is defined analogously (to Def. 3.1(ii)) and denoted by $f_t^{[1]}(x,t)$.

We may now set up a partial differential equation of type (1.1) in L_1^*:

(5.1) $c^2 D_{xx}^{[2]} w - D_{tt}^{[2]} w = 0$

with initial condition

(5.2) $\lim_{t \to 0+} \|w(\cdot,t) - f(\cdot)\| = 0,\ f \in L_1(R_+).$

In order to find a solution in L_1^* we proceed formally as follows: Apply the Walsh-Fourier transform of (2.4) with respect to x, namely $w^{\wedge}(v,t) = \int_0^{\infty} w(x,t)\psi(v,t)dx$, to equation (5.1). In view of Theorem 3.2

(5.3) $(cv)^2 w^\wedge(v,t) = [D_{tt}^{[2]} w(\cdot,t)]^\wedge(v)$

Now, would

(5.4) $[D_{tt}^{[2]} w(\cdot,t)]^\wedge(v) = [w^\wedge(v,t)]_{tt}^{[2]}$,

then the transformed equation would have the form

(5.5) $(cv)^2 w^\wedge(v,t) = [w^\wedge(v,t)]_{tt}^{[2]}$.

For each $v\varepsilon R_+$ this is an "ordinary" dyadic differential equation of second order with respect to t. In view of Theorem 3.1 a solution of (5.5) is given by

(5.6) $w^\wedge(v,t) = A(v)\psi(cv,t)$,

A(v) being independent of t. The initial condition (5.2) yields $A(v) = f^\wedge(v)$.

For all c such that

(5.7) $\psi(cv,t) = \psi(v,ct) \qquad (t,v\varepsilon R_+)$

one therefore has

$$w^\wedge(v,t) = f^\wedge(v)\psi(v,ct)$$

as a solution of (5.5). In order to represent this solution in terms of the original functions, i.e. to find a solution of the original equation (5.1), we must apply a suitable inversion theorem. Indeed, the uniqueness theorem (Corollary 2.1) together with Lemma 2.2 (i) yields

$$w(x,t) = f(x\oplus ct)$$

as a solution of (5.1).

There are two remarks. Whether $f(x\oplus ct)$ is the most general form of the solution of (5.1) depends upon whether (5.6) is the general solution of (5.5). An answer would be possible if the inverse operation to dyadic derivatives would be available. Secondly, equation (5.7) holds for all $c=2^n$, $n\varepsilon Z$ by Lemma 2.1(v). For the other c values the solution w(x,t) is for almost all $t\varepsilon R_+$ equal to the limit for $n\to\infty$ in the $L_1(R_+)$-norm (taken with respect to x) of

$$\int_0^{2^n} \{f^\wedge(v)\psi(cv,t)\}\psi(v,x)dv.$$

This follows since $w\varepsilon L_1^*$ implies that w(x,t) belongs to $L_1^*(R_+)$ with respect to x.

In order to solve equation (5.1) in precise mathematical language, one must reformulate the problem as follows:

Given a function $f\varepsilon L_1(R_+)$, we call for a function $w(x,t)\varepsilon L_1^*$ such that:

(i) $D_x^{[1]}w$, $D_{xx}^{[2]}w$, $D_t^{[1]}w$, $D_{tt}^{[2]}w$ belong L_1^*,

(ii) w(x,t) satisfies equation (4.1), the constant c being of form 2^n, $n\varepsilon Z$,

(iii) $\lim_{t\to 0+} \| w(\cdot,t)-f(\cdot)\| = 0$,

(iv) $\lim_{m_1,m_2\to\infty} \| \frac{1}{4}\sum_{j=-m_1}^{m_2} \frac{[g(\cdot,t)-g(\cdot,t\oplus 2^{-j})]}{2^{-j}} - D_t^{[1]}g(\cdot,t)\| = 0$

for almost all $t\varepsilon R_+$ with g=w as well as $g=D_t^{[1]}w$.

As already mentioned, the Walsh-Fourier transform applied to (5.1) with respect to x yields equation (5.3) by Theorem 3.2 for all $v\varepsilon R_+$ and almost all $r\varepsilon R_+$. Condition (iv) implies that (5.4) is valid on account of Lemma 2.2 (iii). Therefore the transformed equation has for all $v\varepsilon R_+$ and almost $t\varepsilon R_+$ the form (5.5). The rest of the proof is now clear, giving $w(x,t) = f(x\oplus ct)$ a.e. in $x,t\varepsilon R_+$.

Let us now solve equation (5.1) in case the solution w(x,t) is assumed to be of period 1 in the space variable x. Here set

$$\| w\|^{**} = \int_0^\infty \int_0^1 |w(x,t)|dxdt,$$

and let L_1^{**} be the space of all functions w(x,t) which are periodic in x with $\| w\|^{**}$ finite. If one defines the (strong) dyadic partial derivatives with respect to x analogously to Definition 5.1 in the $L_1(0,1)$-norm, and with respect t in the $L_1(R_+)$-norm, then we may set up the equation in L_1^{**}

$$(5.8)\qquad c^2 D_{xx}^{[2]} w - D_{tt}^{[2]} w = 0$$

$$\lim_{t\to 0+} \| w(\cdot,t) - f(\cdot)\|_{L_1(0,1)} = 0, \quad f \in L_1(0,1).$$

To find a solution in L_1^{**} apply the discrete Walsh transform with respect to x to (5.8), yielding by (4.2)

$$(5.9)\quad (ck)^2 w_F^{\wedge}(k,t) - [w_F^{\wedge}(k,t)]_{tt}^{[2]} = 0 \quad (k \in P)$$

for almost all $t \in R_+$ provided

$$[D_{tt}^{[2]} w(\cdot,t)]_F^{\wedge}(k) - [w_F^{\wedge}(k,t)]_{tt}^{[2]} = 0.$$

Now a solution of (5.9) is obviously

$$w_F^{\wedge}(k,t) = A(k)\psi(ck,t),$$

where $A(k) = f_F^{\wedge}(k)$ in view of the initial condition.

For all c such that $\psi(ck,t) = \psi(k,ct)$ $t \in R_+$, $k \in P$, it follows that $w(x,t) = f(x \oplus ct)$ is a solution of (5.8) by (4.1).

6. Concluding Remarks

It is possible to set up the dyadic wave equation in terms of pointwise derivatives in the form

$$(6.1)\quad [w(x,t)]_{xx}^{[2]} - [w(x,t)]_{tt}^{[2]} = 0.$$

This would also allow $\psi(v, x \oplus t)$ as a possible solution by Theorem 3.1. It is probable that the general solution of (6.1) together with (5.2) is then also given by $w(x,t) = f(x \oplus t)$. The trouble here is that we have not proved Theorem 3.2 for pointwise derivatives.

Secondly, as mentioned, H.F. Harmuth interpreted our solution of (5.1) in particle physics. An essential problem would be an interpretation of the dyadic wave equation, better still, of the dyadic derivative itself.

The contribution of Dr. Wagner was carried out as an associate of the research group "Informatik Nr.7" at the Technological University of Aachen.

References

[1] P.L. Butzer - H.J. Wagner: Walsh-Fourier series and the concept of a derivative. Applicable Anal. 3(1973), 29-46.

[2] P.L. Butzer - H.J. Wagner: Approximation by Walsh polynomials and the concept of a derivative. In: Proc. Sympos. Applications of Walsh Functions, Washington, D.C., March 1972, 388-392.

[3] P.L. Butzer - H.J. Wagner: On a Gibbs-type derivative in Walsh-Fourier analysis with applications. In:Proc.National Electronics Conf., Chicago,Oct.1972, 393-398.

[4] N.J. Fine: On the Walsh functions. Trans.Amer.Math.Soc.,65(1949),372-414.

[5] N.J. Fine: The generalized Walsh functions.Trans.Amer.Math.Soc.,69 (1950), 66-77.

[6] J.E. Gibbs: Functions that are solutions of a logical differential equation.NPL (National Physical Laboratory),Middlesex, England, DES Rept.No. 4 (1970).

[7] J.E. Gibbs - M.J. Millard: Walsh functions as solutions of a logical differential equation.NPL:DES Rept. No. 1 (1969).

[8] J.E. Gibbs - J.S. Mortimore - D.R. Perkins: On an important orthogonal system of functions.NPL:DES Rept. No. 12 (1972).

[9] H.F. Harmuth: Transformation of Information by Orthogonal Functions. Springer, Berlin, New York 1969.

[10] H.F. Harmuth: Real numbers versus dyadic group as basis for models of time and space. (To appear.)

[11] F. Pichler: Walsh functions and linear system theory.In:Proc.Sympos Applications of Walsh Functions, Washington,D.C.,April 1970,175-182.

[12] F. Pichler: Walsh Functions-Introduction to the theory. Nato Adv. Study Inst.on Signal Processing,Aug. 1972, U.of Technology, Loughborough, U.K.

[13] R.G. Selfridge: Generalized Walsh transform. Pacific J. Math., 5 (1955), 451-480.

[14] H.J. Wagner: Ein Differential- und Integralkalkül in der Walsh-Fourier-Analysis mit Anwendungen. (Forschungsber. des Landes Nordrhein-Westfalen), Westdeutsch. Verlag, Köln-Opladen (to appear).

[15] J.L. Walsh: A closed set of normal orthogonal functions. Amer.J.Math 45 (1923), 5-24.

Chapter 8
Hardy Spaces in the Theory of Dyadic Derivative

Ferenc Weisz

8.1 Introduction

In this survey paper we will consider the dyadic version of the classical theorem of Lebesgue. Schipp [13] proved that the dyadic derivative of the dyadic integral of a function $f \in L_1[0,1)$ is almost everywhere f. The theory of Hardy spaces can be well applied in harmonic analysis as well as in the theory of dyadic derivative (see e.g. [19, 24] and the references therein). In this paper we will apply this modern approach for the dyadic derivative. We introduce the dyadic martingale Hardy spaces $H_p[0,1)$ and prove ([22]) that the maximal operator of the dyadic derivative of the dyadic integral is bounded from $H_p[0,1)$ to $L_p[0,1)$, whenever $1/2 < p < \infty$. For $p = 1$ we easily obtain the weak type $(1,1)$ inequality for the maximal operator by interpolation, which implies the dyadic version of Lebesgue's theorem mentioned above. The almost everywhere convergence and the weak type inequality are proved usually with the help of a Calderon-Zygmund type decomposition lemma. However, this lemma does not work in higher dimensions. Our method, that can be applied in higher dimension, too, can be regarded as a new method to prove the almost everywhere convergence and weak type inequalities.

In the multi-dimensional case two types of convergence and maximal operators are considered, the restricted (convergence over the diagonal or over a band around the diagonal), and the unrestricted (convergence over $\mathbb{N}^d$). For these two cases, we have to use two different multi-dimensional dyadic Hardy spaces. We ([20, 21, 23]) can prove again that the maximal operator of the multi-dimensional dyadic derivative of the dyadic integral is bounded from $H_p[0,1)^d$ to $L_p[0,1)^d$. As a consequence, we obtain a weak type inequality for $p = 1$, which implies the multi-dimensional version of the theorem about the dyadic derivative of the dyadic integral. In this survey paper we summarize the results appeared in this topic in the last 20 years.

Ferenc Weisz
Department of Numerical Analysis, Eötvös L. University
H-1117 Budapest, Pázmány P. sétány 1/C., Hungary, e-mail: weisz@inf.elte.hu

R.S. Stanković et al. (eds.), *Dyadic Walsh Analysis from 1924 Onwards Walsh-Gibbs-Butzer Dyadic Differentiation in Science Volume 1 Foundations*, Atlantis Studies in Mathematics for Engineering and Science 12, DOI 10.2991/978-94-6239-160-4_8

8.2 One-dimensional Hardy Spaces

We briefly write $L_p[0,1)$ instead of the space $L_p([0,1),\lambda)$, where λ is the Lebesgue measure. By a *dyadic interval* we mean one of the form $[k2^{-n},(k+1)2^{-n})$ for some $k,n \in \mathbb{N}$, $0 \le k < 2^n$. Given $n \in \mathbb{N}$ and $x \in [0,1)$ let $I_n(x)$ be the dyadic interval of length 2^{-n} which contains x. The σ-algebra generated by the dyadic intervals $\{I_n(x) : x \in [0,1)\}$ will be denoted by $\mathscr{F}_n$ $(n \in \mathbb{N})$.

We investigate the class of *martingales* $f = (f_n, n \in \mathbb{N})$ with respect to $(\mathscr{F}_n, n \in \mathbb{N})$. The *maximal function* of a martingale f is defined by

$$f^* := \sup_{n\in\mathbb{N}} |f_n|.$$

There is a close relation between martingales and Walsh-Fourier series. To explain this let us introduce the Walsh system. The *Rademacher functions* are defined by

$$r(x) := \begin{cases} 1, & \text{if}\, x \in [0,\frac{1}{2}) \\ -1, & \text{if}\, x \in [\frac{1}{2},1) \end{cases}$$

and

$$r_n(x) := r(2^n x) \qquad (x \in [0,1), n \in \mathbb{N}).$$

The product system generated by the Rademacher functions is the *one-dimensional Walsh system*:

$$w_n := \prod_{k=0}^{\infty} r_k^{\,n_k},$$

where

$$n = \sum_{k=0}^{\infty} n_k 2^k \qquad (0 \le n_k < 2).$$

If $f \in L_1[0,1)$, then the number

$$\hat{f}(n) := \int_{[0,1)} f w_n \, d\lambda \qquad (n \in \mathbb{N})$$

is said to be the nth *Fourier coefficient* of f. Denote by $s_n f$ the nth *partial sum* of the Walsh-Fourier series of a martingale f, namely,

$$s_n f(x) := \sum_{k=0}^{n-1} \hat{f}(k) w_k(x) = f * D_n(x) = \int_0^1 f(t) D_n(x \dot{+} t)\, dt,$$

where the *Walsh-Dirichlet kernels*

$$D_n := \sum_{k=0}^{n-1} w_k$$

satisfy

$$D_{2^n}(x) = \begin{cases} 2^n & \text{if } x \in [0, 2^{-n}), \\ 0 & \text{if } x \in [2^{-n}, 1). \end{cases}$$

If $f \in L_1[0,1)$, then $s_{2^n} f = f_n = E_n f$ $(n \in \mathbb{N})$ is the martingale generated by f, where E_n denotes the conditional expectation operator with respect to the σ-algebra $\mathscr{F}_n$.

For $0 < p \leq \infty$ the *martingale Hardy space* $H_p[0,1)$ consists of all one-parameter martingales for which

$$\|f\|_{H_p} := \|f^*\|_p < \infty.$$

Recall that the Hardy and L_p spaces are equivalent, if $p > 1$, in other words,

$$H_p[0,1) \sim L_p[0,1) \qquad (1 < p \leq \infty).$$

Moreover, the martingale maximal function is of weak type $(1,1)$:

$$\|f\|_{H_{1,\infty}} := \sup_{\rho > 0} \rho \lambda(f^* > \rho) \leq C\|f\|_1 \qquad (f \in L_1[0,1))$$

and $H_1[0,1) \subset L_1[0,1)$ (see e.g. Weisz [19]).

A first version of the *atomic decomposition* was introduced by Coifman and Weiss [4] in the classical case and by Herz [10] in the martingale case. The proof of the next theorem can be found in Weisz [19].

A function $a \in L_\infty$ is called a *p-atom* if

(a) supp $a \subset I$, $I \subset [0,1)$ is a dyadic interval,
(b) $\|a\|_\infty \leq |I|^{-1/p}$,
(c) $\int_I a(x)\,dx = 0$.

The basic result of atomic decomposition is the following one.

Theorem 8.1 *A martingale f is in $H_p[0,1)$ $(0 < p \leq 1\})$ if and only if there exist a sequence $(a^k, k \in \mathbb{N})$ of p-atoms and a sequence $(\mu_k, k \in \mathbb{N})$ of real numbers such that*

$$\sum_{k=0}^{\infty} \mu_k a^k = f \quad \textit{in the sense of martingales,}$$
$$\sum_{k=0}^{\infty} |\mu_k|^p < \infty. \tag{8.1}$$

Moreover,

$$\|f\|_{H_p} \sim \inf \Big(\sum_{k=0}^{\infty} |\mu_k|^p \Big)^{1/p},$$

where the infimum is taken over all decompositions of f of the form (8.1).

If I is a dyadic interval then let $I^r = 2^r I$ be a dyadic interval, for which $I \subset I^r$ and $|I^r| = 2^r |I|$ $(r \in \mathbb{N})$.

The following result gives a sufficient condition for V to be bounded from $H_p[0,1)$ to $L_p[0,1)$ (see Weisz [24, 25]). For $p_0 = 1$ it can be found in Schipp, Wade, Simon and Pál [16].

Theorem 8.2 *For each $n \in \mathbb{N}$ let $V_n : L_1[0,1) \to L_1[0,1)$ be a bounded linear operator and let*

$$V_* f := \sup_{n \in \mathbb{N}} |V_n f|.$$

Suppose that

$$\int_{[0,1)\setminus I^r} |V_* a|^{p_0} \, d\lambda \le C_{p_0} \tag{8.2}$$

for all p_0-atoms a and for some fixed $r \in \mathbb{N}$ and $0 < p_0 \le 1$, where the cube I is the support of the atom. If V_* is bounded from $L_{p_1}[0,1)$ to $L_{p_1}[0,1)$ for some $1 < p_1 \le \infty$, then

$$\|V_* f\|_p \le C_p \|f\|_{H_p} \qquad (f \in H_p[0,1) \cap L_1[0,1)) \tag{8.3}$$

for all $p_0 \le p \le p_1$. Moreover, if $p_0 < 1$, then the operator V_* is of weak type $(1,1)$, i.e., if $f \in L_1[0,1)$, then

$$\sup_{\rho > 0} \rho \lambda(|V_* f| > \rho) \le C \|f\|_1. \tag{8.4}$$

Note that (8.4) can be obtained from (8.3) by interpolation. For the basic definitions and theorems on interpolation theory of dyadic Hardy spaces see Weisz [19, 24]. Theorem 8.2 can be regarded also as an alternative tool to the Calderon-Zygmund decomposition lemma for proving weak type $(1,1)$ inequalities. In many cases this theorem can be applied better and more simply than the Calderon-Zygmund decomposition lemma.

8.3 The One-dimensional Dyadic Derivative

It is well-known that the derivative of the trigonometric function $e^{2\pi \imath kx}$ is $2\pi \imath k e^{2\pi \imath kx}$. Moreover, the one-dimensional classical differentiation theorem due to Lebesgue (see e.g. Zygmund [27]) says that

$$f(x) = \lim_{h \to 0} \frac{1}{h} \int_x^{x+h} f(t)\, dt \qquad \text{a.e.} \qquad (f \in L_1[0,1)).$$

In this section the dyadic analogue of this result will be formulated. Gibbs [7], Butzer and Wagner [2, 3] introduced the concept of the *dyadic derivative* as follows. For each function f defined on $[0,1)$ set

$$(\mathbf{d}_n f)(x) := \sum_{j=0}^{n-1} 2^{j-1} (f(x) - f(x \dot{+} 2^{-j-1})) \qquad (x \in [0,1)),$$

where $\dot{+}$ denotes the dyadic addition.

Then f is said to be *dyadically differentiable* at $x \in [0,1)$ if $(\mathbf{d}_n f)(x)$ converges as $n \to \infty$. It was verified by Butzer and Wagner [3] that every Walsh function is differentiable and

$$\lim_{n\to\infty} (\mathbf{d}_n w_k)(x) = k w_k(x) \qquad (x \in [0,1), k \in \mathbb{N}).$$

To define the dyadic integral set

$$W := 1 + \sum_{n=1}^{\infty} \frac{w_n}{n}.$$

The *dyadic integral* of $f \in L_1[0,1)$ is introduced by

$$\mathbf{I}f(x) := f * W(x) = \int_0^1 f(t) W(x \dot{+} t)\, dt.$$

Notice that $W \in L_2[0,1) \subset L_1[0,1)$, so $\mathbf{I}$ is well defined on $L_1[0,1)$.

Let the *maximal operator* be defined by

$$\mathbf{I}_* f := \sup_{n\in\mathbb{N}} |\mathbf{d}_n(\mathbf{I}f)|.$$

The boundedness of $\mathbf{I}_*$ from $L_p[0,1)$ to $L_p[0,1)$ $(1 < p \leq \infty)$ is due to Schipp [13]:

Theorem 8.3 *If* $1 < p \leq \infty$ *then*

$$\|\mathbf{I}_* f\|_p \leq C_p \|f\|_p \qquad (f \in L_p[0,1)).$$

Schipp and Simon [14] verified that $\mathbf{I}_*$ is bounded from $L\log L[0,1)$ to $L_1[0,1)$. Recall that $L\log L[0,1) \subset H_1[0,1)$. These results can be extended to dyadic Hardy spaces as follows (see Weisz [22]).

Theorem 8.4 *If* $f \in H_p[0,1) \cap L_1[0,1)$, *then*

$$\|\mathbf{I}_* f\|_p \leq C_p \|f\|_{H_p}$$

for all $1/2 < p < \infty$.

For the proof let us introduce the function

$$W_K := \sum_{n=2^K}^{\infty} \frac{w_n}{n}.$$

The main point of the proof ([22]) is the following very sharp estimation:

$$|\mathbf{d}_n W_K(x)| \leq C \sum_{i=1}^{5} F_{K,n}^i(x),$$

where

$$F^1_{K,n}(x) := \frac{1}{2^{K-n} \vee 1} \sum_{j=0}^{n-1} \sum_{i=j+1}^{n-1} (n-i) 2^{j-n} D_{2^i}(x \dot{+} 2^{-j-1}),$$

$$F^2_{K,n}(x) := \frac{1}{2^{K-n} \vee 1} \sum_{i=0}^{n-1} (n-i) 2^{i-n} D_{2^i}(x),$$

$$F^3_{K,n}(x) := \sum_{j=0}^{n-1} 2^j \sum_{i=n}^{\infty} 2^{-i} \frac{D_{2^i}(x \dot{+} 2^{-j-1})}{2^{K-i} \vee 1},$$

$$F^4_{K,n}(x) := \sum_{k=0}^{\infty} 2^{-k} \frac{D_{2^{n+k}}(x)}{2^{K-n-k} \vee 1}$$

and

$$F^5_{K,n}(x) := D_{2^K}(x) 1_{\{n>K\}}.$$

Using this estimation, we ([22]) can prove that (8.2) holds and that $\|\mathbf{d}_n W\|_1 \le C$ for all $n \in \mathbb{N}$. Since

$$\mathbf{d}_n(\mathbf{I}f)(x) = \mathbf{d}_n \left(\int_0^1 f(t) W(x \dot{+} t)\, dt \right) = \int_0^1 f(t) \mathbf{d}_n W(x \dot{+} t)\, dt,$$

we obtain

$$\|\mathbf{d}_n(\mathbf{I}f)\|_1 \le \|f\|_1 \, \|\mathbf{d}_n W\|_1$$

and

$$\|\mathbf{d}_n(\mathbf{I}f)\|_\infty \le \|f\|_\infty \, \|\mathbf{d}_n W\|_1 .$$

Thus Theorem 8.2 can be applied, which finishes the proof.

Theorem 8.2 can be extended to $p = 1/2$: Goginava [9] verified, that $\mathbf{I}_*$ is bounded from $H_{1/2}[0,1)^d$ to the weak $L_{1/2}[0,1)$ space. However, $\mathbf{I}_*$ is not bounded from $H_p[0,1)^d$ to $H_p[0,1)^d$ if $0 < p \le 1$ (see Goginava [8] and Zhang [26]).

We get from Theorem 8.2 by interpolation (see [22])

Corollary 8.1 *If $f \in L_1[0,1)$, then*

$$\sup_{\rho>0} \rho\, \lambda(\mathbf{I}_* f > \rho) \le C \|f\|_1.$$

The dyadic analogue of the Lebesgue's differentiation theorem follows easily from the preceding weak type inequality:

Corollary 8.2 *If $f \in L_1[0,1)$ and*

$$\int_0^1 f(x)\, dx = 0, \tag{8.5}$$

then

$$\lim_{n\to\infty} \mathbf{d}_n(\mathbf{I}f) = f \quad a.e.$$

Since the dyadic derivative has 0 integral, it is reasonable to assume (8.5) in the preceding corollary. Schipp [13] proved Corollaries 8.1 and 8.2 with another method (for the Vilenkin system see Pál, Simon and Weisz [11, 12, 18]).

To formulate the multi-dimensional versions of these results, we have to introduce three types of dyadic Hardy spaces for higher dimensions.

8.4 d-dimensional Hardy Spaces

For a set $\mathbb{X} \neq \emptyset$ let $\mathbb{X}^d$ be its Cartesian product $\mathbb{X} \times \ldots \times \mathbb{X}$ taken with itself d-times. We denote the space $L_p([0,1)^d, \lambda)$ by $L_p[0,1)^d$ $(d \geq 1)$.

By a *dyadic rectangle* we mean a Cartesian product of d dyadic intervals. For $n \in \mathbb{N}^d$ and $x \in [0,1)^d$, let $I_n(x) := I_{n_1}(x_1) \times \ldots \times I_{n_d}(x_d)$, where $n = (n_1, \ldots, n_d)$ and $x = (x_1, \ldots, x_d)$. The σ-algebra generated by the dyadic rectangles $\{I_n(x) : x \in [0,1)^d\}$ will be denoted again by $\mathscr{F}_n$ $(n \in \mathbb{N}^d)$.

For d-parameter martingales $f = (f_n, n \in \mathbb{N}^d)$ with respect to $(\mathscr{F}_n, n \in \mathbb{N}^d)$, we introduce three kinds of maximal functions and Hardy spaces. The *maximal functions* are defined by

$$f^{\square} := \sup_{n \in \mathbb{N}} |f_{\mathbf{n}}|, \qquad f^* := \sup_{n \in \mathbb{N}^d} |f_n|,$$

where $\mathbf{n} := (n, \ldots, n) \in \mathbb{N}^d$ for $n \in \mathbb{N}$. In the first maximal function we have taken the supremum over the diagonal, in the second one over $\mathbb{N}^d$. Let E_n denote the conditional expectation operator with respect to $\mathscr{F}_n$. Obviously, if $f \in L_1[0,1)^d$, then $(E_n f, n \in \mathbb{N}^d)$ is a martingale. In the third maximal function the supremum is taken over $d-1$ indices: for fixed x_i we define

$$f^i(x) := \sup_{n_k \in \mathbb{N}, k=1,\ldots,d; k \neq i} |E_{n_1} \ldots E_{n_{i-1}} E_{n_{i+1}} \ldots E_{n_d} f(x)|.$$

For $0 < p \leq \infty$ the *martingale Hardy spaces* $H_p^{\square}[0,1)^d$, $H_p[0,1)^d$ and $H_p^i[0,1)^d$ consists of all d-parameter martingales for which

$$\|f\|_{H_p^{\square}} := \|f^{\square}\|_p < \infty, \qquad \|f\|_{H_p} := \|f^*\|_p < \infty, \qquad \|f\|_{H_p^i} := \|f^i\|_p < \infty,$$

respectively. One can show (see Weisz [19]) that $L(\log L)^{d-1}[0,1)^d \subset H_1^i[0,1)^d \subset H_{1,\infty}[0,1)^d$ $(i = 1, \ldots, d)$, more exactly,

$$\|f\|_{H_{1,\infty}} := \sup_{\rho > 0} \rho \lambda(f^* > \rho) \leq C \|f\|_{H_1^i} \qquad (f \in H_1^i[0,1)^d)$$

and

$$\|f\|_{H_1^i} \leq C + C \| |f| (\log^+ |f|)^{d-1} \|_1 \qquad (f \in L(\log L)^{d-1}[0,1)^d)$$

where $\log^+ u = 1_{\{u>1\}} \log u$. Moreover, it is known that

$$H_p^{\square}[0,1)^d \sim H_p[0,1)^d \sim H_p^i[0,1)^d \sim L_p[0,1)^d \qquad (1 < p \le \infty).$$

8.4.1 The Hardy Spaces $H_p^{\square}[0,1)^d$

To obtain some convergence results of the multi-dimensional dyadic derivative over the diagonal or over a band around the diagonal, we consider the Hardy space $H_p^{\square}[0,1)^d$. Now the situation is similar to the one-dimensional case.

A function $a \in L_\infty[0,1)^d$ is a *cube p-atom* if

(a) $\operatorname{supp} a \subset I$, $I \subset [0,1)^d$ is a dyadic cube,
(b) $\|a\|_\infty \le |I|^{-1/p}$,
(c) $\int_I a(x)\,dx = 0$.

The atomic decomposition theorem reads as follows (see Weisz [19, 24]).

Theorem 8.5 *A d-parameter martingale f is in $H_p^{\square}[0,1)^d$ $(0 < p \le 1)$ if and only if there exist a sequence $(a^k, k \in \mathbb{N})$ of cube p-atoms and a sequence $(\mu_k, k \in \mathbb{N})$ of real numbers such that*

$$\begin{aligned} &\sum_{k=0}^{\infty} \mu_k a^k = f \quad \textit{in the sense of martingales,} \\ &\sum_{k=0}^{\infty} |\mu_k|^p < \infty. \end{aligned} \tag{8.6}$$

Moreover,

$$\|f\|_{H_p^{\square}} \sim \inf\Big(\sum_{k=0}^{\infty} |\mu_k|^p\Big)^{1/p},$$

where the infimum is taken over all decompositions of f of the form (8.6).

For a rectangle $R = I_1 \times \ldots \times I_d \subset \mathbb{R}^d$ let $R^r := I_1^r \times \ldots \times I_d^r$ $(r \in \mathbb{N})$. The following result generalizes Theorem 8.2.

Theorem 8.6 *For each $n \in \mathbb{N}^d$ let $V_n : L_1[0,1)^d \to L_1[0,1)^d$ be a bounded linear operator and let*

$$V_* f := \sup_{n \in \mathbb{N}^d} |V_n f|.$$

Suppose that

$$\int_{[0,1)\setminus I^r} |V_* a|^{p_0}\, d\lambda \le C_{p_0}$$

for all cube p_0-atoms a and for some fixed $r \in \mathbb{N}$ and $0 < p_0 \le 1$, where the cube I is the support of the atom. If V_* is bounded from $L_{p_1}[0,1)^d$ to $L_{p_1}[0,1)^d$ for some $1 < p_1 \le \infty$, then

$$\|V_* f\|_p \le C_p \|f\|_{H_p^{\square}} \qquad (f \in H_p^{\square}[0,1)^d \cap L_1[0,1)^d)$$

for all $p_0 \leq p \leq p_1$. *Moreover, if* $p_0 < 1$, *then the operator* V_* *is of weak type* $(1,1)$, *i.e. if* $f \in L_1[0,1)^d$ *then*

$$\sup_{\rho>0} \rho \lambda(|V_* f| > \rho) \leq C\|f\|_1.$$

8.4.2 The Hardy Spaces $H_p[0,1)^d$

In the investigation of the convergence in the Prighheim's sense (i.e. over all $n \in \mathbb{N}^d$), we use the Hardy spaces $H_p[0,1)^d$. The atomic decomposition for $H_p[0,1)^d$ is much more complicated. One reason of this is that the support of an atom is not a rectangle but an open set. Moreover, here we have to choose the atoms from $L_2[0,1)^d$ instead of $L_\infty[0,1)^d$. This atomic decomposition was proved by the author (see [24] and the references therein). For an open set $F \subset [0,1)^d$, denote by $\mathscr{M}(F)$ the maximal dyadic subrectangles of F.

A function $a \in L_2[0,1)^d$ is a *p-atom* if

(a) supp $a \subset F$ for an open set $F \subset [0,1)^d$,
(b) $\|a\|_2 \leq |F|^{1/2-1/p}$,
(c) a can be further decomposed into the sum of "elementary particles" $a_R \in L_2$, $a = \sum_{R \in \mathscr{M}(F)} a_R$ in L_2, satisfying
  (d) supp $a_R \subset R \subset F$,
  (e) for all $i = 1, \ldots, d$ and $R \in \mathscr{M}(F)$ we have

$$\int_{[0,1)} a_R(x)\, dx_i = 0,$$

  (f) for every disjoint partition $\mathscr{P}_l$ $(l = 1, 2, \ldots)$ of $\mathscr{M}(F)$,

$$\Big(\sum_l \|\sum_{R \in \mathscr{P}_l} a_R\|_2^2\Big)^{1/2} \leq |F|^{1/2-1/p}.$$

Theorem 8.7 *A d-parameter martingale* f *is in* $H_p[0,1)^d$ $(0 < p \leq 1)$ *if and only if there exist a sequence* $(a^k, k \in \mathbb{N})$ *of p-atoms and a sequence* $(\mu_k, k \in \mathbb{N})$ *of real numbers such that*

$$\begin{aligned} &\sum_{k=0}^{\infty} \mu_k a^k = f \quad \textit{in the sense of martingales,} \\ &\sum_{k=0}^{\infty} |\mu_k|^p < \infty. \end{aligned} \tag{8.7}$$

Moreover,

$$\|f\|_{H_p} \sim \inf \Big(\sum_{k=0}^{\infty} |\mu_k|^p\Big)^{1/p},$$

where the infimum is taken over all decompositions of f of the form (8.7).

The corresponding results to Theorems 8.2 and 8.6 for the $H_p[0,1)^d$ spaces are much more complicated. Since the definition of the p-atom is very complex, to obtain a usable condition about the boundedness of the operators, we have to introduce simpler atoms. A function $a \in L_2[0,1)^d$ is called a *simple p-atom*, if there exist dyadic intervals $I_i \subset [0,1)$, $i = 1,\dots,j$ for some $1 \le j \le d-1$ such that

(a) supp $a \subset I_1 \times \dots I_j \times A$ for some measurable set $A \subset [0,1)^{d-j}$,
(b) $\|a\|_2 \le (|I_1| \cdots |I_j||A|)^{1/2-1/p}$,
(c) $\int_{I_i} a(x) x_i \, dx_i = \int_A a \, d\lambda = 0$ for $i = 1,\dots,j$.

Of course if $a \in L_2[0,1)^d$ satisfies these conditions for another subset of $\{1,\dots,d\}$ than $\{1,\dots,j\}$, then it is also called simple p-atom.

Note that $H_p[0,1)^d$ cannot be decomposed into simple p-atoms, a counterexample can be found in Weisz [19]. However, the following result, which is due to the author [24], says that for an operator V to be bounded from $H_p[0,1)^d$ to $L_p[0,1)^d$ $(0 < p \le 1)$ it is enough to check V on simple p-atoms and the boundedness of V on $L_2[0,1)^d$. Let H^c denote the complement of the set H.

Theorem 8.8 *Let $0 < p_0 \le 1$ and for each $n \in \mathbb{N}^d$ let $V_n : L_1[0,1)^d \to L_1[0,1)^d$ be a bounded linear operator and*

$$V_* f := \sup_{n \in \mathbb{N}^d} |V_n f|.$$

Suppose that there exist $\eta_1,\dots,\eta_d > 0$ such that for every simple p-atom a and for every $r_1 \dots, r_d \ge 1$

$$\int_{(I_1^{r_1})^c \times \dots \times (I_j^{r_j})^c} \int_A |V_* a|^{p_0} \, d\lambda \le C_{p_0} 2^{-\eta_1 r_1} \cdots 2^{-\eta_j r_j},$$

where $I_1 \times \dots \times I_j \times A$ is the support of a. If $j = d-1$ and $A = I_d$ is an interval, then we also assume that

$$\int_{(I_1^{r_1})^c \times \dots \times (I_{d-1}^{r_{d-1}})^c} \int_{(I_d)^c} |V_* a|^{p_0} \, d\lambda \le C_{p_0} 2^{-\eta_1 r_1} \cdots 2^{-\eta_{d-1} r_{d-1}}.$$

If V_* is bounded from $L_2[0,1)^d$ to $L_2[0,1)^d$, then

$$\|V_* f\|_p \le C_p \|f\|_{H_p} \qquad (f \in H_p[0,1)^d \cap H_1^i[0,1)^d)$$

for all $p_0 \le p \le 2$ and $i = 1,\dots,d$. In particular, if $p_0 < 1$ and $f \in H_1^i[0,1)^d$ for some $i = 1,\dots,d$, then

$$\sup_{\rho > 0} \rho \lambda(|V_* f| > \rho) \le C \|f\|_{H_1^i}.$$

In some sense in the multi-dimensional case the space $H_1^i[0,1)^d$ plays the role of the one-dimensional $L_1[0,1)$ space.

8.5 More-dimensional dyadic derivative

The multi-dimensional version of Lebesgue's differentiation theorem reads as follows:

$$f(x) = \lim_{h\to 0} \frac{1}{\prod_{j=1}^d h_j} \int_{x_1}^{x_1+h_1} \cdots \int_{x_d}^{x_d+h_d} f(t)\,dt \qquad \text{a.e.}$$

if $f \in L(\log L)^{d-1}[0,1)^d$. If $\tau^{-1} \le |h_i/h_j| \le \tau$ for some fixed $\tau \ge 0$ and all $i,j = 1,\ldots,d$, then it holds for all $f \in L_1[0,1)^d$ (see Zygmund [27]).

To present the dyadic version of this result we introduce first the *multi-dimensional dyadic derivative* ([1]) by the limit of $\mathbf{d}_n f$, where

$$(\mathbf{d}_n f)(x) := \sum_{j_1=0}^{n_1-1}\sum_{j_2=0}^{n_2-1} 2^{j_1+j_2-2}\Big(f(x_1,x_2) - f(x_1,x_2\dot{+}2^{-j_2-1})$$
$$- f(x_1\dot{+}2^{-j_1-1},x_2) + f(x_1\dot{+}2^{-j_1-1},x_2\dot{+}2^{-j_2-1})\Big)$$

if $d = 2$ and

$$(\mathbf{d}_n f)(x) := \sum_{j_1=0}^{n_1-1} \cdots \sum_{j_d=0}^{n_d-1} 2^{j_1+\ldots+j_d-d} \sum_{\varepsilon_i=0}^{1} (-1)^{\varepsilon_1+\ldots+\varepsilon_d} f(x_1\dot{+}\varepsilon_1 2^{-j_1-1},\ldots,x_d\dot{+}\varepsilon_d 2^{-j_d-1})$$

if $d > 3$, $n = (n_1,\ldots,n_d) \in \mathbb{N}^d$, $x = (x_1,\ldots,x_d) \in [0,1)^d$.

The Kronecker product $(w_n, n \in \mathbb{N}^d)$ of d Walsh systems is said to be a *d-dimensional Walsh system*. Thus

$$w_n(x) := w_{n_1}(x_1)\cdots w_{n_d}(x_d).$$

Every d-dimensional Walsh function is dyadically differentiable and

$$\lim_{n\to\infty} (\mathbf{d}_n w_k)(x) = \left(\prod_{j=1}^d k_j\right) w_k(x) \qquad (x \in [0,1)^d, k \in \mathbb{N}^d).$$

The *d-dimensional dyadic integral* is defined by

$$\mathbf{I}f(x) := f * (W \times \ldots \times W)(x) = \int_0^1 \cdots \int_0^1 f(t) W(x_1\dot{+}t_1)\cdots W(x_d\dot{+}t_d)\,dt.$$

8.5.1 Restricted dyadic derivative

In this section we investigate the dyadic derivative $\mathbf{d}_n(\mathbf{I}f)$ if n is in a band around the diagonal, more exactly, if for a given $\tau \geq 0$, n is in the region

$$\left\{n \in \mathbb{N}^d : |n_i - n_j| \leq \tau, i,j = 1,\ldots,d\right\}.$$

Define the *restricted maximal operator* by

$$\mathbf{I}_\square f := \sup_{|n_i - n_j| \leq \tau, i,j=1,\ldots,d} |\mathbf{d}_n(\mathbf{I}f)|.$$

This operator is bounded from $L_p[0,1)^d$ to $L_p[0,1)^d$ if $1 < p \leq \infty$ and from $H_p^\square[0,1)^d$ to $L_p[0,1)^d$ if $d/(d+1) < p \leq \infty$ (see [20]). More exactly, applying Theorem 8.6 we can prove

Theorem 8.9 *Suppose that* $f \in H_p^\square[0,1)^d \cap L_1[0,1)^d$ *and*

$$\int_0^1 f(x)\,dx_i = 0 \quad (i = 1,\ldots,d). \tag{8.8}$$

Then

$$\|\mathbf{I}_\square f\|_p \leq C_p \|f\|_{H_p^\square}$$

for all $d/(d+1) < p \leq \infty$.

The next two corollaries follow again from interpolation and from a density argument.

Corollary 8.3 *If* $f \in L_1[0,1)^d$ *satisfies (8.8), then*

$$\sup_{\rho > 0} \rho\, \lambda(\mathbf{I}_\square f > \rho) \leq C \|f\|_1.$$

Corollary 8.4 *If* $\tau \geq 0$ *is arbitrary and* $f \in L_1[0,1)^d$ *satisfies (8.8), then*

$$\lim_{n\to\infty, |n_i - n_j| \leq \tau} \mathbf{d}_n(\mathbf{I}f) = f \quad a.e.$$

Theorem 8.9 and Corollaries 8.3 and 8.4 are due to the author [20, 24, 17]. The two corollaries were also shown by Gát [5] and Gát and Nagy [6].

8.5.2 Unrestricted dyadic derivative

Here we consider the dyadic derivative $\mathbf{d}_n(\mathbf{I}f)$ if $n \in \mathbb{N}^d$. Now we introduce the *unrestricted maximal operator* by

$$\mathbf{I}_* f := \sup_{n \in \mathbb{N}^d} |\mathbf{d}_n(\mathbf{I} f)|.$$

With the help of Theorem 8.8 we proved in Weisz [21, 23, 24], that $\mathbf{I}_*$ is bounded from $L_p[0,1)^d$ to $L_p[0,1)^d$ if $1 < p \leq \infty$ and from $H_p[0,1)^d$ to $L_p[0,1)^d$ if $1/2 < p \leq \infty$.

Theorem 8.10 *If (8.8) is satisfied,* $1/2 < p \leq \infty$ *and* $f \in H_p[0,1)^d \cap H_1^i[0,1)^d$, *then*

$$\|\mathbf{I}_* f\|_p \leq C_p \|f\|_{H_p} \qquad (f \in H_p[0,1)^d).$$

Corollary 8.5 *If* $f \in H_1^i[0,1)^d$ $(i = 1, \ldots, d)$ *satisfies (8.8), then*

$$\sup_{\rho > 0} \rho \, \lambda(\mathbf{I}_* f > \rho) \leq C \|f\|_{H_1^i}.$$

Corollary 8.6 *If* $f \in H_1^i[0,1)^d (\supset L(\log L)^{d-1}[0,1)^d)$ $(i = 1, \ldots, d)$ *satisfies (8.8), then*

$$\lim_{n \to \infty} \mathbf{d}_n(\mathbf{I} f) = f \quad a.e.$$

Note that this result for $f \in L \log L[0,1)^2$ is due to Schipp and Wade [15] in the two-dimensional case.

References

1. P. L. Butzer and W. Engels. Dyadic calculus and sampling theorems for functions with multi-dimensional domain. *Information and Control.*, 52:333–351, 1982.
2. *P. L. Butzer and H. J. Wagner. Walsh-Fiurier series and the concept of a derivative. *Applic. Anal.*, 3:29–46, 1973.
3. *P. L. Butzer and H. J. Wagner. On dyadic analysis based on the pointwise dyadic derivative. *Anal. Math.*, 1:171–196, 1975.
4. R. R. Coifman and G. Weiss. Extensions of Hardy spaces and their use in analysis. *Bull. Amer. Math. Soc.*, 83:569–645, 1977.
5. G. Gát. On the two-dimensional pointwise dyadic calculus. *J. Appr. Theory*, 92:191–215, 1998.
6. G. Gát and K. Nagy. The fundamental theorem of two-parameter pointwise derivate on Vilenkin groups. *Anal. Math.*, 25:33–55, 1999.
7. J. E. Gibbs. Some properties of functions of the non-negative integers less than 2^n. Technical Report DES 3, NPL National Physical Laboratory Middlesex, England, 1969.
8. U. Goginava. The Hardy type inequality for the maximal operator of the one-dimensional dyadic derivative. *Acta Math. Sci., Ser. B, Engl. Ed.*, 31(4):1489–1493, 2011.
9. U. Goginava. Weak type inequality for the one-dimensional dyadic derivative. *Math. Inequal. Appl.*, 14(4):839–848, 2011.
10. C. Herz. Bounded mean oscillation and regulated martingales. *Trans. Amer. Math. Soc.*, 193:199–215, 1974.
11. J. Pál and P. Simon. On a generalization of the concept of derivative. *Acta Math. Hungar.*, 29:155–164, 1977.
12. J. Pál and P. Simon. On the generalized Butzer-Wagner type a.e. differentiability of integral function. *Annales Univ. Sci. Budapest, Sectio Math.*, 20:157–165, 1977.

13. F. Schipp. Über einen Ableitungsbegriff von P.L. Butzer and H.J. Wagner. *Mat. Balkanica.*, 4:541–546, 1974.
14. F. Schipp and P. Simon. On some (H, L_1)-type maximal inequalities with respect to the Walsh-Paley system. In *Functions, Series, Operators, Proc. Conf. in Budapest, 1980*, volume 35 of *Coll. Math. Soc. J. Bolyai*, pages 1039–1045. North Holland, Amsterdam, 1981.
15. F. Schipp and W. R. Wade. A fundamental theorem of dyadic calculus for the unit square. *Applic. Anal.*, 34:203–218, 1989.
16. F. Schipp, W. R. Wade, P. Simon, and J. Pál. *Walsh Series: An Introduction to Dyadic Harmonic Analysis*. Adam Hilger, Bristol, New York, 1990.
17. P. Simon and F. Weisz. On the two-parameter Vilenkin derivative. *Math. Pannonica*, 12:105–128, 2000.
18. P. Simon and F. Weisz. Hardy spaces and the generalization of the dyadic derivative. In L. Leindler, F. Schipp, and J. Szabados, editors, *Functions, Series, Operators, Alexits Memorial Conference, Budapest, (Hungary, 1999)*, pages 367–388, 2002.
19. F. Weisz. *Martingale Hardy Spaces and their Applications in Fourier Analysis*, volume 1568 of *Lecture Notes in Math.* Springer, Berlin, 1994.
20. F. Weisz. (H_p, L_p)-type inequalities for the two-dimensional dyadic derivative. *Studia Math.*, 120:271–288, 1996.
21. F. Weisz. Some maximal inequalities with respect to two-parameter dyadic derivative and Cesàro summability. *Applic. Anal.*, 62:223–238, 1996.
22. F. Weisz. Martingale Hardy spaces and the dyadic derivative. *Anal. Math.*, 24:59–77, 1998.
23. F. Weisz. The two-parameter dyadic derivative and the dyadic Hardy spaces. *Anal. Math.*, 26:143–160, 2000.
24. F. Weisz. *Summability of Multi-dimensional Fourier Series and Hardy Spaces*. Mathematics and Its Applications. Kluwer Academic Publishers, Dordrecht, Boston, London, 2002.
25. F. Weisz. Boundedness of operators on Hardy spaces. *Acta Sci. Math. (Szeged)*, 78:541–557, 2012.
26. X. Zhang, X. Yu, and C. Zhang. The boundedness of two-dimensional maximal operator of dyadic derivative. *J. Math., Wuhan Univ.*, 31(3):395–400, 2011.
27. A. Zygmund. *Trigonometric Series*. Cambridge Press, London, 3rd edition, 2002.

STUDIA MATHEMATICA 120 (3) (1996)

(H_p, L_p)-type inequalities for the two-dimensional dyadic derivative

by

FERENC WEISZ (Budapest)

Abstract. It is shown that the restricted maximal operator of the two-dimensional dyadic derivative of the dyadic integral is bounded from the two-dimensional dyadic Hardy–Lorentz space $H_{p,q}$ to $L_{p,q}$ ($2/3 < p < \infty$, $0 < q \leq \infty$) and is of weak type (L_1, L_1). As a consequence we show that the dyadic integral of a two-dimensional function $f \in L_1$ is dyadically differentiable and its derivative is f a.e.

1. Introduction. It is known that

$$f(x) = \lim_{h\to 0} \frac{1}{h} \int_x^{x+h} f(s)\,ds \quad \text{a.e.}$$

if $f \in L_1[0,1)$. The dyadic analogue of this result can be formulated as follows. Butzer and Wagner [5] introduced the dyadic derivative to be the limit of

$$(\mathrm{d}_n f)(x) := \sum_{j=0}^{n-1} 2^{j-1}(f(x) - f(x \dotplus 2^{-j-1})) \qquad (x \in [0,1))$$

as $n \to \infty$ where $\dotplus$ denotes the dyadic addition (see e.g. Schipp, Wade, Simon and Pál [13]). The dyadic integral $\mathbf{I}f$ is defined by the convolution of f and the function W whose kth Walsh–Fourier coefficient is $1/k$ ($k \neq 0$). The boundedness of $\mathbf{I}^* f = \sup_{n\in\mathbb{N}} |\mathrm{d}_n(\mathbf{I}f)|$ from $L_p[0,1)$ to $L_p[0,1)$ ($1 < p \leq \infty$) and the weak type $(L_1[0,1), L_1[0,1))$ inequality

(1) $$\sup_{\gamma>0} \gamma\lambda(\sup_{n\in\mathbb{N}} \mathbf{I}^* f > \gamma) \leq C\|f\|_1 \qquad (f \in L_1[0,1))$$

1991 *Mathematics Subject Classification*: Primary 42C10, 43A75; Secondary 60G42, 42B30.

Key words and phrases: martingale Hardy spaces, p-atom, interpolation, Walsh functions, dyadic derivative.

This research was partly supported by the Hungarian Scientific Research Funds (OTKA) No F019633.

are due to Schipp [9]. The dyadic analogue of the differentiation theorem follows easily from the last weak type inequality:

$$\lim_{n\to\infty} \mathbf{d}_n(\mathbf{I}f) = f \quad \text{a.e.}$$

if $f \in L_1[0,1)$ is of mean zero (see Schipp [9]).

The weak type inequality was extended by the author [15]. We proved that

$$\|\mathbf{I}^* f\|_{p,q} \le C\|f\|_{H_{p,q}} \quad (1/2 < p < \infty,\ 0 < q \le \infty) \tag{2}$$

where $H_{p,q}$ denotes the one-dimensional dyadic Hardy–Lorentz space. As a special case we obtain (1) from this by choosing $p = 1$ and $q = \infty$.

The two-dimensional differentiation theorem

$$f(x,y) = \lim_{h,k\to 0} \frac{1}{hk} \int_x^{x+h} \int_y^{y+k} f(s,t)\, ds\, dt \quad \text{a.e.} \tag{3}$$

if $f \in L \log L[0,1)^2$ can be found in Zygmund [18]. The dyadic analogue of this result is

$$\lim_{n,m\to\infty} \mathbf{d}_{n,m}(\mathbf{I}f) = f \quad \text{a.e.} \quad (f \in L\log L[0,1)^2)$$

where $\mathbf{I}f$ now denotes the convolution of f and $W \times W$ and, moreover,

$$(\mathbf{d}_{n,m}f)(x,y) := \sum_{i=0}^{n-1}\sum_{j=0}^{m-1} 2^{i+j-2}(f(x,y) - f(x, y \dotplus 2^{-j-1}) \tag{4}$$
$$- f(x \dotplus 2^{-i-1}, y) + f(x \dotplus 2^{-i-1}, y \dotplus 2^{-j-1}))$$

(see Schipp and Wade [12] and also Weisz [17]). Recently the author [15] generalized this convergence result for $f \in H_1^\sharp \supset L\log L[0,1)^2$ where $H_1^\sharp$ is the two-dimensional dyadic hybrid Hardy space.

In this paper the Hardy–Lorentz spaces $H_{p,q}$ of dyadic martingales on the unit square are introduced with the $L_{p,q}$ Lorentz norms of the maximal function $\sup_{n\in\mathbb{N}} |f_{n,n}|$. Of course, $H_p = H_{p,p}$ are the usual Hardy spaces $(0 < p \le \infty)$.

We verify here the same results for the two-dimensional dyadic derivative as we proved in [14] for Cesàro means of two-dimensional Walsh–Fourier series. We denote the restricted maximal operator $\sup_{|n-m|\le\alpha} |\mathbf{d}_{n,m}(\mathbf{I}f)|$ for any $\alpha \ge 0$ by $\mathbf{I}_\alpha^* f$ and prove inequality (2) for this operator $(2/3 < p < \infty,\ 0 < q \le \infty)$. The two-dimensional version of (1) follows from this with $p = 1$ and $q = \infty$. Note that the unrestricted maximal operator is investigated in Weisz [17].

It is known that if $\alpha^{-1} \le |h/k| \le \alpha$ for any $\alpha > 0$ then (3) holds for all $f \in L_1[0,1)^2$. The dyadic analogue of this follows from the two-dimensional version of (1):

$$\lim_{\substack{n,m\to\infty \\ |n-m|\le\alpha}} \mathbf{d}_{n,m}(\mathbf{I}f) = f \quad \text{a.e.} \quad (f \in L[0,1)^2).$$

This convergence is also proved by Gát [7] with another method.

2. Martingales and Hardy–Lorentz spaces. In this paper the unit square $[0,1)^2$ and the Lebesgue measure λ are considered. By a *dyadic interval* we mean one of the form $[k2^{-n},(k+1)2^{-n})$ for some $k,n \in \mathbb{N}$, $0 \le k < 2^n$. Given $n \in \mathbb{N}$ and $x \in [0,1)$ let $I_n(x)$ denote the dyadic interval of length 2^{-n} which contains x. If I_1 and I_2 are dyadic intervals and $\lambda(I_1) = \lambda(I_2)$ then the set

$$I := I_1 \times I_2$$

is a *dyadic square*. Clearly, the dyadic square of area 2^{-2n} containing $(x,y) \in [0,1)^2$ is given by

$$I_{n,n}(x,y) := I_n(x) \times I_n(y).$$

The σ-algebra generated by the dyadic squares $\{I_{n,n}(x) : x \in [0,1)^2\}$ will be denoted by $\mathcal{F}_{n,n}$ $(n \in \mathbb{N})$, more precisely,

$$\mathcal{F}_{n,n} = \sigma\{[k2^{-n},(k+1)2^{-n}) \times [l2^{-n},(l+1)2^{-n}) : 0 \le k < 2^n,\ 0 \le l < 2^n\}$$

where $\sigma(\mathcal{H})$ denotes the σ-algebra generated by an arbitrary set system $\mathcal{H}$. We will investigate martingales of the form $f = (f_{n,n}, n \in \mathbb{N})$ with respect to $(\mathcal{F}_{n,n}, n \in \mathbb{N})$. We briefly write L_p instead of the real $L_p([0,1)^2,\lambda)$ space while the norm (or quasinorm) of this space is defined by $\|f\|_p := (\int_{[0,1)^2} |f|^p \, d\lambda)^{1/p}$ $(0 < p \le \infty)$.

The distribution function of a Borel-measurable function f is defined by

$$\lambda(\{|f| > \gamma\}) := \lambda(\{x : |f(x)| > \gamma\}) \qquad (\gamma \ge 0).$$

The *weak* L_p space L_p^* $(0 < p < \infty)$ consists of all measurable functions f for which

$$\|f\|_{L_p^*} := \sup_{\gamma>0} \gamma[\lambda(\{|f| > \gamma\})]^{1/p} < \infty,$$

while we set $L_\infty^* = L_\infty$.

The spaces L_p^* are special cases of the more general Lorentz spaces $L_{p,q}$. In their definition another concept is used. For a measurable function f the *non-increasing rearrangement* is defined by

$$\tilde{f}(t) := \inf\{\gamma : \lambda(\{|f| > \gamma\}) \le t\}.$$

The Lorentz space $L_{p,q}$ is defined as follows: for $0 < p < \infty$ and $0 < q < \infty$,

$$\|f\|_{p,q} := \left(\int_0^\infty \tilde{f}(t)^q t^{q/p} \frac{dt}{t} \right)^{1/q},$$

while for $0 < p \le \infty$,

$$\|f\|_{p,\infty} := \sup_{t>0} t^{1/p}\tilde{f}(t).$$

Let

$$L_{p,q} := L_{p,q}([0,1)^2, \lambda) := \{f : \|f\|_{p,q} < \infty\}.$$

One can show the following equalities:

$$L_{p,p} = L_p, \quad L_{p,\infty} = L_p^* \quad (0 < p \le \infty)$$

(see e.g. Bennett and Sharpley [1] or Bergh and Löfström [2]).

The *maximal function* of a martingale $f = (f_{n,n}, n \in \mathbb{N})$ is defined by

$$f^* := \sup_{n\in\mathbb{N}} |f_{n,n}|.$$

It is easy to see that, in case $f \in L_1$, the maximal function can also be given by

$$f^*(x,y) = \sup_{n\in\mathbb{N}} \frac{1}{\lambda(I_{n,n}(x,y))}\left|\int_{I_{n,n}(x,y)} f\,d\lambda\right|.$$

For $0 < p,q \le \infty$ the *martingale Hardy-Lorentz space* $H_{p,q}$ consists of all martingales $f = (f_{n,n}, n \in \mathbb{N})$ for which

$$\|f\|_{H_{p,q}} := \|f^*\|_{p,q} < \infty.$$

Note that in case $p = q$ the usual definition of Hardy space $H_{p,p} = H_p$ is obtained.

It is well known that for a martingale $f = (f_{n,n}, n \in \mathbb{N})$,

$$\sup_{\gamma>0} \gamma\lambda(f^* > \gamma) \le \sup_{n\in\mathbb{N}} \|f_{n,n}\|_1 \tag{5}$$

and

$$\|f^*\|_p \le \frac{p}{p-1}\|f\|_p \quad (1 < p \le \infty),$$

hence $H_p \sim L_p$ whenever $1 < p \le \infty$ (see Neveu [8]), where $\sim$ denotes the equivalence of the norms and spaces. Moreover, it is proved in Weisz [16] that

$$H_{p,q} \sim L_{p,q} \quad (1 < p \le \infty,\ 0 < q \le \infty).$$

A bounded measurable function a is a *p-atom* if $a = 1$ or there exists a dyadic square Q such that

(i) $\int_Q a\,d\lambda = 0$,
(ii) $\|a\|_\infty \le \lambda(Q)^{-1/p}$,
(iii) $\{a \ne 0\} \subset Q$.

Using the atomic decomposition we verified the next theorem in [14].

THEOREM A. *Suppose that the operator T is sublinear and, for each $p_0 \le p \le 1$, there exists a constant $C_p > 0$ such that*

$$\int_{[0,1)^2\setminus Q} |Ta|^p \, d\lambda \le C_p \tag{6}$$

for every p-atom a where the support of a is contained in Q as in (i)–(iii). If T is bounded from L_∞ to L_∞ then for every $p_0 \le p \le 1$,

$$\|Tf\|_p \le C_p \|f\|_{H_p} \qquad (f \in H_p \cap L_1).$$

The following interpolation result concerning Hardy-Lorentz spaces will be used in this paper (see Weisz [16]).

THEOREM B. *If a sublinear operator T is bounded from H_{p_0} to L_{p_0} and from L_∞ to L_∞ then it is also bounded from $H_{p,q}$ to $L_{p,q}$ if $p_0 < p < \infty$ and $0 < q \le \infty$.*

3. The two-dimensional dyadic derivative. First we introduce the Walsh system. Every point $x \in [0,1)$ can be written in the following way:

$$x = \sum_{k=0}^{\infty} \frac{x_k}{2^{k+1}}, \qquad 0 \le x_k < 2, \; x_k \in \mathbb{N}.$$

In case there are two different forms, we choose the one for which $\lim_{k\to\infty} x_k = 0$.

The functions

$$r_n(x) := \exp(\pi x_n \sqrt{-1}) \qquad (n \in \mathbb{N})$$

are called *Rademacher functions*. The product system generated by these functions is the *one-dimensional Walsh system*:

$$w_n(x) := \prod_{k=0}^{\infty} r_k(x)^{n_k}$$

where $n = \sum_{k=0}^{\infty} n_k 2^k$, $0 \le n_k < 2$ and $n_k \in \mathbb{N}$.

The Kronecker product $(w_{n,m}; n, m \in \mathbb{N})$ of two Walsh systems is said to be the *two-dimensional Walsh system*. Thus

$$w_{n,m}(x,y) := w_n(x) w_m(y).$$

Recall that the *Walsh-Dirichlet kernels*

$$D_n := \sum_{k=0}^{n-1} w_k$$

satisfy

$$D_{2^n}(x) = \begin{cases} 2^n & \text{if } x \in [0, 2^{-n}), \\ 0 & \text{if } x \in [2^{-n}, 1), \end{cases} \tag{7}$$

for $n \in \mathbb{N}$ (see e.g. Schipp, Wade, Simon and Pál [13]).

For each function f defined on $[0,1)^2$ Butzer and Engels [3] introduced the concept of the two-dimensional dyadic derivative by (4). Then f is said to be *dyadically differentiable* at $x, y \in [0,1)$ if $(\mathbf{d}_{n,m}f)(x,y)$ converges as $n, m \to \infty$. It was verified by Butzer and Wagner [4] that every Walsh function is dyadically differentiable and

$$\lim_{\min(n,m)\to\infty} \mathbf{d}_{n,m}(w_k \times w_l)(x,y) = kl(w_k \times w_l)(x,y)$$

for all $x, y \in [0,1)$ and $k, l \in \mathbb{N}$. Let W be the function whose Walsh–Fourier coefficients satisfy

$$\widehat{W}(k) := \int_0^1 W w_k \, d\lambda := \begin{cases} 1 & \text{if } k = 0, \\ 1/k & \text{if } k \in \mathbb{N},\ k \neq 0. \end{cases}$$

The two-dimensional dyadic integral of $f \in L_1$ is introduced by

$$\mathbf{I}f(x,y) := f * (W \times W)(x,y) := \int_0^1\int_0^1 f(t,u) W(x \dot{+} t) W(y \dot{+} u) \, dt \, du.$$

Notice that $W \in L_2 \subset L_1$, so $\mathbf{I}$ is well defined on L_1.

Set

$$W_K := \sum_{n=2^K}^{\infty} \frac{w_n}{n}$$

and let us estimate $\mathbf{d}_n W$ and $\mathbf{d}_n W_K$. The following theorem can be proved with the help of the ideas in Schipp, Wade, Simon and Pál [13] (pp. 272–275) and in Weisz [15].

THEOREM 1. *For all $n, K \in \mathbb{N}$ we have*

$$|\mathbf{d}_n W(x) + 1| \leq C \sum_{i=1}^{4} F_{0,n}^i(x) \quad \textit{and} \quad |\mathbf{d}_n W_K(x)| \leq C \sum_{i=1}^{5} F_{K,n}^i(x)$$

where

$$F_{K,n}^1(x) := \frac{1}{2^{K-n} \vee 1} \sum_{j=0}^{n-1} \sum_{i=j+1}^{n-1} (n-i) 2^{j-n} D_{2^i}(x \dot{+} 2^{-j-1}),$$

$$F_{K,n}^2(x) := \frac{1}{2^{K-n} \vee 1} \sum_{i=0}^{n-1} (n-i) 2^{i-n} D_{2^i}(x),$$

$$F_{K,n}^3(x) := \sum_{j=0}^{n-1} 2^j \sum_{i=n}^{\infty} 2^{-i} \frac{D_{2^i}(x \dot{+} 2^{-j-1})}{2^{K-i} \vee 1},$$

$$F^4_{K,n}(x) := \sum_{k=0}^{\infty} 2^{-k} \frac{D_{2^{n+k}}(x)}{2^{K-n-k} \vee 1}$$

and

$$F^5_{K,n}(x) := D_{2^K}(x) 1_{\{n>K\}}.$$

4. Inequalities concerning the two-dimensional dyadic derivative. Before considering the operator

$$\mathbf{I}^*_\alpha f := \sup_{|n-m|\le\alpha} |\mathbf{d}_{n,m}(\mathbf{I}f)| \quad (f \in L_1)$$

for any $\alpha \geq 0$ let us modify slightly the dyadic derivative. Set

$$\delta_{n,m} f(x,y) := \int_0^1\int_0^1 f(t,u)[\mathbf{d}_n W(x \dot{+} t) + 1][\mathbf{d}_m W(y \dot{+} u) + 1]\, dt\, du$$

and

$$\mathbf{J}^*_\alpha f := \sup_{|n-m|\le\alpha} |\delta_{n,m} f| \quad (f \in L_1).$$

First we can prove that $\mathbf{J}^*_\alpha$ is bounded from H_p to L_p.

THEOREM 2. *There exist constants C_p depending only on p and α such that for each $2/3 < p \le 1$,*

$$\|\mathbf{J}^*_\alpha f\|_p \le C_p \|f\|_{H_p} \quad (f \in H_p)$$

where $\mathbf{J}^*_\alpha f$ will be defined for $f \in H_p \setminus L_1$ in the proof.

Proof. First assume that $f \in H_p \cap L_1$. By Theorem A the proof of Theorem 2 will be complete if we show that the operator $\mathbf{J}^*_\alpha$ satisfies (6) and is bounded from L_∞ to L_∞.

Since $\|D_{2^n}\|_1 = 1$, we can show that

$$\|F^i_{0,n}\|_1 \le C \quad (i = 1, \ldots, 4;\ n \in \mathbb{N}). \tag{8}$$

From this it follows that $\|\mathbf{d}_n W + 1\|_1 \le C$ for all $n \in \mathbb{N}$, which verifies that $\mathbf{J}^*_\alpha$ is bounded on L_∞.

If $a = 1$ then the left hand side of (6) is zero. Let $a \neq 1$ be an arbitrary p-atom with support $Q = I \times J$ and $\lambda(I) = \lambda(J) = 2^{-K}$ $(K \in \mathbb{N})$. Without loss of generality we can suppose that $I = J = [0, 2^{-K})$. If $k < 2^K$ and $l < 2^K$ then $w_{k,l}$ is constant on Q and so

$$\int_0^1\int_0^1 a(t,u) w_k(x \dot{+} t) w_l(y \dot{+} u)\, dt\, du = 0.$$

Since

$$\mathbf{d}_n(w_{i2^n+k}) = k w_{i2^n+k} \quad (0 \le k < 2^n;\ i, n \in \mathbb{N})$$

(see Schipp, Wade, Simon and Pál [13], p. 272) it is not hard to see that

$$\delta_{n,m}a(x,y) = \int_0^1\int_0^1 a(t,u)[\mathrm{d}_n W_K(x \dot+ t)(\mathrm{d}_m W(y \dot+ u) + 1)$$
$$+ (\mathrm{d}_n W(x \dot+ t) + 1)\mathrm{d}_m W_K(y \dot+ u)$$
$$- \mathrm{d}_n W_K(x \dot+ t)\mathrm{d}_m W_K(y \dot+ u)]\,dt\,du.$$

By the fact that $F^i_{K,n} \le F^i_{0,n}$ $(i = 1,\ldots,4; n, K \in \mathbb{N})$ and by Theorem 1 we obtain

$$\mathbf{J}^*_\alpha a \le \sup_{|n-m|\le\alpha} \sup_{\substack{i=1,\ldots,5\\ j=1,\ldots,4}} |a| * F^i_{K,n} \times F^j_{0,m}$$
$$+ \sup_{|n-m|\le\alpha} \sup_{\substack{i=1,\ldots,4\\ j=1,\ldots,5}} |a| * F^i_{0,n} \times F^j_{K,m}$$
$$+ \sup_{|n-m|\le\alpha} \sup_{\substack{i=1,\ldots,5\\ j=1,\ldots,5}} |a| * F^i_{K,n} \times F^j_{K,m}$$
$$\le 2 \sup_{|n-m|\le\alpha} \sup_{\substack{i=1,\ldots,5\\ j=1,\ldots,4}} |a| * F^i_{K,n} \times F^j_{0,m}$$
$$+ 2 \sup_{|n-m|\le\alpha} \sup_{\substack{i=1,\ldots,4\\ j=1,\ldots,5}} |a| * F^i_{0,n} \times F^j_{K,m} + \sup_{|n-m|\le\alpha} |a| * F^5_{K,n} \times F^5_{K,m}.$$

Now we investigate the first term, the integral of $[\sup_{|n-m|\le\alpha} |a| * F^i_{K,n} \times F^j_{0,m}]^p$ over $[0,1)^2 \setminus Q$ for all $i = 1,\ldots,5$ and $j = 1,\ldots,4$.

Step 1: *Integrating over* $([0,1) \setminus I) \times J$. We proved in [15] that for all $n, K \in \mathbb{N}$ and $i = 1,\ldots,5$,

$$\int_{I^c} \Big(\sup_{n\in\mathbb{N}} \int_I F^i_{K,n}(x \dot+ t)\,dt\Big)^p dx \le C_p 2^{-K} \tag{9}$$

where $I^c := [0,1) \setminus I$. Taking into account (8) and the definition of the p-atom, we can establish that, for all $i = 1,\ldots,5$ and $j = 1,\ldots,4$,

$$\int_{I^c}\int_J \Big(\sup_{n,m\in\mathbb{N}} \int_I\int_J |a(t,u)| F^i_{K,n}(x \dot+ t) F^j_{0,m}(y \dot+ u)\,dt\,du\Big)^p dx\,dy \tag{10}$$
$$\le C_p 2^{2K} \int_{I^c}\int_J \Big(\sup_{n\in\mathbb{N}} \int_I F^i_{K,n}(x \dot+ t)\,dt\Big)^p dx\,dy \le C_p.$$

Step 2: *Integrating over* $I \times ([0,1) \setminus J)$. If $j < K$ and $x \in I$ then $x \dot+ 2^{-j-1} \notin I$. Hence, it follows from (7) that

$$\int_I D_{2^i}(x \dot+ t \dot+ 2^{-j-1})\,dt = 0$$

whenever $x \in I$ and $i > j$. Using this and (7) we can calculate the integrals $\int_I F^i_{K,n}(x \dotplus t)\,dt$ if $x \in I$ and $i = 1, \ldots, 5$:

$$\int_I F^1_{K,n}(x \dotplus t)\,dt = \frac{1}{2^{K-n} \vee 1} \sum_{j=0}^{n-1} \sum_{i=j+1}^{n-1} (n-i) 2^{j-n} \int_I D_{2^i}(x \dotplus t \dotplus 2^{-j-1})\,dt$$

$$\leq \begin{cases} 0 & \text{if } n \leq K, \\ C & \text{if } n > K, \end{cases}$$

$$\int_I F^2_{K,n}(x \dotplus t)\,dt = \frac{1}{2^{K-n} \vee 1} \sum_{i=0}^{n-1} (n-i) 2^{i-n} \int_I D_{2^i}(x \dotplus t)\,dt$$

$$\leq \begin{cases} C 2^{n-K} & \text{if } n \leq K, \\ C & \text{if } n > K, \end{cases}$$

$$\int_I F^3_{K,n}(x \dotplus t)\,dt = \sum_{j=0}^{n-1} 2^j \sum_{i=n}^{\infty} 2^{-i} \frac{1}{2^{K-i} \vee 1} \int_I D_{2^i}(x \dotplus t \dotplus 2^{-j-1})\,dt$$

$$\leq \begin{cases} 0 & \text{if } n \leq K, \\ C & \text{if } n > K, \end{cases}$$

$$\int_I F^4_{K,n}(x \dotplus t)\,dt = \sum_{k=0}^{\infty} 2^{-k} \frac{1}{2^{K-n-k} \vee 1} \int_I D_{2^{n+k}}(x \dotplus t)\,dt$$

$$\leq \sum_{k=0}^{K-n-1} 2^{n-K} 2^{n+k-K} + \sum_{k=(K-n) \vee 0}^{\infty} 2^{-k}$$

$$\leq \begin{cases} C 2^{n-K} & \text{if } n \leq K, \\ C & \text{if } n > K, \end{cases}$$

$$\int_I F^5_{K,n}(x \dotplus t)\,dt = 1_{\{n>K\}} \int_I D_{2^K}(x \dotplus t)\,dt = \begin{cases} 0 & \text{if } n \leq K, \\ 1 & \text{if } n > K. \end{cases}$$

Let $r \in \mathbb{N}$ satisfy $r - 1 < \alpha \leq r$ and observe that

$$\begin{aligned} \sup_{|n-m| \leq \alpha} |a| * F^i_{K,n} \times F^j_{0,m} \leq & \sup_{\substack{|n-m| \leq r \\ n,m \leq K}} |a| * F^i_{K,n} \times F^j_{0,m} \\ & + \sup_{n,m \geq K-r} |a| * F^i_{K,n} \times F^j_{0,m} \\ & =: (A_{i,j}) + (B_{i,j}). \end{aligned}$$

for all $i = 1, \ldots, 5$ and $j = 1, \ldots, 4$. Of course, $(A_{i,j})(x, y) = 0$ if $i = 1, 3, 5$ and $x \in I$. So suppose that $i = 2, 4$ and $j = 1, \ldots, 4$. It is easy to see that

$$(11) \qquad 2^{m-K} F^j_{0,m} \leq F^j_{K,m} \quad (m \leq K;\ j = 1, \ldots, 4).$$

Consequently,

$$\begin{aligned}(A_{i,j})(x,y) &= \sup_{\substack{|n-m|\le r\\ n,m\le K}} \int_I\int_J |a(t,u)| F^i_{K,n}(x\dot+t)F^j_{0,m}(y\dot+u)\,dt\,du\\ &\le C_p 2^{2K/p} \sup_{\substack{|n-m|\le r\\ n,m\le K}} 2^{n-K}\int_J F^j_{0,m}(y\dot+u)\,du\\ &\le C_p 2^{2K/p}2^r \sup_{m\le K} 2^{m-K}\int_J F^j_{0,m}(y\dot+u)\,du\\ &\le C_p 2^{2K/p}\sup_{m\le K}\int_J F^j_{K,m}(y\dot+u)\,du.\end{aligned}$$

Then the inequality

$$\int_I\int_{J^c} (A_{i,j})^p\,d\lambda \le C_p \tag{12}$$

can be proved as in (10) where $i=1,\ldots,5$ and $j=1,\ldots,4$.

Since $F^j_{0,m}=F^j_{K,m}$ for $m>K$, (11) yields that

$$F^j_{0,m}\le 2^r F^j_{K,m} \quad (m\ge K-r;\ j=1,\ldots,4). \tag{13}$$

Then, for each $i=1,\ldots,5$ and $j=1,\ldots,4$,

$$\int_I\int_{J^c}(B_{i,j})^p\,d\lambda \le C_p 2^{2K}\int_I\int_{J^c}\Big(\sup_{m\ge K-r}\int_J 2^r F^j_{K,m}(y\dot+u)\,du\Big)^p\,dx\,dy \le C_p \tag{14}$$

as we have seen in (10).

Step 3: *Integrating over* $([0,1)\setminus I)\times([0,1)\setminus J)$. By (7) it is easy to verify that, for $x\notin I$,

$$\int_I D_{2^i}(x\dot+t\dot+2^{-j-1})\,dt = 2^{i-K}1_{[2^{-j-1},2^{-j-1}\dot+2^{-i})}(x) \tag{15}$$

if $j<i\le K-1$,

$$\int_I D_{2^i}(x\dot+t)\,dt = 2^{i-K}1_{[2^{-K},2^{-i})}(x) \tag{16}$$

if $i\in\mathbb{N}$ and

$$\int_I D_{2^i}(x\dot+t\dot+2^{-j-1})\,dt = 1_{[2^{-j-1},2^{-j-1}\dot+2^{-K})}(x) \tag{17}$$

if $i\ge K$.

Now we modify slightly the kernel functions $F^i_{K,n}$ $(i=1,\ldots,4)$ and calculate their integrals like (9). By (15),

$$\int_{I^c}\Big(\sup_{n\le K}\int_I 2^{n/2}\sum_{j=0}^{n-1}\sum_{i=j+1}^{n-1}(n-i)2^{j-n}D_{2^i}(x\dot+t\dot+2^{-j-1})\,dt\Big)^p\,dx$$

$$= \int_{I^c} \Big(\sup_{n\le K} \sum_{j=0}^{n-1} \sum_{i=j+1}^{n-1} (n-i) 2^{j-n/2} 2^{i-K} 1_{[2^{-j-1}, 2^{-j-1} \dot{+} 2^{-i})}(x) \Big)^p \, dx$$

$$= C_p 2^{-Kp} \int_{I^c} \Big(\sup_{n\le K} \sum_{j=0}^{n-1} \sum_{i=j+1}^{n-1} \frac{n-i}{2} 2^{(i-n)/2} 2^{i/2+j} 1_{[2^{-j-1}, 2^{-j-1} \dot{+} 2^{-i})}(x) \Big)^p \, dx.$$

Since the function $f(n) := (n/2)2^{-n/2}$ is decreasing for $n \ge 3$, we obtain

$$\text{(18)} \quad \int_{I^c} \Big(\sup_{n\le K} \int_I 2^{n/2} \sum_{j=0}^{n-1} \sum_{i=j+1}^{n-1} (n-i) 2^{j-n} D_{2^i}(x \dot{+} t \dot{+} 2^{-j-1}) \, dt \Big)^p \, dx$$

$$\le C_p 2^{-Kp} \int_{I^c} \sum_{j=0}^{K-1} \sum_{i=j+1}^{K-1} 2^{ip/2+jp} 1_{[2^{-j-1}, 2^{-j-1} \dot{+} 2^{-i})}(x) \, dx$$

$$\le C_p 2^{-Kp} \sum_{j=0}^{K-1} \sum_{i=j+1}^{K-1} 2^{jp} 2^{i(p/2-1)}$$

$$\le C_p 2^{-Kp} \sum_{j=0}^{K-1} 2^{j(3p/2-1)} \le C_p 2^{Kp/2-K}$$

provided that $2/3 < p \le 1$.

Using (16) we get

$$\text{(19)} \quad \int_{I^c} \Big(\sup_{n\le K} \int_I 2^{n/2} \sum_{i=0}^{n-1} (n-i) 2^{i-n} D_{2^i}(x \dot{+} t) \, dt \Big)^p \, dx$$

$$\le C_p \int_{I^c} \Big(\sup_{n\le K} \sum_{i=0}^{n-1} \frac{n-i}{2} 2^{(i-n)/2} 2^{i/2} 2^{i-K} 1_{[0, 2^{-i})}(x) \Big)^p \, dx$$

$$\le C_p 2^{-Kp} \sum_{i=0}^{K-1} 2^{i(3p/2-1)} \le C_p 2^{Kp/2-K}.$$

It follows from (15) and (17) that

$$\text{(20)} \quad \int_{I^c} \Big(\sup_{n\le K} \int_I \sum_{j=0}^{n-1} 2^j \sum_{i=n}^{\infty} 2^{-i/2} D_{2^i}(x \dot{+} t \dot{+} 2^{-j-1}) \, dt \Big)^p \, dx$$

$$\le \int_{I^c} \Big(\sup_{n\le K} \sum_{j=0}^{n-1} 2^j \sum_{i=j+1}^{K-1} 2^{-i/2} 2^{i-K} 1_{[2^{-j-1}, 2^{-j-1} \dot{+} 2^{-i})}(x) \Big)^p \, dx$$

$$+ \int_{I^c} \Big(\sup_{n\le K} \sum_{j=0}^{n-1} 2^j \sum_{i=K}^{\infty} 2^{-i/2} 1_{[2^{-j-1}, 2^{-j-1} \dot{+} 2^{-K})}(x) \Big)^p \, dx$$

$$\leq \sum_{j=0}^{K-1} 2^{jp} \sum_{i=j+1}^{K-1} 2^{i(p/2-1)} 2^{-Kp} + \sum_{j=0}^{K-1} 2^{jp} \sum_{i=K}^{\infty} 2^{-ip/2} 2^{-K}$$

$$\leq C_p 2^{-Kp} \sum_{j=0}^{K-1} 2^{j(3p/2-1)} + C_p 2^{-K} 2^{Kp} 2^{-Kp/2} \leq C_p 2^{Kp/2-K}.$$

Similarly,

$$\text{(21)} \qquad \int_{I^c} \left(\sup_{n \leq K} \int_I \sum_{j=0}^{n-1} 2^{j/2} \sum_{i=n}^{\infty} 2^{-i} \frac{D_{2^i}(x \dot{+} t \dot{+} 2^{-j-1})}{2^{K-i} \vee 1} \, dt \right)^p dx$$

$$\leq \int_{I^c} \left(\sup_{n \leq K} \sum_{j=0}^{n-1} 2^{j/2} \sum_{i=j+1}^{K-1} 2^{-K} 2^{i-K} 1_{[2^{-j-1}, 2^{-j-1} \dot{+} 2^{-i})}(x) \right)^p dx$$

$$+ \int_{I^c} \left(\sup_{n \leq K} \sum_{j=0}^{n-1} 2^{j/2} \sum_{i=K}^{\infty} 2^{-i} 1_{[2^{-j-1}, 2^{-j-1} \dot{+} 2^{-K})}(x) \right)^p dx$$

$$\leq \sum_{j=0}^{K-1} 2^{jp/2} \sum_{i=j+1}^{K-1} 2^{i(p-1)} 2^{-2Kp} + \sum_{j=0}^{K-1} 2^{jp/2} \sum_{i=K}^{\infty} 2^{-ip} 2^{-K}$$

$$\leq C_p 2^{-2Kp} \sum_{j=0}^{K-1} 2^{j(3p/2-1)} + C_p 2^{-K} 2^{Kp/2} 2^{-Kp} \leq C_p 2^{-Kp/2-K}$$

whenever $p < 1$. If $p = 1$ then

$$\sum_{j=0}^{K-1} 2^{jp/2} \sum_{i=j+1}^{K-1} 2^{i(p-1)} 2^{-2Kp} = \sum_{j=0}^{K-1} 2^{(j-K)/2} (K-j) 2^{-3K/2} \leq C 2^{-3K/2}$$

and (21) is true in this case, too.

Obviously, if $x \notin I$ and $i \geq K$ then

$$\text{(22)} \qquad \int_I D_{2^i}(x \dot{+} t) \, dt = 0.$$

This implies that

$$\text{(23)} \qquad \int_{I^c} \left(\sup_{n \leq K} \int_I 2^{n/2} \sum_{k=0}^{\infty} 2^{-k} D_{2^{n+k}}(x \dot{+} t) \, dt \right)^p dx$$

$$= \int_{I^c} \left(\sup_{n \leq K} \sum_{k=0}^{K-n-1} 2^{n/2-k} 2^{n+k-K} 1_{[2^{-K}, 2^{-n-k})}(x) \right)^p dx$$

$$= 2^{-Kp} \int_{I^c} \Big(\sum_{k=0}^{K-1} \sum_{n=0}^{K-k-2} 2^{3n/2} 1_{[2^{-n-k-1},2^{-n-k})}(x)$$

$$+ \sum_{k=0}^{K-1} 2^{3(K-k-1)/2} 1_{[2^{-K},2^{-K+1})}(x) \Big)^p \, dx$$

$$\leq C_p 2^{-Kp} \sum_{k=0}^{K-1} \sum_{n=0}^{K-k-2} 2^{n(3p/2-1)} 2^{-k} + C_p 2^{-Kp} \sum_{k=0}^{K-1} 2^{3Kp/2} 2^{-3kp/2} 2^{-K}$$

$$\leq C_p 2^{-Kp} \sum_{k=0}^{K-1} 2^{(K-k)(3p/2-1)} 2^{-k} + C_p 2^{Kp/2-K} \leq C_p 2^{Kp/2-K}.$$

In the same way we conclude that

$$(24) \qquad \int_{I^c} \Big(\sup_{n \leq K} \int_I 2^{-n/2} \sum_{k=0}^{\infty} 2^{-k} \frac{D_{2^{n+k}}(x \dot{+} t)}{2^{K-n-k} \vee 1} \, dt \Big)^p \, dx$$

$$= \int_{I^c} \Big(\sup_{n \leq K} \sum_{k=0}^{K-n-1} 2^{-n/2} 2^{n-K} 2^{n+k-K} 1_{[2^{-K},2^{-n-k})}(x) \Big)^p \, dx$$

$$\leq 2^{-2Kp} \int_{I^c} \Big(\sum_{k=0}^{K-1} \sum_{n=0}^{K-k-2} 2^{3n/2+k} 1_{[2^{-n-k-1},2^{-n-k})}(x)$$

$$+ \sum_{k=0}^{K-1} 2^{3(K-k-1)/2+k} 1_{[2^{-K},2^{-K+1})}(x) \Big)^p \, dx$$

$$\leq C_p 2^{-Kp/2-K}.$$

Now we are ready to deal with the integrals of $(A_{i,j})^p$ over $I^c \times J^c$ $(i = 1, \ldots, 5; j = 1, \ldots, 4)$. We investigate only three terms, $(A_{1,1})$, $(A_{1,3})$ and $(A_{3,1})$, because the others are all similar. Applying (18) twice we obtain

$$(25) \qquad \int_{I^c} \int_{J^c} (A_{1,1})^p \, d\lambda$$

$$= \int_{I^c} \int_{J^c} \Big(\sup_{\substack{|n-m| \leq r \\ n,m \leq K}} \int_I \int_J |a(t,u)| 2^{n-K} \sum_{j=0}^{n-1} \sum_{i=j+1}^{n-1} (n-i) 2^{j-n} D_{2^i}(x \dot{+} t \dot{+} 2^{-j-1})$$

$$\times \sum_{k=0}^{m-1} \sum_{l=k+1}^{m-1} (m-l) 2^{k-m} D_{2^l}(y \dot{+} u \dot{+} 2^{-k-1}) \, dt \, du \Big)^p \, dx \, dy$$

$$\leq C_p 2^{2K-Kp} \int_{I^c}\int_{J^c}\Big(\sup_{n,m\leq K}\int_I 2^{n/2}\sum_{j=0}^{n-1}\sum_{i=j+1}^{n-1}(n-i)2^{j-n}D_{2^i}(x\dot{+}t\dot{+}2^{-j-1})\,dt$$

$$\times\int_J 2^{r/2}2^{m/2}\sum_{k=0}^{m-1}\sum_{l=k+1}^{m-1}(m-l)2^{k-m}D_{2^l}(y\dot{+}u\dot{+}2^{-k-1})\,du\Big)^p\,dx\,dy$$

$$\leq C_p 2^{2K-Kp}2^{Kp/2-K}2^{Kp/2-K} = C_p.$$

By (18) and (20),

$$(26)\qquad \int_{I^c}\int_{J^c}(A_{1,3})^p\,d\lambda$$

$$= \int_{I^c}\int_{J^c}\Big(\sup_{\substack{|n-m|\leq r\\ n,m\leq K}}\int_I\int_J |a(t,u)|2^{n-K}\sum_{j=0}^{n-1}\sum_{i=j+1}^{n-1}(n-i)2^{j-n}D_{2^i}(x\dot{+}t\dot{+}2^{-j-1})$$

$$\times\sum_{k=0}^{m-1}2^k\sum_{l=m}^{\infty}2^{-l}D_{2^l}(y\dot{+}u\dot{+}2^{-k-1})\,dt\,du\Big)^p\,dx\,dy$$

$$\leq C_p 2^{2K-Kp}\int_{I^c}\int_{J^c}\Big(\sup_{n,m\leq K}\int_I 2^{n/2}\sum_{j=0}^{n-1}\sum_{i=j+1}^{n-1}(n-i)2^{j-n}D_{2^i}(x\dot{+}t\dot{+}2^{-j-1})\,dt$$

$$\times\int_J 2^{r/2}\sum_{k=0}^{m-1}2^k\sum_{l=m}^{\infty}2^{-l/2}D_{2^l}(y\dot{+}u\dot{+}2^{-k-1})\,du\Big)^p\,dx\,dy$$

$$\leq C_p 2^{2K-Kp}2^{Kp/2-K}2^{Kp/2-K} = C_p.$$

Similarly, using (18) and (21) we can see that

$$(27)\qquad \int_{I^c}\int_{J^c}(A_{3,1})^p\,d\lambda$$

$$= \int_{I^c}\int_{J^c}\Bigg(\sup_{\substack{|n-m|\leq r\\ n,m\leq K}}\int_I\int_J |a(t,u)|\sum_{j=0}^{n-1}2^j\sum_{i=n}^{\infty}2^{-i}\frac{D_{2^i}(x\dot{+}t\dot{+}2^{-j-1})}{2^{K-i}\vee 1}$$

$$\times\sum_{k=0}^{m-1}\sum_{l=k+1}^{m-1}(m-l)2^{k-m}D_{2^l}(y\dot{+}u\dot{+}2^{-k-1})\,dt\,du\Bigg)^p\,dx\,dy$$

$$\leq C_p 2^{2K}\int_{I^c}\int_{J^c}\Bigg(\sup_{n,m\leq K}\int_I\sum_{j=0}^{n-1}2^{j/2}\sum_{i=n}^{\infty}2^{-i}\frac{D_{2^i}(x\dot{+}t\dot{+}2^{-j-1})}{2^{K-i}\vee 1}\,dt$$

$$\times\int_J 2^{r/2}2^{m/2}\sum_{k=0}^{m-1}\sum_{l=k+1}^{m-1}(m-l)2^{k-m}D_{2^l}(y\dot{+}u\dot{+}2^{-k-1})\,du\Bigg)^p\,dx\,dy$$

$$\leq C_p 2^{2K}2^{-Kp/2-K}2^{Kp/2-K} = C_p.$$

Observe that $(A_{5,j})(x,y) = 0$ $(j = 1, \ldots, 4)$ follows from the definition.

(13) and (9) imply that

$$(28)\qquad \int_{I^c}\int_{J^c} (B_{i,j})^p \, d\lambda \le C_p 2^{2K} \int_{I^c}\int_{J^c} \Big(\sup_{n,m \ge K-r} \int_I F^i_{K,n}(x \dotplus t)\, dt \times \int_J 2^r F^j_{K,m}(y \dotplus u)\, du \Big)^p \, dx\, dy \le C_p$$

for each $i = 1, \ldots, 5$ and $j = 1, \ldots, 4$.

Notice that the terms $|a| * F^i_{0,n} \times F^j_{K,m}$ $(i = 1, \ldots, 4; j = 1, \ldots, 5)$ can be handled similarly. Finally, by (22), $|a| * F^5_{K,n} \times F^5_{K,m}(x,y) = 0$ when $x \in I^c$ or $y \in J^c$. Combining this and (10), (12), (14) and (25)–(28) we can establish that

$$\int_{[0,1)^2 \setminus Q} (\mathbf{J}^*_\alpha a)^p \, dx \le C_p,$$

which proves the theorem for $f \in H_p \cap L_1$.

If $f \in H_p$ $(2/3 < p < 1)$ then $f_{k,k} \in L_1$ and $f_{k,k} \to f$ in H_p norm as $k \to \infty$. We have

$$\|\mathbf{J}^*_\alpha f_{j,j} - \mathbf{J}^*_\alpha f_{k,k}\|_p \le \|\mathbf{J}^*_\alpha (f_{j,j} - f_{k,k})\|_p \le C_p \|f_{j,j} - f_{k,k}\|_{H_p} \to 0$$

as $j, k \to \infty$. For $f \in H_p$ we define $\mathbf{J}^*_\alpha f \in L_p$ by

$$\mathbf{J}^*_\alpha f := \lim_{k \to \infty} \mathbf{J}^*_\alpha f_{k,k} \quad \text{in } L_p \text{ norm},$$

which finishes the proof of the theorem. ∎

The next corollary follows from (5) and from Theorems B and 2.

COROLLARY 1. *There are absolute constants C_1 and $C_{p,q}$ such that*

$$\|\mathbf{J}^*_\alpha f\|_{p,q} \le C_{p,q} \|f\|_{H_{p,q}} \qquad (f \in H_{p,q})$$

for every $2/3 < p < \infty$ and $0 < q \le \infty$. In particular, $\mathbf{J}^*_\alpha$ is of weak type (L_1, L_1), i.e. if $f \in L_1$ then

$$\|\mathbf{J}^*_\alpha f\|_{1,\infty} = \sup_{\gamma > 0} \gamma \lambda(\mathbf{J}^*_\alpha f > \gamma) \le C_1 \|f\|_{H_{1,\infty}} = C_1 \sup_{\gamma > 0} \gamma \lambda(f^* > \gamma) \le C_1 \|f\|_1.$$

Now we can state our main result.

COROLLARY 2. *Suppose that for a martingale $f = (f_{n,n}, n \in \mathbb{N}) \in H_{p,q}$ we have $\int_0^1 f_{n,n}(x, y_0)\, dx = \int_0^1 f_{n,n}(x_0, y)\, dy = 0$ for each $n \in \mathbb{N}$ and almost every $x_0, y_0 \in [0,1)$. Then*

$$\|\mathbf{I}^*_\alpha f\|_{p,q} \le C_{p,q} \|f\|_{H_{p,q}}$$

for every $2/3 < p < \infty$ *and* $0 < q \leq \infty$. *In particular,* $\mathbf{I}_\alpha^*$ *is of weak type* (L_1, L_1), *i.e. if* $f \in L_1$ *such that* $\int_0^1 f(x, y_0)\,dx = \int_0^1 f(x_0, y)\,dy = 0$ *for almost every* $x_0, y_0 \in [0,1)$ *then*

$$\sup_{\gamma>0} \gamma\lambda(\mathbf{I}_\alpha^* f > \gamma) \leq C_1 \|f\|_{H_{1,\infty}} \leq C_1 \|f\|_1.$$

P r o o f. By the proof of Theorem 2 it is enough to verify the corollary for integrable functions. Let $f \in L_1$ such that $\int_0^1 f(x, y_0)\,dx = \int_0^1 f(x_0, y)\,dy = 0$ for almost every $x_0, y_0 \in [0,1)$. Then it is easy to see that

$$\begin{aligned}\mathbf{d}_{n,m}(\mathbf{I}f)(x,y) &= \mathbf{d}_{n,m}\Big(\int_0^1\int_0^1 f(t,u) W(x \dotplus t) W(y \dotplus u)\,dt\,du\Big) \\ &= \int_0^1\int_0^1 f(t,u)\mathbf{d}_n W(x \dotplus t)\mathbf{d}_m W(y \dotplus u)\,dt\,du \\ &= \delta_{n,m}(x,y).\end{aligned}$$

Hence $\mathbf{I}_\alpha^* f = \mathbf{J}_\alpha^* f$ and the result follows from Corollary 1. ∎

The next corollary follows from the weak type inequality in Corollary 2 and from the fact that the Walsh polynomials are dense in L_1.

COROLLARY 3. *If* $\alpha \geq 0$ *is arbitrary and if* $f \in L_1$ *is such that*

$$\int_0^1 f(x, y_0)\,dx = \int_0^1 f(x_0, y)\,dy = 0$$

for almost every $x_0, y_0 \in [0,1)$ *then*

$$\mathbf{d}_{n,m}(\mathbf{I}f) \to f \quad \text{a.e. as } n, m \to \infty \text{ and } |n - m| \leq \alpha.$$

We remark that this corollary is also proved by Gát [7].

Finally, we note that without the condition $\int_0^1 f(x, y_0)\,dx = \int_0^1 f(x_0, y)\,dy = 0$ we can prove Corollary 2 only for $p \geq 1$; more exactly:

THEOREM 3. *There are absolute constants* C_1 *and* $C_{p,q}$ *such that*

$$\|\mathbf{I}_\alpha^* f\|_1 \leq C_1 \|f\|_{H_1} \qquad (f \in H_1)$$

and

$$\|\mathbf{I}_\alpha^* f\|_{p,q} \leq C_{p,q} \|f\|_{H_{p,q}} \qquad (f \in H_{p,q})$$

for every $1 < p < \infty$ *and* $0 < q \leq \infty$.

P r o o f. We can apply only the second inequality of Theorem 1. That is to say, we have to investigate the terms $\sup_{|n-m|\leq\alpha} |a| * F_{K,n}^i \times F_{0,m}^j$

$(i, j = 1, \ldots, 5)$. If $j \neq 5$ then they are considered in the proof of Theorem 2. If $j = 5$ then

$$|a| * F^i_{K,n} \times F^5_{0,m}(x,y) = \int_I \int_J |a(t,u)| F^i_{K,n}(x \dotplus t) F^5_{0,m}(y \dotplus u)\, dt\, du$$
$$\leq 2^{2K} 2^{-K} \int_I F^i_{K,n}(x \dotplus t)\, dt$$

where a is a 1-atom with support $Q = I \times J$, $I = J = [0, 2^{-K})$. Applying (9), we get

$$\int_{I^c} \int_{J^c} \sup_{|n-m| \leq \alpha} |a| * F^i_{K,n} \times F^5_{0,m}(x,y)\, dx\, dy$$
$$\leq 2^K \int_{I^c} \sup_{n \in \mathbb{N}} \int_I F^i_{K,n}(x \dotplus t)\, dt\, dx \leq C.$$

We get the same result if we integrate over $I^c \times J$ or $I \times J^c$. Hence the condition (6) is verified for $p = 1$; this means that the first inequality in Theorem 3 is proved. The second inequality follows by interpolation. ∎

References

[1] C. Bennett and R. Sharpley, *Interpolation of Operators*, Pure Appl. Math. 129, Academic Press, New York, 1988.

[2] J. Bergh and J. Löfström, *Interpolation Spaces. An Introduction*, Springer, Berlin, 1976.

[3] P. L. Butzer and W. Engels, *Dyadic calculus and sampling theorems for functions with multidimensional domain*, Inform. and Control 52 (1982), 333–351.

[4] P. L. Butzer and H. J. Wagner, *On dyadic analysis based on the pointwise dyadic derivative*, Anal. Math. 1 (1975), 171–196.

[5] —, —, *Walsh series and the concept of a derivative*, Appl. Anal. 3 (1973), 29–46.

[6] A. M. Garsia, *Martingale Inequalities. Seminar Notes on Recent Progress*, Math. Lecture Notes Ser., Benjamin, New York, 1973.

[7] Gy. Gát, *On the two-dimensional pointwise dyadic calculus*, J. Approx. Theory, to appear.

[8] J. Neveu, *Discrete-Parameter Martingales*, North-Holland, 1971.

[9] F. Schipp, *Über einen Ableitungsbegriff von P. L. Butzer und H. J. Wagner*, Math. Balkanica 4 (1974), 541–546.

[10] —, *Über gewissen Maximaloperatoren*, Ann. Univ. Sci. Budapest. Sect. Math. 18 (1975), 189–195.

[11] F. Schipp and P. Simon, *On some (H, L_1)-type maximal inequalities with respect to the Walsh–Paley system*, in: Functions, Series, Operators, Budapest 1980, Colloq. Math. Soc. János Bolyai 35, North-Holland, Amsterdam, 1981, 1039–1045.

[12] F. Schipp and W. R. Wade, *A fundamental theorem of dyadic calculus for the unit square*, Appl. Anal. 34 (1989), 203–218.

[13] F. Schipp, W. R. Wade, P. Simon and J. Pál, *Walsh Series: An Introduction to Dyadic Harmonic Analysis*, Adam Hilger, Bristol, 1990.

[14] F. Weisz, *Cesàro summability of two-dimensional Walsh-Fourier series*, Trans. Amer. Math. Soc. (1996), to appear.
[15] —, *Martingale Hardy spaces and the dyadic derivative*, Anal. Math., to appear.
[16] —, *Martingale Hardy Spaces and Their Applications in Fourier-Analysis*, Lecture Notes in Math. 1568, Springer, Berlin, 1994.
[17] —, *Some maximal inequalities with respect to two-parameter dyadic derivative and Cesàro summability*, Appl. Anal., to appear.
[18] A. Zygmund, *Trigonometric Series*, Cambridge Univ. Press, London, 1959.

Department of Numerical Analysis
Eötvös L. University
Múzeum krt. 6-8
H-1088 Budapest, Hungary
E-mail: weisz@ludens.elte.hu

Received January 18, 1996
Revised version April 16, 1996
(3632)

Chapter 9
Term by Term Dyadic Differentiation of Walsh Series

William R. Wade

9.1 Introduction

Let $N = \{0,1,2,\dots\}$ denote the nonnegative integers. Each $n \in N$ can be represented uniquely in binary notation using its *binary coefficients* $n_k \in \{0,1\}$, i.e.,

$$n = \sum_{k=0}^{\infty} n_k 2^k.$$

Let G denote the *dyadic group*, i.e.,

$$G := \{x = (x_0, x_1, \dots) : x_k \in \{0,1\} \quad \text{for all } k \in N\}.$$

These numbers x_k are called the *components* of x. They define the group operation on G by

$$x \dotplus y := (|x_0 - y_0|, |x_1 - y_1|, \dots).$$

Rademacher-like functions r_n are defined on G by

$$r_n(x) = \prod_{k=0}^{\infty} (-1)^{x_k},$$

where the x_k's are the components of $x \in G$. The character group of G is a product system generated by the r_k's, i.e.,

$$w_n(x) := \prod_{k=0}^{\infty} r_k^{n_k}(x),$$

William R. Wade
Mathematics and Computer Science, Biola University, La Mirada CA 90639, USA, e-mail: william.wade@biola.edu

R.S. Stanković et al. (eds.), *Dyadic Walsh Analysis from 1924 Onwards Walsh-Gibbs-Butzer Dyadic Differentiation in Science Volume 1 Foundations*, Atlantis Studies in Mathematics for Engineering and Science 12, DOI 10.2991/978-94-6239-160-4_9

where the n_k's are the binary coefficients of n.

For each $j \in N$, let $e_j := (0,0,\dots,0,1,0,\dots) \in G$ denote the element whose components are all zero except for the jth component which is 1. The group G can be identified with the interval $[0,1]$ by the map

$$(x_0, x_1, \dots) \longmapsto \sum_{k=0}^{\infty} x_k 2^{-(k+1)}$$

which for each $j \in N$ takes the elements $e_j \in G$ to the dyadic rational $2^{-(j+1)}$. This map identifies the characters w_n with the Walsh functions and the generators r_k with the *Rademacher functions*, all defined on the interval $[0,1)$. To streamline the presentation, we will use the same notation for functions defined on the group G as we do for the functions defined on the interval $[0,1)$. In particular, e_j will denote the dyadic rationals $2^{-(j+1)}$, $j \in N$, and r_k will denote the Rademacher functions defined on $[0,1]$ by

$$r_k(x) = \begin{cases} 1, & x \in [0,0.5), \\ -1, & x \in [0.5,1). \end{cases}$$

For details and better notation, see the monograph [18].

9.2 The original problem

In 1969, Gibbs and Millard [10] introduced a derivative for discrete functions. In 1973, Butzer and Wagner [1] modified this definition to apply to functions defined on the group G. Their definition introduced the differential partial sums

$$d_n(f,x) := \sum_{j=0}^{n-1} 2^{j-1} \left(f(x) - f(x \dot{+} 2^{-j-1}) \right).$$

Specifically, they said that a function $f : G \longmapsto \mathbf{R}$ is *strongly dyadically differentiable* in some Banach space $\boldsymbol{X}$ if d_n converges to some function g in the norm of $\boldsymbol{X}$, in which case the limit g is called the *strong dyadic derivative* of f, here denoted by $Df := g$. They proved that $Dw_k = kw_k$ for all $k \in N$, i.e., that the eigensolutions of the dyadic derivative operator are the Walsh functions. They also proved a good many other results in this ground breaking paper which showed that the dyadic derivative plays the same role for the Walsh system that the classical derivative plays for the trigonometric system, e.g., that the differential operator D is closed, that any function in Lip_α for $\alpha > 1$ is dyadically differentiable, and that any dyadically differentiable function must belong to Lip_α for $0 < \alpha \leq 1$. These, and many other results, will undoubtedly appear in other papers in this volume, but we want to focus our attention on one problem: term-by-term dyadic differentiation.

The pointwise dyadic derivative was introduced by Butzer and Wagner [2] in 1975. They said that a function $f : [0,1] \longmapsto \mathbf{R}$ is *dyadically differentiable* at a point x if $d_n(f,x)$ converges to some value $g(x)$ in which case the limit $g(x)$, is called the *(pointwise) dyadic derivative* of f at x, here denoted by $f^{[1]}(x) := g(x)$. They proved that $w_k^{[1]} = kw_k$ for all $k \in N$ and that for a large class of functions, the strong dyadic derivative and the pointwise dyadic derivative agree almost everywhere.

Let x be a point in $[0,1)$. A Walsh series $\sum_{k=0}^{\infty} a_k w_k$ is said to be *term-by-term dyadically differentiable* at x if

$$f(t) := \sum_{k=0}^{\infty} a_k w_k(t), \tag{9.1}$$

exists and is finite for $t = x, x \dotplus 2^{-l}$ and for all $l \in P$, and if the dyadic derivative of f exists at x and satisfies

$$f^{[1]}(x) = \sum_{k=1}^{\infty} k a_k w_k(x). \tag{9.2}$$

Butzer and Wagner [2] asked what conditions on the a_k's would make the function (9.1) term-by-term dyadically differentiable at some point x, i.e., would make (9.2) hold. They answered this question with the following result.

Theorem 9.1 (Butzer-Wagner) *If the sequences $\{a_k\}$ and $\{ka_k\}$ are quasiconvex and if $ka_k \to 0$, then $f = \sum_{k=0}^{\infty} a_k w_k$ is almost everywhere term-by-term dyadically differentiable.*

They also proved that if the derived series (9.2) converges absolutely, then f is term-by-term dyadically differentiable everywhere on $[0,1]$. In view of this last result, they made the following conjecture: If ka_k decreases monotonically to zero (i.e., if $a_1 \geq 2a_2 \geq 3a_3 \geq \cdots$ and $ka_k \to 0$ as $k \to \infty$), then f is term-by-term dyadically differentiable almost everywhere on $[0,1]$. This conjecture was proved by Schipp [16] in 1976. Indeed, he proved that term-by-term dyadic differentiability happened not just almost everywhere, but at all but countably many points in $[0,1]$.

Theorem 9.2 (Schipp) *If ka_k decreases monotonically to zero, then $f = \sum_{k=0}^{\infty} a_k w_k$ is term-by-term dyadically differentiable at every $x \in (0,1)$ except possibly $x = e_j$, $j \in N$.*

In 1979 Skvortsov and Wade [19] showed that it was enough for the a_k's to be monotone decreasing in dyadic blocks. In fact, they proved the following more general result.

Theorem 9.3 (Skvortsov-Wade) *Suppose that $a_k \in \mathbf{R}$, $k \in N$, and that a sequence of numbers R_n is defined by $R_0 := 0$ and*

$$R_n := \sum_{k=2^n}^{2^{n+1}-2} |a_k - a_{k+1}| \qquad n > 0.$$

If $a_n \to 0$ and $2^n R_n \to 0$ as $n \to \infty$, then a necessary and sufficient condition that $f := \sum_{k=0}^{\infty} a_k w_k$ be dyadically differentiable at some $x \in (0,1)$, $x \neq e_j$ for $j \in N$, is that

$$S_{2^n}^{[1]}(x) := \sum_{k=0}^{2^n-1} k a_k w_k(x)$$

converge to a finite real number $g(x)$, as $n \to \infty$, in which case

$$f^{[1]}(x) = g(x).$$

Since convergence of dyadic partial sums is weaker than convergence of the series $\sum_{k=0}^{\infty} k a_k w_k$, this result generalizes Schipp's theorem cited above.

The paper by Skvortsov and Wade also contained another interesting result. Butzer and Wagner [2] had shown that if f is continuous and if $f^{[1]}$ exists everywhere on $[0,1]$, then f is constant. Skvortsov [19] showed the reason for this is a connection between the dyadic derivative and the classical Dini derivates: If f is dyadically differentiable at x, then the upper right and lower left Dini derivates satisfy $D^+ f(x) \geq 0 \geq D_- f(x)$. In particular, if f is continuous on $[0,1]$ and dyadically differentiable off a countable set, then f must be constant. It doesn't matter whether $f^{[1]}$ is continuous or not. There is a fundamental incompatibility between dyadic differentiability and classical continuity.

This theme was further emphasized by Engels in [6], who proved the following result.

Theorem 9.4 (Engels) *Suppose that f is bounded on $[0,1)$ and has at most countably many discontinuity points, all of the first kind. If $f^{[1]}$ exists off a countable set, then f is piecewise constant.*

The question of term-by-term dyadic differentiation in rearranged Walsh series remains almost unexplored, partly because these results are very difficult to obtain. In a long, complicated paper published in 2003, Gát [9] examined dyadic differentiation for Walsh series in the Kaczmarz arrangement. He proved the following result.

Theorem 9.5 (Gát) *Let κ_k, $k \in N$, represent the Walsh-Kaczmarz system. If a_k is a decreasing sequence of real numbers that converges monotonically to zero, if*

$$a_k = o\left(\frac{1}{\log^{\alpha} k}\right)$$

as $k \to \infty$ for some $\alpha > 1$, then the function

$$f = \sum_{k=0}^{\infty} \frac{a_k}{k} \kappa_k$$

is term-by-term dyadically differentiable and $f^{[1]} = \sum_{k=0}^{\infty} a_k \kappa_k$.

9.3 Gap series

It is easy to see by Dirichlet's Test, that the condition that ka_k decreases montonically to zero implies that the derived series $\sum_{k=0}^{\infty} ka_k w_k$ converges everywhere on $[0,1]$. It is natural to ask

Question 9.1 *If the derived series $\sum_{k=0}^{\infty} ka_k w_k$ converges everywhere on $[0,1]$, is $f(x) = \sum_{k=0}^{\infty} a_k w_k$ term-by-term differentiable almost everywhere on $[0,1]$?*

This question is still open for general Walsh series. In 1977 Onneweer [12] solved it completely for Rademacher series as follows.

Theorem 9.6 (Onneweer) *A function $f = \sum_{k=0}^{\infty} a_k r_k$ is dyadically differentiable at some point $x \in G$ if and only if the derived series $\sum_{k=0}^{\infty} 2^k a_k r_k$ converges at x, in which case the function f is term-by-term dyadically differentiable at x.*

In fact, he obtained this result for the much more general situation when G is a Vilenkin group.

Onneweer's theorem answers Question 9.1 for Rademacher series, and shows that the convergence of the derived series is both necessary and sufficient for dyadic differentiability. This leads us to another open question.

Question 9.2 *If $f = \sum_{k=0}^{\infty} a_k w_k$ is dyadically differentiable everywhere on $[0,1)$, does $\sum_{k=0}^{\infty} ka_k w_k$ converge almost everywhere on $[0,1)$?*

For Walsh series with coefficients monotone in dyadic blocks, this question has been answered (see the theorem by Skvortsov and Wade cited above). It is still open for general Walsh series.

Of course, since $r_k = w_{2^k}$, a Rademacher series is just a Walsh series with gaps. It is natural to ask whether Onneweer's theorem holds for other Walsh series with gaps, say $\sum_{k=0}^{\infty} a_{\lambda_k} w_{\lambda_k}$ where λ_k is a subsequence of N.

A sequence of indices λ_k is called a *Paley set* if the number of λ_k's between 2^N and 2^{N+1} is bounded for all $N \in N$, and *lacunary* if there is a number $q > 1$ such that $\lambda_{k+1}/\lambda_k \geq q$ for all $k \in N$. It is easy to see that the sequence $\lambda_k = 2^k$ is lacunary, and that every lacunary sequence is a Paley set. Thus the following result, found in [15], answers Question 9.1 for lacunary series.

Theorem 9.7 (Powell-Wade) *Suppose that $f = \sum_{k=0}^{\infty} a_k w_k$ has Paley gaps, i.e., that there is a Paley set λ_k such that $a_j = 0$ for all $j \neq \lambda_k$, $k \in N$. If the derived series $\sum_{k=0}^{\infty} ka_k w_k$ converges at some $x \in G$, then f is term-by-term dyadically differentiable at x.*

Recently, a new kind of gap series has been introduced [20] which treats non-Paley gaps. A *filtered* Walsh series generated by an increasing sequence of integers $m_0 < m_1 < \cdots$ is a series of the form

$$\sum_{k=0}^{\infty} a_{j_k} w_{j_k},$$

where $j_0 < j_1 < \cdots$ are all those integers j which can be written in the form $j = \sum_{\nu=0}^{N} \eta_\nu 2^{m_\nu}$ for some $N \geq 0$ and some $n_\nu \in \{0,1\}$.

The convergence properties of filtered series are similar to those of the older lacunary series. This leads us to the following question.

Question 9.3 *If $f = \sum_{k=0}^{\infty} a_k w_k$ is a filtered Walsh series, and if the derived series $\sum_{k=0}^{\infty} k a_k w_k$ converges everywhere, is f term-by-term dyadically differentiable almost everywhere on $[0,1]$?*

9.4 Rapidly converging series

Since Question 9.1 proved difficult to answer for non-gap series, and since if the derived series converges, then $ka_k \to 0$ as $k \to \infty$, it was natural to ask that if $ka_k \to 0$ as $k \to \infty$, what additional conditions must be imposed on the a_k's so that the function $f = \sum_{k=0}^{\infty} a_k w_k$ is term-by-term dyadically differentiable.

These additional conditions have come in two types:

1. "Type 1:" The sequence a_k converges to zero so rapidly that $ka_k \to 0$ is a simple consequence.
2. "Type 2:" The derived series $\sum_{k=0}^{\infty} k a_k w_k$ converges rapidly.

The first result of type 1, obtained by Powell and Wade [13] in 1981, proved that if $\alpha > 1$ and if $\sum_{k=0}^{\infty} k^\alpha a_k w_k(x)$ converges at some point $x \in [0,1)$, then $f = \sum_{k=0}^{\infty} a_k w_k$ is term-by-term dyadically differentiable at x. This result has been generalized several times and the best result along these lines was obtained in 1991 (see [14]):

Theorem 9.8 (Powell-Wade) *If $\beta_0 \leq \beta_1 \leq \cdots$ is an increasing sequence of positive real numbers, if*

$$\sum_{k=0}^{\infty} \frac{1}{\beta_k} < \infty, \quad \textit{and if} \quad \sum_{k=0}^{\infty} \beta_k a_k w_k(x) < \infty$$

for some $x \in [0,1)$ then $f := \sum_{k=0}^{\infty} a_k w_k$ is term by term dyadically differentiable at x.

As far as rate of growth is concerned, a result in [8] shows that this theorem is best possible.

Theorem 9.9 (Fridli) *If $\beta_0 \leq \beta_1 \leq \cdots$ is an increasing sequence of positive real numbers, if $x \in [0,1)$, and if*

$$\sum_{k=0}^{\infty} \frac{1}{\beta_k} = \infty,$$

then there is an increasing sequence of numbers Γ_k and a sequence of coefficients a_k such that $\beta_k \le \Gamma_k$, $\sum \Gamma_k a_k w_k(x)$ converges to a finite real number, but $f = \sum_{k=0}^{\infty} a_k w_k$ is not dyadically differentiable at x.

Despite all this progress, the following question, first posed in 1989 [20], is still open.

Question 9.4 *If $\sum_{k=0}^{\infty} k \log k a_k w_k$ converges everywhere on $[0,1)$, is $f = \sum_{k=0}^{\infty} a_k w_k$ almost everywhere dyadically differentiable on $[0,1)$?*

The first result of type 2 was obtained in 1991 by Powell and Wade [14]. The best results along these lines, which appeared in [8], are as follows.

Theorem 9.10 (Fridli-Wade) *Suppose that $x \in [0,1)$ and that $\beta_0 \le \beta_1 \le \cdots$ is an increasing sequence of positive real numbers which satisfies*

$$\sum_{k=0}^{\infty} \frac{1}{\beta_k} < \infty.$$

If

$$\sum_{k=m}^{\infty} a_k w_k(x) = O\left(\frac{1}{\beta_m}\right)$$

as $m \to \infty$, then $f = \sum_{k=0}^{\infty} a_k w_k$ converges absolutely everywhere on $[0,1)$ and is term-by-term dyadically differentiable at x.

Theorem 9.11 (Fridli-Wade) *Suppose that $x \in [0,1)$ and that $\beta_0 \le \beta_1 \le \cdots$ is an increasing sequence of positive real numbers which satisfies*

$$\sum_{k=0}^{\infty} \frac{1}{k\beta_k} < \infty.$$

If

$$\sum_{k=m}^{\infty} k a_k w_k(x) = O\left(\frac{1}{\beta_m}\right)$$

as $m \to \infty$, then $f = \sum_{k=0}^{\infty} a_k w_k$ converges absolutely everywhere on $[0,1)$ and is term-by-term dyadically differentiable at x.

The first of these results is best possible (see [8]):

Theorem 9.12 (Fridli) *If $\beta_0 \le \beta_1 \le \cdots$ is an increasing sequence of positive real numbers, if $x \in [0,1)$, and if*

$$\sum_{k=0}^{\infty} \frac{1}{\beta_k} = \infty,$$

then there is a sequence of coefficients a_k *such that* $\sum_{k=0}^{\infty} a_k w_k(x \dotplus e_j)$ *converges to a finite real number for all* $j \in N$*, but* $f = \sum_{k=0}^{\infty} a_k w_k$ *is not dyadically differentiable at* x*.*

It is not known whether the second of these results is best possible. In particular, the following question is open.

Question 9.5 *Is there a Walsh series* $f = \sum_{k=0}^{\infty} a_k w_k$ *which converges at* x *and* $x \dotplus e_j$ *for all* $j \in N$ *and some* $x \in [0,1)$*, and satisfies*

$$\sum_{k=m}^{\infty} k a_k w_k(x) = O\left(\frac{1}{\log m}\right)$$

as $m \to \infty$*, such that the function* f *is not dyadically differentiable at* x*?*

Finally, we notice that all the positive results of types 1 and 2 depend on the following Dirichlet-like test which deserves to be better known.

Lemma 9.1 (Powell) *Suppose* $(\omega_m, m \in N)$ *is a non-decreasing sequence of positive real numbers and* $\sum_{k=0}^{\infty} x_k$ *is a convergent series of real numbers such that*

$$\sum_{k=m}^{\infty} x_k = o\left(\frac{1}{\omega_m}\right), \quad as \quad m \to \infty.$$

Suppose further that $(b_k, k \in P)$, $(b_k^{(n)}, k \in P)$ *are sequences of real numbers such that*

$$\lim_{n \to \infty} b_k^{(n)} = b_k \qquad (k \in P).$$

If there is an absolute constant $M > 0$ *such that*

$$\sum_{k=1}^{\infty} \left(\frac{1}{\omega_k}\right) \left| b_k^{(n)} - b_{k+1}^{(n)} \right| \leq M \qquad (n \in P),$$

then the series $\sum_{k=0}^{\infty} b_k x_k$ *and* $\sum_{k=0}^{\infty} b_k^{(n)} x_k$ *converge to finite real numbers for* $n \in P$*, and*

$$\lim_{n \to \infty} \sum_{k=1}^{\infty} b_k^{(n)} x_k = \sum_{k=1}^{\infty} b_k x_k.$$

This result first appeared in [13], but the simple, direct proof which appears in [8] is due to Fridli.

9.5 Term-by-term strong dyadic differentiation

A Walsh series $f = \sum_{k=0}^{\infty} a_k w_k$ is said to be *term-by-term strongly dyadically differentiable* in some linear space X if the strong dyadic derivative of f exists in X and satisfies

$$Df = \sum_{k=1}^{\infty} k a_k w_k. \tag{9.3}$$

Butzer and Wagner [1] were first to consider term-by-term strong dyadic differentiation.

To state their results, recall that *dyadic convolution* of two functions $f, g \in L^1[0,1)$ is defined by

$$(f * g)(x) = \int_0^1 f(x \dot{+} y) g(y)\, dy.$$

They proved the following result.

Theorem 9.13 (Butzer-Wagner) *Suppose that X is a translation invariant, L^1 embeddable Banach algebra, i.e., that X is a Banach space of functions in L^1, on G or $[0,1)$, whose norm satisfies*

$$\|\tau_j(f)\|_X = \|f\|_X \quad \textit{for all } j \in N, \textit{ where } \tau_j(f)(x) = f(x \dot{+} e_j),$$

$$\|f * g\|_X \leq \|f\|_X \|g\|_1 \quad \textit{for all } f \in X \textit{ and } g \in L^1[0,1),$$

and

$$\|f\|_1 \leq \|f\|_X \quad \textit{for all } f \in X.$$

Then f is strongly dyadically differentiable in X if and only if there is a function $g \in X$ such that the Walsh-Fourier coefficients of f and g satisfy $\widehat{g}(k) = k\widehat{f}(k)$, in which case f is term-by-term strongly dyadically differentiable in X and $g = Df$.

Of course, $L^p[0,1)$ and $H^p[0,1)$ are translation invariant, L^1 embeddable Banach algebras for $1 \leq p < \infty$. Thus it is natural to ask whether this result holds for L^p and H^p when $0 < p < 1$. This was first discussed in [14]. Before we cite known results which address this question, we need to introduce some terminology.

A quasi-normed linear space is a vector space X together with a positive-definite function $\|\cdot\| : X \longmapsto [0,\infty)$ which satisfies the following four conditions for all $f, g \in X$ and all scalars α: $\|f+g\| \leq \|f\| + \|g\|$, $\|-f\| = \|f\|$, $\|\alpha_k f\| \to 0$ as the scalars $\alpha_k \to 0$, and $\|\alpha f_k\| \to 0$ as $f_k \to 0$ in X. X is called Ψ-monotone if there is an increasing subadditive function $\Psi : [0,\infty) \longmapsto (-\infty,\infty)$ such that $\Psi(\alpha) \to 0$ as $\alpha \downarrow 0$, $\Psi(|\alpha\beta|) \leq \Psi(|\alpha|)\Psi(|\beta|)$, and $\|\alpha f\| \leq \Psi(|\alpha|)\,\|f\|$ for all scalars α, β and all $f \in X$. And, X is called appropriate for dyadic differentiation if every element of X is a function almost everywhere defined on $[0,1)$, if $w_k \in X$ for all $k \in N$, if $\tau_k f \in X$ for every $f \in X$ and every $k \in N$, and $\|f\| = 0$ if and only if $f = 0$ almost everywhere.

Notice that every translation invariant, L^1 embeddable Banach algebra is a quasi-normed, monotone linear space with $\Psi(\alpha) = \alpha$. Since it is well known that $\mathrm{BMO}[0,1)$, $L^p[0,1)$, $H^p[0,1)$, and the block spaces of Taibleson and Weiss, $\mathscr{B}_q$, $q \geq 1$, are all quasi-normed, monotone linear spaces (see [14]), the following results, obtained in 1991 [14] and 1993 [8], answer the question raised above.

Theorem 9.14 (Powell-Wade) *Suppose that X is a Ψ-monotone, quasi-normed linear space appropriate for dyadic differentiation, that $f = \sum_{k=0}^{\infty} a_k w_k$ converges in X, and that β_k is an increasing sequence of positive numbers. If*

$$\sum_{k=0}^{\infty} \Psi\left(\frac{1}{\beta_k}\right) < \infty$$

and if the series

$$\sum_{k=0}^{\infty} \beta_k a_k w_k$$

converges in X, then f is term-by-term strongly dyadically differentiable in X.

Theorem 9.15 (Fridli-Wade) *Suppose that X is a Ψ-monotone, quasi-normed linear space appropriate for dyadic differentiation, that $f = \sum_{k=0}^{\infty} a_k w_k$ converges in X, and that β_k is an increasing sequence of positive numbers.*

a) If

$$\sum_{k=0}^{\infty} \Psi\left(\frac{1}{\beta_k}\right) < \infty$$

and if

$$\left\| \sum_{k=m}^{\infty} a_k w_k \right\| = O\left(\Psi\left(\frac{1}{\beta_m}\right)\right)$$

as $m \to \infty$, then f is term-by-term strongly dyadically differentiable in X.

b) If

$$\sum_{k=0}^{\infty} \Psi\left(\frac{1}{k\beta_k}\right) < \infty$$

and if

$$\left\| \sum_{k=m}^{\infty} k a_k w_k \right\| = O\left(\Psi\left(\frac{1}{\beta_m}\right)\right)$$

as $m \to \infty$, then f is term-by-term strongly dyadically differentiable in X.

There are several open questions associated with this last result. For example, applying part a) to L^p, $0 < p < 1$, and the Block spaces $\mathscr{B}_q$, $q \geq 1$, shows that for any $\alpha > 1$, the function $f = \sum_{k=0}^{\infty} a_k w_k$ is term-by-term dyadically differentiable in $L^p[0,1)$ provided f converges in $L^p[0,1)$ and

$$m^{\alpha/p} \left\| \sum_{k=m}^{\infty} a_k w_k \right\|_p^p = O(1)$$

as $m \to \infty$; and that f is term-by-term dyadically differentiable in $\mathscr{B}_q$ provided f converges in $\mathscr{B}_q$ and

$$\frac{m^\alpha}{\log m}\left\|\sum_{k=m}^{\infty} a_k w_k\right\|_{\mathscr{B}_q} = O(1)$$

as $m \to \infty$. Do these hold when $\alpha = 1$? Similar questions about corollaries to part b) and the Powell-Wade result are also open.

Finally, there are other unexplored topics for term-by-term dyadic differentiation. F. Weisz has several papers (see [22], [23], and [24], for example) which explore the dyadic derivative for multiple Walsh series. Nothing is known about term-by-term dyadic differentiation on the unit square $\mathbf{Q} = [0,1) \times [0,1)$, either for the pointwise derivative, or the strong derivative, even in the Banach spaces $L^p(\mathbf{Q})$, $p \geq 1$.

Question 9.6 *If the derived Walsh series of* $f(x,y) = \sum_{j=1}^{\infty}\sum_{k=0}^{\infty} a_{k,j} w_k(x) w_j(y)$ *converges in* $L^p(\mathbf{Q})$, *or if* $\sum_{j=1}^{\infty}\sum_{k=0}^{\infty} k^2 j^2 a_{k,j} w_k(x) w_j(y)$ *converges everywhere on* $\mathbf{Q}$, *is* f *strongly or pointwise term-by-term dyadically differentiable?*

Schipp (see [17]) introduced the hybrid Hardy spaces $H_p^{\#}(\mathbf{Q})$ and proved that the maximal dyadic integral operator, I^*, takes positive functions in $H_1^{\#}$ to $L^1(\mathbf{Q})$. Weisz [23] proved that this result holds even if f takes negative values. In a later paper (see [24]), he proved that I^* is of type H^p, L^p for $1/2 < p \leq 1$, of weak type $H_1^{\#}, L^1$ and of type $H^{p,q}, L^{p,q}$ for all $1/2 < p < \infty$ and $0 < q \leq \infty$. Further progress has been made by Nie, Li, and Lou [11], who showed that I^* is of type $H_p^{\#}(\mathbf{Q}), H_p^{\#}(\mathbf{Q})$ for $0 < p < 1$ and by Chen and Zhang [3], who showed I^* is a bounded linear operator from the weak martingale Hardy spaces $wH^p(\mathbf{Q})$ to the weak Lebesgue spaces $wL^p(\mathbf{Q})$ for $1/2 < p < 1$.

Question 9.7 *What can be said about term-by-term dyadic differentiation, either pointwise or strong, in the hybrid Hardy spaces* $H_p^{\#}(\mathbf{Q})$ *or the weak Hardy spaces* $wH^p(\mathbf{Q})$ *for* $1/2 < p < 1$?

Finally, Chen and Liu (see [4] and [5]) have used the dyadic derivative and the dyadic integral to introduce norms on some martingale Banach spaces and prove inequalities including the boundedness of the Cesáro means in the spaces which are reminiscent of results in $L^p[0,1)$.

Question 9.8 *What can be said about term-by-term dyadic differentiation, either pointwise or strong, in these martingale Banach spaces?*

References

1. ∗P. L. Butzer, H. J. Wagner, "Walsh series and the concept of a derivative", *Appl Anal.*, Vol. 3, 1973, 29-46.
2. ∗P. L. Butzer, H. J. Wagner, "On dyadic analysis based on the pointwise dyadic derivative", *Analysis Math.*, Vol. 1, 1975, 171-196.
3. Li-Hong Chen, Chuan-Zhou Zhang, "The dyadic derivative and weak martingale Hardy spaces", *J. Math. (Wuhan)*, Vol. 27 , 2007, 961-964.
4. Li-Hong Chen, Peide Liu, "The dyadic derivative and Cesáro means of Banach-valued martingales", *Acta Math. Sci. Ser. B (Eng. Ed.)*, Vol. 29, 2009, 265-275.
5. Li-Hong Chen, Peide Liu, "Boundedness of the'dyadic derivative and Cesáro means on some *B*-valued martingale spaces", *Acta Math. Sci. Ser. B (Eng. Ed.)*, Vol. 31, 2011, 268-280.
6. ∗W. Engels, "On the characterization of the dyadic derivative", *Acta Math. Hungar.*, Vol. 46, 1985, 47-56.
7. N. J. Fine, "On the Walsh functions", *Trans. Amer. Math. Soc.*, Vol. 65, 1949, 372-414.
8. S. Fridli, W. R. Wade, "Rate of convergence and term by term differentiability of Walsh series", *Jour. d'Anal. Math.*, 1993, 287-305.
9. G. Gát, "On term by term dyadic differentiability of Walsh-Kaczmarz series", *Anal. Theory Appl.*, Vol. 19, 2003, 55-75.
10. ∗J. E. Gibbs, M. S. Millard, "Walsh functions as solutions of logical differential equations", *National Physical Lab*, Middlesex, England, DES Report No. 1, 1969.
11. Jian Ying Nie, Xing Guo Li, Guo Wei Lou, "The Martingale Hardy type inequalities for the dyadic derivative and integral", *Acta Math. Sin. (Eng. Ser.)*, Vol. 21, 2005, 1465-1474.
12. C. W. Onneweer, "Differentiability for Rademacher series on groups", *Acta Sci. Math. (Szeged)*, Vol. 39, 1977, 121-128.
13. C. H. Powell, W. R. Wade, "Term by term dyadic differentiation", *Can. J. Math.*, Vol. 33, 1981, 247-256.
14. C. H. Powell, W. R. Wade, "Term by term dyadic differentiation of rapidly convergent Walsh series", *Approx. Theory and Appl.*, Vol. 7, 1991, 20-40.
15. C. H. Powell, W. R. Wade, "Paley sets and term by term differentiation of Walsh series", *Acta Math. Hung.*, Vol. 62, No. 1-2, 1993, 89-96.
16. F. Schipp, "On term by term dyadic differentiability of Walsh series", *Analysis Math.*, Vol. 2, 1976, 149-154.
17. F. Schipp, W. R. Wade, "A fundamental theorem of dyadic calculus for the unit square", *Applic. Anal.*, Vol. 34, 1989, 203-218.
18. F. Schipp, P. Simon, W. R. Wade, *Walsh Series, An Introduction to Dyadic Harmonic Analysis*, Adam Hilger, 1989.
19. ∗V. A. Skvortsov, W. R. Wade, "Generalization of some results concerning Walsh series and the dyadic derivative", *Analysis Math.*, Vol. 5, 1979, 249-255.
20. W. R. Wade, "The Gibbs derivative and term by term dyadic differentiation of Walsh series", *Theory and Applications of Gibbs Derivatives*, P. L. Butzer, R. S. Stanković, (Eds.), Mathematical Institute, Belgrade, 1989, 59-72.
21. W. R. Wade, "Convergence of filtered Walsh-Fourier series", *Anales Univ. Ser. Sci. Budapest, Sect. Comp.*, 2010, 1-11.
22. F. Weisz, "Some maximal inequalities with respect to two-parameter dyadic differentiation and Cesáro summability", *Appl. Anal.*, Vol. 62, 1996, 223-238.
23. F. Weisz, "Martingale Hardy spaces and the dyadic derivative", *Analysis Math.*, Vol. 24, 1998, 58-77.
24. F. Weisz, "The two-parameter dyadic derivative and dyadic Hardy spaces", *Analysis Math.*, Vol. 26, 2000, 143-160.

The ∗ sign in the citations indicates that the paper is reprinted in this book.

Chapter 10
Why I got Interested in Dyadic Differentiation

William R. Wade

My story begins in 1969. I spent my first year after graduate school proving that the Walsh-Fourier series of an Lp function converges almost everywhere when $p > 1$. I submitted it for publication and waited on the referee's report. About three months later the report came back. The result was already known, proved by P. Sjolin a few months earlier.

This shook my world. At the University of Tennessee, I was supposed to produce a steady stream of research. I had just wasted almost a year proving something that was known. If I didn't do any better the next year, they would not retain me. The job market for Ph D's was brutal in those days, and I feared that I would never get another chance to work in the academic world, a world I already loved and hoped to stay in for the rest of my career.

Naturally, when a crisis occurs, one tries to think of ways to avoid such crises in the future. My first reaction was that I decided to spend much time in the library reading journals and poring over the *Mathematical Reviews* and *Zentralblatt*. I wanted to know exactly what was out there so that I would never again spend time proving a known result. By following this practice over the years, I ended up with a broader knowledge of what was known for the Walsh and Haar systems than almost anyone else in the business and was one of the main reasons that F. Schipp asked me to join his project for writing a treatise on Dyadic Harmonic Analysis. I also sent requests for reprints to authors and filed them away for future reference, an invaluable tool for research. In fact, at one time I had a copy of about 90% of the articles ever published in this field.

But it would take more than just knowing the literature to avoid the disappointment I had faced when I received that referee's report. Two things happened within weeks that would changed the way I did mathematics and give me a life-long strategy for conducting business.

William R. Wade
Mathematics and Computer Science, Biola University, La Mirada CA 90639, USA, e-mail: william.wade@biola.edu

R.S. Stanković et al. (eds.), *Dyadic Walsh Analysis from 1924 Onwards Walsh-Gibbs-Butzer Dyadic Differentiation in Science Volume 1 Foundations*, Atlantis Studies in Mathematics for Engineering and Science 12, DOI 10.2991/978-94-6239-160-4_10

The first one occurred during lunch with my colleagues at the University of Tennessee. I and three others often went to lunch at the school cafeteria together: the late William T. Eaton, who became a Sloan Fellow and had a distinguished career at the University of Texas, Robert J. Daverman, who had a distinguished career at the University of Tennessee and currently holds the prestigious office of Secretary to the American Mathematical Society, and J. S. Bradley, who after becoming head of the mathematics department at the University of Tennessee, ended up at the National Science Foundation exerting much influence on mathematical education in the process. The cafeteria was more crowded than usual and we were having difficulty finding a table. Bill led the way, with Bob, John, and me bringing up the rear, snaking through the tables single file. Suddenly, Bob said, "Well, we won't get a table following each other," and headed off in a direction normal to that in which the rest of us were heading. Almost immediately, he found an empty table. Coming so close to my disappointment, I thought: "This is the way I should conduct my research. Look for problems away from the crowd."

The next incident that happened was a visit from my academic "grandfather" Zygmund, who directed V. L. Shapiro's dissertation, who in turn directed mine, had come down to the University of Tennessee at my invitation to give a talk. While there, he told me that the Walsh system had always interested him since R. E. A. C. Paley had published the original article in 1932. He thought that the similarity of the Walsh system to the trigonometric system (they are the best piecewise constant approximation to the trigonometric system) coupled with their greater simplicity might prove useful in that problems might first be solved in the Walsh case that would lead to further advances in the trigonometric case. This has, indeed, proved to be true for several results. But what he said next helped me change the arc of my career. When I complained that research on Walsh functions did not get much attention, he said that the same had been true for multiple trigonometric series. "The reason for this was that we were busy generalizing older one-dimensional results, and had not yet asked the right questions, the questions that uniquely arose from the new setting." I decided that not only would I look for results far from the crowd, I would also start trying to ask questions and look at situations that had no trigonometric analogues.

About seven years later, during one of my weekly visits to the library, I discovered the first article F. Schipp wrote in 1976 on the dyadic derivative. This subject seemed to fit all my criteria. It had no trigonometric analogue, and was far from the crowd. Further stimulus was given me next year when I was awarded a Fulbright Senior Lectureship to spend a semester at Moscow State. The Russians had done so much for Walsh-Fourier analysis, but not one of them had a result on the dyadic derivative. So I began to prepare lectures on dyadic differentiation in hopes of teaching them something new while I worked in Moscow.

As a result of those lectures, V. A. Skvortsov and I wrote an important paper on the incompatibility of dyadic differentiation and classical continuity for nonconstant functions. Ironically, the effect of dyadic differentiation did not linger in Moscow. To this day, they have produced only two or three papers on the subject. In fact, when you read the word "binary derivative" in a Russian paper or book (e.g., the book by Golubov, Efimov, and Skvortsov), it refers to a symmetric derivative taken through dyadic rationals, not the dyadic derivative, i.e., it refers to a phenomenon which is determined only by the local behavior of the function, quite unlike dyadic differentiation where the value of the dyadic derivative of a function at a specific point is influenced by remote values of that function.

However, my efforts were not wasted. Working hard to present a comprehensive and enlightening set of lectures, I really understood the new concept and posed several problems which helped guide my research and that of my Ph.D. students for the next decade.

АКАДЕМИЯ
НАУК
СССР

MAGYAR
TUDOMÁNYOS
AKADÉMIA

ANALYSIS
MATHEMATICA

TOMUS 5 FASCICULUS 3 1979

Separatum

AKADÉMIAI KIADÓ, BUDAPEST

Analysis Mathematica is an international quarterly sponsored jointly by the Academy of Sciences of the USSR and the Hungarian Academy of Sciences.

It is dedicated primarily to such problems of classical mathematical analysis as differentiation and integration of functions, measure theory, analytic and harmonic functions, Fourier analysis and orthogonal expansions, approximation of functions and quadrature formulae, function spaces, extremal problems, inequalities, etc.

The quarterly publishes research papers containing new essential results with complete proofs, and occasionally also survey papers prepared to the initiative of the Editorial Board.

The four issues published per year make up a volume of about 320 printed pages.

Papers are published according to the author's choice in English, Russian and occasionally in French or German.

Manuscripts for publication should be submitted to any of the two branches of the Editorial Board of Analysis Mathematica:

Analysis Mathematica — международный журнал, издаваемый совместно Академией наук СССР и Венгерской Академией наук.

Журнал посвящен в первую очередь проблемам классического математического анализа, таким как теория дифференцирования и интегрирования функций, теория меры, аналитические и гармонические функции, анализ Фурье и ортогональные разложения, приближение функций и квадратурные формулы, пространства функций, экстремальные задачи, неравенства и т. д.

В журнале публикуются оригинальные научные статьи, содержащие новые существенные результаты с полными доказательствами, а также обзорные статьи, которые подготавливаются по заказу редколлегии.

В год выходят четыре номера журнала, которые составляют том объемом примерно 320 печатных страниц.

Статьи по усмотрению авторов публикуются на русском, английском и в отдельных случаях на немецком или французском языках.

Рукописи статей следует направлять в одну из секций редколлегии журнала:

Bolyai Institute
Aradi vértanúk tere 1
6720 Szeged, Hungary.

СССР, Москва 117 333, ул. Вавилова 42
Математический институт
им. В. А. Стеклова АН СССР.

Распространитель в СССР:

СОЮЗПЕЧАТЬ, СССР, Москва 123 308, Новохорошевское шоссе 4.

Distributors:

KULTURA, Budapest 1389, P.O. Box 149, Hungary
(for Albania, Bulgaria, China, Cuba, Czechoslovakia, German Democratic Republic, Hungary, Korean People's Republic, Mongolia, Poland, Rumania, Vietnam, and Yugoslavia);

PERGAMON PRESS, Oxford OX3 0BW, Headington Hill Hall, England, or
MAXWELL HOUSE, Elmsford, N.Y. 10 523, Fairview Park, USA
(for all other countries). Subscription rate per volume US Dollars 50.00 including postage and insurance.

Analysis Mathematica, 5 (1979), 249—255

Generalization of some results concerning Walsh series and the dyadic derivative

V. A. SKVORCOV and W. R. WADE

In this paper we present generalizations of some recent results concerning Walsh series, the dyadic derivative introduced by GIBBS, BUTZER and WAGNER (see [1], [2], [5], [9]), and the possibility of dyadically differentiating a Walsh series term by term. Our proofs, instead of being modifications of the earlier proofs, are new and less complicated.

In section 2 we prove a theorem which gives conditions for a Walsh series to be dyadically differentiable. A consequence of this theorem (see Corollary 2 below) is the result of SCHIPP [7] which settled a problem left open by BUTZER and WAGNER [2] concerning whether a Walsh series $\sum a_k w_k(x)$ could be dyadically differentiated term by term if its coefficients satisfied $ka_k \downarrow 0$ as $k \to \infty$.

In section 3 we show that a non-constant continuous function cannot be dyadically differentiable at all but countable many points in the interval (0, 1). This contains the celebrated result of Butzer and Wagner (see Theorem A below) and emphasizes once again the basic incompatibility of the classical concept of smoothness and dyadic differentiability.

By combining these two theorems, it is possible to obtain certain conditions on the Walsh—Fourier coefficients of a continuous function sufficient to conclude that the function is constant. As an example, we state Corollary 3 below, which was obtained by COURY [3] by another method.

1. In this section we briefly review basic definitions and well-known properties of Walsh functions and the dyadic derivative.

For each positive integer k and each point $x \in [0, 1)$ we shall represent their respective dyadic expansions by

$$(1) \qquad k = \sum_{j=0}^{\infty} k_j 2^j, \quad k_j = 0 \text{ or } 1 \quad \text{and} \quad x = \sum_{j=1}^{\infty} x_j 2^{-j}, \quad x_j = 0 \text{ or } 1.$$

To avoid ambiguity in (1), we shall use the finite expansion for each dyadic rational x.

Received July 1, 1978.

The *Walsh—Paley functions* $w_0, w_1, \ldots$ can be defined by the expression

$$w_k(x) = \exp\left(\pi i \sum_{j=0}^{\infty} k_j x_{j+1}\right).$$

The *dyadic sum* of two points $x, y \in [0, 1)$ is given in terms of their dyadic expansions as follows:

$$x \dotplus y = \sum_{j=1}^{\infty} |x_j - y_j| 2^{-j}.$$

It is well-known (see [4]) that under this operation the interval $[0, 1)$ is essentially a compact group whose character group is the set of Walsh functions.

The *dyadic derivative* $df(x)$ of a function $f(x)$ at a point x is defined by the equations

$$\text{(2)} \qquad df(x) = \lim_{n\to\infty} d_n f(x), \quad \text{where} \quad d_n f(x) = \sum_{j=0}^{n-1} 2^{j-1}[f(x) - f(x \dotplus 2^{-j-1})].$$

It is not difficult to verify (see e.g. [9]) that

$$\text{(3)} \qquad dw_k(x) = d_n w_k(x) = k w_k(x) \quad \text{for} \quad k < 2^n,$$

and for each $x \in [0, 1)$. Also, it is well-known (see [4] or [8]) that if $D_k(t) = \sum_{j=0}^{k-1} w_j(t)$ represents the Walsh—Dirichlet kernel, then

$$\text{(4)} \qquad |D_k(t)| \le \frac{\text{const}}{t}$$

for $t \in (0, 1)$ and $k = 1, 2, \ldots$.

2. In this section we shall prove the following

Theorem 1. *Let $\{a_k\}$ be a sequence of real numbers and set*

$$\text{(5)} \qquad K_n = |a_{2^n}| + |a_{2^{n+1}-1}| + \sum_{k=2^n}^{2^{n+1}-2} |a_k - a_{k+1}| \quad \left(n = 0, 1, \ldots, \sum_{k=1}^{0} = 0\right).$$

If

$$\text{(6)} \qquad 2^n K_n \to 0 \quad as \quad n \to \infty,$$

then the sum

$$\text{(7)} \qquad f(x) = \sum_{k=0}^{\infty} a_k w_k(x)$$

is defined for all $x \in (0, 1)$; moreover, for each point $x \in (0, 1)$, $x \ne 2^{-j-1}$, $j = 0, 1, \ldots$, a necessary and sufficient condition that $df(x)$ exist is the convergence of the 2^nth *partial sums of the series* $\sum_{k=0}^{\infty} k a_k w_k(x)$. *In either case, the following equation holds:*

$$df(x) = \lim_{n\to\infty} \sum_{k=0}^{2^n-1} k a_k w_k(x),$$

where the existence of the right-hand side implies the existence of the left-hand side, and visa versa.

To begin the proof of this theorem, we observe by Abel's transformation that

$$(8) \quad \sum_{k=2^m}^{2^{m+1}-1} a_k w_k(x) = \sum_{k=2^m}^{2^{m+1}-2} (a_k - a_{k+1}) D_{k+1}(x) + a_{2^{m+1}-1} D_{2^{m+1}}(x) - a_{2^m} D_{2^m}(x)$$

for all $x \in (0, 1)$ and for $m = 0, 1, \ldots$. In particular, since $a_k \to 0$ follows from hypothesis (6), we conclude from (4), (5), and (6) that the series represented by (7) converges for all $x \in (0, 1)$, and consequently, that $f(x)$ is defined for all $x \in (0, 1)$.

Next, set

$$R_n(x) = \sum_{k=2^n}^{\infty} a_k w_k(x)$$

and notice by (3) that

$$d_n f(x) = \sum_{k=0}^{2^n-1} k a_k w_k(x) + d_n(R_n(x)).$$

In view of (2), then, Theorem 1 will be established if we can show that $d_n(R_n(x)) \to 0$, as $n \to \infty$, for every point $x \in (0, 1)$ with $x \neq 2^{-j-1}$ for any integer $j \geq 0$.

Toward this, apply (8) for each $m \geq n$ to verify that

$$d_n(R_n(x)) \equiv \sum_{m=n}^{\infty} d_n\left(\sum_{k=2^m}^{2^{m+1}-1} a_k w_k(x)\right) =$$

$$= \sum_{m=n}^{\infty} \left\{ \sum_{k=2^m}^{2^{m+1}-2} (a_k - a_{k+1}) d_n(D_{k+1}(x)) + a_{2^{m+1}-1} d_n(D_{2^{m+1}}(x)) - a_{2^m} d_n(D_{2^m}(x)) \right\}.$$

In particular, we can estimate $d_n(R_n(x))$ as follows

$$|d_n(R_n(x))| \leq W_n(x) \sum_{m=n}^{\infty} K_m,$$

where for each $x \in (0, 1)$ we have set

$$W_n(x) = \sup_{k \geq 1} |d_n(D_k(x))|.$$

Now, by hypothesis (6), we know that

$$\sum_{m=n}^{\infty} K_m = o\left(\frac{1}{2^n}\right) \quad \text{as} \quad n \to \infty,$$

and therefore it remains to show that $W_n(x)=O(2^n)$, as $n\to\infty$, for each point $x\in(0, 1)$ with $x\neq 2^{-j-1}$.

Fix such an x and observe by (4) that

$$|D_k(x\dotplus 2^{-j-1})| < C(x) < \infty \quad (x \neq 2^{-k},\ k = 1, 2, \dots) \tag{9}$$

for all $k=1, 2, \dots$ and $j=0, 1, \dots$. Moreover, by definition we have

$$d_n(D_k(x)) = \sum_{j=0}^{n-1} 2^{j-1}[D_k(x)-D_k(x\dotplus 2^{-j-1})].$$

Therefore, by (4) and (9) we conclude that

$$|d_n(D_k(x))| \leqq C_1(x) \sum_{j=0}^{n-1} 2^{j-1} < 2^n C_1(x),$$

i.e., $W_n(x)=O(2^n)$. This completes the proof of Theorem 1.

A sequence $\{a_k\}$ is said to be *monotone in dyadic blocks* if $a_{2^n}\geqq a_{2^n+1}\geqq \dots \geqq a_{2^{n+1}-1}\geqq 0$ for n sufficiently large. For such sequences, the numbers K_n defined by (5) obviously satisfy $K_n=2a_{2^n}$ for n sufficiently large. Hence Theorem 1 contains the following result.

Corollary 1. *Suppose that $\{a_k\}$ is monotone in dyadic blocks. If $2^k a_{2^k}\to 0$ as $k\to\infty$, then the function*

$$f(x) = \sum_{k=0}^{\infty} a_k w_k(x) \tag{10}$$

is dyadically differentiable at a point $x\in(0, 1)$, $x\neq 2^{-j-1}$, $j=0, 1, \dots$, if and only if the 2^nth partial sums of the series $\sum_{k=0}^{\infty} ka_k w_k(x)$ converge in which case

$$df(x) = \lim_{n\to\infty} \sum_{k=0}^{2^n-1} ka_k w_k(x).$$

Since it is quite easy to see that the series $\sum_{k=0}^{\infty} ka_k w_k(x)$ converges when $ka_k\downarrow 0$, as $k\to\infty$, Theorem 1 contains the following result as well.

Corollary 2 (Schipp [7]). *If $ka_k\downarrow 0$ as $k\to\infty$, then the function* (10) *is dyadically differentiable at each point $x\neq 2^{-j-1}$, $j=0, 1, \dots$, and in fact,*

$$df(x) = \sum_{k=0}^{\infty} ka_k w_k(x).$$

3. BUTZER and WAGNER [2, Theorem 6.7] have established the following

Theorem A. *If $f(x)$ is continuous on $[0, 1]$, and if $df(x)$ exists everywhere on $[0, 1]$ and is continuous everywhere on $[0, 1]$, then $f(x)$ is constant.*

In this section we show that the hypothesis in Theorem A concerning the continuity of df can be discarded. In fact, the following is true.

Theorem 2. *If $f(x)$ is continuous on $(0, 1)$ and if $df(x)$ exists and is finite for all but countably many points in $(0, 1)$, then $f(x)$ is constant.*

To prove Theorem 2, let E denote the countable set of points at which the dyadic derivative of f fails to exist. By (2), then, the series

$$\sum_{j=0}^{\infty} 2^{j-1}[f(x)-f(x\dotplus 2^{-j-1})]$$

converges for each $x\in(0, 1)\setminus E$. In particular, we have that

$$\lim_{j\to\infty} \frac{f(x\dotplus 2^{-j})-f(x)}{2^{-j}} = 0 \quad \text{for} \quad x\notin E. \tag{11}$$

Next, we observe that given $x\in[0, 1)$ we can choose integers $j_1<j_2<\dots$ such that $x_{j_n}=0$ for $n=1, 2, \dots$, where the coefficients x_j are defined by (1). Hence, by the definition of the dyadic sum $\dotplus$, we have that $x\dotplus 2^{-j_n}=x+2^{-j_n}$ for $n=1, 2, \dots$. It follows from (11), then, that

$$\lim_{n\to\infty} \frac{f(x+2^{-j_n})-f(x)}{2^{-j_n}} = 0 \quad \text{for} \quad x\notin E.$$

We have thus established that the upper right and lower right Dini derivates of f satisfy the inequality

$$D^+f(x) \geqq 0 \geqq D_-f(x) \quad \text{for} \quad x\in(0, 1)\setminus E.$$

Since the set E is countable, and the function f is continuous, we conclude (see [6]) that $f(x)$ is constant on $(0, 1)$. The proof of Theorem 2 is complete.

Suppose now that the a_k's in Theorem 1 are the Walsh—Fourier coefficients of some continuous function f. If they satisfy condition (6) and the 2^nth partial sums of the series $\sum_{k=0}^{\infty} ka_k w_k(x)$ converge for all but countably many points in $(0, 1)$, then by Theorem 1 f is dyadically differentiable at all but countably many points in $(0, 1)$. Thus by Theorem 2, the function f must be constant. A special case of this result is the following

Corollary 3 (Coury [3]). *If f is a continuous function whose Walsh—Fourier coefficients a_k satisfy $ka_k \downarrow 0$ as $k \to \infty$, then f is constant.*

Notice that if the function f in Theorem 2 is not continuous, then even the condition

(12) $$df(x) \equiv 0 \quad \text{for} \quad x \in [0, 1]$$

is not sufficiently powerful to conclude that f is constant. Indeed, divide the interval [0, 1] into classes of rationality and consider a function f which is constant on each class.

However, if f is continuous on the group, then condition (12) does imply that f is constant. Indeed, Wagner [9] has shown that any function which satisfies (12) and which assumes its maximum at a point x_0, necessarily satisfies $f(x_0 \dotplus \varrho) = $ $=$const for all dyadic rationals ϱ. Since the group ([0, 1], $\dotplus$) is compact (see [4]), we know that any of its continuous functions assumes its maximum, and therefore, if such a function also satisfies (12), then there is a point x_0 such that $f(x_0 \dotplus \varrho) = $ $=$const for all dyadic rationals ϱ. But given $x \in [0, 1)$ there is a sequence $\{\varrho_n\}$ of dyadic rationals such that $x_0 \dotplus \varrho_n \to x$ on the group as $n \to \infty$. Since f is continuous on the group, it follows that f is constant throughout [0, 1).

The following is still an open question. If $f(x)$ is continuous on the group and if $df(x) = 0$ for all but countably many points $x \in (0, 1]$, is $f(x)$ constant?

References

[1] P. L. Butzer and H. J. Wagner, On a Gibbs-type derivative in Walsh—Fourier analysis with applications, *Proc. of Electronics Conference, Chicago, 1972;* 393—398 (Oak Brook, Illinois, 1972).

[2] P. L. Butzer and H. J. Wagner, On dyadic analysis based on the pointwise dyadic derivative, *Analysis Math.*, **1** (1975), 171—196.

[3] J. E. Coury, Walsh series with coefficients tending monotonically to zero, *Pacific J. Math.*, **54** (1974), 1—16.

[4] N. J. Fine, On the Walsh functions, *Trans. Amer. Math. Soc.*, **65** (1949), 372—414.

[5] J. E. Gibbs, *Some properties of functions on the nonnegative integers less than* 2^n, NPL (National Physical Laboratory), Middlessex, England, DES Rept., no. 3 (1969).

[6] S. Saks, *Theory of the integral*, Dover (New York, 1964) — С. Сакс, *Теория интеграла*, Иностранная литература (Москва, 1949).

[7] F. Schipp, On term by term dyadic differentiability of Walsh series, *Analysis Math.*, **2** (1976), 149—154.

[8] А. А. Шнейдер, О сходимости рядов Фурье по функциям Уолша, *Матем. сб.*, **34** (1954), 441—472.

[9] H. J. Wagner, *Ein Differential- und Integralkalkül in der Walsh—Fourier Analysis mit Anwendungen*, Westdeutscher Verlag (Köln—Opladen, 1973).

Обобщение некоторых результатов, касающихся рядов Уолша и двоичной производной

В. А. СКВОРЦОВ и У. Р. УЭЙД

Обобщаются некоторые недавние результаты, связанные с поточечным дифференцированием, введенным Гибсом, Бутцером и Вагнером и предназначенным для почленного дифференцирования рядов Уолша. Наряду с обобщением, целью статьи является дать существенно более простые доказательства обобщаемых результатов.

Устанавливается, что для рядов Уолша $\sum_{k=0}^{\infty} a_k w_k(x)$, коэффициенты которых удовлетворяют условию

$$2^n\left\{|a_{2^n}|+|a_{2^{n+1}-1}|+\sum_{k=2^n}^{2^{n+1}-2}|a_k-a_{k+1}|\right\}\to 0 \quad \text{при} \quad n\to\infty,$$

двоичная дифференцируемость суммы ряда Уолша в произвольной точке $x\in(0, 1)$, $x\neq 2^{-j}$, $j=1, 2, \ldots$, эквивалентна сходимости в этой точке последовательности частичных сумм с номерами 2^n ряда $\sum_{k=0}^{\infty} ka_k w_k(x)$.

Тем самым обобщается теорема Шиппа о почленном двоичном дифференцировании ряда Уолша, коэффициенты которого удовлетворяют условию $ka_k\downarrow 0$ при $k\to\infty$.

Доказывается также, что если непрерывная на (0, 1) функция имеет конечную двоичную производную всюду, кроме, быть может, счетного множества, то она является постоянной. Этот результат обобщает теорему Бутцера и Вагнера, где дополнительно предполагалась непрерывность производной.

В. А. СКВОРЦОВ
СССР, МОСКВА 117 234
МОСКОВСКИЙ ГОСУДАРСТВЕННЫЙ
УНИВЕРСИТЕТ ИМ. М. В. ЛОМОНОСОВА
МЕХАНИКО-МАТЕМАТИЧЕСКИЙ ФАКУЛЬТЕТ

W. R. WADE
UNIVERSITY OF TENNESSEE
DEPARTMENT OF MATHEMATICS
KNOXVILLE 37 916, USA

Acta Math. Hung.
46 (1—2) (1985), 47—56.

ON THE CHARACTERIZATION OF THE DYADIC DERIVATIVE

W. ENGELS (Aachen)

1. Introduction

Two essential open problems in dyadic analysis, as initiated by J. E. Gibbs (and his collaborators M. J. Millard and B. Ireland) [9, 10, 11] and further developed by P. L. Butzer and H. J. Wagner [2, 3, 4, 5], F. Schipp (and his collaborators J. Pál and P. Simon) [19, 20, 21, 25, 26], N. R. Ladhawala [14], R. Penney [23], C. W. Onneweer [15, 16, 17], Zheng-Wei-xing and Su-Wei-yi [32], and He Zelin [13] are the characterization and interpretation of the dyadic derivative. Gibbs and Ireland [12] gave a first, but rather abstract interpretation in the realm of locally compact abelian groups. However, just as the classical derivative may be associated with the slope of a tangent to a curve, or with the speed of an object — thus associated with basic geometric or physical notions — there is still no intuitive interpretation of the dyadic derivative (which may lie in the setting of these modern sciences such as information or signal theory which make use of dyadic Walsh analysis). Nevertheless, as a further step in this direction Skvorcov and Wade [28] derived a first characterization of the class of $f\in C[0, 1)$; they improved some earlier results due to Bočkarev [1], Butzer and Wagner [5] as well as to Schipp [27].

The aim of this paper is to give a rather complete characterization of the class of functions that are dyadic differentiable in dependence upon the discontinuities of the first kind (which means only jumps) of the function in question. The cases that the function has a finite number of jumps or an infinite number of jumps under the additional constraint that this set of discontinuities has only a finite number of cluster points are distinguished. Our main results state roughly that a function of either case is dyadic differentiable if and only if it is a piecewise constant.

Although this is a rather restrictive condition, the dyadic derivative is especially adapted to functions that have only a few or small intervals of constancy. It is even applicable to functions having a denumerable set of discontinuities like the well-known Dirichlet-function.

This paper will deal with the situation of functions defined on [0, 1). The corresponding material for functions defined on the positive real axis $\mathbf{R}^+$ will be treated in a further paper.

The paper is divided into four sections. Section 2 is concerned with a summary of the fundamental properties of dyadic analysis, including dyadic representation, dyadic addition, the basic Walsh functions, some elements of Walsh—Fourier transforms, as well as the dyadic derivative. Section 3 deals with two characterization theorems. Finally, Section 4 is devoted to some representative examples which are worked out in detail.

2. Preliminaries

2.1. Dyadic addition. In the following, let $\mathbf{N}:=\{1,2,3,\dots\}$, $\mathbf{P}:=\mathbf{N}\cup\{0\}$ and $\mathbf{Z}:=\{0,\pm 1,\pm 2,\dots\}$. Each $k\in\mathbf{P}$ has a unique dyadic expansion

$$k=\sum_{j=0}^{\infty}k_j 2^j \quad (k_j\in\{0,1\}). \tag{2.1}$$

Likewise each $x\in[0,1)$ has a unique dyadic representation

$$x=\sum_{j=0}^{\infty}x_j 2^{-j} \quad (x_j\in\{0,1\}) \tag{2.2}$$

if the finite expansion is chosen in case x belongs to the dyadic rationals (=D. R.), i.e., the set of all numbers of the form $x=p2^{-q}\in[0,1)$, $p\in\mathbf{P}$, $q\in\mathbf{N}$. The dyadic sum of $x=\sum_{j=1}^{\infty}x_j 2^{-j}$ and $y=\sum_{j=1}^{\infty}y_j 2^{-j}$ is defined by

$$x\oplus y=\sum_{j=1}^{\infty}h_j 2^{-j}, \tag{2.3}$$

$$h_j:=x_j\dot{+}y_j:=(x_j+y_j)_{\mathrm{mod}\,2}=|x_j-y_j|. \tag{2.4}$$

In view of the uniqueness of the representation, dyadic addition is only defined for almost all $y\in[0,1)$. For example, one easily sees that

$$x\oplus 2^{-j}=x-2^{-j}(x_j-(x_j\dot{+}1)), \tag{2.5}$$

and, setting

$$J_0:=\{j\in\mathbf{P},\ x_{j+1}=0\},\quad J_1:=\{j\in\mathbf{P},\ x_{j+1}=1\}, \tag{2.6}$$

$$x\oplus 2^{-j-1}=\begin{cases}x+2^{-j-1}, & j\in J_0\\ x-2^{-j-1}, & j\in J_1.\end{cases} \tag{2.7}$$

Formulas (2.5)—(2.7) will often be used later on in this paper.

2.2. Walsh functions. The functions which are taken as a basis for dyadic analysis are the Walsh functions $\psi_k(x)$ [31]. For $x\in[0,1)$ — using Paley's enumeration [22] — they are given by

$$\psi_k(x)=\exp\Big\{\pi i\sum_{j=0}^{\infty}k_j x_{j+1}\Big\}=(-1)^{\sum_{j=0}^{\infty}k_j x_{j+1}},\quad k\in\mathbf{P}, \tag{2.8}$$

and on $\mathbf{R}$ (set of all reals) by periodic extension. The ψ_k, $k\in\mathbf{P}$, form a complete, orthonormal system, and possess the important property

$$\psi_k(x\oplus y)=\psi_k(x)\psi_k(y)\quad (k\in\mathbf{P}) \tag{2.9}$$

for fixed $x\in[0,1)$ and almost all $y\in[0,1)$.

Denote by $L^p(0,1)$, $1\le p\le\infty$, the set of all functions f of period 1 which are p-th power Lebesgue integrable and endowed with norm

$$\|f\|_p:=\Big\{\int_0^1|f(x)|^p\,dx\Big\}^{1/p}, \tag{2.10}$$

$L^\infty(0, 1)$ that of all essentially bounded functions f of period 1 with norm

$$\|f\|_\infty := \operatorname*{ess\,sup}_{x\in[0,1)} |f(x)|, \tag{2.11}$$

and finally $C^\oplus[0, 1]$ that of all dyadic continuous functions f of period 1, endowed with the usual sup norm, thus

$$C^\oplus[0,1] := \{f;\ \lim_{h\to 0} \|f(\cdot\oplus h)-f(\cdot)\|_{C^\oplus} := \lim_{h\to 0} \sup_{x\in[0,1)} |f(x\oplus h)-f(x)| = 0\}. \tag{2.12}$$

In the following $X=X(0, 1)$ always stands for one of the (Banach) spaces $L^p(0, 1)$, $1\leq p<\infty$, $L^\infty(0, 1)$, and $C^\oplus[0, 1]$, with norm $\|f\|_X$.

Denoting the Walsh—Fourier coefficients of $f\in X(0, 1)$ by

$$f^\wedge(k) = \int_0^1 f(u)\psi_k(u)\,du \quad (k\in\mathbf{P}), \tag{2.13}$$

the formal series

$$\sum_{k=0}^{\infty} f^\wedge(k)\psi_k(x) \tag{2.14}$$

is called the Walsh—Fourier series of f.

2.3. Dyadic differentiation. Further, the concept of dyadic differentiation, as defined by Butzer and Wagner [2], [4], is basic.

DEFINITION 3.1. a) Let $f\in X(0, 1)$. If there exists $g\in X(0, 1)$ such that

$$\lim_{m\to\infty} \Big\| \sum_{j=0}^{m} 2^{j-1}[f(\cdot)-f(\cdot\oplus 2^{-j-1})]-g(\cdot)\Big\|_X = 0, \tag{2.15}$$

then g is called the first strong dyadic derivative of f, denoted by $g=D^{[1]}f$. Derivatives of higher order are defined successively by $(D^{[0]}f:=f)$

$$D^{[r]}f = D^{[1]}(D^{[r-1]}f), \quad r\in\mathbf{N}. \tag{2.16}$$

b) Let f be defined on $[0, 1)$. If

$$\sum_{j=0}^{\infty} 2^{j-1}[f(x)-f(x\oplus 2^{-j-1})] = c < \infty \tag{2.17}$$

for $x\in[0, 1)$, then c is called the first pointwise dyadic derivative of f at x, denoted by $f^{[1]}(x)$. Setting $f^{[0]}(x)=f(x)$, derivatives of higher order are given by

$$f^{[r]}(x) = (f^{[r-1]})^{[1]}(x) \quad (r\in\mathbf{N}). \tag{2.18}$$

Some of the most important properties of the dyadic derivative are (for the proofs see [2, 3, 4]) listed below:

(i) $D^{[r]}$ is a closed, linear operator in $X(0, 1)$.

(ii) The Walsh functions ψ_k are infinitely differentiable, and

$$D^{[r]}\psi_k = \psi_k^{[r]} = k^r\psi_k. \tag{2.19}$$

(iii) If $f, D^{(r)}f\in X(0, 1)$ for some $r\in\mathbf{N}$, then

$$[D^{[r]}f]^\wedge(k) = k^r f^\wedge(k) \quad (k\in\mathbf{P}). \tag{2.20}$$

The pointwise counterpart of (2.20) can be found in [5]. Moreover, one can construct antidifferentiation operators to (2.15) and (2.17), respectively. This leads to the fundamental theorem of dyadic analysis; it is similar to the fundamental theorem of classical differential and integral calculus. For the strong version of the dyadic derivative the fundamental theorem is due to Butzer and Wagner [4], in the more difficult pointwise sense it is due to Schipp [25], [26]. The dyadic derivative gives information about the smoothness of a function. It allows one to study the rate of approximation of dyadic differentiable functions by partial sums of Walsh—Fourier series, the best degree of approximation in the Walsh setting, partial differential equations with the derivatives being understood in the dyadic sense, etc. One of the more important application lies, for instance, in the field of digital signal processing, cf. [6], [7], [8], [29].

3. Characterization theorems

Though the dyadic derivative has attained significance in various fields of dyadic (Walsh)-analysis, an interpretation of it or even a full characterization of the class of functions which are dyadic differentiable is still lacking. In this respect, however, a first result, due to Skvorcov and Wade [28], which generalizes an earlier one by Butzer and Wagner [5], states:

Let f be continuous on $[0, 1)$, *and let $f^{[1]}$ exist for all but countably many points $x \in (0, 1)$. Then f is constant.*

This result tells us that it is not reasonable to begin with functions that are continuous on the whole interval of definition $[0, 1)$. So we will deal with functions f that have infinitely many discontinuities. It seems convenient to divide this class of functions into two parts, namely those functions the set of discontinuity-points of which possesses either a finite or an inifinite number of cluster-points in $[0, 1)$. The first group of functions will be treated in our main theorem. For the second group it is clearly not possible to obtain an equivalent statement between dyadic differentiability and piecewise constancy, because the members of this class of functions cannot consist totally of piecewise (non-degenerate) constant functions. One could even say that in this event the intervals of constancy may degenerate to points. Nevertheless even those functions can be dyadic differentiable. In the next section we will show that Dirichlet's function, which is an important member of this 'exotic' class, is indeed dyadic differentiable. However, in case of f having a countable set of discontinuities which have at most a *finite* number of cluster-points we are able to prove the following

THEOREM. *Let f be defined and bounded on* $[0, 1)$, *possessing a countable set of discontinuities $x^{(k)}$, $k \in \mathbf{N}$, exclusively of first kind, which have at most a finite number of cluster-points in* $[0, 1)$. *Then f is pointwise dyadic differentiable except on a countable set in* $[0, 1)$ *if and only if the function is a piecewise constant on* $[0, 1)$.

PROOF. In the following it is assumed that the function is righthand continuous at the points of discontinuity. Although our results are also valid in case of left-hand continuity, they are slightly more complicated to establish.

At frist we will deal with the case that f has only one discontinuity, namely $x^{(1)} \in [0, 1)$. Suppose that f is dyadic differentiable on $[0, 1)$ except on a countable set $H\subset[0, 1)$. Then the series (2.17) converges for all $x\in[0, 1)\setminus H$, and therefore

$$\lim_{j\to\infty} 2^{j-1}[f(x)-f(x\oplus 2^{-j-1})] = 0. \tag{3.1}$$

If $x^{(1)}=0$, then f is continuous on $(0, 1)$, and with the help of the result of Skvorcov and Wade [28] cited above, one immediately concludes that f is constant on $(0, 1)$. Now suppose that $x^{(1)}>0$. Then f is continuous on $(0, x^{(1)})\cup(x^{(1)}, 1)$. If $x\in(0, x^{(1)})$, one can choose a sequence of integers $(j_n)_{n=1}^{\infty}$ with $j_n<j_{n+1}$, $n\in\mathbf{N}$, such that $x_{j_n}=0$ for all $n\in\mathbf{N}$. Noting (2.5) and (3.1), one deduces for all $x\notin H$

$$\lim_{n\to\infty}\frac{f(x)-f(x\oplus 2^{-j_n})}{2^{-j_n}} = \lim_{n\to\infty}\frac{f(x)-f(x+2^{-j_n})}{2^{-j_n}} = 0.$$

Likewise, if $x\notin$D.R., then there exists a sequence of integers $(\bar{j}_n)_{n=1}^{\infty}$ with $\bar{j}_n<\bar{j}_{n+1}$, $n\in\mathbf{N}$ such that $x_{\bar{j}_n}=1$ for all $n\in\mathbf{N}$. So similarly, one concludes for all $x\notin H\cup$D.R.

$$\lim_{n\to\infty}\frac{f(x)-f(x\oplus 2^{-\bar{j}_n})}{2^{-\bar{j}_n}} = \lim_{n\to\infty}\frac{f(x)-f(x-2^{-\bar{j}_n})}{2^{-\bar{j}_n}} = 0.$$

Therefore the upper and lower right Dini derivatives of f exist and are equal for $x\notin H\cup$D.R. (cf. [24], pp. 155). Since $H\cup$D.R. is countable, and f is continuous on $[0, x^{(1)})$, it follows as in [28] (using [24]) that $f=$const. on $(0, x^{(1)})$. Analogously one can verify that f is also constant on $(x^{(1)}, 1)$.

Conversely, suppose that f is a piecewise constant function on $[0, 1)$, possessing a jump at $x^{(1)}\in(0, 1)$. Obviously this function can be expressed as

$$f(x) = \begin{cases} A, & x\in[0, x^{(1)}) \\ B, & x\in[x^{(1)}, 1) \end{cases} \quad (A, B\in\mathbf{R}). \tag{3.2}$$

If $A=B$, the matter is trivial, because in this event f equals $A\psi_0$, which is clearly dyadic differentiable. So, let $A\neq B$; with J_0, J_1 defined as in (2.6) one has by (2.5), (2.7)

$$f(x\oplus 2^{-j-1}) = \begin{cases} f(x+2^{-j-1}), & j\in J_0 \\ f(x-2^{-j-1}), & j\in J_1. \end{cases} \tag{3.3}$$

It can be easily shown that for $x\in[0, x^{(1)})$ there exist $\bar{J}, \bar{\bar{j}}\in\mathbf{N}$ such that for the difference $f(x)-f(x\oplus 2^{-j-1})$ the following three relations are valid:

$$f(x)-f(x\oplus 2^{-j-1}) = \begin{cases} 0, & j\in J_1 \\ A-B, & j\in J_0;\ j\leqq \bar{J}, \quad x\in[0, x^{(1)}) \\ 0, & j\in J_0;\ j>\bar{J}, \end{cases} \tag{3.4}$$

$$f(x)-f(x\oplus 2^{-j-1}) = \begin{cases} B-A, & j\in J_1; \\ 0, & j\in J_0, \end{cases} \quad x = x^{(1)} \tag{3.5}$$

$$f(x)-f(x\oplus 2^{-j-1}) = \begin{cases} B-A, & j\in J_1,\ j<\bar{\bar{j}} \\ 0, & j\in J_1;\ j\geqq\bar{\bar{j}}, \quad x\in[x^{(1)}, 1) \\ 0, & j\in J_0. \end{cases} \tag{3.6}$$

In view of these, the pointwise dyadic derivative (2.17) clearly exists for all $x\in[0, 1)\setminus\{x^{(1)}\}$. If $x^{(1)}$ is a dyadic rational, f is dyadic differentiable there too. Hence f is dyadic differentiable almost everywhere on $[0, 1)$. It is now obvious that both directions of the proof can easily be extended to the case of a finite number of discontinuities $x^{(k)}$, $1\leq k\leq n$.

Let us now assume that f possesses infinitely many discontinuities having in the first instance only one cluster point. Then, clearly, the limit $\lim_{k\to\infty} x^{(k)}=x_0\in[0, 1]$ exists. Furthermore, the sequence may without loss of generality be assumed to be monotone, namely $x^{(k)}>x^{(k+1)}$. Now construct a sequence of intervals $I_k=(x^{(k+1)}, x^{(k)})$ with $I_k\cap I_j=\emptyset$, $k\neq j$, so that the interval $[0, 1)$, apart from the jumps $x^{(k)}$, is representable as a union of pointwise disjoint intervals:

$$[0, x_0)\cup \bigcup_{k=1}^{\infty} (x^{(k+1)}, x^{(k)})\cup(x^{(1)}, 1) = [0, 1)\setminus\bigcup_{k=1}^{\infty} \{x^{(k)}\}. \tag{3.7}$$

We can proceed as in the case of finitely many discontinuities. Suppose f is pointwise dyadic differentiable on $[0, 1)$ except on a countable set $H\subset[0, 1)$. Then one has (3.1) again for all $x\in[0, 1)\setminus H$, where H is again the set of all points for which the derivative fails to exist. Now apply the above arguments to each of the intervals of the decomposition (3.7). It turns out that f is continuous on each of the intervals $[0, x_0)$, $(x^{(1)}, 1)$, I_k, $k\in\mathbf{N}$; similarly upper as well as lower right Dini-derivatives of f exist for $x\notin H\cup\mathrm{D.R.}$, and they are as well equal to one another. So one can conclude that f is a constant on $[0, x_0)$, $(x^{(1)}, 1)$ as well as on each I_k, $k\in\mathbf{N}$.

Conversely, f is assumed to be a piecewise constant function possessing the discontinuities $x^{(k)}$, $k\in\mathbf{N}$, which are only jumping-points of f. Consequently f is constant on each of the intervals $[0, x_0)$, $(x^{(1)}, 1)$, and I_k, $k\in\mathbf{N}$, respectively. Now we are principally in the position to use the methods of the converse direction of the proof of the corresponding finite case. Suppose for example that $x\in I_k$, $k\in\mathbf{N}$. Obviously there exists $j_0\in\mathbf{N}$ such that for all $j>j_0$ the dyadic sum $x\oplus 2^{-j-1}$ belongs to I_k, I_{k+1} or I_{k-1}, respectively. (For $k=1$ define $I_0:=(x^{(1)}), 1)$.) Hence for $j>j_0$ we have the same situation as in the above converse direction. For example, if $x\oplus 2^{-j-1}$ belongs to I_k or I_{k+1}, respectively, one obtains eight relations for the difference $x\pm 2^{-j-1}$, quite similar to those in (3.4)—(3.6). Thus the difference $f(x)-f(x\oplus 2^{-j-1})$ vanishes for some $j>j^*$, since f is constant on $(0, x_0)$, $(x^{(1)}, 1)$ and on I_k, $k\in\mathbf{N}$. Consequently we have for all $x\in(0, x_0)\cup(x^{(1)}, 1)\cup I_k$, $k\in\mathbf{N}$

$$\sum_{j=0}^{\infty} 2^{j-1}[f(x)-f(x\oplus 2^{-j-1})] < \infty.$$

Therefore the dyadic derivative can only fail to exist at the points $x_0, x^{(k)}$, $k\in\mathbf{N}$. Since $(x^{(k)})_{k=1}^{\infty}$ is countable, f is dyadic differentiable on $[0, 1)$, except the countable set.

Let us finally discuss the case that the sequence of discontinuities $(x^{(k)})_{k=1}^{\infty}$ has a finite number of cluster points in $[0, 1]$, say x_i, $1\leq i\leq n$. Then we can split up the sequence $(x^{(k)})_{k=1}^{\infty}$ into the subsequences $(x^{(i_k)})_{k=1}^{\infty}$, $1\leq i\leq n$, having the property that $\lim_{k\to\infty} x^{(i_k)}=x_i$, $1\leq i\leq n$. The proof of the result then follows along the same lines as the above proof concerning one cluster point.

For the particular case of a *finite* number of discontinuities of the function f it is quite clear that the statement of the Theorem remains valid if the pointwise derivative is replaced by the *strong* dyadic derivative. In this regard we have

COROLLARY. *Let f be a function on $[0, 1)$ possessing a finite number of discontinuities $x^{(k)}$, $1 \leqq k \leqq n$, in $[0, 1)$ and no points of discontinuity of the second kind (poles, etc.). Then f is dyadic differentiable on $[0, 1)$ in the strong sense if and only if $f=\varphi$ except on a countable set, φ being a piecewise constant function on $[0, 1)$.*

4. Examples

In this section we will present some examples which illustrate the applicability as well as the limits of the dyadic derivative. We begin our considerations with an example of a function that is piecewise constant and consequently is dyadic differentiable in both senses. Nevertheless this simple example will point out what the derivative actually effects. Note that the general theorems treated do not give any information about the real nature of the dyadic derivative.

Let g_1 be defined by

$$g_1(x) = \begin{cases} 3, & x \in [0, 1/4) \\ -1, & x \in [1/4, 1). \end{cases}$$

According to our theorem as well as the corollary, g_1 is differentiable in the strong as well as in the pointwise sense and, since the point of discontinuity is a dyadic rational, g_1 is dyadic differentiable everywhere on $[0, 1)$. Evaluating the derivative gives

$$D^{[1]}g_1(x) = g_1^{[1]}(x) = \begin{cases} 6, & x \in [0, 1/4) \\ -4, & x \in [1/4, 1/2) \\ -2, & x \in [1/2, 3/4) \\ 0, & x \in [3/4, 1). \end{cases}$$

This example shows that the derivative of a piecewise constant function can possess more points of discontinuity than the original function does have; this clearly results from the global character of the derivative.

Let us now consider a piecewise linear function, for example $g_2(x)=x\psi_1(x)$. Clearly the theorem tells us that g_2 cannot be dyadic differentiable since it is not piecewise constant. It is even *nowhere* dyadic differentiable on $[0, 1)$. Indeed, the terms of the series (2.17) are

$$2^{j-1}[x\psi_1(x)-(x\oplus 2^{-j-1})\psi_1(x\oplus 2^{-j-1})] = \begin{cases} (1/2)\psi_1(x)[x\oplus 1/2]+x, & j=0 \\ -(1/4)\psi_1(x), & j\in J_0 \quad j \geqq 1 \\ (1/4)\psi_1(x), & j\in J_1 \end{cases}$$

noting that $\psi_1(2^{-j-1})=-1$, $j=0$, $\psi(2^{-j-1})=1$, $j\geqq 1$, as well as formula (2.9). So the series (2.17) does not converge for any $x\in[0, 1)$, since the necessity condition (3.1) is not satisfied. Thus g_2 is nowhere dyadic differentiable on $[0, 1)$.

Let us now present an example of a function the discontinuities of which have one cluster point in [0, 1), namely

$$g_3(x)=\begin{cases}1/2^n, & x\in[2^{-n-1}, 2^{-n})\\ 0, & x=0\end{cases} \quad (n\in\mathbf{N}\cup\{0\}).$$

Since g_3 is bounded with cluster point $x=0$, g_3 is dyadic differentiable on [0, 1) except on a countable set. Its dyadic derivative at $x=1/2$, for example, is given by $g_3^{[1]}(1/2)=1/2$; at $x=0$ it does not exist.

One may conjecture that a function which possesses a countable number of discontinuities, which lie *dense* in [0, 1], is not dyadic differentiable. On the contrary, the dyadic derivative is especially suited for such exotic functions, although the general theorem is not applicable. One such example is Dirichlet's function

$$g_4(x)=\begin{cases}0, & x\in[0,1)\cap\mathbf{Q}\\ 1, & \text{elsewhere on } [0,1),\end{cases}$$

where $\mathbf{Q}$ denotes the set of all rationals. g_4 is dyadic differentiable everywhere on [0, 1) with $D^{[1]}g_4=g_4^{[1]}=0$. In fact, if $x\in[0,1)\setminus\mathbf{Q}$, then $g_4(x\oplus 2^{-j-1})=0$ by (2.7). If $x\in[0,1)\cap\mathbf{Q}$, then (2.7) and the definition of g_4 deliver $g_4(x\oplus 2^{-j-1})=0$. So $g_4(x)-g_4(x\oplus 2^{-j-1})=0$, for all $x\in[0,1)$ and $j\in\mathbf{P}$. Hence $d^{[1]}(x)=0$, $x\in[0,1)$.

For the strong dyadic derivative (2.15) with respect to the spaces $L^p(0,1)$, $1\leqq p\leqq\infty$ one has on account of

$$\Big\|\sum_{j=0}^{n-1}2^{j-1}[g_4(\cdot)-g_4(\cdot\oplus 2^{-j-1})]\Big\|_{L^p(0,1)}=$$

$$=\begin{cases}\Big\{\displaystyle\int_{[0,1)\setminus Q}\Big|\sum_{j=0}^{n-1}2^{j-1}[g_4(x)-g_4(x\oplus 2^{-j-1})]\Big|^p dx\Big\}^{1/p}, & 1\leqq p<\infty,\\[2ex] \operatorname*{ess\,sup}_{x\in[0,1)}\Big|\displaystyle\sum_{j=0}^{n-1}2^{j-1}[g_4(x)-g_4(x\oplus 2^{-j-1})]\Big|, & p=\infty\end{cases}$$

and in connection with the preceding remarks that $D^{[1]}g_4(x)=0$. Note that the space $C^{\oplus}[0,1]$ is not permissible here since $g_4\notin C^{\oplus}[0,1]$.

A more complicated function dyadic differentiable at every $x\in[0,1)$ is

$$g_5(x)=\begin{cases}g_4(x)+1, & x\in[0,1/2)\\ g_4(x)+3, & x\in[1/2,3/4)\\ g_4(x)-1, & x\in[3/4,1).\end{cases}$$

Another exotic function is $g_6(x)=1+\sum_{k=1}^{\infty}(1/k^2)\psi_k(x)$, $x\in[0,1)$. Although it has discontinuities at each dyadic rational, it is still dyadic differentiable with $g_6^{[1]}(x)=$ $=D^{[1]}g_6(x)=\sum_{k=1}^{\infty}(1/k)\psi_k(x)$ (see [6] for exact evaluation, [8] for its graph).

Acknowledgement. Thanks are due to Prof. P. L. Butzer as well as to Priv.-Doz. Dr. W. Splettstößer, both Aachen, who read the manuscript and gave helpful advice and criticism.

References

[1] S. V. Bočkarev, On the Walsh—Fourier coefficients, *Izv. Akad. Nauk SSSR, Ser. Math.*, 34 (1970), 203—208.

[2] P. L. Butzer—H. J. Wagner, Approximation by Walsh polynomials and the concept of a derivative. In: *Applications of Walsh-Functions* (Proc. Naval. Res. Lab., Washington, D. C., 27—29 March 1972; Eds. R. W. Zeek—A. E. Showalter) Washington, D. C. 1972, xi+401 pp.; pp. 388—392.

[3] P. L. Butzer—H. J. Wagner, On a Gibbs type derivative in Walsh—Fourier analysis with applications. In: *Proceedings of the 1972 National Electronics Conference* (Chicago, 9—10 Oct. 1972; Ed. R. E. Horton) Oak Brook, Illinois, 1972, xxvi+457 pp.; pp. 393—398.

[4] P. L. Butzer—H. J. Wagner, Walsh—Fourier series and the concept of a derivative, *Appl. Anal.*, 3 (1973), 29—46.

[5] P. L. Butzer—H. J. Wagner, On dyadic analysis based on the pointwise dyadic derivative, *Analysis Math.*, 1 (1975), 171—196.

[6] P. L. Butzer—W. Splettstößer, Sampling principle for duration limited signals and dyadic Walsh analysis, *Inf. Sci.*, 14 (1978), 93—106.

[7] P. L. Butzer—W. Engels, Dyadic calculus and sampling theorems for functions with multidimensional domain. Part I: General Theory. Part II: Applications to sampling theorems, *Information and Control* (1983), 333—351; 351—363.

[8] W. Engels—W. Splettstößer, On Walsh differentiable dyadically stationary random processes, *IEEE Trans. on Inf. Theory*, IT—28 (1982), 612—619.

[9] J. E. Gibbs—M. J. Millard, Walsh functions as solutions of a logical differential equation, *NPL: DES Rept.* No. 1 (1969).

[10] J. E. Gibbs, Some properties of functions on the non-negative integers less than 2^n, *NPL (National Physical Laboratory)*, Middlesex, England, DES Rept. No. 3 (1969).

[11] J. E. Gibbs—B. Ireland, Some generalizations of the logical derivative, *NPL: DES Rept.* No. 8 (1971).

[12] J. E. Gibbs—B. Ireland, Walsh functions and differentiation. In: *Proceedings of the Symposium and Workshop on Applications of Walsh-Functions.* Naval Research Laboratory Washington, D. C. 1974, (A. Bass, Edit.). (Springfield Va. 2251: National Technical Information Service); 1—29.

[13] He Zelin, Derivatives and integrals of fractional order in generalized Walsh—Fourier analysis with application to approximation theory, *J. Approximation Theory*, in print.

[14] N. R. Ladhawala, Absolute summability of Walsh—Fourier series, *Pac. J. Math.*, 65 (1976), 103—108.

[15] C. W. Onneweer, Fractional differentiation on the group of integers of the *p*-adic or *p*-series field, *Analysis Math.*, 3 (1977), 119—130.

[16] C. W. Onneweer, Differentiation on a *p*-adic or *p*-series field. In: *Proceedings of the Conference „Linear Spaces and Approximation"*, in Oberwolfach (Schwarzwald), Herausgeber: P. L. Butzer und B. Sz.-Nagy, Birkhäuser, ISNM 40, Basel 1978, 685 pp. 187—198.

[17] C. W. Onneweer, On the definition of dyadic differentiation, *Applicable Math.*, 9 (1979), 267—278.

[18] C. W. Onneweer, Fractional differentiation and Lipschitz spaces on local fields, *Trans. Amer. Math. Soc.*, 258 (1980), 155—165.

[19] J. Pál—P. Simon, On a generalization of the concept of derivative, *Acta Math. Acad. Sci. Hungar.*, 29 (1977), 155—164.

[20] J. Pál, On a concept of a derivative among functions defined on a dyadic field, *SIAM J. Math. Anal.*, 8 (1977), 375—391.

[21] J. Pál, On almost everywhere differentiation of dyadic integral function on R_+. In: *Fourier Analysis and Approximation Theory*, Proc. Colloq., Budapest, 1976, Colloq. Math. Soc. János Bolyai, 19, North Holland, Amsterdam 1978, vol. II, 591—601.

[22] R. E. A. C. Paley, A remarkable series of orthogonal functions, *Proc. London Math. Soc.*, 34 (1932), 241—279.

[23] R. Penney, On the rate of growth of the Walsh antidifferentiation operator, *Proc. Amer. Math. Soc.*, 55 (1976), 57—61.

[24] L. Saks, *Théorie de l'Integrale* (Warschau, 1933).

[25] F. Schipp, Über einen Ableitungsbegriff von P. L. Butzer und H. J. Wagner, *Math. Balkanica*, 4 (1974), 541—546.

[26] F. Schipp, Über gewisse Maximaloperatoren, *Annales Univ. Sci. Budapestinensis de Rolando Eötvös Nominatae. Sec. Math.*, **28** (1976), 145—152.
[27] F. Schipp, On term by term dyadic differentiability of Walsh series, *Analysis Math.*, **2** (1976), 149—154.
[28] V. A. Skvorcov—W. R. Wade, Generalizations of some results concerning Walsh series and the dyadic derivative, *Analysis Math.*, **5** (1979), 249—255.
[29] W. Splettstößer, Error analysis in the Walsh sampling theorem. In: *1980 IEEE International Symposium on Electromagnetic Compatibility*, Baltimore. The Institute of Electrical and Electronics Engineers, Piscataway, N. Y., 409 pp.; 366—370.
[30] H. J. Wagner, Ein Differential- und Integralkalkül in der Walsh—Fourier Analysis mit Anwendungen (*Forschungsbericht des Landes Nordrhein-Westfalen* Nr. 2334), Westdeutscher Verlag, Köln-Opladen 1973, 71 pp.
[31] J. L. Walsh, A closed set of normal orthogonal functions, *Amer. J. Math.*, **55** (1923), 5—24.
[32] Zheng-Wei-xing—Su-Wei-yi, The logical derivative and integral, *Jour. of Math. Research and Exposition*, **1** (1981), 79—90.

(Received April 7, 1983; revised June 21, 1983)

LEHRSTUHL A FÜR MATHEMATIK
AACHEN UNIVERSITY OF TECHNOLOGY
AACHEN, FEDERAL REPUBLIC OF GERMANY

ANNALES
UNIVERSITATIS SCIENTIARUM
BUDAPESTINENSIS
DE ROLANDO EÖTVÖS NOMINATAE

SEPARATUM

SECTIO MATHEMATICA

TOMUS XVIII.

1975

ÜBER GEWISSEN MAXIMALOPERATOREN

Von

F. SCHIPP

II. Lehrstuhl für Analysis der Eötvös Loránd Universität, Budapest

(Eingegangen am 2. Oktober 1974)

Es sei $B=(b_{ij})_{i,j\in N}$ eine Matrix, $(\lambda_i)_{i\in N}$[1] eine positive Zahlenfolge mit folgenden Eigenschaften:

$$\text{(B)} \qquad 0\leq b_{ij}\leq 2^{-\lambda_i} \quad (i,j\in N).$$

Es sei weiterhin $A=(a_{ij}^n)_{i,j,n\in N}$ und für $i,j\in N$ setzen wir

$$\alpha_{ij}=\sup_{n<j} a_{ij}^n, \quad \beta_{ij}=\sup_{n\geq j} 2^n a_{ij}^n \quad (i,j\in N).$$

In dieser Arbeit wird vorausgesetzt, daß A den folgenden Bedingungen genügt:

$$1^\circ \qquad a_{ij}^n\geq 0 \quad (i,j,n\in N),$$

$$\text{(A)} \quad 2^\circ \qquad \sup_n \sum_{i,j=0}^{\infty} a_{ij}^n = K_1<\infty,$$

$$3^\circ \qquad \sup_m \sum_{\lambda_i\leq m} \sum_{j=m}^{\infty} (\alpha_{ij}+2^{-j}\beta_{ij}) = K_2<\infty.$$

Führen wir noch die Bezeichnungen

$$f_n(x)=\sum_{i,j=0}^{\infty} a_{ij}^n D_{2^j}(x\dotplus b_{ij}) \quad (n\in N)$$

ein, wobei D_{2^j} die 2^j-te Walsh-Dirichletsche Kernfunktion und $\dotplus$ die Fine-sche Operation bezeichnet (vergl [1]), d.h.:

$$D_{2^j}(x)=\sum_{n=0}^{2^j-1}\psi_n(x)=\prod_{k=0}^{j-1}(1+\psi_{2^k}(x))=\begin{cases}2^j & (0\leq x<2^{-j}),\\ 0 & (2^{-j}\leq x<1).\end{cases}$$

[1] $N=\{0,1,2,\dots\}$.

In dieser Arbeit beschäftigen wir uns mit der Konvergenz der Operatorfolge

$$(T_n f)(x) = (f_n \ast f)(x) = \int_0^1 f(t) f_n(x \dotplus t)\, dt \quad (n \in N).$$

Wir beweisen den folgenden

SATZ. *Genügen A und B den Bedingungen* (A) *und* (B), *so ist der Operator*

$$(Tf)(x) = \sup_n |(T_n f)(x)|$$

vom Typ (∞, ∞) *und von schwachem Typ* (1, 1), *d.h.*

a) *für* $f \in L^\infty(0, 1)$ *gilt*

$$\|Tf\|_\infty \leq K_1 \|f\|_\infty,^2$$

b) *für* $f \in L(0, 1)$ *und* $y > 0$ *gilt*

(1) $$\operatorname{mes}\{x : (Tf)(x) > y\} \leq M \|f\|_1 / y,$$

wobei $M = (4K_2 + 1) 2K_1$ *ist und* $K_i\ (i = 1, 2)$ *die Konstante der Bedingung* (A) *bedeutet.*

Aus dem Satz, durch Anwendung eines Satzes von MARCINKIEWICZ (vergl. [2], Seite 111) ergibt sich

KOROLLAR 1. Für $f \in L^p(0, 1)$ gilt

$$\|Tf\|_p \leq K(p) \|f\|_p \quad (1 < p \leq \infty),$$

wobei $K(p)$ eine von f unabhängige Konstante ist. Durch Anwendung des Satzes ergibt sich das folgende

KOROLLAR 2. Es bezeichne $\sigma_n(f)$ die n-te $(C, 1)$-Mittel der Walsh-Fourier-Entwicklung von $f \in L(0, 1)$. Dann

a) ist der Operator

(2) $$(\sigma f)(x) = \sup_n |(\sigma_n f)(x)|$$

vom Typ (∞, ∞) und von schwachem Typ (1, 1),

b) für $f \in L^p(0, 1)$ $(1 < p \leq \infty)$ gilt

$$\|\sigma f\|_p \leq C(p) \|f\|_p,$$

wobei $C(p)$ eine von f unabhängige Konstante ist,

c) für $f \in L(0, 1)$ gilt in fast allen $x \in (0, 1)$

$$\lim_{n \to \infty} (\sigma_n f)(x) = f(x).$$

[2] $\|\ \|_p$ bezeichnet die Norm in $L^p(0, 1)$ $(1 \leq p \leq \infty)$.

Die Behauptung *c)* hat N. J. FINE [3] bewiesen. Eine andere Anwendung des Satzes ist in [4] gegeben.

Beweis des Satzes

Zum Beweis des Satzes benutzen wir einen Hilfssatz von Calderon-Zygmund Typ (Vergl. [5], Hilfssatz III.).

HILFSSATZ 1. Es sei $f \in L(0,1)$ und $y > \|f\|_1$. Dann gibt es ein System $(I_k, k = 1, 2, \dots)$ von dyadischen Intervallen und zwei funktionen φ_0 und φ derart, daß die folgenden Relationen gelten:

$$\text{a)} \quad f = \varphi_0 + \varphi, \quad \|\varphi_0\|_\infty \leq 2y,$$

$$\text{(3) b)} \quad I_i \cap I_j = \emptyset \ (i \neq j), \quad \sum_{i=1}^{\infty} \text{mes } I_i \leq \|f\|_1 / y,$$

$$\text{c)} \quad \varphi = \sum_{i=1}^{\infty} \varphi_i, \ \text{supp } \varphi_i \subset I_i, \ \int_0^1 \varphi_i = 0, \ \|\varphi\|_1 \leq 4 \|f\|_1 .$$

BEWEIS DES SATZES. *a)* Da $\|T_n f\|_\infty \leq \|f_n\|_1 \|f\|_\infty$ ist, deshalb gilt

$$\|Tf\|_\infty \leq (\sup_n \|f_n\|_1) \|f\|_\infty$$

und so genügt es zu zeigen, daß

$$\text{(4)} \qquad \sup_n \|f_n\|_1 \leq K_1$$

ist. Auf Grund der Relationen $D_{2^j}(x) \geq 0$, $\int_0^1 D_{2^j}(t)\,dt = 1$ $(x \in (0,1), j \in N)$ und nach (A) 2° ergibt sich

$$\|f_n\|_1 = \sum_{i,j=0}^{\infty} a_{ij}^n \leq K_1 \quad (n \in N),$$

womit (4) bewiesen ist.

Zum beweis der Ungleichung (1) benutzen wir Hilfssatz 1. Es sei mes $I_k = 2^{-m_k}$ $(m_k \in N)$, $H = \bigcup_{k=1}^{\infty} I_k$, und $H' = [0,1) - H$. Ist $j \leq m_k$, oder $x \notin I_k$ so gilt auf Grund von (3) c) $(\varphi_k * D_{2^j})(x) = 0$. Da im Falle $x \notin I_k$ und $\lambda_i > m_k$ auch $x \dotplus b_{ij} \notin I_k$ ist, deshalb gilt

$$(\varphi_k * D_{2^j})(x \dotplus b_{ij}) = 0 \quad (j \leq m_k, \text{ oder } x \notin I_k \text{ und } \lambda_i > m_k)$$

und so ist für $x \notin I_k$

$$|(f_n * \varphi_k)(x)| = \left| \sum_{\lambda_i \leq m_k} \sum_{j=m_k}^{\infty} a_{ij}^n (D_{2^j} * \varphi_k)(x \dotplus b_{ij}) \right| =$$

$$= \left| \sum_{\lambda_i \leq m_k} \left(\sum_{j=m_k}^{n} + \sum_{j=n+1}^{\infty} \right) (a_{ij}^n (D_{2^j} * \varphi_k)(x \dotplus b_{ij})) \right| \leq$$

$$\leq \sum_{\lambda_i \leq m_k} \sum_{j=m_k}^{\infty} (\beta_{ij} (\chi_j * |\varphi_k|)(x \dotplus b_{ij}) + \alpha_{ij} (D_{2^j} * |\varphi_k|)(x \dotplus b_{ij})) \quad (n \in N)$$

wobei $\chi_i = 2^{-j} D_{2^j}$ gesetzt wird.

Daraus folgt

$$(T\varphi_k)(x) \leq \sum_{\lambda_i \leq m_k} \sum_{j=m_k}^{\infty} (\beta_{ij} (\chi_j * |\varphi_k|)(x \dotplus b_{ij}) + \alpha_{ij} (D_{2^j} * |\varphi_k|)(x \dotplus b_{ij})) \quad (x \notin I_k)$$

und so gilt auf Grund der Ungleichungen $\|f * g\|_1 \leq \|f\|_1 \|g\|_1$, $\|D_{2^j}\| = 1$, $\|\chi_j\|_1 = 2^{-j}$ und nach (A)

$$\|\chi_{H'} T \varphi_k\|_1 \leq \|\varphi_k\|_1 \sum_{\lambda_i \leq m_k} \sum_{j=m_k}^{\infty} (\alpha_{ij} + 2^{-j} \beta_{ij}) \leq K_2 \|\varphi_k\|_1 .\text{[3]}$$

Daraus nach (3) *b), c)* ergibt sich

$$\|\chi_{H'} T \varphi\|_1 \leq \sum_{k=1}^{\infty} \|\chi_{H'} T \varphi_k\|_1 \leq K_2 \sum_{k=1}^{\infty} \|\varphi_k\|_1 \leq 4K_2 \|f\|_1 .$$

Hiraus bekommen wir

$$\text{mes}\{x \in H' : (T\varphi)(x) > y\} \leq 1/y \, \|\chi_{H'} T \varphi\|_1 \leq (4K_2/y) \|f\|_1$$

und so gilt nach (3) b)

$$\text{mes}\{x : (T\varphi)(x) > y\} \leq \text{mes}\, H + \text{mes}\{x \in H' : (T\varphi)(x) > y\} \leq$$
$$\leq (4K_2 + 1)/y \|f\|_1 .$$

Da nach (3) a) und nach Teil *a)* des Satzes $\text{mes}\{x : (T\varphi_0)(x) > 2K_1 y\} = 0$ ist, deshalb gilt

$$\text{(5)} \quad \text{mes}\{x : (Tf)(x) > (2K_1 + 1) y\} \leq \text{mes}\{x : (T\varphi_0)(x) > 2K_1 y\} +$$
$$+ \text{mes}\{x : (T\varphi)(x) > y\} \leq (4K_2 + 1)/y \|f\|_1 .$$

Offensichtlich ist (5) für $y < \|f\|_1$ auch richtig. Die Behauptung (1) folgt nun schon aus (5) (mit der Konstante $M = (4K_2 + 1)(2K_1 + 1)$).

Damit haben wir den Satz bewiesen.

[3] $\chi_{H'}(x) = 0 \ (x \in H)$, $\chi_{H'}(x) = 1 \ (x \in H')$.

Beweis von Korollar 2

Zum beweis von Korollar 2 führen wir den Operator (vergl. [6])

$$(d_n f)(x) = \sum_{k=0}^{n-1} 2^{k-1}(f(x) - f(x \dotplus e_k))$$

$$(e_k = 2^{-k-1},\ f \in L(0,1),\ x \in (0,1))$$

ein. Da für die m-te Walsh-Paley-Funktion

$$\psi_m(x) = (-1)^{\sum_{k=0}^{\infty} m_k x_k}$$

gilt, für $m < 2^n$ ist

$$(d_n \psi_m)(x) = \sum_{k=0}^{n-1} 2^{k-1}(\psi_m(x) - \psi_m(x \dotplus e_k)) =$$

$$= \psi_m(x) \sum_{k=0}^{n-1} 2^k (1 - (-1)^k)/2 = \psi_m(x)\left(\sum_{k=0}^{n-1} m_k 2^k\right) = m\,\psi_m(x).$$

Daraus für die Walsh-Fejérsche Kernfunktion K_m im Falle $2^{n-1} \leq m < 2^n$ ergibt sich die folgende Darstellung:

$$K_m = \frac{1}{m}\sum_{k=0}^{m-1}(m-k)\psi_k = D_m - \frac{1}{m}\sum_{k=0}^{m-1} k\,\psi_k = D_m - \frac{1}{m} d_n D_m, \tag{6}$$

wobei

$$D_m = \sum_{k=0}^{m-1} \psi_k$$

die m-te Walsh-Dirichletsche Kernfunktion bedeutet. Wir benutzen den folgenden

HILFSSATZ 2. Für $2^{n-1} \leq m < 2^n$ gilt

$$|K_m(x)| \leq \sum_{i=0}^{n-1} 2^{i-n} \sum_{i=j}^{n-1} (D_{2^j}(x) + D_{2^j}(x \dotplus e_i)) \quad (x \in (0,1)).$$

BEWEIS VON HILFSSATZ 2. Hat $k \in N$ die dyadische Darstellung

$$k = \sum_{j=0}^{\infty} k_j 2^j (k_j \in \{0,1\}),$$

so gilt (vergl. [7])

$$D_k = \psi_k \sum_{j=0}^{\infty} k_j \psi_{2^j} D_{2^j},$$

woraus sich für $m K_m$ auf Grund von

$$m K_m(x) = \psi_m(x)\left(\sum_{i=0}^{n-1} m_i 2^i\right)\left(\sum_{j=0}^{n-1} m_j \psi_{2^j}(x) D_{2^j}(x)\right) -$$

$$-\psi_m(x)\sum_{i=0}^{n-1} 2^i \sum_{j=0}^{n-1} m_j \psi_{2^j}(x)\left(D_{2^j}(x) - \psi_m(e_i)\psi_{2^j}(e_i) D_{2^j}(x \dotplus e_i)\right) =$$

$$= \psi_m(x)\sum_{i=0}^{n-1} 2^i \sum_{j=0}^{n-1} c_{ij}(x)$$

ergibt. Da

$$D_{2^j}(x \dotplus e_i) = D_{2^j}(x)\ (i \geq j),\quad \psi_{2^j}(e_i) = 1\ (i > j)$$

und

$$(1-\psi_m(e_i))/2 = \frac{1}{2}(1-(-1)^{m_i}) = m_i$$

sind, deshalb ist $c_{ij}(x) = 0 \quad (j < i,\ x \in (0,1))$ und so gilt

$$m\,|K_m(x)| \leq 1/2 \sum_{i=0}^{n-1} 2^i \sum_{j=i}^{n-1} \left(D_{2^j}(x) + D_{2^j}(x \dotplus e_i)\right).$$

Daraus folgt die Behauptung.

Beweis von Korollar 2. Es sei $a_{ij}^n = 2^{i-n}$ für $0 \leq i \leq j < n$ und $a_{ij}^n = 0$ sonst, $b_{ij} = 0$, und $c_{ij} = e_j\ (i, j \in N)$ und führen wir die Bezeichnungen

$$f_n^1(x) = \sum_{i,j=0}^{\infty} a_{ij}^n D_{2^j}(x \dotplus b_{ij}),\quad f_n^2(x) = \sum_{i,j=0}^{\infty} a_{ij}^n D_{2^j}(x \dotplus c_{ij})$$

ein. Dann gilt nach Hilfssatz 2 für $2^{n-1} \leq m < 2^n$

$$|K_m(x)| \leq \sum_{i=0}^{n-1} 2^{i-n} \sum_{j=i}^{n-1} \left(D_{2^j}(x) + D_{2^j}(x \dotplus e_i)\right) = f_n^1(x) + f_n^2(x).$$

Die Matrizen $(b_{ij})_{i,j\in N'}$ $(c_{ij})_{i,j\in N}$ mit $\lambda_i = i\ (i \in N)$ genügen der Bedingung (B). Da

$$\sum_{i,j=0}^{\infty} a_{ij}^n = \sum_{i=0}^{n-1} 2^{i-n} \sum_{j=i}^{n-1} 1 = \sum_{i=0}^{n-1} (n-i)\,2^{-n+i} < \sum_{k=0}^{\infty} k\,2^{-k} = K_1 < \infty$$

ist genügt A der Bedingung (A) 2°. Da endlich $\alpha_{ij} = 0$, $\beta_{ij} = 2^i\ (i, j \in N)$ und

$$\sum_{i=0}^{m} \sum_{j=m}^{\infty} 2^{-j} \beta_{ij} = \sum_{i=0}^{m} 2^i \sum_{j=m}^{\infty} 2^{-j} = 2^{-m+1} \sum_{i=0}^{m} 2^i < 4 \quad (m \in N)$$

ist, sind die Bedingungen (A) 1° und 3° auch erfüllt. Auf Grund des Satzes sind die Operatoren

$$(T^i f)(x) = \sup_n |(f_n^i * f)(x)| \quad (i = 1, 2)$$

vom Typ (∞, ∞) und von schwachem Typ $(1,1)$ und auf Grund der Ungleichung

$$(\sigma f)(x) = \sup_m |(f * K_m(x)| \leq T^1(|f|)(x) + T^2(|f|)(x)$$

dasselbe gilt für σ.

Damit ist die Behauptung *a)* bewiesen. Die Behauptungen *b)* und *c)* kann man auf Grund von *a)* mit wohlbekannten Methoden beweisen.

Literaturverzeichnis

[1] N. J. Fine, On the Walsh functions, *Trans. Amer. Math. Soc.*, **65** (1949), 372–414.
[2] A. Zygmund, *Trigonometric series*, Vol. II. (Cambridge, 1959).
[3] N. J. Fine, Cesàro summability of Walsh-Fourier series, *Proc. Nat. Acad. Sci. USA*, **41** (1955), 588–591.
[4] F. Schipp, Über einen Ableitungsbegriff von P. L. Butzer und H. J. Wagner *Mat. Balkanika*, **4** (1974), 541–546.
[5] F. Schipp, Über die starke Summation von Walsh-Fourierreihen, *Acta Sci. Math.*, **30** (1969), 77–87.
[6] P. L. Butzer – H. J. Wagner, Walsh-Fourier series and the concept of a derivative, *Applicable Analysis*, **3** (1973), 29–46.
[7] F. Schipp, Über die Größenordnung der Partialsummen der Entwicklung integrierbarer Funktionen nach *W*-Systemen, *Acta Sci. Math.*, **28** (1967), 123–134.

Annales Univ. Sci. Budapest., Sect. Comp. **8** (1987) 91-108

ON THE DYADIC DIFFERENTIABILITY OF DYADIC INTEGRAL FUNCTIONS ON R+

J. PÁL and F. SCHIPP

Dedicated to Prof. I. Kátai
on the occasion of his fiftieth birthday

In their paper [1] P. L. Butzer and H. J. Wagner have introduced the concept of dyadic derivative for functions defined on the dyadic field $\mathbf{R}^+$. Furthermore, Wagner [5] has defined the notion of dyadic integral as the inverse of the dyadic derivative and investigated, among others, the strong dyadic differentiability of dyadic integrals. In this paper we shall prove some estimates from which the result of Wagner easily follows. Moreover, these estimates can be used also for the proof of the almost everywhere dyadic differentiability of dyadic integrals. We shall concern ourselves with this question in a forthcoming paper. In connection with this see also [2] and [3] .

Let $f : \mathbf{R}^+ \to \mathbf{C}$ be a function defined on the dyadic field and let for every $n \in \mathbf{N}$

$$d_n f := \frac{1}{2} \sum_{j=-n}^{n} 2^j (f - \tau_{2^{-(j+1)}} f) \tag{1}$$

be the nth dyadic "differential quotient" of f, where τ_h $(h \in \mathbf{R}^+)$ are the dyadic translation operators:

$$(\tau_h f)(x) := f(x \dot{+} h) \quad (x, h \in \mathbf{R}^+). \tag{2}$$

If for some point $x \in \mathbf{R}^+$ the limit

$$\lim_{n\to\infty} (d_n f)(x) =: f^{[1]}(x) \tag{3}$$

exists, then we say that f is dyadically differentiable at $x \in \mathbf{R}^+$ and $f^{[1]}(x)$ is the dyadic derivative of f at $x \in \mathbf{R}^+$. If $f \in L^1(\mathbf{R}^+)$ is an integrable function and there exists a function $g \in L^1(\mathbf{R}^+)$ for which

$$\lim_{n\to\infty} \| d_n f - g \|_1 = 0 \tag{4}$$

holds, then f is said to be strongly dyadically differentiable in $L^1(\mathbf{R})^+$ and $Df := g$ is the strong dyadic derivative of f.

Let $n \in \mathbf{N}$ and define the function W_n by its Walsh-Fourier transform $\widehat{W}_n$ as follows:

$$\widehat{W}_n(y) := \begin{cases} 0, & y \in [0, 2^{-n}) \\ \frac{1}{y}, & y \in [2^{-n}, +\infty)\ . \end{cases} \tag{5}$$

Wagner has proved (see [5]) that there exists uniquely a function $W_n \in L^1(\mathbf{R}^+)$ for which (5) holds; moreover,

$$W_n(x) = \lim_{k\to\infty} \int_{2^{-n}}^{2^k} \frac{1}{y} w_x(y)dy \quad (x \in \mathbf{R}^+) \tag{6}$$

and the limit can be taken either in the $L^1(\mathbf{R}^+)$-norm or in the pointwise sense. Here and in the sequel the symbols w_x $(x \in \mathbf{R}^+)$ denote the generalized Walsh functions.

In the following we introduce the inverse operation of the dyadic derivative by the following definition (see [5]): if for a function $f \in L^1(\mathbf{R}^+)$ there exists a function $g \in L^1(\mathbf{R}^+)$ such that

$$\lim_{n\to\infty} \| W_n * f - g \|_1 = 0, \tag{7}$$

then g is called the strong dyadic integral of f and is denoted by If ($*$ denotes dyadic convolution) .

For this notion of dyadic integral Wagner proved that the following assertions are equivalent for $f, g \in L^1(\mathbf{R}^+)$:

$$\text{(8)} \quad \begin{array}{ll} i) & g = If, \\ ii) & \hat{g}(y) = \begin{cases} 0, & y = 0 \\ \frac{1}{y}\hat{f}(y), & y > 0 . \end{cases} \end{array}$$

In the following we investigate the strong dyadic differentiability of the dyadic integral $If \in L^1(\mathbf{R}^+)$ for $f \in L^1(\mathbf{R}^+)$. We remark that if $f \in L^1(\mathbf{R}^+)$ then If is not necessarily defined. For example, if $f := \chi[0,1)$ is the characteristic function of the interval $[0,1)$ then If is not defined (see [5]). Therefore, in the following we suppose that $f \in L^1(\mathbf{R}^+)$ and the dyadic integral $If \in L^1(\mathbf{R}^+)$ of f exists. First we compute the dyadic "differential quotients" $d_n(If)$ $(n \in \mathbf{N})$ of If . In connection with this we shall show the following

Lemma 1. *If for a function $f \in L^1(\mathbf{R}^+)$ the dyadic integral $If \in L^1(\mathbf{R}^+)$ exists, then*

$$\text{(9)} \qquad d_n(If) = d_n W_n * f \quad (n \in \mathbf{N}).$$

Proof. Since τ_h $(h \in \mathbf{R}^+)$ are isometries in $L^1(\mathbf{R}^+)$, from the definition of d_n it follows that

$$d_n(If) = \lim_{m\to\infty} d_n(W_m * f) = \lim_{m\to\infty} (d_n W_m) * f = (\lim_{m\to\infty} d_n W_m) * f,$$

where the limit is taken in the $L^1(\mathbf{R}^+)$-norm sense. Furthermore,

$$\text{(10)} \qquad d_n W_m = \lim_{k\to\infty} d_n W_{m,k}$$

and the limit can be taken either in the $L^1(\mathbf{R}^+)$-norm or in the pointwise sense, where

$$(11)\qquad W_{m,k}(x) := \int_{2^{-m}}^{2^k} \frac{1}{y} w_x(y)dy \quad (x \in \mathbf{R}^+,\ m,k \in \mathbf{N})\ .$$

For $m \geq n$ and $k \in \mathbf{N}$ we have

$$(d_n W_{m,k})(x) = \frac{1}{2} \sum_{j=-n}^{n} \int_{2^{-m}}^{2^k} \frac{2^j}{y}\big(w_x(y) - w_{x \dotplus 2^{-(j+1)}}(y)\big)dy =$$

$$= \int_{2^{-m}}^{2^k} \frac{1}{y}\alpha_n(y) w_x(y)dy \quad (x \in \mathbf{R}^+),$$

where

$$(12)\qquad \begin{gathered} \alpha_n(y) := \frac{1}{2}\sum_{j=-n}^{n} 2^j\big(1 - w_{2^{-(j+1)}}(y)\big) = \sum_{j=-n}^{n} y_j 2^{-j} \\ \Big(y = \sum_{j=-\infty}^{\infty} y_j 2^{-j} \in \mathbf{R}^+,\ y_j \in \{0,1\},\ n \in \mathbf{N}\Big). \end{gathered}$$

Since $\alpha_n(y) = 0$ if $y \in [0, 2^{-n})$, we have that for any $m \geq n$ and $k \in \mathbf{N}$

$$d_n W_{m,k} = d_n W_{n,k},$$

and, consequently, our lemma is proved. □

In the following our aim is to give an estimate for the functions $d_n W_n$ $(n \in \mathbf{N})$. To this end, we define $\beta_n(y)$ for $y \in \mathbf{R}^+$ and $n \in \mathbf{N}$ as follows:

$$(13)\qquad \beta_n(y) := \sum_{j=-n}^{0} y_j 2^{-j}.$$

It is easy to see that

$$2^n\alpha_n(2^{-n}y) = \beta_{2n}(y) \quad (y \in \mathbf{R}^+,\ n \in \mathbf{N}). \tag{14}$$

Let us introduce the functions V_n $(n \in \mathbf{N})$ by the following equality:

$$V_n(x) := \lim_{k\to\infty} \int_0^{2^k} \frac{1}{y}\beta_n(y)w_x(y)dy \quad (x \in \mathbf{R}^+,\ n \in \mathbf{N}) \tag{15}$$

(it is easy to show that this limit exists). With the functions V_n $(n \in \mathbf{N})$ we can express the functions d_nW_n $(n \in \mathbf{N})$. Namely, the following lemma is true.

Lemma 2. *For every $n \in \mathbf{N}$ and $x \in \mathbf{R}^+$*

$$(d_nW_n)(x) = 2^{-n}V_{2n}(2^{-n}x). \tag{16}$$

Proof. Using in the integral (15) the transformation $z := 2^{-n}y$ and (14), we get

$$(d_nW_n)(x) = \lim_{k\to\infty} \int_0^{2^k} \frac{1}{z}\alpha_n(z)w_x(z)dz =$$

$$= \lim_{k\to\infty} 2^{-n} \int_0^{2^{k+n}} \frac{1}{2^{-n}y}\alpha_n(2^{-n}y)w_x(2^{-n}y)dy =$$

$$= 2^{-n} \lim_{k\to\infty} \int_0^{2^k} \frac{1}{y}\beta_{2n}(y)w_{2^{-n}x}(y)dy =$$

$$= 2^{-n}V_{2n}(2^{-n}x) \quad (x \in \mathbf{R}^+,\ n \in \mathbf{N}). \ \square$$

In the following we give an estimate for the functions V_n $(n \in \mathbf{N})$. For this we define the functions L and J as follows:

$$L(u) := \begin{cases} 2^{-n-s-2}, & u \in [2^n + 2^s + k,\ 2^n + 2^s + k + 1) \\ & (0 \le k < 2^s,\ 0 \le s < n,\ n \in \mathbf{P}) \\ 0, & \text{otherwise,} \end{cases} \tag{17}$$

$$J(u) := \begin{cases} 1, & u \in [0,1) \\ 2^{-n-1}, & u \in [2^n,\ 2^n + 1),\ n \in \mathbf{N} \\ 0, & \text{otherwise.} \end{cases} \tag{18}$$

It is easy to see that $L, J \in L^1(\mathbf{R}^+)$ and

$$\| J \|_1 = 2, \qquad \| L \|_1 = \frac{1}{2}\,. \tag{19}$$

Using the functions L and J we can give an estimate for the functions V_n $(n \in \mathbf{N})$. For this we denote by $d_n W$ $(n \in \mathbf{N})$ the functions defined on the interval $[0,1)$ in [4]. In this case the following assertion is true.

Theorem 1. *For every $n \in \mathbf{N}$ we have the estimate*

$$| V_n | \le 7(L + J) + 2J \sum_{k=0}^{\infty} 2^{-k} \overline{D}_{2^k} + \chi[0,1)\,(| d_n W | + 1)\ , \tag{20}$$

where $\overline{D}_{2^k}$ $(k \in \mathbf{N})$ is the periodic extension of the Walsh–Dirichlet kernel D_{2^k} from the interval $[0,1)$ to $\mathbf{R}^+$ with period 1.

Proof. Since

$$\beta_n(y) = k \quad (y \in [i2^n + k, i2^n + k + 1)) \tag{21}$$

for $i, n \in \mathbf{N}$, $0 \le k < 2^n$, $V_n(x)$ can be written in the following form:

$$V_n(x) = \sum_{k=1}^{2^n - 1} w_k(x) \int_k^{k+1} \frac{k}{y} w_{[x]}(y)\,dy +$$

(22)

$$+ \sum_{i=1}^{\infty} \Big(\sum_{k=0}^{2^n - 1} w_{i2^n + k}(x) \int_{i2^n + k}^{i2^n + k + 1} \frac{k}{y} w_{[x]}(y)\,dy \Big) \quad (x \in \mathbf{R}^+,\ n \in \mathbf{N}).$$

In the following we use the notation

$$A_{i2^n+k}(x) := \int_{i2^n+k}^{i2^n+k+1} \left(\frac{k}{y} - \frac{k}{i2^n+k}\right) w_{[x]}(y)dy \quad (x \in \mathbf{R}^+), \tag{23}$$

where for $i \in \mathbf{P}$, $n \in \mathbf{N}$ and $0 \leq k < 2^n$ and for $i = 0$, $1 \leq k < 2^n$. Using this notation we have

$$\begin{aligned} V_n = \sum_{k=1}^{2^n-1} A_k w_k + \sum_{i=1}^{\infty} \left(\sum_{k=0}^{2^n-1} A_{i2^n+k} w_{i2^n+k} \right) + \\ + \chi[0,1)(d_n W - 1) \qquad (n \in \mathbf{N}). \end{aligned} \tag{24}$$

Let us introduce the following notations:

$$\begin{aligned} V_n^1 &:= \sum_{k=1}^{2^n-1} A_k w_k, \\ V_n^2 &:= \sum_{i=1}^{\infty} \left(\sum_{k=0}^{2^n-1} A_{i2^n+k} w_{i2^n+k} \right) \quad (n \in \mathbf{N}). \end{aligned} \tag{25}$$

Firstly we investigate the functions V_n^2 $(n \in \mathbf{N})$ and we shall show that

$$| V_n^2 | \leq 3(L+J) \quad (n \in \mathbf{N}). \tag{26}$$

Using the second mean value theorem of integral calculus and integration by parts, we get that for every $x \in [1, +\infty)$, $i \in \mathbf{P}$ and $0 \leq k < 2^n$ we have

$$\begin{aligned} A_{i2^n+k}(x) = \int_{i2^n+k}^{i2^n+k+1} \frac{k}{y^2} J_{[x]}(y)dy = \\ = \frac{k}{(i2^n+k)^2} \int_0^{\xi} J_{[x]}(y)dy + \frac{k}{(i2^n+k+1)^2} \int_{\xi}^{1} J_{[x]}(y)dy, \end{aligned} \tag{27}$$

where $\xi \in (0,1)$ is an appropriate number, and $J_{[x]}$ $(x \in \mathbf{R}^+)$ denotes the integral function of $w_{[x]}$ vanishing at 0 :

$$J_{[x]}(t) := \int_0^t w_{[x]}(u)du \quad (t, x \in \mathbf{R}^+). \tag{28}$$

In the following we shall need some estimates for the functions

$$L(x,\omega) := \int_0^\omega J_{[x]}(t)dt \quad (x, \omega \in \mathbf{R}^+). \tag{29}$$

It can be shown that

$$| J_{2^n+k}(t) | \leq 2^{-n-1} \ (t \in \mathbf{R}^+, \ 0 \leq k < 2^n, \ n \in \mathbf{N}) \tag{30}$$

and $\omega \to L(x,\omega)$ is a 1-periodic function, if $[x] \neq 2^l$ for some $l \in \mathbf{N}$ and $x \geq 1$.

Furthermore, it is easy to check that

$$\begin{gathered} | L(2^n + 2^s + k, \omega) | \leq 2^{-n-s-2} \\ (\omega \in \mathbf{R}^+, \ 0 \leq k < 2, \ 0 \leq s < n, \ n \in \mathbf{P}). \end{gathered} \tag{31}$$

Using the definition of the functions L and J (see(17), (18)), from (30) and (31) we get the following estimate:

$$| L(x,\omega) - L(x,[\omega]) | \leq L(x) + J(x) \quad (x, \omega \in \mathbf{R}^+). \tag{32}$$

From this and (27) the following estimate follows:

$$\begin{aligned} | A_{i2^n+k}(x) | &\leq \frac{k}{i^2 \cdot 2^{2n}} (2 \mid L(x,\xi) \mid + \mid L(x,1) \mid) \leq \\ &\leq \frac{3k}{i^2 \cdot 2^{2n}} (L(x) + J(x)) \quad (x \in [1, +\infty)). \end{aligned} \tag{33}$$

Using this, we get that on the interval $[1,+\infty)$

$$
\begin{aligned}
&| V_n^2 | \leq \sum_{i=1}^{\infty}\Big(\sum_{k=0}^{2^n-1} | A_{i2^n+k} |\Big) \leq \\
(34)\qquad &\leq \frac{3}{2}(L+J)\sum_{i=1}^{\infty}\frac{1}{i^2} < 3(L+J) \quad (n \in \mathbf{N}).
\end{aligned}
$$

For $x \in [0,1)$, $i \in \mathbf{P}$ and $0 \leq k < 2^n$ we have that

$$(35) \qquad A_{i2^n+k}(x) = k \cdot \left[\ln\left(1+\frac{1}{i2^n+k}\right) - \frac{1}{i2^n+k}\right].$$

From this using the inequality

$$| \ln(1+t) - t | \leq \frac{1}{2}t^2 \ (0 \leq t < 1),$$

it follows that on the interval $[0,1)$

$$(36) \quad | A_{i2^n+k} | \leq \frac{k}{2(i2^n+k)^2} \leq \frac{k}{2i^2 \cdot 2^{2n}} \quad (i \in \mathbf{P},\ 0 \leq k < 2^n)$$

and, since $L=0$ and $J=1$ on the interval $[0,1)$, we get that on the interval $[0,1)$

$$(37) \ | V_n^2 | \leq \sum_{i=1}^{\infty}\Big(\sum_{k=0}^{2^n-1} | A_{i2^n+k} |\Big) \leq \frac{1}{2} < 3(L+J) \quad (n \in \mathbf{N}).$$

Consequently, for the functions V_n^2 $(n \in \mathbf{N})$ we have the estimate (26).

In the following we investigate the functions V_n^1 $(n \in \mathbf{N})$. Namely, we shall show that

$$(38) \qquad | V_n^1 | \leq 4(L+J) + 2J\Big(\sum_{k=0}^{\infty} 2^{-k}\overline{D}_{2^k}\Big) \ (n \in \mathbf{N}).$$

For $1 \le k < 2^n$ and $x \in [1, +\infty)$, using integration by parts, we get that

$$A_k(x) = \int_k^{k+1} (\frac{k}{y} - 1) w_{[x]}(y)dy =$$

$$(39) \quad = \int_0^1 \left(\frac{1}{1 + \frac{y}{k} - 1} \right) w_{[x]}(y)dy = \frac{1}{k} \int_0^1 \frac{1}{(1+\frac{y}{k})^2} J_{[x]}(y)dy =$$

$$= \frac{1}{k} \int_0^1 \left(\frac{1}{(1+\frac{y}{k})^2} - 1 \right) J_{[x]}(y)dy + \frac{1}{k} \int_0^1 J_{[x]}(y)dy :=$$

$$:= A_k^1(x) + A_k^2(x).$$

Using the second mean value theorem of integral calculus, we have

$$(40) \quad A_k^1(x) = \frac{1}{k} \left(\frac{1}{(1+\frac{1}{k})^2} - 1 \right) \cdot \int_\xi^1 J_{[x]}(y)dy$$

$$(x \in [1, +\infty), \ 1 \le k < 2^n)$$

with an appropriate number $\xi \in (0,1)$. Using this,we get the following estimate on the interval $[1, +\infty)$:

$$(41) \quad \begin{aligned} &| \sum_{k=1}^{2^n-1} A_k^1 w_k | \le \sum_{k=1}^{2^n-1} | A_k^1 | \le \\ &\le \sum_{k=1}^{2^n-1} \frac{2}{k^2}(L+J) \le 4(L+J) \quad (n \in \mathbf{N}). \end{aligned}$$

From the definition of A_k^2 we have

$$(42) \quad \sum_{k=1}^{2^n-1} A_k(x)w_k(x) = \left(\sum_{k=1}^{2^n-1} \frac{w_k(x)}{k} \right) \cdot \int_0^1 J_{[x]}(y)dy$$

$$(x \in [1, +\infty), \ n \in \mathbf{N}).$$

By the definition of J and an estimate for the first factor in (42) (see [4]), we get

$$| \sum_{k=1}^{2^n-1} A_k^2(x) w_k(x) | \leq 4 \left(\sum_{k=0}^{\infty} 2^{-k} \overline{D}_{2^k}(x) \right) \cdot \int_0^1 J_{[x]}(y) dy = \tag{43}$$

$$= 2J(x) \left(\sum_{k=0}^{\infty} 2^{-k} \overline{D}_{2^k}(x) \right) \quad (x \in [1, +\infty), \ n \in \mathbf{N}).$$

For $x \in [0, 1)$ and $1 \leq k < 2^n$ we can calculate $A_k(x)$ simply as follows:

$$A_k(x) = \int_k^{k+1} (\frac{k}{y} - 1) dy = k \ln(1 + \frac{1}{k}) - 1. \tag{44}$$

Write the sum

$$\sum_{k=1}^{2^n-1} A_k w_k$$

on the interval $[0, 1)$ in the form

$$\sum_{k=1}^{2^n-1} A_k w_k = \tag{45}$$

$$= \sum_{k=1}^{2^n-1} (k \ln(1 + \frac{1}{k}) - 1 + \frac{1}{2k}) w_k - \frac{1}{2} \sum_{k=1}^{2^n-1} \frac{w_k}{k} \quad (n \in \mathbf{N}).$$

Since

$$| \ln(1 + t) - t + \frac{t^2}{2} | \leq \frac{1}{3} t^3 \ (0 \leq t < 1),$$

from (45) we get the following estimate on the interval $[0,1)$:

$$
(46)\qquad \begin{aligned} \left|\sum_{k=1}^{2^n-1} A_k w_k\right| &\leq \frac{1}{3}\sum_{k=1}^{2^n-1}\frac{1}{k^2}+\frac{1}{2}\left|\sum_{k=1}^{2^n-1}\frac{w_k}{k}\right| \leq \\ &\leq \frac{2}{3}+2\sum_{k=0}^{\infty}2^{-k}\overline{D}_{2^k} \quad (n\in \mathbf{N}). \end{aligned}
$$

Summarizing our results, we get the desired estimate (38) for the functions V_n^1 $(n \in \mathbf{N})$ and on the basis of (24), (26) and (38) we have proved the theorem. □

In the following we shall show that the estimate (20) for V_n $(n \in \mathbf{N})$ implies among others the strong dyadic differentiability of the dyadic integrals (see also [5]). Namely, the following theorem is true.

Theorem 2. i) $\sup_{n\in\mathbf{N}} \| V_n \|_1 < +\infty$.

ii) *If for a function* $f \in L^1(\mathbf{R}^+)$

$$
(47)\qquad \widehat{f}(0) = \int_{\mathbf{R}^+} f = 0
$$

and the dyadic integral $If \in L^1(\mathbf{R}^+)$ *exists, then* If *is strongly dyadically differentiable and*

$$
(48)\qquad D(If) = f.
$$

Proof. i) We know that $L, J \in L^1(\mathbf{R}^+)$ and on the basis of the definition of J we have

$$
(49)\qquad \begin{aligned} &\int_{\mathbf{R}^+} J\left(\sum_{k=0}^{\infty} 2^{-k}\overline{D}_{2^k}\right) = \\ &= \int_0^1 \left(\sum_{k=0}^{\infty} 2^{-k}\overline{D}_{2^k}\right) + \sum_{j=0}^{\infty} 2^{-j-1}\int_{2^j}^{2^j+1}\left(\sum_{k=0}^{\infty} 2^{-k}\overline{D}_{2^k}\right) \leq 4. \end{aligned}
$$

Furthermore, using a result in [4], one has

$$\sup_{n\in\mathbf{N}} \int_0^1 | d_n W | < +\infty. \tag{50}$$

Thus, by the estimate (20) for V_n $(n \in \mathbf{N})$ part i) of the theorem is proved.

ii) Define the operators T_n $(n \in \mathbf{N})$ on $L^1(\mathbf{R}^+)$ as follows:

$$T_n f := d_n W_n * f \quad (f \in L^1(\mathbf{R}^+),\ n \in \mathbf{N}). \tag{51}$$

The operators T_n $(n \in \mathbf{N})$ are uniformly bounded. In fact, by (16) we have

$$\begin{aligned} \| T_n \| \leq \| d_n W_n \|_1 &= \int_{\mathbf{R}^+} 2^{-n} | V_{2n}(2^{-n}x) | \, dx = \\ &= \| V_{2n} \|_1 = O(1) \quad (n \to \infty). \end{aligned} \tag{52}$$

Thus, by (9) and Banach–Steinhaus theorem it is enough to show (48) for the elements of a dense subset in the dyadically integrable functions satisfying (47). To this end, let us define the set of functions

$$S := \{\chi[0, 2^s) w_{2^{-s}m} : \ s \in \mathbf{N},\ m \in \mathbf{P}\}. \tag{53}$$

Then it can be shown that S is a dense subset in the set of dyadically integrable functions.

Firstly we shall show that if $f \in L^1(\mathbf{R}^+)$ is a dyadically integrable function, then

$$\lim_{n\to\infty} 2^n \int_0^{2^n} f = 0. \tag{54}$$

Let $h := \chi[0, 2^{-n})$ $(n \in \mathbf{N})$. Then its Walsh–Fourier transform is

$$\widehat{h} = 2^{-n}\chi[0, 2^n)$$

and by the equality

$$\int_{\mathbf{R}^+} f\widehat{h} = \int_{\mathbf{R}^+} \widehat{f}h$$

we get

$$2^{-n}\int_0^{2^n} f = \int_0^{2^{-n}} \widehat{f}. \tag{55}$$

Putting $g := If$, by (8) we have

$$\int_0^{2^{-n}} \widehat{f} = \int_0^{2^{-n}} y\widehat{g}(y)dy. \tag{56}$$

Since $\widehat{g}(0) = 0$ and $\widehat{g}$ is w-continuous, we can estimate $\widehat{g}$ by its modulus of continuity as follows:

$$|\, \widehat{g}(y) \,|=|\, \widehat{g}(y) - \widehat{g}(0) \,|\leq \omega(\widehat{g}, 2^{-n}) \quad (y \in [0, 2^{-n})). \tag{57}$$

From this we get

$$|\int_0^{2^{-n}} y\widehat{g}(y)\, dy \,|\leq \omega(\widehat{g}, 2^{-n}) \int_0^{2^{-n}} y\, dy = \frac{1}{2}2^{-2n}\omega(\widehat{g}, 2^{-n}),$$

and, consequently, by (55) and (56) the inequality (54) follows.

In the following let $f \in L^1(\mathbf{R}^+)$ be a dyadically integrable function and ε an arbitrary positive number. Let us decompose the function f in the form

$$f = \chi[0, 2^N)f + \chi[2^N, +\infty)f := f_1 + f_2, \tag{58}$$

where $N \in \mathbf{N}$ is natural number for which

$$\| f_2 \|_1 = \int_{2^N}^{\infty} | f | < \varepsilon, \quad 2^N \left| \int_0^{2^N} f \right| < \varepsilon \tag{59}$$

hold. Let us define an element P in the linear hull of S as follows:

$$P := S_{2^n} f_1 - \left(\int_0^{2^N} f\right) \chi[0, 2^N), \tag{60}$$

where $S_{2^N} f_1$ denotes the 2^N-th partial sum of f_1 on the interval $[0, 2^N)$. Then

$$\lim_{N \to \infty} \| f_1 - S_{2^N} f_1 \|_1 = 0 \tag{61}$$

and, consequently, by (59) and (61)

$$\begin{aligned} &\| f - P \|_1 \leq \| f_1 - P \|_1 + \| f_2 \|_1 \leq \\ &\leq \| f_1 - S_{2^N} f_1 \|_1 + \| \left(\int_0^{2^N} f\right) \chi[0, 2^N) \|_1 + \| f_2 \|_1 < 3\varepsilon, \end{aligned} \tag{62}$$

if N is large enough. Thus we have proved that S is a dense subset in the set of dyadically integrable functions.

In the following we shall show that for elements of S (48) is true. Let

$$f := \chi[0, 2^s) w_{2^{-s}m} \quad (s \in \mathbf{N}, \ m \in \mathbf{P}) \tag{63}$$

be an arbitrary element of S. We must show that

$$\lim_{n \to \infty} \| d_n W_n * f - f \|_1 = 0. \tag{64}$$

By an easy calculation we get that

$$(d_nW_n * f)(x) - f(x) =$$

(65)
$$= \int_{\mathbf{R}^+} \frac{\alpha_n(y) - y}{y} \widehat{f}(y) w_x(y) dy \quad (x \in \mathbf{R}^+,\ n \in \mathbf{N})$$

and

(66)
$$\widehat{f} = 2^s \chi[\frac{m}{2^s}, \frac{m+1}{2^s}).$$

Consequently,

$$(d_nW_n * f)\,(x) - f(x) =$$

(67)
$$= 2^s \int_{m/2^s}^{(m+1)/2^s} \frac{\alpha_n(y) - y}{y} w_x(y) dy \quad (x \in \mathbf{R}^+,\ n \in \mathbf{N}).$$

Let us introduce the following notation:

$$F_n(x) := (d_nW_n * f)(2^s x) - f(2^s x) \quad (x \in \mathbf{R}^+,\ n \in \mathbf{N}).$$

It is enough to show that

(69)
$$\lim_{n \to \infty} \| F_n \|_1 = 0.$$

Applying a simple transformation we can show that

(70)
$$F_n(x) = 2^s w_m(x) \int_0^1 \gamma_n(u)\gamma(u) w_{[x]}(u) du$$
$$(x \in \mathbf{R}^+,\ n \in \mathbf{N}),$$

where

$$(71)\qquad \begin{aligned} \gamma_n(u) &:= \alpha_n(2^{-s}u) - 2^{-s}u, \\ \gamma(u) &:= \frac{1}{m+u} \quad (0 \le u < 1). \end{aligned}$$

From (70) we get that

$$(72)\qquad \begin{aligned} &\| F_n \|_1 = \| (\gamma_n \gamma)^\wedge \|_{l^1} = \\ &= \| \widehat{\gamma}_n * \widehat{\gamma} \|_{l^1} \le \| \widehat{\gamma}_n \|_{l^1} \| \widehat{\gamma} \|_{l^1} \quad (n \in \mathbf{N}), \end{aligned}$$

where $\widehat{\varphi}$ denotes the sequence of Walsh–Fourier coefficients for any function φ defined $[0,1)$. Integrating by parts one can show that

$$(73)\qquad \| \widehat{\gamma} \|_{l^1} < +\infty.$$

Moreover, since

$$(74)\quad \gamma_n(u) = \sum_{k=n+1}^{\infty} \frac{w_{2^{k-s}}(u)}{2^{k+1}} - \frac{1}{2^{n+1}} \quad (0 \le u < 1,\ n \in \mathbf{N}),$$

we get

$$(75)\qquad \| \gamma_n \|_{l^1} = \sum_{k=n+1}^{\infty} \frac{1}{2^{k+1}} + \frac{1}{2^{n+1}} = \frac{1}{2^n} \quad (n \in \mathbf{N}).$$

Consequently, by (72) we have

$$\lim_{n\to\infty} \| F_n \|_1 = 0. \ \square$$

References

[1] BUTZER, P. L. and WAGNER, H. J., A calculus for Walsh functions defined on $\mathbf{R}^+$. In: *Proc. Symp. on Applications of Walsh Functions.* Washington D. C., 1973, 75-81.

[2] PÁL, J. On almost everywhere differentiability of dyadic integral functions on $\mathbf{R}^+$. In: *Colloq. Math. Soc. J. Bolyai, 19. Fourier Analysis and Approximation Theory,* Vol. II. Budapest, 1978, 591-601.

[3] PÁL, J., The almost everywhere differentiability of Wagner's dyadic integral function on $\mathbf{R}^+$. In: *Colloq. Math. Soc. J. Bolyai, 49. Alfred Haar Memorial Conference.* Budapest, 1985, 695-701.

[4] SCHIPP, F., Über einen Ableitungsbegriff von P. L. Butzer und H. J. Wagner. *Math. Balkanica* **4** (1974) 541-546.

[5] WAGNER, H. J., On the dyadic calculus for functions defined on $\mathbf{R}^+$. In: *Theory and Applications of Walsh Functions and Other Nonsinusoidal Functions.* Hatfield Polytechnic, Hatfield, 1975, 101-129.

(Receieved December 28, 1987)

JENŐ PÁL FERENC SCHIPP

Dept. of Numerical Analysis
Eötvös Loránd University
H-1088 Budapest, Múzeum krt. 6-8.

HUNGARY

ON THE ROLE OF WALSH AND HAAR FUNCTIONS IN DYADIC ANALYSIS

P.L. Butzer, W. Splettstößer, H.J. Wagner
Lehrstuhl A für Mathematatik, Technological University of Aachen
Aachen, Western Germany

1. Summary

In a series of papers Gibbs [6,7], Pichler [11] Butzer-Wagner [3,4,5], Schipp [12,13], Penney [10] and others studied a differential and integral calculus based upon the Walsh system, the so-called dyadic calculus, a good portion of which has been developed. For further references to this subject see [6].

The aim of this talk is to introduce, in analogy with the dyadic calculus, a derivative and integral concept with respect to the Haar system and to build it up into a calculus suitable for Haar analysis. Its application potentialities are examined in Sec. 5; these are to the theory of Haar series, namely to the order of magnitude of the Haar coefficients, to best approximation by Walsh polynomials. In Sec. 6 the comparisons between the Haar calculus and the dyadic calculus are stressed. For example, the derivatives in both cases are infinite sums, and the corresponding integrals are defined via dyadic convolution.

Details to the material sketched in the talk are to be found in [14].

2. Preliminaries; Haar-Translation

In the following let $N=\{1,2,\dots\}$, $P=\{0,1,2,\dots\}$, $R=(-\infty,+\infty)$ and let $L^p(0,1)$, $1\leq p<\infty$, be the set of all functions f defined on R with period 1 which are pth power (Lebesgue) integrable with norm

$$\|f\|_p = \int_0^1 |f(x)|^p dx \; ^{1/p} \qquad (2.1).$$

This lecture is concerned with an orthonormal system of functions introduced by Haar [9] in 1909. These Haar functions are defined on $[0,1)$ by

$$\chi_0(x)=1$$

$$\chi_{2^n+j}(x) = \begin{cases} \sqrt{2^n}, & x\in[\frac{2j}{2^{n+1}}, \frac{2j+1}{2^{n+1}}) \\ -\sqrt{2^n}, & x\in[\frac{2j+1}{2^{n+1}}, \frac{2j+2}{2^{n+1}}) \\ 0, & \text{otherwise on } [0,1), \end{cases} \qquad (2.2)$$

where $n\in P$, $0\leq j<2^n$, and on R by periodic extension.

For $f\in L^p(0,1)$, $k\in P$, the real numbers

$$f^\wedge(k)=\int_0^1 f(x)\chi_k(x)dx$$

are called the Haar (Fourier) coefficients of f,

$$\sum_{k=0}^{\infty} f^\wedge(k)\chi_k(x)$$

the Haar (Fourier) series of f, and

$$(s_m f)(x)=\sum_{k=0}^{m-1} f^\wedge(k)\chi_k(x)$$

the mth partial sum of the Haar series.

These concepts allow one to define a Haar translation, needed to define a Haar derivative. Recall that in the classical derivative

$f'(x)=\lim_{h\to 0} h^{-1}[f(x+h)-f(x)]$ the translation $f(x+h)$ plays the basic rôle, and in the dyadic derivative

$$f^{\{1\}}(x) = \tfrac{1}{2}\sum_{j=0}^{\infty} 2^j[f(x)-f(x\oplus 2^{-j-1})] \qquad (2.3)$$

the dyadic translation $f(x\oplus h)$ (see Sec. 6). A suitable translation or shift concept $f_h(x)$ for the Haar system is defined by:

Definition 1. With $h\in[2^{-m-1}, 2^{-m})$, $m\in P$ and $f\in L^p(0,1)$,

$$f_h(x) = 2(s_{2^m}f)(x)-(s_{2^{m+1}}f)(x) \qquad (2.4)$$

is called the Haar translation of f by h.

The function $f_h(x)$ has the following typical properties:

i) If $f\in L^p(0,1)$, then $f_h\in L^p(0,1)$ with

$$\|f_h\|_p\leq\|f\|_p;$$

ii) For each $f\in L^p(0,1)$ one has

$$\lim_{h\to 0+} \|f_h-f\|_p = 0;$$

iii) The Haar coefficients of f are with $n\in P$, $0\leq j<2^n$,

$$f_h^\wedge(0) = f^\wedge(0)$$

$$f_h^\wedge(2^n+j)=\sqrt{2^{-n}}\chi_{2^n}(h)f^\wedge(2^n+j);$$

iv) The Haar functions, shifted by h, have for $n\in P$, $0\leq j<2^n$ the form

$$(\chi_0)_h(x) = \chi_0(x)$$

$$(\chi_{2^n+j})_h(x)=\sqrt{2^{-n}}\chi_{2^n}(h)\chi_{2^n+j}(x).$$

Concerning the proof of (ii) one needs the known fact that $f\in L^p(0,1)$ implies

$$\lim_{n\to\infty} \|s_{2^n}f-f\|_p = 0 \qquad (2.5).$$

For complete proofs see Splettstößer-Wagner [14].

3. The Haar-Derivative

A derivative concept adapted to the Haar system and given via translation (2.4) is

Definition 2. For $f\in L^p(0,1)$, $1\le p<\infty$, the first Haar derivative is defined by

$$(D^{[1]}f)(x)=\frac{1}{3}\sum_{j=0}^{\infty}2^j[f(x)-f_{2^{-j-1}}(x)] + \frac{1}{3}[f(x)-\int_0^1 f(u)du]. \tag{3.1}$$

The rth derivative of f is defined successively by $D^{[r]}f=D^{[1]}(D^{[r-1]}f)$, $r=2,3,\dots$.

The convergence of the sum in (3.1) is to be understood in the norm (2.1) of the space $L^p(0,1)$. The derivative index is written in square brackets to distinguish it from the classical derivative $f^{(r)}(x)$ and the dyadic derivative $f^{\{r\}}(x)$.

The Haar derivative has the following basic properties:

i) $D^{[r]}$ is a closed, linear operator;

ii) $D^{[1]}f=0 \iff f=\text{const.}$;

iii) The Haar functions are arbitrarily often differentiable in the sense of Def. 2 with $(n\in P,\ 0\le j<2^n)$

$$(D^{[r]}\chi_0)(x)=0$$

$$(D^{[r]}\chi_{2^n+j})(x)=2^{nr}\chi_{2^n+j}(x);$$

iv) If f, $D^{[r]}f\in L^p(0,1)$ for some $r\in N$, then for $n\in P$, $0\le j<2^n$,

$$[D^{[r]}f]^\wedge(0)=0$$

$$[D^{[r]}f]^\wedge(2^n+j)=2^{nr}f^\wedge(2^n+j).$$

Essential in the proofs of properties (ii)-(iv) is the fundamental identity $(n\in P)$

$$3\cdot 2^n=\sum_{j=0}^{\infty}2^j[1-\sqrt{2^{-n}}\chi_{2^n}(2^{-j-1})]+1 .$$

which follows by (2.2) together with $(n,j\in P)$

$$\chi_{2^n}(2^{-j-1})=\begin{cases}\sqrt{2^n}, & j>n\\ -\sqrt{2^n}, & j=n\\ 0, & j<n.\end{cases}$$

The proof of (iv) for $r=1$ also makes use of a result of Ul'janov [15], namely that if $f\in L^p(0,1)$ then

$$|f^\wedge(k)|\le 2\,k^{1/p-1/2}\|f\|_p .$$

For complete proofs see again [14].

Note that it would be possible to define the Haar derivative differently and still be able to obtain the same applications as those to be treated in Sec. 5. One could, for example, take the dyadic derivative for the Haar setting, but this would be difficult technically.

4. The Haar Anti-derivative

In order to define the inverse operator to the Haar derivative $D^{[1]}f$, namely the anti-differentiation operator, one needs the concepts of dyadic addition and dyadic convolution.

Each $x\in[0,1)$ with $x\ne p\cdot 2^{-q}$, $p\in P, q\in N$, has the unique expansion

$$x=\sum_{j=1}^{\infty}x_j 2^{-j},\quad x_j\in\{0,1\}.$$

For $x=p\cdot 2^{-q}$ there being two expansions, the finite one is chosen. The dyadic addition of two numbers $x,y\in[0,1)$ is then defined by

$$x\oplus y=\sum_{j=1}^{\infty}|x_j-y_j|2^{-j}.$$

If $x,y\in R$, $[x]$ being the largest integer $\le x$, then $x\oplus y=(x-[x])\oplus(y-[y])$.

The dyadic convolution of $f\in L^p(0,1)$ with a function $g\in L^1(0,1)$ is now defined by

$$(f*g)(x)=\int_0^1 f(u)g(x\oplus u)du.$$

One has

$$\|f*g\|_p\le\|g\|_1\|f\|_p \tag{4.1}.$$

This leads to the concept of integration with respect to the Haar system.

Definition 3. If $f\in L^p(0,1)$, $1\le p<\infty$, then

$$(I_{[r]}f)(x)=(f*K_r)(x) \tag{4.2}$$

with

$$K_r(x)=1+\sum_{k=0}^{\infty}2^{-k(r-1/2)}\chi_{2^k}(x)\quad\text{a.e.} \tag{4.3}$$

is called the rth Haar anti-derivative of f.

The function $K_r(x)$ is well-defined since the sum in (4.3) is convergent for each $x\in R$ in case $r\ge 2$, and for each non-integer in case $r=1$. $K_r(x)$ belongs to $L^1(0,1)$ for each $r\in N$, so that dyadic convolution (4.2) is also well-defined.

The Haar anti-differentiation operator $I_{[r]}$ has the following properties:

i) $I_{[r]}$ is a closed, linear operator;

ii) If $f\in L^p(0,1)$, then $I_{[r]}f\in L^p(0,1)$, $r\in N$;

iii) For the Haar functions one has for $n\in P$, $0\le j<2^n$

$$(I_{[r]}\chi_0)(x)=\chi_0(x)$$

$$(I_{[r]}\chi_{2^n+j})(x)=2^{-nr}\chi_{2^n+j}(x);$$

iv) If $f\in L^p(0,1)$, $1\le p<\infty$, then

$[I_{[r]}f]^\wedge(0) = f^\wedge(0)$

$[I_{[r]}f]^\wedge(2^n+j)=2^{-nr}f^\wedge(2^n+j)$.

The proofs of i) and ii) follow directly from Def. 3 and (4.1); for iii) and iv) see [14].

A comparison of the properties iii) and iv) with those of iii) and iv) for the operator $D^{[r]}$ suggest that the operations $D^{[r]}$ and $I_{[r]}$ are actually inverse to another. One can indeed show (see [14]):

Theorem 1. Let $f \in L^p(0,1), f^\wedge(0)=0$. One has:

i) If $D^{[r]}f \in L^p(0,1)$, then

$I_{[r]}(D^{[r]}f)=f$,

ii) $D^{[r]}(I_{[r]}f)=f$.

With this result, which is the counterpart of the fundamental theorem of the classical differential and integral calculus, we have developed a calculus with respect to the Haar system, the so-called Haar calculus.

5. Applications of the Haar Calculus to the Theory of Haar Series

5.1 Haar Modulus of Continuity.

The Haar calculus set up above allows new applications to Haar series. In order to examine the rate of convergence of these series the classical modulus of continuity

$$\omega(f;\delta) = \sup_{0<|h|<\delta} \|f(\cdot+h)-f(\cdot)\|_p \qquad (5.1)$$

plays an important rôle. If one replaces the translation $f(x+h)$ in (5.1) by the Haar shift $f_h(x)$, one obtains the Haar modulus of continuity

$$\omega_H(f;\delta) = \sup_{0<h<\delta} \|f_h-f\|_p.$$

For $f \in L^p(0,1)$, $n \in N$, these two concepts are related by

$$\omega_H(f;2^{-n}) < 8\omega(f;2^{-n}) \qquad (5.2),$$

the proof of which depends on (2.4) and an estimate of Ul'janov [15]:

$$\|f-s_{2^n}\|_p < 4\omega(f;2^{-n}).$$

Defining the Haar Lipschitz class $Lip_H\ \alpha$, $\alpha>0$, by $(\delta \to 0+)$

$$Lip_H\ \alpha = \{f \in L^p(0,1); \omega_H(f;\delta) = O(\delta^\alpha)\} \quad (5.3),$$

and comparing it with the classical Lipschitz class

$$Lip\ \alpha = \{f \in L^p(0,1); \omega(f;\delta) = O(\delta^\alpha)\},$$

then (5.2) gives a sufficient criterion when f belongs to $Lip_H\ \alpha$, namely:

If $f \in Lip\ \alpha$, $\alpha>0$, then $f \in Lip_H\ \alpha$.

If a function $f \in L^p(0,1)$ has a classical derivative, the modulus of continuity of f can be estimated by the modulus of $f^{(1)}$. The counterpart for the Haar derivative is

Theorem 2. If f, $D^{[r]}f \in L^p(0,1)$, $n \in N$, then

$$\omega_H(f;2^{-n}) < 2^{-nr}\omega_H(D^{[r]}f;\ 2^{-n}).$$

This result enables one to obtain new estimates for the Haar coefficients as well as for the rate of convergence of Haar series, found below.

5.2 Order of Magnitude of Haar Coefficients

Ul'janov [15] showed that for $f \in L^p(0,1)$ one has for $k \to \infty$

$$|f^\wedge(k)| = O(k^{1/p-1/2}\omega(f;k^{-1})).$$

This estimate remains valid when replacing the modulus of continuity here by the Haar one. Indeed,

$$|f^\wedge(k)| = O(k^{1/p-1/2}\omega_H(f;k^{-1})).$$

Together with Thm. 2 this yields

Theorem 3. If f, $D^{[r]}f \in L^p(0,1)$, then for $k \to \infty$

$$|f^\wedge(k)| = O(k^{1/p-1/2-r}\omega_H(D^{[r]}f;k^{-1})).$$

Note that a corresponding assertion involving the rth classical derivative would not hold. Indeed, Golubov [8] has shown that for functions whose rth classical derivative is integrable one has for $k \to \infty$:

$$f^\wedge(k) = o(k^{-3/2}) \Longrightarrow f = \text{const.}$$

5.3 Approximation by Haar Polynomials

Letting P_n be the set of all Haar polynomials of degree $<n$, namely $p_n(x)=\sum_{k=0}^{n-1}a_k\chi_k(x)$ where $a_k \in R$, then

$$E_n(f) = \inf_{p_n \in P_n} \|f-p_n\|_p = \|f-p_n^*\|_p$$

is called the best approximation of $f \in L^p(0,1)$ by Haar polynomials, and p_n^* the corresponding polynomial of best approximation. It is obvious that $E_n(f) > E_{n+1}(f)$, $n \in N$, and that $\lim_{n\to\infty}E_n(f)=0$ by (2.5).

To treat the rate of convergence of $\{E_n(f)\}_{n=1}^\infty$ to zero, the modulus (5.1) is again needed. Ul'janov [15] showed that for $n \in N$

$$E_n(f) < 6\omega(f;n^{-1}).$$

Employing now the Haar modulus one has

$$(3/2)E_{2^n}(f)\leq\omega_H(f;2^{-n})\leq 4E_{2^n}(f).$$

Concerning the desired rate one has

Theorem 4. The following five assertions are equivalent for $f\in L^p(0,1)$, fixed $r\in P$, $\alpha\in R$, $\alpha>0$, $n\to\infty$:

i) $E_n(f)=O(n^{-r-\alpha})$,

ii) $\omega_H(D^{[r]}f;n^{-1})=O(n^{-\alpha})$,

iii) $D^{[r]}f\in Lip_H\ \alpha$,

iv) $D^{[m]}f\in L^p(0,1)$, $m\in P$, $0\leq m\leq r$, with

$$\|D^{[m]}f-D^{[m]}p_n^*\|_p=O(n^{-r-\alpha+m}),$$

v) $\|D^{[m]}p_n^*\|_p=O(n^{m-r-\alpha})$, $0<r+\alpha<m$.

The equivalence ii)$\Longleftrightarrow$ iii) is Def. (5.3). The proof of the other assertions follows from a general theorem on best approximation in Banach spaces due to Butzer-Scherer [2]. The latter is based upon the verification of corresponding Jackson and Bernstein type inequalities. In our present frame these are

a) For $p_n\in P_n$, $r,n\in N$, one has

$$\|D^{[r]}p_n\|_p\leq An^r\|p_n\|_p;$$

b) If f, $D^{[r]}f\in L^p(0,1)$, then

$$E_n(f)\leq Bn^{-r}\|D^{[r]}f\|_p,$$

A and B being constants independent of n,p_n and f.

Note that the assertions i)-v) above remain valid if the polynomial $p_n^*(f;x)$ of best approximation to f is replaced by the partial sum $(s_{2^n}f)(x)$.

6. Comparison of the Haar Calculus with the Walsh Dyadic Calculus

As mentioned in the summary, dyadic analysis is based upon the Walsh system $\{\psi_k(x)\}_{k=0}^{\infty}$. For $f\in L^p(0,1)$, $f^{\{1\}}(x)$ as defined in (2.3) is the first dyadic derivative, $f^{\{r\}}(x)$, $r\in N$, being defined successively. The corresponding inverse operator, the rth dyadic integral of f, is given by

$$(I_{\{r\}}f)(x) = (f*W_r)(x) \qquad (6.1)$$

where

$$W_r(x) = 1+\sum_{k=1}^{\infty}\psi_k(x)/k^r \text{ a.e. }.$$

This leads to a fundamental theorem of the calculus based upon Walsh functions and to a so-called Walsh dyadic calculus. As a simple example, the Walsh functions are arbitrarily often dyadic differentiable and integrable. For this theory and its many applications see [3, 4, 16].

A comparison of the Haar derivative (3.1) with the dyadic derivative (2.3) and of the Haar anti-derivative (4.2) with (6.1) reveals that the Haar calculus is closely connected with that of Walsh dyadic calculus, both with respect to their structures and applications. The reason of course is that the Haar functions may be written as certain linear combinations of Walsh functions, and conversely. The methods of proof of the two theories are however very different. For the Walsh, in distinction to the Haar functions, build a so-called periodic multiplicative system; i.e. the former can be defined as the characters of the dyadic group (see survey paper [1] and references cited there). In other words, in developing the dyadic calculus one has $\psi_k(x\oplus y)=\psi_k(x)\psi_k(y)$ a.e. in $x,y\in R$ with $k\in P$, at one's disposal, a similar equation not being valid for Haar functions.

It must be pointed out that one could principally introduce a calculus for any orthogonal system using the methods of this talk, the resulting calculus being different for each system, however. The multiplicity of possible derivative and integral concepts would naturally be disturbing, and a unified embracing calculus would be desirable. Whereas one can embed the classical and dyadic derivative in the unified frame of locally compact abelian groups, this is improbable for the Haar system.

However, if one restricts the matter to orthogonal systems consisting of piecewise constant functions, then it may perhaps be possible to build up a unified differential and integral calculus. This talk is a first attempt in this direction. This may possibly help one in finding an appropriate intuitive interpretation of the dyadic or Haar derivative in terms of the modern sciences making use of Walsh or Haar analysis, one of the basic open problems here.

This contribution was carried out in membership with the research group "Informatik Nr. 14" at Aachen.

Table

$\{\chi_k(x)\}_{k=0}^{\infty}$ Haar system		$\{\psi_k(x)\}_{k=0}^{\infty}$ Walsh system
$f^\wedge(k)=\int_0^1 f(u)\chi_k(u)du \qquad (k\in P)$		$f_W^\wedge(k)=\int_0^1 f(u)\psi_k(u)du \qquad (k\in P)$
$s_{2^n}(f,x)=\sum_{k=0}^{2^n-1} f^\wedge(k)\chi_k(x)$	=	$s_{2^n}^{(W)}(f,x)=\sum_{k=0}^{2^n-1} f_W^\wedge(k)\psi_k(x)$
$f_h(x)=2s_{2^n}(f,x)-s_{2^{n+1}}(f,x),\ h\in[\frac{1}{2^{n+1}},\frac{1}{2^n})$		$f(x\oplus h)$
$(D^{[1]}f)(x)=\frac{1}{3}\{\sum_{j=0}^{\infty}2^j[f(x)-f_{2^{-j-1}}(x)] + [f(x)-\int_0^1 f(u)du]\}$		$f^{\{1\}}(x)=\frac{1}{2}\sum_{j=0}^{\infty}2^j[f(x)-f(x\oplus 2^{-j-1})]$
$(I_{[1]}f)(x)=(f*K_1)(x)$		$(I_{\{1\}}f)(x)=(f*W_1)(x)$
$K_1(x)=1+\sum_{k=0}^{\infty}2^{-k/2}\chi_{2^k}(x)$ a.e.		$W_1(x)=1+\sum_{k=1}^{\infty}\frac{\psi_k(x)}{k}$ a.e.
$f(x)\sim\sum_{k=0}^{\infty}f^\wedge(k)\chi_k(x)$		$f(x)\sim\sum_{k=0}^{\infty}f_W^\wedge(k)\psi_k(x)$
$f_h(x)\sim f^\wedge(0)+\sum_{k=0}^{\infty}\sum_{j=0}^{2^k-1}f^\wedge(2^k+j)[\chi_{2^k+j}(x)]_h$ $=f^\wedge(0)+\sum_{k=0}^{\infty}\sum_{j=0}^{2^k-1}f^\wedge(2^k+j)\frac{1}{\sqrt{2^k}}\chi_{2^k}(h)\chi_{2^k+j}(x)$		$f(x\oplus h)\sim\sum_{k=0}^{\infty}f_W^\wedge(k)\psi_k(x\oplus h)$ $=\sum_{k=0}^{\infty}f_W^\wedge(k)\psi_k(h)\psi_k(x)$
$(D^{[1]}f)(x)\sim\sum_{k=0}^{\infty}2^k\sum_{j=0}^{2^k-1}f^\wedge(2^k+j)\chi_{2^k+j}(x)$		$f^{\{1\}}(x)\sim\sum_{k=1}^{\infty}k f_W^\wedge(k)\psi_k(x)$
$(I_{[1]}f)(x)\sim f^\wedge(0)+\sum_{k=0}^{\infty}2^{-k}\sum_{j=0}^{2^k-1}f^\wedge(2^k+j)\chi_{2^k+j}(x)$		$(I_{\{1\}}f)(x)\sim f_W^\wedge(0)+\sum_{k=1}^{\infty}\frac{f_W^\wedge(k)}{k}\psi_k(x)$

References

[1] Balašov, L.A. - Rubinšteǐn, A.I.: Series with respect to the Walsh system and their generalizations. Itogi Nauki, Ser. Mat. (Mat. Anal.), 147-202 (1970) = J.Soviet Math., 1, No. 6, 727-763 (1973).

[2] Butzer, P.L. - Scherer, K.: Jackson and Bernstein-type inequalities for families of commutative operators in Banach spaces. J. Approximation Theory 5, 308-342 (1972).

[3] Butzer, P.L. - Wagner, H.J.: Walsh-Fourier series and the concept of a derivative. Applicable Anal. 3, 29-46 (1973).

[4] Butzer, P.L. - Wagner, H.J.: On a Gibbs-type derivative in Walsh-Fourier analysis with applications. In: Proc. of the 1972 National Electronics Conference (Chicago, Illinois, October 9-11, 1972; Ed. R.E. Horton) Oak Brook, Illinois, 1972, xxvi + 427 pp.; pp. 393-398.

[5] Butzer, P.L. - Wagner, H.J.: On dyadic analysis based on the pointwise dyadic derivative. Analysis Matematica 1, No. 3 (1975). (To appear).

[6] Gibbs, J.E. - Ireland, B.: Walsh functions and differentiation. In: Applications of Walsh Functions and Sequency Theory (New York: IEEE; Ed. Schreiber, H.H. - Sandy, G.F.), Order No. 74 CH 0861-5 EMC (1974), vi + 460 pp; 147-176.

[7] Gibbs, J.E. - Millard M.J.: Walsh functions as solutions of a logical differential equation. National Physical Laboratory,

Teddington, Middlesex, England, DES Report No. 1 (1969).

[8] Golubov, B.I.: Series with respect to the Haar system. Itogi Nauki, Ser. Mat. (Mat. Anal. 1970), 109-146 (1971).

[9] Haar, A.: Zur Theorie der orthogonalen Funktionensysteme. Math. Ann. 69, 331-371 (1910).

[10] Penney, R.: On the rate of growth of the Walsh anti-differentiation operator. (To appear).

[11] Pichler, F.R.: Walsh functions and linear system theory. In: Applications of Walsh Functions (Proc. Sympos. Naval Res. Lab., Washington, D.C. March 31 - April 3, 1970; Ed. C.A. Bass) Washington, D.C. 1970, 273 pp., pp. 175-182 (Springfield, Va 22151: National Technical Information Service: AD 707431).

[12] Schipp, F.: Über einen Ableitungsbegriff von P.L. Butzer und H.J. Wagner. Matematica Balkanica, 4 (1974).

[13] Schipp, F.: On dyadic derivative. (To appear).

[14] Splettstößer, W. - Wagner, H.J.: Ein Infinitesimalkalkül für Haarfunktionen. (To appear).

[15] Ul'janov, P.L.: On Haar series (Russian). Mat. Sb. (N.S.) 63 (105), 356-391 (1964).

[16] Wagner, H.J.: Ein Differential- und Integralkalkül in der Walsh-Fourier-Analysis mit Anwendungen. Forschungsber. des Landes Nordrhein-Westfalen Nr. 2334, Westdeutscher Verlag, Köln-Opladen, 71 pp. (1973).

ON DYADIC CALCULUS FOR FUNCTIONS DEFINED ON R_+

by

H.J. Wagner

Lehrstuhl A für Mathematik
Technological University of Aachen
Fed. Rep. of Germany

Theory and Applications of Walsh Functions
(and other nonsinusoidal functions)

Hatfield, Herts., July 1-3, 1975

Abstract

Already in earlier papers did P.L. Butzer and the author examine the concept of a dyadic derivative for functions defined on $[0,\infty)$. In this paper a corresponding inverse operator is defined for such functions, namely the dyadic integral operator, and a dyadic calculus is developed. It is shown that the analogue of the fundamental theorem of the calculus holds for dyadic differentiation and integration. The application potentialities of this calculus are demonstrated with examples.

1. Introduction

Beginning with work by J.E. Gibbs (and his collaborators) [8,9,10], P.L. Butzer and H.J. Wagner developed in a series of papers [1,2,3,15] a differential and integral calculus which is applicable to functions defined on the dyadic group, or to functions defined on $(-\infty,+\infty)$ having period 1. This dyadic calculus is based upon the system of Walsh functions $\{\psi_k(x)\}_{k=0}^{\infty}$ and is therefore particularly suitable for Walsh-Fourier analysis.

In [4,5] did P.L. Butzer and the author first examine a dyadic differentiation concept (denoted by $D^{[1]}f$) for non-periodic functions f defined on $[0,\infty)$. In this situation this concept is based upon the generalized Walsh functions $\psi(y,x)$ with $y,x \in [0,\infty)$ (which coincide with $\psi_k(x)$ for $y = k \in P$).

Starting off with these two papers, J. Pál [11] introduced an operator U_a such that $U_a f$ defines the inverse operation to $D^{[r]}f$ under the rather strong additional condition

$$(1.1) \qquad f^{\wedge}(v) := \int_0^{\infty} f(u)\psi(v,u)du = 0$$

for $0 \leq v < a$.

In the present paper there is defined the anti-differentation operator $I_{[r]}$, and it is shown that it is the inverse to $D^{[r]}$, $r \in N$, provided f satisfies

$$(1.2) \qquad \int_0^{\infty} f(x)dx = f^{\wedge}(0) = 0$$

- 2 -

The latter condition is essentially weaker than (1.1).

In § 2 the class of functions is considered which characterizes (1.2). In § 3 the dyadic differentiation and integration operators $D^{[r]}$ and $I_{[r]}$ are introduced and it is shown that

$$I_{[r]}(D^{[r]}f) = f$$

$$D^{[r]}(I_{[r]}f) = f$$

provided f satisfies condition (1.2). Here the basic differential and integral calculus for Walsh functions defined on $[0,\infty)$ is developed. In § 4 there are several general remarks concerning the dyadic calculus as well as applications to estimates involving the dyadic derivative in Walsh analysis.

This paper was carried out while the author was a member of the research group "Informatik Nr. 14" at the Technological University of Aachen.

- 3 -

2. Preliminary Results

In the following let $N = \{1,2,3,\ldots\}$, $P = \{0,1,2,\ldots\}$ and $Z = \{0,\pm 1,\pm 2,\ldots\}$; furthermore set $R_+ = [0,\infty)$ and let D_+ be the set of all non-negative numbers of the form $p2^q$ with $p \in P$ and $q \in Z$. Each $y \in R_+$ has the dyadic representation

$$y = \sum_{j=-K}^{\infty} y_j 2^{-j} \qquad (y_j \in \{0,1\},\ K \in P),$$

which is unique in case $y \notin D_+$. If $y \in D_+$ there are two representations, of which we choose the finite one, i.e. the expansion with $y_j = 0$ for all $j \geq j_0$. If $x = \sum_{j=-L}^{\infty} x_j 2^{-j}$ $(x_j \in \{0,1\},\ L \in P)$, then dyadic addition is defined by

$$x \oplus y = \sum_{j=-M}^{\infty} |x_j - y_j| 2^{-j} \qquad (M = \max\{K,L\}).$$

For an integrable function f defined on R_+ one has

$$(2.1) \qquad \int_0^{\infty} f(x \oplus y)\, dx = \int_0^{\infty} f(x) dx \qquad (y \in R_+).$$

For the proof of this property see N.J. Fine [7]. According to Fine the generalized Walsh functions are then given by

$$\psi(y,x) = \exp\{\pi i \sum_{j=-K}^{L+1} y_j x_{1-j}\} \qquad (x,y \in R_+).$$

Some of their more important properties are:

For each $y \in R_+$ and $v \in R_+$ and almost all $x \in R_+$

$$(2.2) \qquad \psi(y, v \oplus x) = \psi(y,v)\psi(y,x).$$

- 4 -

For fixed $n \in Z$ and $h \in [0,2^{-n})$, $y \in [0,2^{n})$ and almost all $x \in R_+$

(2.3) $$\psi(y,x \oplus h) = \psi(y,x).$$

The problems to be investigated are mainly carried out for functions belonging to the space $L^1(R_+)$ of functions that are absolutely Lebesgue integrable over R_+ with norm

$$\|f\| = \|f(\cdot)\| = \int_0^\infty |f(x)|\,dx < +\infty.$$

Furthermore let $L^2(R_+) = \{f| \{\int_0^\infty f(x)|^2 dx\}^{1/2} < +\infty\}$.
The Walsh-Fourier transform $f^\wedge$ of the function $f \in L^1(R_+)$ is then given by

$$f^\wedge(v) = \int_0^\infty f(u)\psi(v,u)du \qquad (v \in R_+)$$

and one has

(2.4) $$[f(\cdot \oplus h)]^\wedge(v) = \psi(v,h)f^\wedge(v) \qquad (v,h \in R_+),$$

(2.5) $$|f^\wedge(v)| \leq \|f\| \qquad (v \in R_+).$$

If one defines the dyadic convolution of $f,g \in L^1(R_+)$ by

$$(f*g)(x) = \int_0 f(x \oplus u)g(u)du = (g*f)(x),$$

then

(2.6) $$\|f*g\| \leq \|f\|\,\|g\|,$$

and

(2.7) $$(f*g)^\wedge(v) = f^\wedge(v)g^\wedge(v) \qquad (v \in R_+).$$

- 5 -

An important instrument in the proofs are the Dirichlet partial sums. If, following Fine [7] , one defines Dirichlet's kernel by

$$J(\omega,x) = \int_0^{\omega} \psi(x,v)dv \qquad (\omega,x \in R_+),$$

then one has for $n \in Z$

$$(2.8) \qquad J(2^n,x) = \begin{cases} 2^n, & x \in [0,2^{-n}) \\ 0, & x \in [2^{-n},\infty), \end{cases}$$

and

$$(2.9) \qquad \|J(2^n,\cdot)\| = 1.$$

Defining the Dirichlet sums by

$$(2.10) \qquad S(f;\omega,x) = \int_0^{\omega} f^{\wedge}(v)\psi(x,v)dv$$

or in the transformed state by

$$(2.11) \qquad [S(f;\omega,\cdot)]^{\wedge}(v) = \begin{cases} f^{\wedge}(v), & v \in [0,\omega) \\ 0, & v \in [\omega,\infty), \end{cases}$$

one obtains for almost all $x \in R_+$

$$(2.12) \qquad S(f;\omega,x) = \int_0^{\infty} f(x \oplus u)J(\omega,u)du = (f*J(\omega,\cdot))(x).$$

It is known that for each $f \in L^1(R_+)$

$$(2.13) \qquad \lim_{n\to\infty} \|f(\cdot) - S(f;2^n,\cdot)\| = 0.$$

(2.13) together with (2.10) immediately yield the uniqueness theorem for the Walsh transform:

(2.14) If $f \in L^1(R_+)$ and $f^\wedge(v) = 0$ a.e., then $f = 0$ a.e. in R_+.

For the proofs of the properties mentioned the reader is referred to N.J. Fine [7], R.G. Selfridge [14] or P.L. Butzer - H.J. Wagner [4].

Of special interest below are the Dirichlet sums of the form $S(f;2^{-n},x)$ with $n \in N$. On account of (2.8) and (2.12) one has for almost all $x \in R_+$

$$S(f;2^{-n},x) = 2^{-n} \int_0^{2^n} f(x \oplus u)du. \tag{2.15}$$

Since $f \in L^1(R_+)$,

$$\lim_{n\to\infty} S(f;2^{-n},x) = 0. \tag{2.16}$$

In contrast to (2.16) one has for norm-convergence

<u>Theorem 2.1</u> The following assertions are equivalent for $f \in L^1(R_+)$:

i) $\int_0^\infty f(x)dx = f^\wedge(0) = 0$,

ii) $\lim_{n\to\infty} \| 2^{-n} \int_0^{2^n} f(\cdot \oplus u)du\| = \lim_{n\to\infty}\| S(f;2^{-n},\cdot)\| = 0.$

Proof: Concerning ii)$\Rightarrow$i), assume $\int_0^\infty f(x)dx \neq 0$. Since $f \in L^1(R_+)$ this implies

$$\left|\int_0^\infty f(x)dx\right| = A > 0.$$

By (2.1) and $2^{-n} \int_0^{2^n}\int_0^\infty | f(x \oplus u)| dxdu < +\infty$, Fubini's theorem yields

- 7 -

$$0 < A = \left| 2^{-n} \int_0^{2^n} \int_0^{\infty} f(x)dxdu \right| = \left| 2^{-n} \int_0^{2^n} \int_0^{\infty} f(x \oplus u)dxdu \right|$$

$$= \int_0^{\infty} \left| 2^{-n} \int_0^{2^n} f(x \oplus u)du \right| dx,$$

whence $\lim_{n\to\infty} \left\| 2^{-n} \int_0^{2^n} f(\cdot \oplus u)du \right\| = 0.$

For the proof of i)$\Rightarrow$ii) we need the following facts. According to N.J.Fine [7] one has for each $x \in [0,2^n)$

$$\int_0^{2^n} f(x \oplus u)du = \int_0^{2^n} f(u)du. \tag{2.17}$$

By (2.1) and (2.17) we obtain for $u \in [0,2^n)$ likewise

$$\int_{2^n}^{\infty} f(x \oplus u)dx = \int_{2^n}^{\infty} f(x)dx. \tag{2.18}$$

But the assumption $\int_0^{\infty} f(x)dx = 0$ implies

$$\lim_{n\to\infty} \int_0^{2^n} f(x)dx = 0 \tag{2.19}$$

and from $f \in L^1(R_+)$ we have

$$\lim_{n\to\infty} \int_{2^n}^{\infty} |f(u)| du = 0 \tag{2.20}$$

With the results (2.17) - (2.20) one can now establish the implication i)$\Rightarrow$ii) of Theorem 2.1. Indeed,

$$\left\| 2^{-n} \int_0^{2^n} f(\cdot \oplus u)du \right\| = \int_0^{2^n} \left| 2^{-n} \int_0^{2^n} f(x \oplus u)du \right| dx$$

$$+ \int_{2^n}^{\infty} \left| 2^{-n} \int_0^{2^n} f(x \oplus u)du \right| dx$$

- 8 -

By (2.17) and (2.19) the integral

$$I_1 = \int_0^{2^n} | 2^{-n} \int_0^{2^n} f(u)du| \, dx = | \int_0^{2^n} f(u)du|$$

tends to zero for $n\to\infty$. Since

$$2^{-n}\int_0^{2^n} \int_{2^n}^{\infty} | f(x \oplus u)| \, dxdu \leqslant \| f\| < +\infty$$

and by (2.18)

$$I_2 \leqslant 2^{-n} \int_0^{2^n} \int_{2^n}^{\infty} | f(x \oplus u)| \, dxdu = \int_{2^n}^{\infty} | f(x)| \, dx.$$

This expression also converges to zero for $n\to\infty$ on account of (2.20). This completes the proof.

<u>Remarks to Theorem 2.1</u>

1) If instead of (2.15) on considers the sequence of functions

$$2^{-n} \int_0^{2^n} f(x + u)du$$

in the sense of classical analysis, then one even has for any $f \in L^1(R_+)$

$$\lim_{T\to\infty} \frac{1}{T}\| \int_0^T f(\cdot + u)du\| = 0$$

(without any additional condition such as $\int_0^\infty f(x)dx = 0$). Therefore Theorem 2.1 is a peculiarity in Walsh analysis and caused by the operation $\oplus$ of addition.

2) Naturally there exist functions $f \in L^1(R_+)$ such that

$$\lim_{n\to\infty} \| 2^{-n} \int_0^{2^n} f(\cdot \oplus u)du \| \neq 0.$$

For if $f \in L^1(R_+)$ with $f(x) \geqslant 0$, all $x \in R_+$, and $\|f\| > 0$, then by (2.1) and the Fubini theorem

$$\| 2^{-n} \int_0^{2^n} f(\cdot \oplus u)du \| = \int_0^{\infty} 2^{-n} \int_0^{2^n} f(x \oplus u)dudx$$

$$= 2^{-n} \int_0^{2^n} \int_0^{\infty} f(x \oplus u)dxdu = \int_0^{\infty} f(x)dx = \|f\| > 0.$$

3. Dyadic Differentiation and Integration

Beginning with a definition of F. Pichler [12], P.L. Butzer - H.J. Wagner [4] studied the concept of a dyadic derivative for functions defined on R_+. As mentioned, the main purpose of this paper is to introduce an inverse operator to the dyadic differentiation operator. Let us first define the latter and recall some of its most important properties.

Definition 3.1 If for $f \in L^1(R_+)$ there exists $g \in L^1(R_+)$ such that

$$\lim_{m\to\infty} \| \frac{1}{2} \sum_{j=-m}^{m} 2^j [f(\cdot) - f(\cdot \oplus 2^{-j-1})] - g(\cdot) \| = 0,$$

then g is called the strong dyadic derivative of f, denoted by $D^{[1]}f$. The derivatives of order r are defined successively by

$$D^{[r]}f = D^{[1]}(D^{[r-1]}f) \qquad (r \in N).$$

- 10 -

The designation "strong" derivative refers to the fact that the sum in question converges in the norm. If the sum in (3.1) converges in the pointwise sense to g(x), then one speaks of the pointwise dyadic derivative, denoted by $f^{[1]}(x)$.

In [4] it was shown that

<u>Lemma 3.2</u> If f, $D^{[r]}f \in L^1(R_+)$, then

$$[D^{[r]}f]^\wedge(v) = v^r f^\wedge(v) \qquad (v \in R_+).$$

The proof is based upon the identity

$$(3.1) \quad \frac{1}{2}\sum_{j=-\infty}^{m-1} 2^j[1-\psi(v,2^{-j-1})] = v \quad (v \in [0,2^m),\ m \in N).$$

Given a dyadic derivative $D^{[r]}f \in L^1(R_+)$, the question is now to regain the original function $f \in L^1(R_+)$, thus to find the "dyadic" primitive function.

Starting off with paper [4], J. Pál [11] introduced a "restricted" inverse operator U_a to $D^{[1]}$. Indeed he showed that for arbitrary but fixed a > 0 one has under the additional condition

$$(3.2) \qquad f^\wedge(v) = 0 \qquad (0 \leqslant v < a):$$

(3.3) If f, $D^{[1]}f \in L^1(R_+)$, then $U_aD^{[1]}f = f$,

(3.4) If $f \in L^1(R_+)$, then $D^{[1]}(U_af) = f$.

In the following we define the operator $I_{[r]}$ i.e. the inverse to $D^{[r]}$, for which we shall obtain results of the type (3.3) and (3.4), the Pál condition (3.2) now being replaced by the weaker condition

- 11 -

$$\int_0^\infty f(x)dx = f^\wedge(0) = 0.$$

The latter is by Theorem 2.1 equivalent to

$$\lim_{n\to\infty} \| S(f;2^{-n},\cdot)\| = 0.$$

Concerning (3.2), if $f \in L^1(R_+)$, then a weaker condition than (3.2) is that $S(f;a,x) = 0$. If $f \in L^2(R_+)$ then (3.2) is equivalent to

$$\int_0^\infty f^\wedge(v)\psi(v,x)dv = \int_a^\infty f^\wedge(v)\psi(v,x)dv,$$

which is a rather restrictive (and unnatural) condition. On the other hand, we shall make use of estimates carried out by Pál to prove our results.

For the above purpose we need the following lemmas.

Let the function $W_{r,n}(x)$ be defined via its transform

(3.5) $$W^\wedge_{r,n}(v) = \begin{cases} v^{-r}, & v \in [2^{-n},\infty) \\ 0, & v \in [0,2^{-n}) \end{cases} \qquad n \in Z, r \in N.$$

For this function we obtain

<u>Lemma 3.3</u> $W_{r,n}(x) \in L^1(R_+) \cap L^2(R_+)$ for each fixed $n \in Z$.

For the negative integers in (3.5) one has, now using the notation

- 12 -

$$W_r^{n\wedge}(v) := W^{\wedge}_{r,-n}(v)$$

$$(3,6)\quad W_r^{n\wedge}(v) = \begin{cases} v^{-r}, & v \in [2^n,\infty) \\ 0, & v \in [0,2^n) \end{cases} \qquad n \in N,$$

<u>Lemma 3.4</u> $\|W_r^n\| = \mathcal{O}[2^{-nr}]\ (n\to+\infty)$.

The proofs of Lemmas 3.3 and 3.4, formulated in a somewhat different terminology, are to be found in J. Pal [11]. It is to be noted that the proofs use very intricate and lengthy estimates which are well carried out.

With these preparations we may finally introduce the inverse operator to $D^{[r]}$.

<u>Definition 3.5</u> If for $f \in L^1(R_+)$ there exists $g \in L^1(R_+)$ such that

$$\lim_{n\to\infty} \| g(\cdot) - (W_{r,n} * f)(\cdot) \| = 0,$$

then g is called the rth fold strong dyadic integral of f, denoted by $I_{[r]}f$.

The operator $I_{[r]}$ possesses the following properties.

<u>Theorem 3.6</u> The following assertions are equivalent for $f,g \in L^1(R_+)$ and fixed $r \in N$:

(i) $g = I_{[r]}f$,

(ii) $g^{\wedge}(v) = \begin{cases} v^{-r} f^{\wedge}(v) & , v \in (0,\infty) \\ 0 & , v = 0. \end{cases}$

Proof: Concerning (i)$\Rightarrow$(ii), $g = I_{[r]} f$ states that

$$\lim_{n\to\infty} \| g - (W_{r,n} * f) \| = 0.$$

By the properties (2.5) and (2.7) this yields

$$g^\wedge(v) = \lim_{n\to\infty} (W_{r,n} * f)^\wedge(v) = \begin{cases} v^{-r} f^\wedge(v), & v \in (0,\infty) \\ 0 \quad , & v \in 0 . \end{cases}$$

For (ii)$\Rightarrow$(i), first note that in view of the convolution property (2.7) and hypothesis (ii) one has the following identity for $m>n$

$$(3.7) \quad (W_{r,n} * f)(x) - (W_{r,m} * f)(x) = S(g;2^{-n},x) - S(g;2^{-m},x)$$

since the Walsh transforms of both sides are equal. Let us now show that the limit of (3.7) in the $L^1(R_+)$-norm for $m,n\to\infty$ is equal to zero. Since $g^\wedge(o) = \int_0^\infty g(x)dx = 0$ by (ii), one has by Theorem 2.1 that

$$\lim_{n\to\infty} \| S(g;2^{-n},\cdot) \| = 0.$$

This implies that

$$\lim_{n\to\infty} \| S(g;2^{-n},\cdot) - S(g;2^{-m},\cdot) \| = 0.$$

Therefore also the left hand side of (3.7) tends to zero in the $L^1(R_+)$-norm for $m,n\to\infty$. Since the space $L^1(R_+)$ is complete, there exists a function $h \in L^1(R_+)$ such that

$$\lim_{n\to\infty} \| h - (W_{r,n} * f) \| = 0, \text{ i.e. } h = I_{[r]} f$$

which, according to (2.5) and (2.7), yields

$$h^\wedge(v) = v^{-r} f^\wedge(v) = g^\wedge(v) \qquad (v \in R_+).$$

Hence h = g a.e. by the uniqueness assertion (2.14), and so there follows part (i). This establishes the equivalence of (i) with (ii).

As an application of Theorem 3.6

<u>Theorem 3.7</u> If f, $D^{[r]}$ f ε $L^1(R_+)$ with $\int_0^\infty f(x)dx = 0$, then

$$f = I_{[r]} D^{[r]} f.$$

<u>Proof:</u> Lemma 3.2 together with $\int_0^\infty f(x)dx = 0$ tells us that

$$f^\wedge(v) = \begin{cases} v^{-r}[D^{[r]} f]^\wedge(v) & , v \in (0,\infty) \\ 0 & , v = 0 \end{cases} .$$

Therefore condition (ii) of Theorem 3.6 is satisfied, and so $f = I_{[r]} D^{[r]} f$, as was to be shown.

Our next goal is to examine dyadic differentiability of $I_{[r]} f$ for f ε $L^1(R_+)$. At first a remark concerning the dyadic integral:

If f ε $L^1(R_+)$, then $I_{[1]} f$ need in general not belong to $L^1(R_+)$, it need not even be defined.

For this purpose consider the following example:

$$f_1(x) = \begin{cases} 1 & , x \in [0,1) \\ 0 & , x \in [1,\infty). \end{cases}$$

Obviously f_1 ε $L^1(R_+)$. Making use of a property of Walsh

- 15 -

functions, namely

$$(3.8) \qquad \psi(x,y) = \psi([x],y)\psi(x,[y]) \qquad (x,y \in R_+)$$

[x] being the greatest integer $\leq$ x, and

$$(3.9) \qquad \int_0^1 \psi([x],v)\psi([y],v)dv = \begin{cases} 1 & , [x] = [y] \\ 0 & , [x] \neq [y], \end{cases}$$

one has by (3.8) and (3.9)

$$f_1^\wedge(v) = \int_0^\infty f_1(u)\psi(v,u)du = \int_0^1 \psi(v,u)du$$

$$= \int_0^1 \psi([v],u)du = \begin{cases} 1 , & v \in [0,1) \\ 0 , & v \in [1,\infty). \end{cases}$$

Using the convolution property (2.7) this yields

$$(W_{1,n}*f_1)^\wedge(v) = \begin{cases} v^{-1} & , v \in [2^{-n},1) \\ 0 & , v \in [1,\infty), \end{cases}$$

and so one has at least for almost all x ε [0,1)

$$(W_{1,n}*f_1)(x) = \int_{2^{-n}}^1 v^{-1}\psi(v,x)dv$$

$$\int_{2^{-n}}^1 v^{-1}dv = -n \log 2.$$

almost all $x \in [0,1)$, $I_{[1]}f_1$ is not defined.

If the dyadic integral of a function in $L^1(R_+)$ also belongs to $L^1(R_+)$, then

<u>Theorem 3.8</u> If f and g belong to $L^1(R_+)$ such that $g = I_{[r]}f$ for an $r \in N$, then

$$D^{[r]}(I_{[r]}f) = f.$$

<u>Proof:</u> Consider first the case $r = 1$. Setting for $h \in L^1(R_+)$

$$(D_m^{[1]}h)(x) = \frac{1}{2} \sum_{j=-(m-1)}^{m-1} 2^j[h(x)-h(x \oplus 2^{-j-1})],$$

the problem is to show that

$$(3.10) \qquad \lim_{m\to\infty} \| D_m^{[1]} g - f\| = 0 .$$

The proof of (3.10) is carried out in three steps. At first we show that

$$\text{a)} \qquad g(x) = S(g;2^m,x) + (W_1^m * f)(x) \qquad (\text{a.e. in } R_+).$$

This would imply the estimate

$$\| D_m^{[1]} g - f\| < \| D_m^{[1]} S(g;2^m,\cdot)-f(\cdot)\| + \| D_m^{[1]} (W_1^m * f)\| .$$

In the second step we show that

$$\text{b)} \qquad \lim_{m\to\infty} \| D_m^{[1]} S(g;2^m,\cdot) - f(\cdot)\| = 0 ,$$

and thirdly that

- 17 -

c) $\lim_{m\to\infty} \| D_m^{[1]}(W_1^m * f)\| = 0,$

which would yield (3.10).

To prove a), since $g = I_{[1]} f$, by Theorem 3.6 one has

$$(3.11) \qquad g^{\wedge}(v) = \begin{cases} v^{-1} f^{\wedge}(v) & , v \in (0,\infty) \\ 0 & , v = 0, \end{cases}$$

as well $\lim_{n\to\infty} \| g - W_{1,n} f\| = 0$. Hence there exists a subsequence $(W_{1,n_k} * f)(x)$ such that

$$(3.12) \qquad \lim_{k\to\infty} (W_{1,n_k} * f)(x) = g(x) \qquad \text{(a.e. in } R_+),$$

for which we have for $n_k > m$ the decomposition

$$(3.13) \quad (W_{1,n_k} * f)(x) = S(W_{1,n_k} * f; 2^m, x) + (W_1^m * f)(x) \quad \text{(a.e. in } R_+).$$

Indeed, the Walsh transforms of both sides of (3.13) are equal to another by the convolution theorem (2.7) and definitions (2.11), (3.5) (3.6). The uniqueness theorem (2.14) then gives the identity (3.13). But by (3.11) we obtain

$$S(W_{1,n_k} * f; 2^m, x) = \int_{2^{-n_k}}^{2^m} v^{-1} f^{\wedge}(v)\psi(v,x)dv$$

$$= S(g; 2^m, x) - S(g; 2^{-n_k}, x),$$

with

$$\lim_{k\to\infty} S(g; 2^{-n_k}, x) = 0$$

- 18 -

by (2.16). Letting $k\to\infty$ in (3.13) one has for almost all x in R_+

$$\lim_{k\to\infty} (W_{1,n_k} * f)(x) = g(x) = S(g;2^m,x) + (W_1^m * f)(x),$$

which then gives result a).

To establish b), let us first verify the identity

$$(3.14) \qquad S(f;2^m,x) = \frac{1}{2}\sum_{j=-\infty}^{m-1} 2^j \Delta_{j,m}(x) \qquad \text{(a.e. in } R_+),$$

with

$$\Delta_{j,m}(x) = S(g;2^m,x) - S(g;2^m,x \oplus 2^{-j-1}),$$

by comparing the Walsh transforms of both sides of (3.14).

Since

$$\frac{1}{2}\sum_{j=-\infty}^{m-1} \int_0^\infty | 2^j \Delta_{j,m}(x)\psi(v,x)| \, dx < \sum_{j=-\infty}^{m-1} 2^j \| S(g;2^m,\cdot)\|$$

$$< 2^m \| g\| < +\infty ,$$

one has by Fubini's theorem together with (2.4), (3.1) and (3.11) vor $v \in R_+$

$$[\frac{1}{2}\sum_{j=-\infty}^{m-1} 2^j \Delta_{j,m}(\cdot)]^\wedge(v) = \frac{1}{2}\sum_{j=-\infty}^{m-1} 2^j [\Delta_{j,m}(\cdot)]^\wedge(v)$$

$$= \sum_{j=-\infty}^{m-1} 2^j [1 - \psi(v,2^{-j-1})][S(g;2^m,)]^\wedge(v)$$

- 19 -

$$= \begin{cases} f^{\wedge}(v), & v \in [0,2^m) \\ 0, & v \in [2^m,\infty) \end{cases} = [S(f;2^m,\cdot)]^{\wedge}(v) \;.$$

Identity (3.14) now follows by the uniqueness result; at the same time this shows that for each fixed $m \in Z$ the sum $\frac{1}{2}\sum_{j=-\infty}^{m} 2^j \Delta_{j,m}(x)$ is convergent. Together with (3.14) this yields for almost all $x \in R_+$

$$(3.15) \quad S(f,2^m,x) - D_m^{[1]} S(g;2^m,x) = \frac{1}{2}\sum_{j=-\infty}^{-m} 2^j \Delta_{j,m}(x).$$

The estimate

$$\left\| \frac{1}{2}\sum_{j=-\infty}^{-m} 2^j \Delta_{j,m}(\cdot)\right\| < \frac{1}{2}\sum_{j=-\infty}^{-m} 2^j \|\Delta_{j,m}\| < 2^{-m+1}\|g\|$$

together with (3.15) delivers

$$(3.16) \qquad \lim_{m\to\infty} \|D_m^{[1]} S(g;2^m,\cdot) - S(f,2^m,\cdot)\| = 0.$$

Then the triangle inequality

$$\|D_m^{[1]} S(g;2^m,\cdot) - f(\cdot)\| < \|D_m^{[1]} S(g;2^m,\cdot) - S(f;2^m,\cdot)\| + \|S(f;2^m,\cdot) - f(\cdot)\|$$

in conjunction with (3.16) and (2.13) give the desired formula b).

Concerning c), it follows by (2.7) and (2.14) that

$$D_m^{[1]}(W_1^m * f) = (f * D_m^{[1]} W_1^m) = (f * F_m)$$

with

$$F_m(x) = (D_m^{[1]} W_1^m)\ (x).$$

For this function one has by Lemma 3.4 the estimate

$$\|F_m\| = \|\frac{1}{2}\sum_{j=-(m-1)}^{m-1} 2^j[W_1^m(\cdot) - W_1^m(\cdot \oplus 2^{-j-1})]\|$$

$$< \sum_{j=-(m-1)}^{m-1} 2^j\|W_1^m\| < \text{const.},$$

the constant being independent of m. With (2.6) one has for all $k,m \in N$ with $m > k$ the inequality

$$(3.17) \qquad \|F_m*f\| \leq \|F_m*(f-S(f;2^k,\cdot)\| + \|F_m*S(f;2^k,\cdot)\|$$

$$\leq \|F_m\|\,\|f(\cdot) - S(f;2^k,\cdot)\|$$

$$\leq \text{const.}\ \|f(\cdot) - S(f;2^k,\cdot)\|,$$

where one needs the fact that $[F_m*S(f;2^k,\cdot)](x) = 0$ a.e. in R_+ for $m > k$. Indeed, by the convolution property (2.7) and the definition of F_m and W_1^m we have

$$[F_m*S(f,\ 2^k,\cdot)]^\wedge(v) = 0 \qquad (v \in R_+).$$

By (3.17) and (2.13) there now follows c). This establishes the theorem for the case $r = 1$.

If $r > 1$, note that

$$g = I_{[r]}f = I_{[1]}(I_{[r-1]}f).$$

But this yields by the case $r = 1$

$$D^{[1]}g = D^{[1]}(I_{[1]}(I_{[r-1]}f) = I_{[r-1]}f.$$

The proof is then completed by induction.

Let us finally summarize some of the major results of this section, characterising the strongly dyadic differentiable functions in the sense of definition (3.1).

<u>Lemma 3.9</u> The following assertions are equivalent for $f, g \in L^1(R)$ with $\int_0^\infty f(x)dx = 0$ and for fixed $r \in N$:

i) $D^{[r]}f = g$,

ii) $v^r f^\wedge(v) = g^\wedge(v) \qquad (v \in R_+)$,

iii) $f = I_{[r]}g.$.

<u>Proof:</u> The implication i)⇒ii) follows by Theorem 3.2. ii)⇒iii) by Theorem 3.6, and iii)⇒i) by Theorem 3.8.

It is important to note that Theorem 3.8 together with Theorem 3.7 give the analogue of the classical fundamental theorem of the differential and integral calculus, this time in the dyadic sense.

<u>4. Some Remarks Concerning the Dyadic Calculus and its Applications to Walsh Analysis</u>

a) Let us first remark that all of the definitions and results considered in this paper may also be carried over to functions belonging to the space $L^2(R_+)$. Hereby the $L^1(R_+)$-norm is replaced by the $L^2(R_+)$-norm, namely,

- 22 -

$$\|f\|_2 = \{\int_0^\infty |f(x)|^2 dx\}^{1/2}.$$

Instead of the inequality (2.5) which was used a lot one now uses the property

$$\|f\|_2 = \|f_2^\wedge(\cdot)\|_2 ,$$

where the Walsh transform $f_2^\wedge(v)$ of $f \in L^2(R_+)$ is defined by

$$\lim_{\rho\to\infty} \|f_2^\wedge(\cdot) - \int_0^\rho f(u)\psi(\cdot,u)du\|_2 = 0 \qquad (\rho \in R_+).$$

b) If one defines the dyadic derivative not in the principal value sense, as was the case in Definition 3.1, but more generally as the function $\tilde{D}^{[1]}f$ for which

$$\lim_{m,n\to\infty} \|\frac{1}{2}\sum_{j=-n}^{m} 2^j[f(\cdot) - f(\cdot \oplus 2^{-j-1})] - (\tilde{D}^{[1]}f)(\cdot)\| = 0,$$

then one obtains with this definition the same results as we did with $D^{[1]}f$.

c) As already mentioned in Section 3, one could also introduce the dyadic derivative via a pointwise limit, namely

$$(4.1) \qquad f^{[1]}(x) = \frac{1}{2}\sum_{j=-\infty}^{+\infty} 2^j[f(x) - f(x \oplus 2^{-j-1})],$$

and likewise the dyadic integral by

$$(4.2) \qquad (\tilde{I}_{[1]}f)(x) = \lim_{n\to\infty} (W_{1,n}*f)(x).$$

These two definitions have the advantage that the Walsh functions $\psi(y,x)$ themselves are then dyadic differentiable and integrable. Indeed, one has for each $y \in R_+$

$$\psi^{[1]}(y,x) = y\psi(y,x) \qquad \text{(almost all } x \in R_+)$$

(the dyadic derivative being taken with respect to x), and for each $y \in (0,\infty)$

$$[\tilde{I}_{[1]}\psi(y,\cdot)](x) = (1/y)\psi(y,x) \qquad \text{(almost all } x \in R_+).$$

These two results cannot hold for the operators $D^{[1]}$ and $I_{[1]}$ since $\psi(y,x)$ does not belong to $L^1(R_+)$. In contrast, the pointwise definitions (4.1) and (4.2) have the disadvantage that the corresponding developement of the dyadic calculus is more difficult. It may be conjectured that in this case one could also introduce a class of functions playing the role of the absolutely continuous functions in classical analysis in order to establish all the results of this paper also with $f^{[r]}$ and $\tilde{I}_{[r]}$. The corresponding problem for functions having period 1 has recently been solved by P.L. Butzer - H.J. Wagner [6], F. Schipp [13].

We next wish to show that the concepts $D^{[r]}$ and $I_{[r]}$ play more or less the same role in Walsh analysis as do the classical derivative and integral in Fourier analysis by comparing their application potentialities.

If Lip $(\alpha,L^1(R_+))$ is the class of functions with

$$\|f(\cdot) - f(\cdot \oplus h)\| = \mathcal{O}[h^\alpha] \qquad (h\to 0+),$$

and the W-modulus of continuity is defined by

$$\omega_W(f;L^1(R_+),\delta) = \sup_{0<h<\delta} \|f(\cdot) - f(\cdot \oplus h)\|,$$

then

$$f \in \mathrm{Lip}\,(\alpha, L^1(R_+)) \Longleftrightarrow \omega_W(f;L^1(R_+),\delta) = \mathcal{O}[\delta^\alpha] \qquad (\delta \to 0+).$$

With these concepts we have

<u>Lemma 4.1</u> If f, $D^{[r]}f \in L^1(R_+)$, then

$$\omega_W(f;L^1(R_+),\delta) = \mathcal{O}[\delta^r \omega_W(D^{[r]}f;L^1(R_+);\delta)]$$

<u>Proof:</u> One has for almost all $x \in R_+$ and for $h \in [0,2^{-m})$

$$(4.3) \quad f(x) - f(x \oplus h) = (W_r^m * D^{[r]}f)(x) - (W_r^m * D^{[r]}f)\,(x \oplus h),$$

since the Walsh transforms of both sides of (4.3) are equal. Indeed, by (2.4) and (2.3) one has $[f(\cdot) - f(\cdot \oplus h)]^\wedge(v) = f^\wedge(v)[1 - \psi(v,h)]$ for $2^m \leq v < \infty$ and $= 0$ for $v \in [0,2^m)$. By (2.7), Lemma 3.2 and (3.6) the Walsh transform of the right side has the same form, proving (4.3). Together with (2.6) and Lemma 3.4 this implies

$$(4.4) \quad \| f(\cdot) - f(\cdot \oplus h)\| \leq \text{const. } 2^{-mr} \| (D^{[r]}f)(\cdot) - (D^{[r]}f)(\cdot \oplus h)\|,$$

from which the lemma follows immeadiately.

It is possible to estimate the difference $\|f(\cdot) - S(f,2^n,\cdot)\|$ by means of the W-modulus of continuity. One has

- 25 -

(4.5) $$\frac{1}{2}\,\omega_W(f;L^1(R_+);2^{-n}) \leq \|f(\cdot) - S(f;2^n,\cdot)\| \leq \omega_W(f;L^1(R_+);2^{-n})$$

(for a proof compare C. Watari [16]). This allows us to establish

Theorem 4.2. The following assertions are equivalent for $f \in L^1(R_+)$ and $\alpha>0$:

(i) $f \in \mathrm{Lip}(\alpha, L^1(R_+))$,

(ii) $\|f(\cdot) - S(f;2^n,\cdot)\| = \mathcal{O}[2^{-n\alpha}] \qquad (n\to\infty)$,

(iii) $D^{[\nu]}f$ exists for $0\leq\nu<\alpha$ with

$$\|(D^{[\nu]}f)(\cdot) - S(D^{[\nu]}f;2^n,\cdot)\| = \mathcal{O}[2^{-n(\alpha-\nu)}] \qquad (n\to\infty).$$

Proof: The equivalence (i)⇔(ii) follows directly from (4.5). For (iii)⇒(ii) one uses property (4.4). The converse (ii)⇒(iii) is the counterpart of the theorem of Bernstein. The proof follows by the standard procedure (compare also with P.L. Butzer - H.J. Wagner [2]).

Let us finally recall that one could also set up and solve partial dyadic differential equations using the concepts $D^{[r]}$ and $I_{[r]}$, as is already carried out in the paper [4]. This matter is important as it may possibly help one in finding a suitable interpretation of the dyadic calculus, still one of the basic open problems here.

- 26 -

References

[1] P.L. Butzer - H.J. Wagner, Walsh-Fourier series and the concept of a derivative. Applicable Anal. 3 (1973), 29 - 46.

[2] P.L. Butzer - H.J. Wagner, Approximation by Walsh polynomials and the concept of a derivative. In: Applications of Walsh Functions (Proc. Sympos. Naval Res. Lab., Washington, D.C., March 27 - 29, 1972; Ed. R.W. Zeek - A.E. Showalter) Washington, D.C. 1972, xi + 401 pp.; pp. 388 - 392.

[3] P.L. Butzer - H.J. Wagner, On a Gibbs-type derivative in Walsh-Fourier analysis with applications. In: Proc. of the 1972 Electronics Conference (Chicago, Oct. 9-10, 1972; Ed. R.E. Horton) Oak Brook, Illinois, 1972, xxvi + 427 pp.; pp. 393 - 398.

[4] P.L. Butzer - H.J. Wagner, A calculus for Walsh functions defined on R_+. In: Applications of Walsh Functions (Proc. Sympos. Naval Res. Lab., Washington, D.C., April 18 - 20, 1973; Ed. R.W. Zeek - A.E. Showalter) Washington, D.C. 1973, xi + 298 pp.; pp. 75 - 81.

[5] P.L. Butzer - H.J. Wagner, A new calculus for Walsh functions with applications. In: Theory and Applications of Walsh and other non-sinusoidal functions (Proc. Sympos. Hatfield Polytechnic, June 28 - 29, 1973; Ed. P.D. Lines) Hatfield, England, 1973.

[6] P.L. Butzer - H.J. Wagner, On dyadic analysis based on the pointwise dyadic derivative. (To appear).

[7] N.J. Fine, The generalized Walsh functions. Trans. Amer. Math. Soc. 69 (1950), 66 - 77.

[8] J.E. Gibbs, Some properties of functions on the non-negative integers less than 2^n. NPL (National Physical Laboratory), Middlesex, England, DES Rept. no. 3 (1969).

[9] J.E. Gibbs - B. Ireland, Walsh functions and differentiation. In: Applications of Walsh Functions (Proc. Sympos. Naval Res. Lab., Washington, D.C., March 18 - 20, 1974).

[10] J.E. Gibbs - M.J. Millard, Walsh functions as solutions of a logical differential equation. NPL, DES Rept. no. 1 (1969).

[11] J. Pál, Concept of a derivative among functions defined on the dyadic field. (To appear).

[12] F. Pichler, Walsh functions and linear system theory. In: Applications of Walsh Functions (Proc. Sympos. Naval Res. Lab., Washington, D.C., 31 March - 1 April, 1970, Ed. C.A. Bass) Washington, D.C. 1970, viii + 274 pp; pp. 175 - 182

- 27 -

[13] F. Schipp. Über einen Ableitungsbegriff von P.L. Butzer und H.J. Wagner. In: Proceedings 5. Balkan Math. Congress, June 24 - 30, 1974, Belgrad.

[14] R.G. Selfridge, Generalized Walsh transform. Pacific J. Math. 5 (1955), 451 - 480.

[15] H.J. Wagner, Ein Differential- und Integralkalkül in der Walsh-Fourier-Analysis mit Anwendungen. (Forschungsber. des Landes Nordrhein-Westfalen Nr. 2334), Westdeutscher Verlag, Köln-Opladen 1973, 71 pp.

[16] C. Watari, Best approximation by Walsh polynomials. Tôhoku Math. J. 15 (1963), 1 - 5.

Chapter 11
Dyadic Derivative and Walsh-Fourier Transform

Boris I. Golubov

11.1 Introduction

Following to the concept of J.E. Gibbs [1] P.L. Butzer and H.J. Wagner [2] defined the notion of a dyadic strong derivative D. After that they introduced the dyadic strong integral I and dyadic pointwise derivative d (see [3]-[5]). Their definitions concerns functions defined on dyadic group G or dyadic field K. The dyadic group G and dyadic field K are isomorphic to modified segment $[0,1]^*$ and modified positive half-line $R_+^* = [0,\infty)^*$, respectively. The characters of dyadic group G and dyadic field K are Walsh-Paley functions $w_n(\cdot)$, $n \in Z_+ = \{0,1,\dots\}$ and generalized Walsh functions $\psi_y(\cdot)$, $y \in R_+^*$, respectively. P.L. Butzer and H.J. Wagner proved the equalities $Dw_n = nw_n$ and $dw_n(x) = nw_n(x)$ for $n \in Z_{n+}$, $x \in G$, and $d\psi_y(x) = |y|\,\psi_y(x)$, for $x,y \in K$. In [3], for the functions $f \in L(R_+)$, the equality $D(\hat{f})(x) = x\hat{f}(x)$ is proved, where $\hat{f}$ is the Fourier transform of the function f.

C.W. Onneweer [6] introduced modified pointwise and strong dyadic derivatives for functions defined on dyadic group G or dyadic field K. He proved that the characters of the dyadic group G or the dyadic field K are differentiable in his sense and they are eigenfunctions of the modified differential operator δ. He proved the equalities $\delta w_0(x) = 0$ and $\delta w_n(x) = 2^k w_n(x)$, if $2^k \le n < 2^{k+1}$, $k \in Z_+$, $x \in G$. In another article, C.W. Onneweer [7] introduced modified fractional differentiation and integration for the functions defined on compact Vilenkin groups G_p of order $p \ge 2$.

J. Pál [8] proved that if $f \in L(R_+)$ and $xf(x) \in L(R_+)$, then Walsh transform $\hat{f}$ is pointwise differentiable in the sense of P.L. Butzer and H.J. Wagner at each point $x \in R_+$ and $d(\hat{f})(x) = (tf(t))\hat{}(x)$.

Boris I. Golubov
Chair of Higher Mathematics, Moscow Institute of Physics and Technology (State University), 141700 Dolgoproudny, Region, Russia, e-mail: golubov@mail.mipt.ru

R.S. Stanković et al. (eds.), *Dyadic Walsh Analysis from 1924 Onwards Walsh-Gibbs-Butzer Dyadic Differentiation in Science Volume 1 Foundations*, Atlantis Studies in Mathematics for Engineering and Science 12, DOI 10.2991/978-94-6239-160-4_11

In our paper [9], the modified dyadic fractional derivatives and integrals on R_+ were introduced and applied to differentiation and integration of the Walsh-Fourier transform. Below we give a sketch of the results from the paper [9] (see also [10], Ch. 3).

11.2 Notations and Definitions

For a number $x \in R_+ = [0,\infty)$ we consider dyadic expansion $x = \sum_{n=-\infty}^{\infty} 2^{-n-1} x_n$, where x_n equals to 0 or 1. Note that $x_n = 0$ for $n \le n(x) \le 0$. If x is the dyadic rational, then we take its finite expansion, i.e., $x_n = 0$ for $n \ge n(x)$. We define dyadic sum of two numbers $x, y \in R_+$ by the operation $\oplus$ as follows: $x \oplus y = z$, where $z_n = x_n + y_n$ mod 2 for all $n \in Z = \{0, \pm 1, \ldots\}$. Let us set $t(x,y) = \sum_{n=-\infty}^{\infty} x_n y_{-n-1}$ and define the generalized Walsh functions $\psi(x,y) = \psi_y(x) = (-1)^{t(x,y)}$ for $x, y \in R_+$. These functions were introduced by N.J. Fine [11]. It is evident that $\psi(x,y) = \psi(y,x)$ and $\psi(x,y) = \pm 1$ for $x, y \in R_+$.

N.J. Fine [11] (see also [12], Ch. 1 or [13], Ch. 9) for the function $f \in L(R_+)$ introduced the Walsh-Fourier transform by the equality

$$\hat{f}(x) = \int_{R_+} f(y)\psi(x,y)dy.$$

He proved that $\hat{f}(x)$ is uniformly W-continuous on R_+ and $\lim_{x\to\infty} \hat{f}(x) = 0$.

For $f \in L(R_+)$ and $g \in L(R_+)$, $1 \le p < \infty$, we set

$$(f * g)(x) = \int_{R_+} f(x \oplus y) g(y) dy,$$

where $x \in R_+$. The function $(f * g)$ is called the dyadic convolution of the functions f and g. Let us note that $(f * g) \in L^p(R_+)$, if $1 \le p < \infty$.

11.3 Lemmas

For $x > 0$ we set $h(x) = 2^{-n}$ for $2^{-n} \le x < 2^{n-1}$, $n \in Z$. It is evident that $x^{-1} \le h(x) < 2x^{-1}$. Let us introduce the sequence of kernels by means of which we shall define the pointwise fractional modified dyadic integral.

Lemma 11.1 *If $\alpha > 0$ and $n \in Z$, then for each $x > 0$ the following limit*

$$W_n^{\alpha}(x) = \lim_{m\to\infty} \int_{2^{-n}}^{2^m} (h(t))^{\alpha} \psi(x,t) dt,$$

exists and is finite. More precisely, $W_n^{\alpha}(x) = -2^{(\alpha-1)n}$ for $2^{n-1} \le x < 2^n$,

$$W_n^\alpha(x) = -2^{(\alpha-1)n} + 2(1-2^{-\alpha})\sum_{i=0}^{k} 2^{(n-i)(\alpha-1)},$$

for $2^{n-k-2} \le x < 2^{n-k-1}$, $k = 0, 1, \ldots$, *and* $W_n^\alpha(x) = 0$ *for* $x \ge 2^n$.

We shall write $f(x) \approx g(x), x \to a$, if $f(x) = O(g(x)), x \to a$ and simultaneously $g(x) = O(f(x)), x \to a$.

Corollary 11.1 *1. If* $0 < \alpha < 1$, $n \in Z$, *then* $W_n^\alpha(x) \approx x^{\alpha-1}$, $x \to +0$,
2. $W_n^1(x) \approx \log_2(x^{-1})$, $x \to +0$,
3. $W_n^\alpha(x)$ *is bounded on* $(0, +\infty)$, *if* $\alpha > 1$.

Corollary 11.2 *If* $\alpha > 0$, $n \in Z$, *then* $W_n^\alpha \in L(R_+)$.

Let $X_E(x)$ denote the test function of the set E, i.e., $X_E(x) = 1$ for $x \in E$ and $X_E(x) = 0$ for $x \notin E$.

Lemma 11.2 *If* $\alpha > 0$, $n \in Z$, *then* $\hat{W}_n^\alpha(x) = (h(x))^\alpha X_{[2^{-n},+\infty)}(x)$.

Corollary 11.3 *If* $\alpha > 0$, $n \in Z$, *then* $\hat{W}_n^\alpha = 0$ *for* $x \in [0, 2^{-n})$.

Let us introduce the sequence of kernels by means of which we shall define the pointwise fractional modified dyadic derivative.

For $\alpha > 0$, $n \in Z$ we set

$$\Lambda_n^\alpha(x) = \int_0^{2^n} (h(t))^{-\alpha} \psi(x,t)\,dt, \quad x \in R_+.$$

Lemma 11.3 *If* $\alpha > 0$, $n \in Z$, *then*

1. $\Lambda_n^\alpha(x) = (2^{\alpha+1} - 1)^{-1} 2^{(\alpha+1)n}$ *for* $0 \le x < 2^{-n}$,
2. $\Lambda_n^\alpha(x) = -C_\alpha 2^{(\alpha+1)(n-i)}$ *for* $2^{-n+i} \le x < 2^{-n+i+1}$, $i \in Z$, *where* $C_\alpha = (1-2^{-\alpha})(2^{\alpha+1} - 1)^{-1}$.

Corollary 11.4 *For each* $\alpha > 0$, *and* $n \in Z$, *the function* $\Lambda_n^\alpha(x)$ *is bounded on* R_+ *and* $\Lambda_n^\alpha(x) \approx x^{-\alpha-1}$, $x \to +\infty$.

Corollary 11.5 *If* $\alpha > 0$, *and* $n \in Z$, *then* $\Lambda_n^\alpha \in L(R_+)$.

11.4 Dyadic Differentiation and Integration of Walsh-Fourier Transform

Taking in account the Corollaries 11.4 and 11.5 we may introduce the following definition.

Definition 11.1 *If* $\alpha > 0$ *and for the function* $f \in L(R_+) \cup L^\infty(R_+)$ *the following limit* $d^{(\alpha)}(f)(x) = \lim_{n\to\infty}(f * \Lambda_n^\alpha(x))$ *exists and is finite at the point* $x \in R_+$, *then the number* $d^{(\alpha)}(f)(x)$ *is called the modified dyadic derivative (MDD) of order* α *of the function* f *at the point* x.

Example 11.1 *For the function* $\varphi = X_{[0,1)}$ *and* $\alpha > 0$*, we have*

1. *If* $x \in [0,2)$*, then* $d^{(\alpha)}(\varphi)(x) = (2^{\alpha+1} - 1)^{-1}$.
2. *If* $x \geq 1$*, then* $d^{(\alpha)}(\varphi)(x) = -(1 - 2^{-\alpha})\sum_{k=1}^{\infty} 2^{-(\alpha+1)k} X_{[0,2^k)}(x)$.

Theorem 11.1 *Let us assume that* $\alpha > 0$ *and* $f, h^{-\alpha} f \in R_+$. *Then the Walsh transform* $\hat{f}$ *of the function* f *has the MDD of order* α *at each point* $x \in R_+$ *and the equality* $d^{(\alpha)}(\hat{f})(x) = (h^{-\alpha} f)\hat{}(x)$ *holds.*

The proof of this theorem is based on the following lemma.

Lemma 11.4 *The generalized Walsh function* $\psi_y(x)$, $x, y \in R_+$*, has the MDD of any order* $\alpha > 0$ *at each point* $x \in R_+$. *More precisely,* $d^{(\alpha)}(\psi_0)(x) = 0$ *on* R_+ *and* $d^{(\alpha)}(\psi_y)(x) = (h(y))^{-\alpha} \psi_y(x)$ *for* $x \in R_+$, $y > 0$.

Taking in account the Corollary 11.2, we may introduce the following definition.

Definition 11.2 *If* $\alpha > 0$ *and for the function* $f \in L(R_+)$ *the following limit* $j_\alpha(f)(x) = \lim_{n\to\infty}(f * W_n^\alpha(x)$ *exists and is finite at the point* $x \in R_+$*, then the number* $j_\alpha(f)(x)$ *is called the modified dyadic integral (MDI) of order* α *of the function* f *at the point* x.

Example 11.2 *For the function* $\varphi = X_{[0,1)}$ *and* $\alpha > 0$*, we have*

1. *If* $0 < \alpha < 1$*, then the function* φ *has the MDI of order* α *at each point* $x \in R_+$.
2. *If* $\alpha \geq 1$*, then the function* φ *does not have the MDI of order* α *at each point* $x \in R_+$.

Theorem 11.2 *Let us assume that* $\alpha > 0$ *and* $f, h^{\alpha} f \in R_+$. *Then the Walsh transform* $\hat{f}$ *of the function* f *has the MDI of order* α *at each point* $x \in R_+$ *and the equality* $j_\alpha(\hat{f})(x) = (h^{\alpha} f)\hat{}(x)$ *holds.*

The proof of this theorem is based on the following lemma.

Lemma 11.5 *The generalized Walsh function* $\psi_y(x)$, $x, y \in R_+$*, has the MDI of any order* $\alpha > 0$ *at each point* $x \in R_+$. *More precisely,* $d^{(\alpha)}(\psi_0)(x) = 0$ *on* R_+ *and* $j_\alpha(\psi_y)(x) = (h(y))^{\alpha} \psi_y(x)$ *for* $x \in R_+$, $y > 0$.

References

1. ∗Gibbs, J.E., *Walsh spectrometry, a form of spectral analysis well suited to binary digital computation*, Nat. Phys. Lab., Teddington, UK, 1967, 24 pages.
2. ∗Butzer, P.L., Wagner, H.J., "Walsh series and the concept of a derivative", *Applicable Analysis*, Vol. 3, No. 1, 1973, 29-46.
3. ∗Butzer, P.L., Wagner, H.J., "A calculus for Walsh functions defined on R_+, *Proc. Symp. Navel Research Laboratory*, Washington, D.C., April 18-20, 1973, 75-81.
4. ∗Butzer, P.L. and Wagner, H.J., "On dyadic analysis based on pointwise dyadic derivative", *Analysis Math.*, Vol. 1, No. 1, 1975, 171-196.

5. *Wagner, H.J., "On dyadic calculus for functions defined on R_+, *Proc. Symp. Theory and Applications of Walsh functions*, Hatfield Politechnic, 1975, 101-129.
6. *Onneweer, C.W., "On the definition of dyadic differentiation", *Applicable Analysis*, Vol. 9, 1979, 267-278.
7. Onneweer, C.W., "Fractional differentiation on the group of integers of a p-adic or p series field", *Analysis Math.*, Vol. 3, 1977, 119-130.
8. Pál, J., "On the connection between the concept of a derivative defined on the dyadic field and the Walsh-Fourier transform", *Annales Sci. Univ. Budapest. Sect. Math.*, Vol. 18, 1975, 49-54.
9. Golubov, B.I. "Dyadic fractional differentiation and integration of Walsh transform", *Proc. Int. Conf. Mathematics and Its Applications*, April 5-7, 2004, Kuwait Univ., Kuwait, 2004, 274-284.
10. Golubov, B.I., *Elements of Dyadic Analysis*, URSS Publ., Moscow, 2007 (in Russian).
11. Fine, N.J., "The generalized Walsh functions", *Trans. Amer. Math. Soc.*, Vol. 69, 1950, 66-77.
12. Golubov, B., Efimov, A., Skvortsov, V., *Walsh Series and Transforms, Theory and Applications*, Kluwer Academic Publishers, Dordrecht-Boston-London, 1991.
13. Schipp, F., Wade, W.R., Simon, P., *Walsh Series - An Introduction to Dyadic Harmonic Analysis*, Akademiai Kiado, Budapest, 1990.

The $*$ sign in the citations indicates that the paper is reprinted in this book.

Chapter 12
How I started my research in Walsh and dyadic analysis

Boris I. Golubov

My first paper [1] related to Walsh functions was published in 1972. After that in a cooperation with A.V. Efimov and V.A. Skvortsov we published the monograph *Walsh Series and Transforms - Theory and Applications* [2]. My first paper [3] on the dyadic derivative and dyadic integral appeared after I read trough the papers of P.L. Butzer and J. Wagner [4], [5].

In [4], P.L. Butzer and H.J. Wagner introduced the strong dyadic derivative $D(f) \in L(G)$ for functions $f \in L(G)$, where G is the dyadic group which is isomorphic to the modified dyadic segment $[0,1]^*$. They proved that for the functions of the Walsh-Paley system $\{w_n\}_{n=0}^{\infty}$ the equalities $D(w_n) = nw_n$, $n \in Z_+$, where Z_+ is the set of non-negative integers, hold. This means that the Walsh-Paley functions are eigenfunctions of the operator D.

In the same paper for the functions $f \in L(G)$ the strong dyadic integral $I(f) \in L(G)$ was introduced and the equalities $D(I(f)) = I(D(f)) = f$ were proved under condition $\hat{f}(0) \equiv \int_G f(x)d\mu(x) = 0$, where μ is the normalized Haar measure on G , i.e., $\mu(G) = 1$. Thus the fundamental theorem of dyadic integral calculus was established.

In the paper [5] of the same authors the strong dyadic derivative $\overline{D}(f)$ for functions $f \in L(R_+)$ is introduced and the equality $(\overline{D}(f))^{\sim}(x) = x\,\widetilde{f}\,(x)$ is proved, where $\widetilde{f}$ is Walsh-Fourier transform of the function f.

For the functions $f \in L(R_+)$ H.J. Wagner [6] defined the strong dyadic integral $\overline{I}(f)$ and proved the equalities $\overline{D}(\overline{I}(f)) = f$ and $\overline{I}(\overline{D}(f)) = f$. (The last equality is proved under the condition $\widetilde{f}= 0$). In the same paper the following criterion was proved: For a pair of functions $f, g \in L(R_+)$ the equality $g = \overline{I}(f)$ holds iff $\widetilde{g}\,(0) = 0$ and $\widetilde{g}\,(x) = \widetilde{f}\,/x, x > 0$.

Boris I. Golubov
Chair of Higher Mathematics, Moscow Institute of Physics and Technology (State University), 141700 Dolgoproudny, Region, Russia, e-mail: golubov@mail.mipt.ru

R.S. Stanković et al. (eds.), *Dyadic Walsh Analysis from 1924 Onwards Walsh-Gibbs-Butzer Dyadic Differentiation in Science Volume 1 Foundations*, Atlantis Studies in Mathematics for Engineering and Science 12, DOI 10.2991/978-94-6239-160-4_12

The reading of papers [4], [5], and [6] was the impetus for me in order I be engaged in the study of dyadic derivatives and integrals. In the paper [3], I introduced the modified strong dyadic integral (MSDI) $J(f) \in L(R_+)$ and the modified strong dyadic derivative (MSDD) $D(f) \in L(R_+)$ for functions $f \in L(R_+)$. It is necessary to note that the MSDD D is equivalent to the strong dyadic derivative $D^{(1)}$ which was introduced by C.W. Onneweer [7] for functions defined on the group K isomorphic to the modified half-line R_+^*. In [3], the following criterion was proved: A pair of functions $f, g \in L(R_+)$ satisfy the equation $g = J(f)$ iff $\widetilde{g}(0) = 0$ and $\widetilde{g}(x) = h(x)\widetilde{f}(x)$, where $h(x) = 2^{-n}$ for $2^n \leq x < 2^{n+1}$, $n \in Z$ and $\widetilde{f}$ is the Walsh-Fourier transform of the function f. This criterion is an analog of a theorem of J. Pál [8] (see also [9], Ch. 9), who considered the strong dyadic integral $\bar{I}$ of H.J. Wagner [6] instead of the modified strong dyadic integral J. In [3], under the condition $\widetilde{f}(0) = 0$ it was proved the equalities $D(J(f)) = f$ and $J(D(f)) = f$. Moreover, the countable set L of eigenfunctions of the operators D and J was found and it was proved that the linear hull of the set L is dense in the dyadic Hardy space $H(R_+)$. Finally, it was proved that on dyadic Hardy space $H(R_+)$ the modified strong dyadic integral $J(f)$ can be defined as the uniform limit on R_+ of some sequence of dyadic convolutions $f * W_n$, where $W_n \in L(R_+)$, $n \in Z_+$.

In my paper [10], the fractional MSDI $J_\alpha(f)$ and MSDD $D^{(\alpha)}(f)$ of positive order α were introduced for functions $f \in L(R_+)$. Let us note that the fractional MSDD $D^{(\alpha)}(f)$ considered by me for functions $f \in L(R_+)$ is in fact a modification of the derivative $D^{(\alpha)}$ introduced by C.W. Onneweer [11] for functions $f \in L(K)$, where the dyadic group K is isomorphic to the modified half-line R_+^*. In [10], most results of the paper [3] were generalized on fractional modified strong dyadic integrals and derivatives. In particular, the following criterion was proved: A pair of functions $f, g \in L(R_+)$ satisfy the equation $g = J_\alpha(f)$ for a given $\alpha > 0$ iff $\widetilde{g}(0) = 0$ and $\widetilde{g}(x) = (h(x))^\alpha \widetilde{f}(x)$, where $h(x) = 2^{-n}$ for $2^n \leq x < 2^{n+1}$, $n \in Z$, and $\widetilde{f}$ is Walsh-Fourier transform of the function f. A similar criterion for the validity of the equation $g = D^{(\alpha)}(f)$ for a given $\alpha > 0$ was obtained. It was proved that if a function $f \in L(R_+)$ has MSDD $D^{(\alpha)}(f)$ of order $\alpha > 0$ and $\widetilde{f}(0) = 0$, then the function $D^{(\alpha)}(f)$ has MSDI of order α and the equality $J_\alpha(D^\alpha(f)) = f$ is valid. The similar statement is true if we change the symbols $D^{(\alpha)}$ and J_α: If a function $f \in L(R_+)$ has MSDI $J_\alpha(f)$ of order $\alpha > 0$, then the function $J_\alpha(f)$ has MSDD of order α and the equality $D^\alpha(J_\alpha(f)) = f$ holds.

It was proved that for $\alpha, \beta > 0$ the equality $D^{(\alpha)}(D^{(\beta)}(f)) = D^{(\alpha+\beta)}(f)$ holds if both MSDD in the left hand side exist. The similar equality for the fractional MSDI also is valid.

Let us note that the results of the paper [10] mentioned above are analogous to that of He Zelin [12] who considered the strong dyadic integrals and derivatives of fractional order of Butzer-Wagner type.

In the paper [13], P.L. Butzer and H.J. Wagner investigated the properties of pointwise dyadic derivatives of functions defined on the segment $[0, 1]$. In our paper [10], for functions $f \in L(R_+)$ and a point $x \in R_+$, the modified dyadic pointwise derivative (MDPD) $d^{(\alpha)}(f)(x)$ and the modified dyadic pointwise integral (MDPI) $j_\alpha(f)(x)$ were introduced. It was proved that if a function $f \in L(R_+)$ has MSDD $D^{(\alpha)}(f)$ (or MSDI $J_{(\alpha)}(f)$) of order $\alpha > 0$ then at each Lebesgue point $x \in R_+$ of the function f, hence almost everywhere on R_+, the equality $j_\alpha(D^{(\alpha)}(f))(x) = f(x)$ (respectively the equality $d^{(\alpha)}(I_\alpha(f))(x) = f(x)$) holds. The sufficient condition for the validity of the equation $D^{(\alpha)}(f)(x) = d^{(\alpha)}(f)(x)$ at all points $x \in R_+$ was obtained. Namely if a function $f \in L(R_+)$ has MSDD $D^{(\alpha)}(f)$ and $h(\cdot)^{-\alpha}\widetilde{f}(\cdot) \in L(R_+)$, then $d^{(\alpha)}(f)(x)$ and $D^{(\alpha)}(f)(x)$ exist at each point $x \in R_+$ and the equation $D^{(\alpha)}(f)(x) = d^{(\alpha)}(f)(x)$ holds. Here, as above, $\widetilde{f}$ is the Walsh-Fourier transform of the function f. A similar statement for the equation $J_\alpha(f)(x) = j_\alpha(f)(x)$ is also valid: If a function $f \in L(R_+)$ has MSDI $J_\alpha(f)$ and $h(\cdot)^{\alpha}\widetilde{f}(\cdot) \in L(R_+)$, then $j_\alpha(f)(x)$ and $J_\alpha(f)(x)$ exist at each point $x \in R_+$ and the equation $J_\alpha(f)(x) = j_\alpha(f)(x)$ holds.

In the paper [10] besides of the MDPD $d^{(\alpha)}(f)(x)$ and MDPI $j_{(\alpha)}(f)(x)$ also the dyadic pointwise derivative (DPD) $\overline{d}^{(\alpha)}(f)(x)$ for a function $f \in L(R_+) \cup L^\infty(R_+)$ and the dyadic pointwise integral (DPI) $\overline{j}_{(\alpha)}(f)(x)$ for a function $f \in L^\infty(R_+)$ were introduced. These derivatives and integrals are applied to the differentiation and integration of Walsh-Fourier transforms. It was proved that if $\alpha > 0$ and for a function $f \in L^\infty(R_+)$ the inclusion $h(\cdot)^{-\alpha}f(\cdot) \in L(R_+)$ holds, then Walsh-Fourier transform $\widetilde{f}$ has DPD $\overline{d}^{(\alpha)}(f)(x)$ at each point $f \in L(R_+)$ and the equality $\overline{d}^{(\alpha)}(f)(x) = (h(\cdot)^{-\alpha}f(\cdot))^{\sim}(x)$ is true. The similar statement for the DPI $\overline{j}_\alpha(f)(x)$ is valid.

In my paper [14] the strong dyadic integral $I_{\{\alpha\}}(f)$ of fractional order $\alpha > 0$ of H. J. Wagner type for the functions $f \in L(R_+)$ was introduced. The following criterion was proved: For a pair of functions $f, g \in L(R_+)$ the equality $g = I_{\{\alpha\}}(f)$, $\alpha > 0$ holds iff $\widetilde{g}\,(0) = 0$ and $\widetilde{g}\,(x) = \widetilde{f}\,(x)/x^\alpha$, $x > 0$. For $\alpha = 1$, this criterion has been proved by H. J. Wagner [6].

Besides of this for the fractional dyadic pointwise derivative and integral the following analogues of two theorems of Lebesgue were proved: the theorem on differentiation of the indefinite Lebesgue integral of an integrable function at its Lebesgue points and the theorem on reconstruction of an absolutely continuous function by means of its derivative. Dyadic fractional analogues of the formula of integration by parts also were obtained. In addition, some theorems were proved on dyadic fractional differentiation and integration of a Lebesgue integral depending on a parameter. Most of the results are new also for dyadic derivatives and integrals of natural order.

In the paper [15], I introduced the space $D'_d(R_+)$ of dyadic distributions on the base of pointwise dyadic derivatives of the functions defined on the half-line $R_+ = [0, \infty)$. The completeness of this space was proved. Besides of this, the space $S_d(R_+)$ of dyadic tempered distributions was introduced and the completeness of this space was proved.

References

1. Golubov, B.I., "Best approximations of functions in L^p metric by the Haar and Walsh polynomials", *Mat. Sbornik*, Vol. 87, No. 2, 1972, 254-274.
2. Golubov, B.I., Efimov, A.V., Skvortsov, V.A., *Walsh Series and Transforms - Theory and Applications*, Moscow, Nauka, 1987, 344p, (in Russian).
3. Golubov, B.I., "On the modified strong dyadic integral and derivative", *Mat. Sbornik.*, Vol. 193, No 4, 2002, 37-60.
4. Butzer, P. L., Wagner, H. J., "Walsh series and the concept of a derivative", *Applicable Anal.*, Vol. 3, 1973, 29-46.
5. $*$Butzer, P. L., Wagner, H. J., "A calculus for Walsh functions defined on R_+", *Proc. Symp. Applications of Walsh functions*, Washington, D. C., April 18 - 20, 1973, 75-81.
6. $*$Wagner, H. J., "On dyadic calculus for functions defined on R_+, *Proc. Symp. Theory and Applications of Walsh functions*, Hatfield Polytechnic, 1975, 101-129.
7. $*$Onneweer, C.W., "On the definition of dyadic differentiation", *Applicable Anal.*, Vol. 9, 1979, 267-278.
8. Pál J., "On the connection between the concept of a derivative defined on the dyadic field and the Walsh - Fourier transform", *Annales Sci. Univ. Budapest. Sect. Math.*, Vol. 18, 1975, 49-54.
9. Schipp, F., Wade, W.R., Simon, P., *Walsh Series - An Introduction to Dyadic Harmonic Analysis*, Adam Hilger, Bristol-New York, 1990.
10. Golubov, B.I., "Modified dyadic integral and derivative of fractional order on R_+", *Math. Notes*, Vol. 79, No. 2, 2006, 213-233, (in Russian).
11. Onneweer, C.W., "Fractional derivatives and Lipschitz spaces on local fields", *Trans. Amer. Math. Soc.*, Vol. 258, 1980, 923-931.
12. $*$Zelin He, "The derivatives and integrals of fractional order in Walsh-Fourier analysis with applications to approximation theory", *J. Approx. Theory*, Vol. 39, 1983, 361-373.
13. $*$Butzer, P. L., Wagner, H. J., "On dyadic analysis based on pointwise dyadic derivative", *Analysis Math.*, Vol. 1, 1975, 171-196.
14. Golubov, B.I., "On some properties of fractional dyadic derivative and integral", *Analysis Math.*, Vol. 32, 2006, 173-205.
15. Golubov, B.I., "Dyadic distributions", *Math. Sbornik*, Vol. 198, No, 2, 2007, 67-90, (in Russian).
16. Golubov, B.I., "Dyadic fractional differentiation and integration of Walsh transform", *Proc. Int. Conf. Mathematics and its Applications*, Kuwait, 2005, 274-284.

The $*$ sign in the citations indicates that the paper is reprinted in this book.

Index

A

Aachen School, 161, 176
Abelian groups, 90, 95, 226
absolutely continuous functions, 3, 213
AEU, 38
antiderivative, 212
atomic decomposition, 317, 322-323

B

Banach algebras, 355
Bessel Potential spaces, 248
Butzer-Wagner derivative, 89, 94-95, 166

C

$(C,1)$-summability, 210, 216, 221
Calderon-Zygmund type decomposition lemma, 315
characteristic function, 215, 222
coordinatewise addition, 91
cube p-atom, 322

D

d-parameter martingale, 322-323
DDR, 38
diadically continuous functions, 218
Dini derivates, 350
Dirichlet kernels, 1, 221-222
Discrete Fourier transform (DFT), 90
dyadic blocks, 222-223, 349, 351
dyadic convolution, 38, 93, 212, 355, 444
dyadic convolution operator, 212
dyadic difference operator, 216
dyadic differentiation, 93-94, 164-165, 174, 211, 222, 225-226, 247-249, 347-348, 350, 355-361, 445, 447, 452
dyadic group, 3-4, 90-91, 95, 164, 211, 215, 347, 443, 449-450
dyadic Hardy spaces, 221-223, 227, 315, 319, 321, 328, 358
dyadically integrable, 215

E

eigenvalue problem, 91-92

F

Fadeev theorem, 217
filtered Walsh series, 351-352
fractional p-adic derivative, 89

G

Gibbs derivative, 89-95, 358
Gibbs differentiator, 91-92, 224
Group character, 90, 91

R.S. Stanković et al. (eds.), *Dyadic Walsh Analysis from 1924 Onwards Walsh-Gibbs-Butzer Dyadic Differentiation in Science Volume 1 Foundations*, Atlantis Studies in Mathematics for Engineering and Science 12, DOI 10.2991/978-94-6239-160-4

H

Haar measure, 248, 449
harmonic analysis, 96, 162, 174, 236, 315, 328, 358-359, 447, 452
Hungarian school, 209

K

kernel function, 216, 222-223
Kronecker delta, 92

L

L.C. Vilenkin groups, 248
Lebesgue point, 217, 219, 451
Lebesgue points, 210, 217, 219, 451
Lipschitz classes, 218
Lipschitz spaces, 225, 248, 452

M

Marcinkiewicz means, 221
martingale Hardy space, 317
martingale transform, 216
Maximal function, 210, 317, 321
modified dyadic derivative, 445
modified strong dyadic derivative, 450
monotonic functions, 210, 213
multi dimensional dyadic Hardy spaces, 221
multi-dimensional dyadic derivative, 315, 322
Multipliers, 175, 211, 222-223, 249

N

Newton-Leibniz derivative, 90
non-sinusoidal functions, 90, 93, 162, 224

P

p-adic derivative, 89
p-atom, 317, 322-323
Paley enumeration, 211
Paley formula, 216
Paley gaps, 351
Paley set, 351
pointwise dyadic derivative, 176, 223, 248, 327, 349, 358, 446, 452
power-weighted Hardy spaces, 248
power-weighted Lebesgue spaces, 248

Q

quasiconvex, 349

R

Rademacher functions, 2, 348
Rademacher-series, 235
Restricted dyadic derivative, 326

S

sal functions, 37
strong dyadic derivative, 212, 348-349, 355, 449-450
strong dyadic integral, 174, 215, 225, 449-452
Strong summability, 218-219, 224, 226

T

term-by-term dyadic differentiation, 348, 350, 357
the dyadic maximal operator, 213, 217
translation invariant, 223, 355-356
trigonometric system, 1-3, 210-211, 221, 236, 348, 360

U

unrestricted maximal operator, 326

V

Vilenkin groups, 4, 215, 223-225, 247-

249, 327, 443
Vilenkin series, 236
Vilenkin-Fourier series, 4, 237, 247

W

Walsh multiplier operator, 211, 222
Walsh polynomials, 95, 168-170, 174-176, 217, 223, 452
Walsh-Kaczmarz series, 358
Walsh-Kaczmarz system, 350
Walsh-Paley functions, 167, 169, 247, 443, 449